D1087279

ALGORITHM DESIGN

ALGORITHM DESIGN

A Recursion Transformation Framework

MARVIN C. PAULL
Rutgers University

A Wiley-Interscience Publication

JOHN WILEY & SONS

New York / Chichester / Brisbane / Toronto / Singapore

Library of Congress Cataloging in Publication Data:

Paull, Marvin C.
Algorithm design.

"A Wiley-Interscience publication."
Bibliography: p.
Includes index.
1. Recursion theory. 2. Algorithms. I. Title.

QA9.6.P38 1987 511.3 87-21674
ISBN 0-471-81688-4

Printed in the United States of America

10 9 8 7 6 5 4 3 2 1

PREFACE

This book is the result of repeated refinement of notes developed for a first semester graduate course in algorithms. In the early years many students arrived with little academic background in computer science. Though the percentage of well-prepared students has grown over the years, the improvement in preparation has been somewhat offset by the diversity in that preparation. To fairly address this mixed population, the course included many of the algorithms which have earned the reputation of being basic, because they are useful and interesting and because they are almost always found in introductory algorithm texts. This assures that those not previously exposed to these core algorithms become familiar with them. The main emphasis of the course, and the aspect designed to hold the attention of those with a stronger background, is the development of these algorithms as illustrations of design principles likely to be new to them. In the course we develop a structured framework for description of the origin of good algorithms. Every algorithm presented is developed within this framework. The approach can be thought of as generalized recursion removal. It is similar to that found, for example, in [Strong 71; Burstall 77; Darlington 76; Bird 77; Cohen 79], where the origin of algorithms is modeled as a transformation from an initial problem formulation as a recursive definition to a "good" algorithm, where the transformation used depends on properties of the definition. Similar synthesis techniques, not explicitly tied to recursive definitions, are also found in popular texts and in current research—Chapter 4 of [Wells 71] is an early example, Chapter 3 of [Aho, Hopcroft, Ullman 75] is another, and the work in [Paige 81] represents recent research along these lines. Some of this work has been given a recursive definition formulation and included.

It is the choice of properties and transformations which complete our adopted framework of recursive definition to good algorithm. To cover the algorithms whose origins are to be explained, we tried to choose as few simple properties and transformations as possible. Hopefully these properties cover an even larger class of definitions. Because different principles can explain overlapping sets of algorithms, the choice is guided in part by esthetics. The principles chosen lead to an explanation of efficiency which emphasizes its fundamental origins. Starting with a recursive definition formulation which is basically nondeterministic and thus permissive in sequencing, space efficiency depends on choice of sequencing, and time efficiency depends on nonrepetition of similar computations. This foundation produces explanations of the origin of the major efficiencies of many algorithms without requiring a strong focus on the details of data structure representation, although a basic familiarity with data structure design is needed and is assumed.

In addition to requirements imposed by preparation, there are those imposed by the need to prepare for more advanced courses. Our program has had a strong artificial intelligence component. The emphasis on recursive definitions, their design and implementation, as well as the discussion of the depth-first, breadth-first, best-first, backtrack, and alpha–beta approaches to design improvement is aimed particularly at supporting this component. Correlating these approaches with Tarjan's depth-first search, elimination, and other techniques developed in other areas of computer science serves to give a broader, unified overview for all students. Many other algorithms are covered because they are needed in advanced courses. But in every case I have tried to develop it within the framework which unifies the entire presentation.

Although all the algorithms given in this book are well known, their development, as well as that of much of the supporting material, is necessarily novel. Included at the end of Chapter 2 is the development of a class of what are called n-dimensional recursive definitions, for which good implementations can be neatly formulated. In Chapter 3 we discuss the difficult problem of designing recursive definitions. In Section 3.3 we discuss how to obtain recursive definitions from a backtrack form of set definition, and also an interesting approach to obtaining a backtrack form from other forms of set definitions by solving sets of inequalities. The inverse has been recognized as an important property in the design of algorithms based on an initial recursive definition formulation. In Chapter 4 we introduce a strong form of this inverse, called the "uniform inverse," which is significant in explaining a large class of good enumeration algorithms. In Chapter 6 we consider a set of transformations which serve to connect different forms of recursive definition which arise repeatedly in practice. Equivalent forms related by such transformations may suggest different implementations with different complexities.

When presented with a new algorithm containing some elegant improvement, students invariably ask: how was that discovered? Though this question, entwined as it is with that of the nature of creativity and inspiration, will probably provide perpetual puzzlement, one can point to some approaches and techniques which could have been used. As the class of algorithms whose origins are to be explained is narrowed, stronger statements can be made. Thus a kind of well structured story, most likely fiction, explaining the origin of a wide range of good algorithms can be developed. Although this does not give their historical origin, it does provide a coherent structuring of principles helpful in understanding these and other algorithms, and perhaps even in designing new ones. This book attempts to present such a story. There will continue to be algorithms which cannot be explained in a satisfying way with these principles. However, given the framework of principles, we can hope that a student, presented with such an algorithm, will be led to reexamine and refine these principles rather than simply add another entry to a catalogue of "inspired" algorithms.

References

References, mostly to books rather than original sources, are embedded in the text. Almost every algorithm covered here is covered in the standard algorithm texts, including, for example [Aho, Hopcroft, Ullman 75; Reingold, Nievergelt, Deo 77; Horowitz, Sahni 78; Papadimitriou, Steiglitz 82; Sedgewick 83; Hu 82]. Alternative developments of these algorithms can be found there. References for particular algorithms are also given if it seems that further background would be particularly helpful.

Acknowledgments

Thanks are due to many, beyond the neglected Charlene, Chris, and Eric, who have been encouraging and helpful over the long and sometimes hesitant development of this book. K. Kaplan, B. Ryder, and N. Sridharan have taught a course along the lines of this book, while T. Ostrand and B. Nudel have used the book in substantial parts of courses. All have contributed to removing errors and have made many helpful suggestions.

Others have done extensive careful reading and editing and made material contributions to developments needed for the presentation. These include M. Berman with special contributions to the section on binary search trees, M. Carroll who helped particularly with the section on games, S. Marchand who was the first unofficial editor of a more or less complete manuscript, and T. Marlowe whose help was essential to the development of material on iterative algorithms, and in fact in many ways to the entire manuscript.

And in the prehistory of the manuscript's development it was J. Maye who did most of the typing.

MARVIN C. PAULL

New Brunswick, New Jersey
January 1987

CONTENTS

ALGORITHM DESIGN

CHAPTER ONE

INTRODUCTION

1.1. OVERALL OUTLINE

An algorithm is a procedure, or sequence of instructions, used to *evaluate a function*. We write $A(f(X))$ to indicate that algorithm A *implements*, *realizes*, or *evaluates* f at all arguments X in the domain of $f(X)$. The legitimate values of X are the legitimate inputs, or initial values of the input variables, of the algorithm $A(f(X))$. The objective of algorithm design is to implement given function definitions with algorithms frugal in their use of time and space. A primary objective of this book is to explore principles useful in developing such designs. Since there are many excellent algorithms known for a variety of function definitions, there is plenty of data available from which to extract such principles. For each function definition there are many implementing algorithms (Figure 1-1).

Algorithms and function definitions can be expressed in many different ways. To formulate design principles compactly, it is necessary to focus on only a few of these.

Most of the time algorithms will be given as deterministic programs, at roughly the ALGOL language level and style. Written this way they are referred to as *sequential algorithms*. On occasion algorithms are given nondeterministically, that is, permissive in the orderings of some of their operations, expressed, however, in a language similar to that used for the sequential algorithms.

As a basis for defining new functions we assume the existence of a pool of already known or primitive functions. Generally a function definition will be made up of a set of equations, each giving the function over part of its domain, either as a composition of primitive functions only, or recursively as a composition of itself and primitive functions. In the interesting cases this form of definition almost always includes a recursive equation, that is, it is a *recursive definition* (Figure 1-2).

The design principles will show how to go from a recursive definition to efficient sequential algorithms. However, the easy, natural initial definitions of a function may not be recursive. In order to make the design principles

FIGURE 1-1 Algorithm design—general.

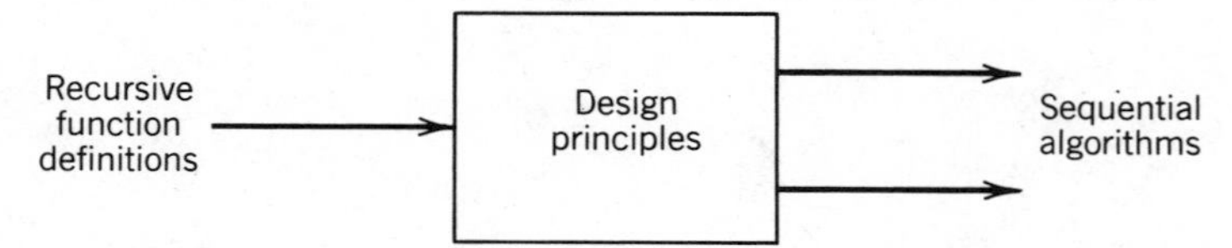

FIGURE 1-2 Algorithm design—Starting with recursive definition.

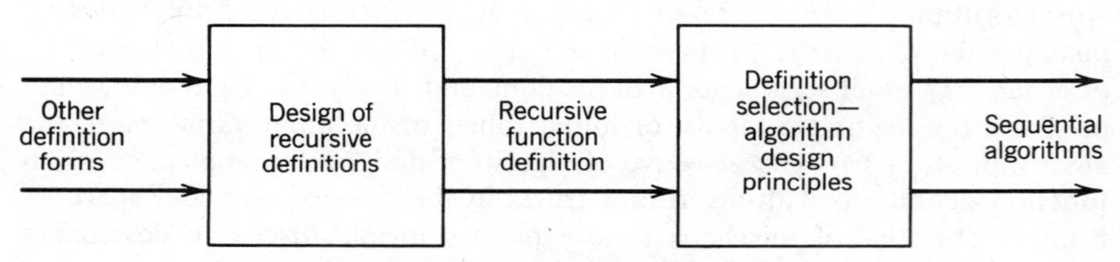

FIGURE 1-3 Algorithm design—two stages.

applicable to other forms of definition, we will develop transformations from a number of such forms to recursive definitions (Figure 1-3). Furthermore, significantly different recursive definitions can define the same function. Each of these may lead to a different sequential algorithm. Principles for selecting the best and transforming among alternatives will also be explored.

Our success in uncovering the principles involved in the design procedure varies widely. For some classes of function definitions a few simple principles explain the origin of all good implementing algorithms. For others, often very small classes, it appears that large numbers of principles are necessary. In any case our classification of algorithms around applicable design principles rather than application areas will in several cases provide a very satisfying unification of function definitions arising out of very different applications.

Certainly the efficiency of an algorithm is sensitive to the design of the data structures it uses. This book, however, will not go deeply into that aspect. Rather we assume that appropriate primitive functions are efficiently implemented and available. In definitions or algorithms involving numbers we assume primitive functions such as addition (+) and subtraction (−); if the definitions or algorithms involve strings or simple vectors, functions such as concatenation (| |) and deconcatenation (−); if trees, parent and child functions. These operate on primitive data structures, that is, numbers, vectors or strings, or linked or vector representation of trees. Such structures are provided either by hardware (usually numbers are), or by software (the usual case for trees). For a collection of data and the set of primitive functions associated with it, detailed data representation or structure can vary, and its form can significantly affect the efficiency of the primitive functions. Whether in hardware or software, the design and access of this structure will be considered as occurring in a different module from that in which the definition or algorithm is executed. There will, however, be occasions for which we will consider the design of the module supporting the primitive functions, mostly to justify complexity claims. The book will nonetheless concentrate on the aspect of algorithm design concerned with sequencing and organization of the primitive functions.

1.1.1. Outline by Chapters

Chapter 2 is devoted to exploring fundamental issues arising in any study of recursive definitions, including the names and meanings of their parts, the conditions and sense in which they actually define functions, and techniques for proving some of their properties. General methods of definition evaluation are also reviewed. These include *substitutional evaluation* (SE) and *equation set evaluation* (ESE). SE is the way one would normally evaluate a recursive definition by hand, with repeated substitution for appearances of the defined function. ESE is a variant of SE which records intermediate results of substitutions for possible reuse in the form of equation sets. Like the definitions themselves, their implementing algorithms can be partitioned into two broad classes—SE and ESE simulations. This division serves to organize our study of algorithm design in subsequent chapters.

Chapter 3 takes up the problem of transforming various types of nonrecursive "natural" definitions into recursive ones. In general only very broad advice is available for the problem of finding a recursive definition for a function originally defined by other means. However, we will explore some classes of definitions, such as backtracking definitions, for which mechanical procedures are available.

For any function f there is a class of equivalent recursive definitions which, though differing from each other, all define f. We explore transformations between equivalent definitions, and the relative value of different equivalent definitions for producing algorithms. We also consider transformations between some definitions which, though not equivalent, are strongly related, as for example between the definition of a set and that of the cardinality of that set.

Chapter 4 examines a variety of implementations of recursive definitions. The algorithms developed in this chapter are all based on SE simulation. The presence of certain properties in a recursive definition will allow particularly efficient implementations. Many such properties, including existence of an inverse and associativity of the primitive functions in the definition, are identified, and consequences in allowing efficiencies in the implementing algorithm are explored. The illustrations in this part of the chapter are largely drawn from enumerations of various sets and sequences, including sets of permutations, combinations, and partitions, as well as Gray code and Tower of Hanoi sequences. The savings made possible by the properties explored here are mainly in storage rather than in time. In an SE simulation there is often some choice in the order in which the necessary operations are to be carried out. Properties such as the inverse and associativity allow new orderings of the operation, but do not eliminate the need to do them all. Temporary storage must be provided for the maximum accumulation of partial results. In order to minimize temporary memory such results will be discarded when no longer needed. Just when information can be discarded depends largely on the choice of operation ordering. The storage requirement is therefore a function of the order of simulation.

In Chapter 5 the property of comparability is introduced. A problem has this property if the same or similar partial results will be produced repeatedly in a straightforward SE simulation. Significant savings in time now become possible if partial results, which may be needed again, are saved and reused, rather than recomputed (as in straightforward simulation). The comparability property and its consequences in making possible significant time-saving modifications in SE simulation algorithms are illustrated with examples of graph and game algorithms.

In Chapter 6, algorithms based on ESE simulation are considered. Here again special properties of the definition make possible efficiencies in the implementing algorithms. These properties are a special, although frequently occurring, example of the concept of comparability introduced in Chapter 5. Thus they also lead to time savings.

As noted above there are often different definitions for the same functions. One definition may be amenable to ESE while another is not. Transformations generating a definition amenable to ESE simulation from one which is not are explored. These transformations and their ESE simulations are illustrated with a variety of examples, including definitions for language recognition and parsing, shortest paths and tours in graphs, and matrix multiplication optimization. Both elimination and iterative ESE are considered.

In Chapter 7 we give examples of problems which fit the framework of the book only if the framework and problem are distorted. This includes definitions for which the explanation of good implementing algorithm requires a large number of or narrowly applicable principles, and those for which it is difficult to see what general properties justify the known "good" implementations.

1.2. SUMMARY OF NOTATION

A set of functions which access and operate on a conceptually unified data collection is an *abstract data structure*. In each definition and algorithm design considered here certain abstract data structures will be considered primitive and available.

The focus of this book is on the use of primitive functions in designing algorithms, rather than on the efficient implementation of the primitive functions. In the following summary we define a large number of primitive functions without discussing implementation. This presents the reader with an abrupt introduction to a large amount of notation. Furthermore, that notation is made compact, though where feasible familiar, to allow compact representation of definitions and algorithms whose representation might otherwise require so much space that the focus on essentials would be blurred. It is probably best to pass through the notation lightly in this summary and actually learn it along with its use in subsequent chapters.

Most of the primitive abstract data structures, their primitive functions, and the notations we will use for them, other than the obvious arithmetic ones, are summarized below. Function domains include vectors, sets, lists, graphs, and grammars. It is followed by a summary of other features of the algorithmic or pseudoprogramming language to be used. The more exotic notation will be defined again in the text on first use. The concepts expressed by the notation in this summary are assumed to be familiar to the reader.

In what follows the notation itself is given on the left while its meaning is briefly described on the right. The notation is organized in the following groups. (Within each grouping, primitive functions are organized according to their domain and range. For example, operations that take a set to a set (set → set) are grouped together.)

- Sets
  - Set → set
  - Sets → set
  - Set → integer
- Vectors, strings, sequences
  - Vectors → vector
  - Vector, scalar → vector
  - Vector → scalar
  - Vector → integer
  - Vectors → set
  - Vector component, vector → index
- Sets of vectors and vectors of vectors
  - Sets of vectors → set of vectors
  - Vectors of vectors → vector of vectors
- Predicates
- Graphs
- Grammars
- Ceiling and floor

Nested function composition
Minimum and maximum
Function multiples
Algorithmic notation

SETS

Assume $A = \{a_1, \ldots, a_n\}$ and $B = \{b_1, \ldots, b_m\}$ are sets.

Membership

$a \in A$ Set membership

Specification

$X = [x_1, \ldots, x_n\}$ Set containing $x_1, \ldots, x_n$
$X = \{x \mid P(x)\}$ or $\{P(x)\}$ Set of x's with property P

Special Sets

$\{\ \} = \{\text{zilch}\}$ Empty set

Note: zilch is the name of the member of the empty set and of the empty vector.

$N = \{1, 2, \ldots\}$ Set of positive integers

$N^b = \{1, 2, \ldots, b\}$
$N_a = \{a, a+1, \ldots\}$
$N_a^b = \{a, a+1, \ldots, b\}$
Sets of integers

Conventions

$$\bigcup_{i=1}^{i=0} [\text{expression in } i] = \{\ \}$$

Note: The union of zero sets, which occurs in settings such as that shown here, is defined to be the empty set.

Primitive Functions

Set → set

A^* Set of subsets of A

Sets → set

$A \cap B = \{c \mid c \in A \text{ and } \in B\}$	Set intersection
$A \cup B = \{c \mid c \in A \text{ or } \in B\}$	Set union
$A - B = \{c \mid c \in A \text{ and } \notin B\}$	Set difference

Set → integers

$\lvert A \rvert$	Cardinality of set (number of members)

Set → vector

$\langle A \rangle$	Vector whose components are the members of A, arbitrarily ordered

VECTORS, STRINGS, SEQUENCES

Assume $V = \langle v_1, \ldots, v_n \rangle$ and $W = \langle w_1, \ldots, w_m \rangle$.

Membership

v_i is ith component of V	Vector membership

Note: If e is an expression whose value is a vector, then $[e]_i$ is the ith component of that vector.

Specification

$X = \langle x_1, \ldots, x_n \rangle$	Vector with components $x_1, \ldots, x_n$
$X = \langle x \mid P(x) \rangle$ or $\langle P(x) \rangle$	Vector with components having property and order given by P; if order is absent, an arbitrary, consistent order applies

Special Vectors

$\langle\ \rangle$ or $\langle \text{zilch} \rangle$	Empty vector
0–1 vector	Components $\in \{0, 1\}$
$x^n = \underbrace{\langle x, x, \ldots, x \rangle}_{n \text{ components}}$	Single-valued vector

Operations

Vectors → vector

$V \parallel W = \langle v_1, \ldots, v_n, w_1, \ldots, w_m \rangle$	Concatenation

If o_i is a two-argument operation and $|V| = |W| = n$, then

$$Vo_1 \ldots o_n W = \langle v_1 o_1 w_1, \ldots, v_n o_n w_n \rangle \qquad \text{Function Multiple}$$

Vector, scalar → vector

$V - v_j = \langle v_1, \ldots, v_{j-1}, v_{j+1}, \ldots, v_n \rangle$		Component removal
$v_j \leftarrow a$	replace v_j in V with a	Component assignment
$v_j \leftarrow$	same as $V^j - v_j$	Component removal
$v_{j+} \leftarrow a$	insert a just after v_j in V	Component insert
$v_{j-} \leftarrow a$	insert a just before v_j in V	Component insert

Vector → scalar

$$V \cdot W = \langle v_1 * w_1 + \cdots + v_n * w_n \rangle \qquad \text{Dot product}$$

Vector → integer

$|V|$ Number of components in V equals length of V

Vectors → set

$\{V\} = \{v \mid v \text{ a component of } V\}$ Set of components of a vector

Vector component, vector → index

ind(x, V) Index of component x in vector V

SETS OF VECTORS AND VECTORS OF VECTORS

Assume Q and R are sets of vectors, W and V are vectors of vectors

Operations

Sets of vectors → set of vectors

$Q|*|R = \{q \| r \mid q \text{ in } Q, r \text{ in } R\}$ Cross concatenation

Example: $\{\langle 5,1\rangle|*|\{\langle 2,3\rangle, \langle 7\rangle\} = \{\langle 5,1,2,3\rangle, \langle 5,1,7\rangle\}$
$(= \{\langle 5,1,7\rangle, \langle 5,1,2\ 3\rangle\})$

Vectors of vectors → vector of vectors

1. For a single-component vector $W = \langle w \rangle$, where w itself is a vector, and an n-component vector $V = \langle v_1, \ldots, v_n \rangle$, where $\forall j\ v_j$ is a vector,
   $W|*|V = \langle w\|v_1, \ldots, w\|v_n \rangle$ Cross concatenation
   Example: $\langle\langle 5,1\rangle\rangle|*|\langle\langle 2,3\rangle, \langle 7\rangle\rangle = \langle\langle 5,1,2,3\rangle, \langle 5,1,7\rangle\rangle$

2. For an m-component vector $W = \langle w_1, \ldots, w_m \rangle$ where $\forall k$, w_k is itself a vector and $m > 1$,
   $W|*|V = (\langle w_1\rangle|*|V)\| \cdots \|(w_m|*|V)$ Cross concatenation

   Example: $\langle\langle 5\rangle, \langle 1\rangle\rangle|*|\langle\langle 2\rangle, \langle 7\rangle\rangle = \langle\langle 5,2\rangle, \langle 5,7\rangle, \langle 1,2\rangle, \langle 1,7\rangle\rangle$

PREDICATES

Assume P is a predicate.

Predicate → set

$\{x \mid P(x)\}$ or $\{P(x)\}$	Set of all x that satisfy $P(x)$
$\langle x \mid P(x)\rangle$ or $\langle P(x)\rangle$	Vector of all x that satisfy $P(x)$

GRAPHS

Basic. A graph [digraph] $G = (VE)$ is a set of *vertices* V, and a set of *edges* E is a subset of $\{\{a, b\}\ [\langle a, b\rangle]\ | a, b \text{ in } V\}$. $G = (V, E, c)$ is an *(edge) weighted* or *costed* graph [digraph] if (V, E) is a graph [digraph] and includes a *weight* or *cost function*: c: $E \to$ reals. Thus $c(a, b)$ is a number giving the cost of edge $\{a, b\}$ $[\langle a, b\rangle]$.

Defined Primitive Functions. The *neighbor function* [*(outward) neighbor*] of graph [digraph] $G = (V, E)$, n: $V \to V^*$, is defined as follows:

$n(v) = \{w \mid \{v, w\}\ [\langle v, w\rangle] \text{ in } E\}$

$n_i(v)$ is the ith component of $\langle n(v)\rangle$. It is called the ith *neighbor* [ith *(outward) neighbor*] of v.

The inverse and other functions and predicates related to the neighborhood function are

$i(v, x)$ is the index of x in $\langle n(v)\rangle$ $[= \text{ind}(v, n(v))]$.

$n^{-1}(i, x)$ is a vertex v such that the ith component of $\langle n(v)\rangle$ is x. This notation will usually be used in situations in which this inverse is unique.

next-neigh $(v, x) = n_{i(v,x)+1}(v)$

$(x =$ last-neigh$(v))$ = true if x is the last component of $\langle n(v)\rangle$,
= false otherwise.

$N(v) = |n(v)|$

$c_i(v) = c(v, n_i(v))$

Defined Terms. p is a *path between a and b* [path from a to b] in graph [digraph] $G = (V, E)$ $[(V, E, c)]$ if $p = \langle e_1, \ldots, e_{n-1}\rangle$, where for i in N^{n-1}, $e_i = \{v_i, v_{i+1}\}$ $[\langle v_i, v_{i+1}\rangle]$ is an edge in E, and $v_1 = a$, $v_n = b$. No two components of $\langle v_1, \ldots, v_n\rangle$ are equal. Without the constraint of the last sentence, *path* becomes *genpath*.

The path p can be represented by the sequence in which vertices first appear in the path edges [sometimes called a *(vertex) path vector*] $\langle v_1, \ldots, v_n\rangle$, or by the reverse of that sequence [*reverse (vertex) path vector*] $\langle v_n, \ldots, v_1\rangle$. The path vector does not have any recurrence of any vertex; a genpath may contain repeated vertices. For most purposes here it is best to think of a path as being a nonrepeating sequence of vertices. The *length* of p is the number of edges in p.

A *cycle* is defined the same way as path, except that $a = v_1$ and $b = v_n$ are the same vertex. No other pair in $\langle v_1, \ldots, v_n \rangle$ is equal. A graph (digraph) with no cycles is called *acyclic*.

A vertex v is *reachable* from vertex w if there is a path from w to v.

A graph (digraph) is *connected* if there is a path between every pair of vertices in V (from any vertex to any other).

A *tree* is a connected, acyclic graph.

GRAMMARS

Basic. A *context-free grammar* (CFG) consists of a finite set of *nonterminal symbols* (NTS), one of which is designated the *distinguished* (*starting*) *symbol*, a finite set of *terminal symbols* (TS), and a finite set of *production rules*, each of which is given in the form $x \to s$, where x is an NTS and s is a finite string of TSs and NTSs. Note that $|s| \geq 1$.

Defined Primitive Predicates

TS(x) is true if x is a terminal symbol of G.

NTS(x) is true if x is a nonterminal symbol of G.

Defined Primitive Functions. If NTS(x) is true:

$N(x)$ is the number of production rules with x on the left.

$N(x, i)$ is the number of symbols on the right of the ith production rule with x on the left.

$P_i(x)$ is the ith production rule with x on the left.

$R_i(x)$ is the right side of the ith production rule with x on the left.

$R_{ij}(x)$ is the jth symbol on the right side of the ith production rule with x on the left.

Defined Terms. Given any string of NTSs and TSs of $G, q = \langle q_1, \ldots, q_j, \ldots, q_n \rangle$, q_j is an NTS, then $w = \langle q_1, \ldots, q_{j-1} \rangle \| R_i(q_j) \| \langle q_{j+1}, \ldots, q_n \rangle$ is said to be *derived* from q in a single step using production rule $P_i(q_j)$, symbolized $q \xrightarrow{P_i(q_j)} w$.

If q_j is the leftmost NTS in q, then it is a *leftmost derivation step*. Any sequence of (leftmost) derivation steps using rules of G which starts with the string a and ends with the string b is a (leftmost) derivation of b from a in G, symbolized $a \to b$.

Any string b which can be derived from a in G can be derived by a leftmost derivation. The intermediate strings, which appear in a derivation of b from a in G, are called *sentential forms*. The *language of G*, L_G, is the set of all terminal strings that can be produced by (leftmost) derivations from the distinguished symbol of G. If there is more than one leftmost derivation by which the same terminal string in the language of G can be derived, then G is *ambiguous*; otherwise G is *unambiguous*. If there is a derivation in G from the NTS A to a string b whose first symbol is A, then G is *left-recursive*. Notice that if G is not left-recursive, there is some number $b(G) \leq$ the number of NTSs in G such that any leftmost derivation must result, after $b(G)$ steps, in a string that starts with a terminal symbol. Given any grammar G, there is always another which generates the same language and is not left-recursive. For all the parsers and recognizers considered here we assume the CFG involved is not left-recursive. A *parse* of a string s in L_G is the sequence of rules used in deriving s (by leftmost derivation) in G. A *parser* is an algorithm which, when given a grammar G and a string s as input, produces the parse of s in G as output if s is in L_G and a "no" if s is not in L_G. A *recognizer* is an algorithm which, with the same inputs as the parser, produces an output "yes" if s is in L_G and a "no" otherwise.

Miscellaneous.
Ceiling, floor

$\lfloor x \rfloor$	Greatest integer $\leq x$
$\lceil x \rceil$	Least integer $\geq x$

NESTED FUNCTION COMPOSITION

$h^k(X) \equiv \underbrace{h(h(\ldots h(h(X))\ldots))}_{k\,h's}$ — Single function

If $I = \langle i_1, \ldots, i_n \rangle$, then

$o_I(X) \equiv o_{i_n}(o_{i_{n-1}}(\ldots o_{i_2}(o_{i_1}(X))\ldots))$ — Function sequence

$N_{i=\text{start}}^{\text{end}}[h(x, y_i)]$
$\equiv h(h(\ldots h(h(x, y_{\text{start}}), y_{\text{start}+1})\ldots), y_{\text{end}})$ — Two arguments

MINIMUM AND MAXIMUM

min [max] (x, y) or $x \downarrow y$ $[x \uparrow y]$ — Minimum [maximum] of x and y

If X is a set;
MIN [MAX] (X) — Minimum [Maximum] $x \in X$

If X is a vector,
MIN [MAX] (X) — Vector with components minimum [maximum] $x \in X$

min-sel is a function each of whose arguments is a two-component vector. The first component is a number, called the *cost* of the argument. The second can be anything. The definition of min-sel is

$$\text{min-sel}(\langle a_1, b_1 \rangle, \ldots, \langle a_n, b_n \rangle) = \langle a_j, b_j \rangle$$

where j is the smallest index such that $a_j = \text{minimum}(a_1, \ldots, a_n)$.

If X is a set or a vector of pairs, say $X = \langle\langle a_1, b_1\rangle, \ldots, \langle a_n, b_n\rangle\rangle$ as above, then

$$\text{min-sel}(X) = \text{min-sel}(\langle a_1, b_1\rangle, \ldots, \langle a_n, b_n\rangle)$$

min-sel*(X) is a nondeterministic operation on a set or vector of pairs whose value is one of the pairs $\langle a_i, b_i\rangle$ with a_i *any index* for which a_i = minimum $(a_1, \ldots, a_n)$.

FUNCTION MULTIPLES

If $v(A, B)$ and $w(C, D)$ are two functions, each with two arguments, then $[v - w](\langle A, C\rangle, \langle B, D\rangle) = \langle v(A, B), w(C, D)\rangle$

ALGORITHM NOTATION

The pseudolanguage used here to describe algorithms is, for the most part, made up of operations common to all PASCAL or PL/1 type languages; few specialized features of these programming languages are used.

The pseudolanguage includes the usual assignment statements: $X \leftarrow$ *expression*, where, however, in addition to the usual arithmetic operations, the expression can involve any of the operations defined earlier in this section. Also, some shorthand notation related to vector manipulations is used. Let $Z = \langle Z_1, \ldots, Z_n\rangle$ and let there be n components in $\langle A, B, \ldots, L\rangle$. Then

> $\langle A, B, \ldots, L\rangle \leftarrow Z$ is equivalent to the set of assignments. $A \leftarrow Z_1$; $B \leftarrow Z_2; \ldots; L \leftarrow Z_n$.
> $Z \leftarrow \langle A, B, \ldots, L\rangle$ is equivalent to the set of assignments. $Z_1 \leftarrow A$; $Z_2 \leftarrow B: \ldots; Z_n \leftarrow L$.

Instead of referring to a vector component index with an integer subscript, as in the text, an integer enclosed in brackets is used in algorithms. Thus Z_1 in the text becomes $Z[1]$ in an algorithm. In fact, in general, wherever a subscript would be used in text, bracketing is used in the algorithms.

Special assignment operations are defined for pushing, popping, and other list manipulations. Lists, stacks, queues, sequences, vectors, and strings are different names for referring to the same structure, namely, an ordered sequence of variables. The difference in names reflects different sets of operations that will be applied to that structure—different ways in which the contents as well as the number of such variables are changed. A structure identified as a vector in the text may be referred to as a stack in an algorithm if the set of primitives which access it is the appropriately restricted.

Important special members of lists are given names.

$Z[\text{top}] = Z[1]$ is the top, or first, entry of the list, in general.
$Z[\text{top} + i - 1] = Z[i]$ is the top + $(i - 1)$th entry in Z. $Z[\text{last}]$ is the bottom or last entry in Z.

If Z is a list and X a variable whose type is that of the components of Z, the list operations are

$Z \leftarrow$ psh- X	Push contents of X onto Z or $Z \leftarrow \langle X\rangle \| Z$, or equivalently, $Z[1\text{-}] \leftarrow X$.
$X \leftarrow$ pop- Z	Remove top entry from Z and place in X, or equivalently, $X \leftarrow Z[1]$; $Z \leftarrow Z - Z[1]$.
$\leftarrow$ pop- Z	Remove top entry from Z, or equivalently, $Z[1] \leftarrow$.
$Z \leftarrow$ queue- X	Put contents of X at the bottom of list Z, or equivalently, $Z \leftarrow Z \| \langle X\rangle$ or $Z[\text{last} +] \leftarrow X$.

If e is an expression whose value is an integer and W is itself a list, then

$W \leftarrow$ pop.e- Z Pop e members of Z and push them into W or do e repeats of the pair of operations;
$X \leftarrow$ pop- Z; $W \leftarrow$ psh- X.

List entries may be vectors. For example, consider a list whose entries are two-component vectors. Pushing a new entry, whose first component is in A and second component is in B, onto the list Z can be expressed as $Z \leftarrow$ psh- $\langle A, B\rangle$, and popping such an entry into the two variables A and B, $\langle A, B\rangle \leftarrow$ pop- Z.

For output we use assignment notation, as in SNOBOL,

OUT $\leftarrow X$ Printout X

Input statements are represented by assignments of the form $X \leftarrow n$ where n is not a variable but an input. This distinction is indicated by use of the lowercase. The nature of n will ordinarily be indicated.

We use the standard decision statement:

IF condition THEN statement1 ELSE statement2. (The ELSE part may be omitted), and the control statements,

FOR variable = lower limit TO upper limit BY increment DO (The BY part may be omitted)

WHILE (condition) DO...END
UNTIL (condition) DO...END.
FOR ever DO...END is a FOR statement which always repeats (until some exit statement internal to the loop is executed.)

In general the pair DO...END, when not in combination with FOR, WHILE, or UNTIL, is used for grouping.

The statement DONE terminates an algorithm; RETURN terminates a procedure. RETURN (x) passes the value of x to the calling procedure.

The type of a variable is usually given in the text preceding the algorithm in which it appears, or in notes following the algorithm statement. Declaration statements are occasionally used. For example, DCL (V[N] vector) INIT(0) declares V to be a vector with N components all initialized to 0.

1.3. PROBLEMS

1. *Recursive Definition to Recursive Procedure.* The purpose of Problem 1 is to help familiarize you with the notation and, more importantly, to review recursive programming of a given recursive definition, and to demonstrate that a recursive definition, as given in the text, is very close to the corresponding recursive program that implements that definition. There are decisions to be made about data structures, for example, how to represent strings that must grow in length, but in the main, the transformation from definition to procedure is straightforward.

   Write and run a recursive procedure in the language of your choice that implements the following definition for $f(\langle\ \rangle, 0, 0, 6, 3)$.

Definition 1.3.1

$$
\begin{cases}
f(x, y, L, n, m) = \{\ \} & :(y = n) \text{ and } (L < m) \\
f(x, y, L, n, m) = \{x\} & :(y \leq n) \text{ and } (L = m) \\
f(x, y, L, n, m) = f(x \| \langle y + 1\rangle, y + 1, L + 1, n, m) & \\
\quad \cup f(x \| \langle y + 2\rangle, y + 2, L + 1, n, m) \cup \cdots \cup f(x \| \langle y + (n - y)\rangle, & \\
\quad y + (n - y), L + 1, n, m) & :(y < n) \text{ and } (L < m) \\
\quad \text{Initially } x = \langle\ \rangle,\ y = 0,\ L = 0,\ n \text{ in } N,\ m \text{ in } N,\ n > m. &
\end{cases}
$$

The function is intended to enumerate combinations of the first n integers taken m at a time. It may be helpful to hand simulate it. In any case one ought to be able to construct the recursive procedure without knowing what the function does.

2. *Recursive Definition to Nonrecursive Procedure.* An important objective of the book is to develop design techniques which start with recursive definitions and produce nonrecursive implementations. The problem below is given so you may contrast the techniques which you now use with those developed in the text.

   Write but do not run a nonrecursive program for the definition in Problem 1. Assume the existence of all primitive functions used in Definition 1.3.1 of Problem 1.

CHAPTER TWO

FUNDAMENTALS

Every algorithm design in this book starts with a recursive definition. In this chapter we explore the syntax and semantics of such definitions. Definitions are classified according to syntactic features which determine their defining power. Termination and, more generally, the sense in which a recursive definition defines a function are explored. Proof techniques for properties of recursive definitions are studied. We consider proofs that a definition d terminates; that d defines a given primitive function f; that it is equivalent to another definition, recursive or otherwise; that a given algorithm implements d. Fundamental ways of evaluating recursive definitions are considered, including modeling of the evaluation process by computation networks. These are useful in suggesting implementations for evaluating recursive definitions. Finally, as an example of the use of such models, the class of k-dimensional definitions, for which there are efficient implementing algorithms, is studied. The fundamentals discussed have varying practical import. Termination is not often an issue. Its existence is usually easily decided. The same is true in deciding whether a recursive definition actually defines the function we want. The networks developed for modeling evaluation are equivalent to the familiar tree structures widely used for the same purpose. Our basic goal in discussing these questions is to make the connections between the simplest ways of evaluating recursive definitions by hand and good algorithms—through concepts such as network models. Detailed consideration of the sections on termination, proofs, and on the implementation of functions defined on k-dimensional integer space is not essential in following the subsequent material.

2.1 DEFINING THE FUNCTION THAT THE ALGORITHM EVALUATES

To define a function it is necessary to give its domain and range, and a member of the range corresponding to each member of the domain. If D is the domain, R the range, and f the function name, then we say f: $D \rightarrow R$, and if d in D corresponds to r of R, we say $r = f(d)$. There are a variety of ways in which the function-defining information can be given. One way is to specify for each d in D a set of properties that $f(d)$ must satisfy. These properties are given in terms of *primitive functions*, relations, predicates, and data structures. Primitive means "assumed to be available." Functions or data structures can be primitive in the sense that they are both simple and common built-in parts of familiar programming languages, such as addition and concatenation. They may also be considered primitive because they are simple relative to the function for which a definition is being formulated. Such "primitives" may themselves later require definition. Consider the following examples.

Example 2.1.1

$$f(d) \leq 3 * d + 5$$

Example 2.1.2

$$f(d) = 3 * d + 5 \qquad \text{if } d \leq 25$$

$$f(d) = 2 * d \qquad \text{if } d > 25$$

Example 2.1.3: D and R both are the set of all vectors whose components are integers (in N). For $d \in D$, $f(d)$ is the vector $r \in R$ such that:

a. $|r| \leq |d|$.
b. For each component in d there is exactly one in r with the same value.
c. $r_i \leq r_{i+1}$.

[$f(d)$ is the sorted irredundant version of the vector d.]

Example 2.1.4: D is the set of integers > 0 ($\in N$), R is the set of all vectors with components in D. For $d \in D$, $f(d)$ is the set of all vectors $r \in R$ with properties:

1. $|r| = d$.
2. r_i is an integer $\leq d$ (in N^d).
3. $r_i \neq r_j$ if $i \neq j$.

Or, more succinctly,

$$f(d) = \left\{ \langle r_1, \ldots, r_d \rangle | r_i \in N^d \text{ and } r_i \neq r_j \text{ if } i \neq j \right\}$$

[$f(d)$ is the set of all permutations of the first d integers, each represented as a vector.]

Example 2.1.5: Each member of D is a vector with three components $\langle G, i, j\rangle$. G is a digraph with integer costs assigned to its edges, and i and j are vertices in G. The cost of a path is the sum of the costs of the branches on the path. $f(G, i, j)$ is the least cost of a path in G from vertex i to vertex j.

Example 2.1.6: $D = N_0$.

$$f(d) = 1 \qquad \text{if } d = 0 \text{ or } 1$$

$$f(d) = f(d-1) + f(d-2) \qquad \text{if } d \in N_2$$

The specification of $f(d)$ in Examples 2.1.1 and 2.1.2 is given by equalities with compositions of primitive functions $(*, +)$ to be performed on d. In Example 2.1.2 the primitive functions differ, depending on the subset of the domain to which d belongs. The determination of d's subset is specified by a test involving primitive relations $(<\, , >)$. Each line in these definitions then gives a procedure in the form of a composition of primitive functions and is therefore called a *composition line*. These lines are also all *nonrecursive* because the compositions on the right have no occurrence of the name of the defined function, called *nonterminal*. A definition made up of such lines is virtually a program.

On the other hand, Examples 2.1.3–2.1.5 contain lines which, in addition to compositions, involve inequalities and other relations used to constrain the function value rather than to delineate a domain. These definitions cannot be evaluated by simply carrying out the given compositions. They do, however, give enough of the properties of $f(d)$ so that precisely one member of the range of f satisfies those properties. Each property that $f(d)$ must satisfy limits the possible value for a given d to a subset of R. The combination of all properties limits $f(d)$ to the intersection of these sets. Example 2.1.6 shares many of the properties of the previous examples. It consists entirely of equalities with a nonterminal on the left and compositions of primitives and, possibly, the nonterminal f on the right, and its last line is called a *recursive* composition line. According to that line, when d is drawn from a specified subset N_2, $f(d)$ is required to have the relation $=$ to its value at other members of its domain $[f(d-1) + f(d-2)]$. Any definition requiring a defined function to satisfy a relation between values at two or more places in its domain is called a *recursive definition*. We will generally reserve the term for definitions containing only recursive and nonrecursive composition lines. It is on such recursive definitions that this book focuses.

Given a recursive definition, we will show how to design an equivalent definition consisting of nonrecursive composition lines only, thus giving a basic design of actual programs. But function definitions do not necessarily originate in the recursive form. To make use of techniques for transforming recursive definitions to programs, it will be necessary also to develop ways to transform nonrecursive definition, to equivalent recursive definitions.

In the next sections the components of recursive definitions are more completely defined, and aspects of their significance are studied. The classical issues of recursive function theory, the power and limitations of such definitions, are only briefly noted. These issues are covered in [Rogers 67; Peter 51; Yasuhara 71], for example.

2.1.1 Composition Lines — Nonrecursive and Recursive

First consider some additional examples of recursive composition lines.

Example 2.1.7

$$h(r) = h(r/2) + h(r/2) \qquad :r \text{ is a real number}$$

Example 2.1.8

$$f(x, n) = f(x\|\langle 0\rangle, n-1) \cup f(x\|\langle 1\rangle, n-1) \quad :n \in N,\ x \text{ is a 0–1 vector}$$

[This line involves the primitive operators $\|$ (concatenation), and $\cup$ (union). The function argument consists of a string vector x and an integer n. Each possible argument is a point in the domain. Each such point is a vector–integer pair (x, n). In English the line says that the value of the function at the point (x, n) in its domain is equal to the union of its values at (x extended by concatenation of 0, $n-1$) and at (x extended by concatenation of 1, $n-1$).]

Now follows the general definition of a composition line. Let f be a nonterminal, X a set of arguments, P a finite set of primitive functions, and Q a predicate on X. In addition let $E(P, f, X)$ be a composition of the primitive functions in P, together with f and X, whose exact form depends on X. Then a composition line is of the form

$$f(X) = E(P, f, X) \qquad :Q(X) \tag{2.1.1}$$

Note that $E(P, f, X)$ is a composition which, beyond containing occurrences of X, can depend on X in other ways. For example, as illustrated in the following line, its length may depend on X,

$$g(v, n) = \bigcup_{i=1}^{n} [g(v\|\langle i\rangle, n-1)] \qquad :n \in N$$

In this line the expression on the right of the equal sign is $E(P, f, X)$, where P contains $\cup$, $\|$, and $-$ (subtraction); f is g; and X is the pair (v, n). The length of $E(P, f, X)$ in this instance depends on n or on a component of X.

In any case $E(P, f, X)$ must contain only a finite number of occurrences of primitive functions and f. Each occurrence of f in $E(P, f, X)$, or *f call*, has the form $f(e)$, where e is itself an expression with the same characteristics as $E(P, f, X)$.

The values of X for which $Q(X)$ is true constitute the domain of the line. [For such a line to be legitimate, the primitive functions in P must make sense in the domain $Q(X)$.]

If $E(P, f, X)$ contains no f calls, then Eq. (2.1.1) is a nonrecursive line. In this case the sense in which Eq. (2.1.1) defines a function is clear. The value of f at any X in its domain is determined by evaluating the primitive functions on the right of Eq. (2.1.1). If, on the other hand, it is recursive, that is, $E(P, f, X)$ does contain f, then the sense in which Eq. (2.1.1) provides a definition is not quite so obvious. A recursive line is *satisfied* by a primitive function $p(x)$ if substitution of p for f throughout the line yields a line in which the two sides of the equality are demonstrably equal whenever $Q(X)$ is true. If $p(x)$ satisfies a line, then that line *defines* $p(x)$.

A recursive line can thus define many, one, or no primitive functions. (Conversely, some primitive functions may not be definable by a definition containing only one recursive line.) The recursive line in Example 2.1.7 defines many—one for each number, called c, below. Included among the functions defined are those in the set

$$\{h_c(r) = c * r \mid c \text{ is any number}\}$$

because substituting in any of the functions in Example 2.1.7, we get for any number c that $c * r = c * (r/2) + c * (r/2)$.

Example 2.1.8 also defines an entire set of functions—a different one for each possible vector, called z below. Each of the functions has as its value a set of 0–1 vectors.

$$\{f_z(x, n) = \{x \| y \| z \mid y = \text{0–1 vector}, |y| = n\} \mid z = \text{any vector}\}$$

This is shown by substituting the primitive function above all occurrences of f throughout Example 2.1.8, namely, for $f(x, n)$, $f(x\|\langle 0\rangle, n - 1)$, and $f(x\|\langle 1\rangle, n - 1)$, and by showing that the equality is valid. When substituting, changing the letter y throughout the primitive function above to, for example, w, will not change its meaning.

$$\begin{aligned}
&\{x\|y\|z \mid y = \text{0–1 vector}, |y| = n\} \\
&\quad = \{x\|\langle 0\rangle\|w\|z \mid w = \text{0–1 vector}, |w| = n - 1\} \\
&\quad\quad \cup \{x\|\langle 1\rangle\|w\|z \mid w = \text{0–1 vector}, |w| = n - 1\}
\end{aligned}$$

This is a valid equality since the set of all y's (all 0–1 vectors of length n) is clearly the union of:

1. The set of all w's (all 0–1 vectors of length $n - 1$), with 0 concatenated in front of each, together with
2. The set of all w's (all 0–1 vectors of length $n - 1$), with 1 concatenated in front of each.

Similar reasoning holds for the recursive line using the primitive function minimum ($\downarrow$).

Example 2.1.9

$$f(n) = n \downarrow f(n) \qquad :n \in N$$

defines at least the set of primitive functions

$$\{ f_k(n) = n - k \mid k \in N^{n-1} \}$$

because if $k \in N^{n-1}$, then $n - k = n \downarrow (n - k)$.

In contrast to the preceding, the following recursive line uniquely defines the primitive function $f(n) = n$.

Example 2.1.10

$$f(n) = n \downarrow (f(n) + 1) \qquad :n \in N$$

Clearly $f(n)$ equals either n or $f(n) + 1$, and it cannot equal the latter.

Although Example 2.1.10 is uniquely satisfiable (in the range N), in practice definitions containing only recursive lines are rarely uniquely satisfied. Consider a recursive line whose evaluation ends after a finite number of substitutions because the arguments of f in the final expression are outside the domain of that line. This happens in Example 2.1.8 where the second argument of f decreases by 1 on each substitution and so eventually becomes less than 0 and thus outside the domain of the line. The value of a function at arguments in its domain, as defined by such a line, is ultimately dependent on the value of the function at argument values outside its domain. Because its specification is consistent with any value given to the function outside its domain, that line cannot uniquely define a function. Even evaluations of recursive lines which never generate an f argument outside the lines' domain can fail to define a unique function. Consider the following examples.

Example 2.1.11

$$f(x) = x/2 + f(x/2) \qquad :x \text{ real}$$

Repeated substitution yields

$$f(x) = x/2 + x/4 + \cdots + x/2^j + f(x/2^j)$$

This might be expected to converge uniquely to x. However, as is shown by substitution, $f(x) = a + x$ also satisfies Example 2.1.11 with any a. There are in fact infinitely many other functions that satisfy this example. Consider

$$f(x) = \begin{cases} x & :x \text{ rational} \\ x + 1 & :x \text{ irrational} \end{cases}$$

Consider the example

Example 2.1.12

$$f(n) = n \downarrow f(n+1) \qquad :n \in N$$

Again repeated substitution gives

$$f(n) = n \downarrow n + 1 \downarrow \cdots \downarrow n + j \downarrow f(n + j + 1)$$

This appears to make $f(n) = n$ satisfy Example 2.1.12. It does, but so does $f(n) = 0$ as well as other primitive functions.

Finally it is obvious that the definition

$$f(n) = f(n) + 1 \qquad :n \in N$$

has no solution. More precisely, there is no function $f(n)$ mapping integers to integers having the above property because there is no member $x \in N$ with the property $x = x + 1$.

2.1.2 Nested and Nonnested Lines

Recursive lines are either *nested* or *nonnested*. When the right side of a recursive line contains a nonterminal whose argument also contains a nonterminal, that line is nested. None of the previous definition lines are nested, but the next one is.

Example 2.1.13

$$f(x, y) = f(1, f(x, y - 1)) \qquad :x, y \in N$$

To express the functions that satisfy this definition we use notation for repeated nested application of a function, that is,

$$h^k(X) = h(h(\ldots h(h(X))\ldots))$$

(see Section 1.2).

$$f_h(x, y) = \{h^y(x) \mid h(x) \text{ is a function with domain and range } N\}$$

For example, if $h(x) = 1 + x$, then the satisfying function is equal to $x + y$; if $h(x) = 2 * x$, then the satisfying function is equal to $2^y * x$.

In Example 2.1.13 the nested definition line serves as a compact representation for repeating a primitive function n times. The closest we can come with a nonnested definition line is

$$f(x, y, z) = w(f(x, y - 1), z) \qquad :x, y \in N$$

where z is such that $w(x, z) = h(x)$.

In the next example a nested definition line defines a rapidly increasing function.

Example 2.1.14

$$f(n, m) = f(n-1, f(n-1, m)) \qquad : n, m \in N$$

As can be verified by substitution, this definition is satisfied by

$$\{ f_h(n, m) = h^{2^n}(m) \mid h(m) \text{ is a function with domain and range } N \}$$

With the nested line of Example 2.1.14 we are able to define a function whose value increases rapidly with its input. This is a power inherent in the nested definition line, which allows definition of functions that cannot be defined at all with nonnested lines. An example of such a definition is that due to Ackermann and Peter, Definition 2.4.4 [Yasuhara 71; Peter 51]. Three uses of nesting are:

1. To give a compact definition of a function which can be defined with a nonnested line of greater length.
2. To impose an ordering on the evaluation which cannot be imposed with an equivalent nonnested definition.
3. To define functions which cannot be otherwise defined.

The third use usually produces functions which are not of practical interest. Though the preponderance of recursive definitions considered here are nonnested, there will be occasion to use nested lines for each of the first two uses above.

2.2 RECURSIVE DEFINITION

A single composition line is not usually sufficient to define a function uniquely. A *recursive definition* consists of a finite number of composition lines, at least one of which is a recursive line, and which are applicable on mutually disjoint domains. (One can weaken this to insist only that, whenever two lines apply to a point in the domain, they yield the same value.) If the nonterminal is f, the definition is designated DEF.f. The union of all domains associated with nonrecursive lines in DEF.f forms its *terminal domain*; the union of all those associated with nonterminal lines is the *nonterminal domain*. The union of terminal and nonterminal domains is the *total domain*. The *starting*, or *initial*, *domain* is the one for which the function has been designed and is specified in the "Initially..." line. Sometimes the domain is made wider than its intended use in order to make the definition simple. Consider the following extension of the definition in Example 2.1.8.

Definition 2.2.1: Set of Binary n-Tuples

$$\left[\begin{array}{ll} f(x,n) = \{x\} & :n = 0 \\ f(x,n) = f(x\|\langle 0\rangle, n-1) \cup f(x\|\langle 1\rangle, n-1) & :n \in N \\ \text{Initially } x = \langle\ \rangle,\ n \in N. & \end{array}\right.$$

Here x is any string. $n = 0$. (x, n) is the terminal domain. With x being any string and $n \in N$, (x, n) is the nonterminal domain. The starting domain is given in the "Initially..." line.

In another example (Definition 2.2.2), $f(x, m, n)$ equals the set of all strings of 0's and 1's of length n, none of which has more than a total of m 1's, and in which every pair of 1's is separated by at least one 0. Here we are interested in illustrating the syntax of the definition. The fact that it actually has the claimed value will be verified in Section 2.3. Note the multiple recursive lines; the reader should verify that the three domains are disjoint.

Definition 2.2.2: Subset of Binary n-Tuples

$$\left[\begin{array}{ll} f(x,m,n) = \{x\} & :m \in N_0 \text{ and } n = 0 \\ f(x,m,n) = f(\langle 0\rangle\|x, m, n-1) & :(x = 1 \text{ and } m \in N), \\ & \quad \text{or } (x_1 = 0 \text{ and } m = 0) \\ & \quad \text{and } n \in N \\ f(x,m,n) = f(\langle 0\rangle\|x, m, n-1) & \\ \qquad\qquad \cup f(\langle 1\rangle\|x, m-1, n-1) & :x_1 = 0 \text{ or zilch} \\ & \quad \text{and } (m \in N \text{ and } n \in N) \\ \text{Initially } x = \langle\ \rangle,\ m \in N,\ n \in N. & \end{array}\right.$$

Here the terminal domain includes any values of (x, m, n) in which x is a string, $m \in N_0$, $n = 0$; and the nonterminal domain includes all values of (x, m, n) in which x is a string, $m \in N_0$, $n \in N$.

2.2.1 Standard Forms for Nonnested Recursive Definitions

The bulk of the algorithms in this book are based on nonnested definitions. Any such definition can be expressed in a *standard form*.

As previously described [Eq. (2.1.1)], any composition line can be represented thus:

$$f(X) = E(P, f, X) \qquad :Q(X)$$

For any nonnested line a specialization of Eq. (2.1.1) can be used:

$$f(X) = w\big(f(o_1(X)), \ldots, f\big(o_{m(X)}(X)\big)\big) \qquad :\sim T(X) \qquad (2.2.1)$$

where X is called the *argument structure*. The o_i's are called *O functions*. The *m function* gives the number of f calls on the right. The *w function* has the value of the right side f calls as arguments. The O, w, and m functions are primitive; $\sim T(X)$ is a predicate giving a nonterminal condition.

For example, examining the recursive line from Definition 2.2.1 for binary n-tuples,

$$f(x, n) = f(x\|\langle 0\rangle, n - 1) \cup f(x\|\langle 1\rangle, n - 1) \qquad :n \in N$$

X is the vector–integer pair (x, n). $o_1(X) = o_1(x, n)$ is the pair obtained by concatenating 0 with x, and subtracting 1 from n, so $o_1(x, n) = (x\|\langle 0\rangle, n - 1)$; similarly $o_2(X) = (x\|\langle 1\rangle, n - 1)$. More compactly, $o_i(X) = (x\|\langle i - 1\rangle, n - 1)$. $m(X) = m(x, n) = 2$; here a constant, though in general the number of f calls on the right depends on X. w is the union. $\sim T(X) = \sim T(x, n) = n \in N$, which in this case does not depend on x.

Definition 2.2.1 has one nonterminal line in standard form, Definition 2.2.2 has two.

Equation (2.2.1) is a specialization of Eq. (2.1.1) to nonnested recursive lines. For nonrecursive or terminal lines there is the simpler specialization, namely,

$$f(X) = t(X) \qquad :T(X) \tag{2.2.2}$$

$t(X)$ is the *terminal value*, $T(X)$ the terminal condition.

In Definition 2.2.1 there is one such standard form line, namely,

$$f(x, n) = \{x\} \qquad :n = 0$$

Here $t(X) = t(x, n) = \{x\}$; it depends only on x. $T(X) = T(x, n)$ is $n = 0$; it depends only on n.

In general *any* nonnested recursive definition can be given as a collection of lines of the form of Eqs. (2.2.1) and (2.2.2). For the most part we use this form or the variant below.

$$f(X) = w\big(f(o_1(X)), \ldots, f(o_{m(X)}(X)), p(X)\big) \qquad :\sim T(X) \tag{2.2.3}$$

Here the *p function* $p(X)$ is included. It provides convenience but no new defining power. In fact, Eq. (2.2.3) is equivalent to a pair of lines having the forms of Eqs. (2.2.1) and (2.2.2). To get the equivalent form, the argument u is added to the argument structure X. Equation (2.2.3) is changed by adding this second component to each f call, making $u = 1$ in each f call on the right of this line, and replacing $p(X)$ with $f(X, 0)$. This gives one line of the equivalent definition. In the second line $f(X, 0)$ is made equal to $p(X)$. This is a terminal line. We get

$$f(X, u) = w\big(f(o_1(X), 1), \ldots,$$

$$f\big(o_{m(X)}(X), 1\big), f(X, 0)\big) \qquad :\sim T(X) \text{ and } u = 1$$

$$f(X, u) = p(X) \qquad :u = 0$$

Any definition composed of recursive lines of the form of Eq. (2.2.1) or (2.2.3) and terminal lines of the form of Eq. (2.2.2) is said to be in *standard (nonnested) form.*

Here is another form which cannot define any new functions, though it is often convenient,

$$f(X) = w_{j \in N^{m(X)}}\big(f\big(o_j(X)\big)\big) \qquad :\sim T(X) \qquad (2.2.4)$$

Since this form makes no commitment on the order of the f call arguments of the w function, its use is restricted to situations in which all orders of evaluation must give the same answer, that is, w is commutative. (Though any ordering will do when the recursive line is written as in Eq. (2.2.4), a particular one will have to be chosen for the implementation algorithm.)

2.3 THE FUNCTION DEFINED BY A RECURSIVE DEFINITION AND TERMINATION

A recursive definition defines the function or functions that satisfy it. Unfortunately there is no general procedure to determine what these functions are. Recursion is the source of the difficulty—but also the source of the power of these definitions. Despite the general difficulty there are significant classes of definitions for which the defined function or functions can be determined. For example, all recursive definitions that *terminate* provide unique function definitions.

2.3.1 Substitutional Evaluation and Termination

Termination is related to the straightforward substitution procedure one might naturally choose to evaluate a recursive definition DEF.f at an argument d. The procedure starts with the expression $f(d)$ and creates a sequence of such expressions, each obtained from its predecessor by replacing one or more occurrences of the form $f(\beta)$ with the right side of the recursive line for $f(\beta)$. If this procedure eventually produces an expression χ, containing no occurrences of f, then we say that DEF.f *terminates* at d. Clearly $f(d) = \chi$ and, as will be shown later, $f(d)$ can only be χ. For example, in obtaining the value of $f(\langle\ \rangle, 2, 3)$, as defined by Definition 2.2.2, using such a substitutional procedure, the following expressions, separated by $\Rightarrow$, would be developed.

Example 2.3.1

$$
\begin{array}{lllll}
f(\langle\ \rangle, 2, 3) \Rightarrow & & & & \\
\mathrm{n} f(\langle 0\rangle, 2, 2) & & & \cup f(\langle 1\rangle, 1, 2) \Rightarrow & \\
f(\langle 00\rangle, 2, 1) & & \cup f(\langle 10\rangle, 1, 1) & \cup f(\langle 1\rangle, 1, 2) \Rightarrow & \\
f(\langle 000\rangle, 2, 0) & \cup f(\langle 100\rangle, 2, 0) & \cup f(\langle 10\rangle, 1, 1) & \cup f(\langle 1\rangle, 1, 2) \Rightarrow & \\
\{\langle 000\rangle\} & \cup \{\langle 100\rangle\} & \cup f(\langle 10\rangle, 1, 1) & \cup f(\langle 1\rangle, 1, 2) \Rightarrow & \\
\Rightarrow \{\langle 000\rangle & \langle 100\rangle\} & \cup f(\langle 101\rangle, 1, 0) & \cup f(\langle 1\rangle, 1, 2) & \\
\Rightarrow \{\langle 000\rangle & \langle 100\rangle & \langle 010\rangle\} & \cup f(\langle 1\rangle, 1, 2) & \\
\Rightarrow \{\langle 000\rangle & \langle 100\rangle & \langle 101\rangle\} & \cup f(\langle 01\rangle, 1, 1) & \\
\Rightarrow \{\langle 000\rangle & \langle 100\rangle & \langle 010\rangle\} & \cup f(\langle 001, 1, 0) \cup f(\langle 101\rangle, 0, 0) & \\
\Rightarrow \{\langle 000\rangle & \langle 100\rangle & \langle 101\rangle & \langle 001\rangle & \langle 101\rangle\}
\end{array}
$$

Since the procedure sketched above for evaluating recursive definitions is fundamental, it is restated in detail below.

The procedure for a *substitutional evaluation* (SE) of DEF.f at a member of its starting domain, say d, involves the progressive refinement of a "sentential form," designated F_i, as follows.

Process 2.3.1: Substitutional Evaluation, or SE
F_1, the first sentential form, is set to $f(d)$. The $(i+1)$th sentential form, F_{i+1}, is produced from the ith, F_i, as follows:

1. Evaluate all primitive subexpressions in F_i. This gives F_i'.
2. (a) If F_i' does not contain any occurrence of f (f *calls*), then F_i' is terminal, $f(d) = F_i'$, and the evaluation *terminates*.
   (b) Otherwise, as is easy to show inductively, F_i' will contain at least one f call, say $f(\beta)$, denoted by saying: "$F_i' = \ldots f(\beta) \ldots$", whose argument β *does not contain another occurrence of* f nor, being completely evaluated, of any primitive function. (So evaluated, β is called a *primitive argument structure*.) Let $f(X) = E(P, f, X)$ be the form of the composition line in DEF.f whose domain contains β. Replace the noted occurrence of $f(\beta)$ in F_i' with $E(P, f, \beta)$ to obtain F_{i+1}.

If Process 2.3.1 terminates, then DEF.f *terminates at* d. If DEF.f terminates at every member of its starting domain, then DEF.f *terminates*. This way of defining termination is tied to evaluation by SE. It is the termination definition to which we adhere throughout the book, but it is not the only one used in the literature. For nonnested lines it is equivalent to the others, but for nested lines it requires that all nonterminal (f) occurrences be eventually removed under the constraint that substitution is always for the innermost of a group of nested nonterminals before any of the others in that group. If termination fails under this SE order, even though there exists an order of substitution under which all f's would eventually be removed, we say the definition *does not* terminate. Other definitions of termination are more permissive in the substitutional order allowed.

Example 2.3.2: A second example of SE, Process 2.3.1, follows:

$$\left[\begin{array}{ll} f(x) = x - 1 & :x > 5 \\ f(x) = f(f(x+2)) & :x \leq 5 \\ \text{Initially } x \in N. & \end{array}\right.$$

Evaluating $f(1)$ by substitutional evaluation,

$$\begin{aligned} f(1) \Rightarrow f(f(3)) \Rightarrow f(f(f(5))) &\Rightarrow f(f(f(f(7)))) \\ &\Rightarrow f(f(f(6))) \\ &\Rightarrow f(f(5)) \Rightarrow f(f(f(7))) \\ &\qquad\qquad\quad \Rightarrow f(f(6)) \\ &\qquad\qquad\quad \Rightarrow f(5) \Rightarrow f(f(7)) \\ &\qquad\qquad\qquad\qquad\quad \Rightarrow f(6) \\ &\qquad\qquad\qquad\qquad\quad \Rightarrow 5 \end{aligned}$$

Example 2.3.3: Similarly, reusing results obtained above

$$f(2) \Rightarrow f(f(4)) \Rightarrow f(f(f(6))) = 5$$

$$f(3) \Rightarrow f(f(5)) = 5$$

$$f(4) \Rightarrow f(f(6)) = 5$$

$$f(5) \Rightarrow f(f(7)) = 5$$

Therefore DEF.f terminates at $x = 1, 2, 3, 4, 5$, and since for all $x > 5$ the nonrecursive line in the definition applies, DEF.f terminates for all $x \in N$.

As noted, SE is not the only plausible evaluation process. In this example, instead of $f(f(f(5)))$, it would have been legitimate to use the recursive line for $f(3)$ to obtain $f(f(f(f(3) + 2)))$ directly from $f(f(3))$ provided it had been ascertained that $f(3) \leq 5$.

This example is not typical. It was chosen because its proof of termination is straightforward. In general, proof of termination is not so easy. To explore some of the more difficult proofs of termination, as well as other proofs, we need additional definitions. These will be considered next.

2.3.2 The Substitutional Domain of a Recursive Definition

A definition DEF.f can fail to terminate in a number of ways. It can fail because F_i of Process 2.3.1 continues to contain occurrences of f no matter how large i gets. We could ensure termination by changing the definition on an appropriate subset of its total domain to a nonrecursive line. DEF.f can also fail to terminate because it is *incomplete*. Assume $f(A)$ is being substitutionally evaluated and the sentential form $F_i = \ldots f(\beta)\ldots$ is eventually produced. If β is outside the total domain of DEF.f, then SE can never replace $f(\beta)$. Therefore $f(\beta)$ must remain in all subsequent sentential forms, $F_{i+1}, F_{i+2}, \ldots,$ if it is even possible to produce more such forms. Failure to terminate results because the total domain of DEF.f does not include all the arguments of f which can arise in an SE of DEF.f, at A in this case. Unlike the previous reason for failure to terminate, is an incompleteness of definition and should not occur in any "reasonable" definition. Such a "reasonable" definition will be called *complete*. Barring explicit denial, all our definitions are assumed to be complete.

If A is a member of the total domain of DEF.f, the set of all primitive arguments of f, which occur in sentential forms of some SE of DEF.f at A, is denoted by $\Delta_f(A)$ and is called the *substitutional domain of DEF.f at A*. Usually we are interested in $\Delta_f(A)$ when A is a member of the starting domain. $\Delta_f(A)$ is a subset of the total domain of DEF.f (by the assumed completeness of DEF.f). $\Delta_f = \bigcup_{d \text{ in starting domain of } f}[\Delta_f(d)]$ is the *substitutional domain of DEF.f*.

2.3.2.1 Finite Substitutional Domain

The following is a direct consequence of the definition of Δ_f.

Lemma 2.3.1: If DEF.f terminates at A, then $\Delta_f(A)$ is finite.

PROOF: We can form a tree which records all possible sentential forms produced by SEs starting with $f(A)$. $f(A)$ is the sentential form at the root. For a tree node n with sentential form S_n there is one child node associated with each f call in S_n. The sentential form associated with each child is the result of substituting for the corresponding f call in S_n. Each node has a finite number of children. Each sentential form is of finite length and contains a finite number of argument structures, and each path from the root, representing an SE, is finite. Therefore by Koenig's lemma [Yasuhara 71, p. 250] the tree is finite, and so is the number of distinct argument structures. ■

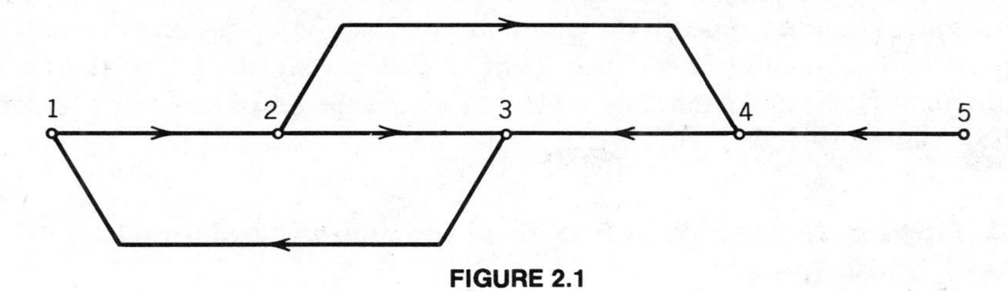

FIGURE 2.1

On the other hand, just because $\Delta_f(A)$ is finite does not mean DEF.f must terminate. In the next definition, $f(x, G)$ is the set of all vertices that can be reached (by a directed path of length 0 or more) from vertex x in the digraph G. It does not terminate if G has any cycles involving a vertex reachable from x. $N(x)$ is the number of outward neighbors of x and $n_i(x)$ is the ith outward neighbor of x.

Definition 2.3.1: Reachable Vertices

$$\left[\begin{array}{ll} f(x,G) = \{x\} & :N(x) = 0 \\ f(x,G) = \bigcup_{i=1}^{N(x)} [f(n_i(x),G)] \cup \{x\} & :N(x) \in N \\ \text{Initially } x \text{ is vertex in } G. & \end{array}\right.$$

For example, consider graph G in Figure 2-1. SE for finding $f(1, G)$ for this graph starts as follows.

Example 2.3.4

$$\begin{aligned} f(1) &\Rightarrow f(2) \cup \{1\} \\ &\Rightarrow f(3) \cup f(4) \cup \{1,2\} \\ &\Rightarrow f(1) \cup f(4) \cup \{1,2,3\} \end{aligned}$$

Since $f(1)$ recurs, the SE procedure will not terminate.

In this example $\Delta_f(1, G)$ is finite $(= \{1, 2, 3, 4\})$, and a set of four equations whose solution will give the answer is:

$$f(1) = f(2) \cup \{1\}$$

$$f(2) = f(3) \cup f(4) \cup \{2\}$$

$$f(3) = f(1) \cup \{3\}$$

$$f(4) = f(3) \cup \{4\}$$

These equations are formed by a variation on SE to be discussed later in this chapter as equation set evaluation (ESE). Though the definition does not terminate at $(1, G)$ and cannot be evaluated by SE, the set of equations formed by ESE can be solved.

2.3.3 Types of Arguments of Recursive Definitions — Information, Control, and Invariant

The arguments of the nonterminal f in a recursive definition, their type and number, constitute its *argument structure*. The values the argument structure takes during SE are the members of Δ_f. The term argument structure serves to emphasize the fact that the arguments are often complex. Typically an argument is an entity such as a sequence, a graph, a matrix, a grammar, or similar structure, and there are usually a number of such arguments in a definition. All the argument structures appearing within a definition, ex., X and $o_i(X)$ in the standard form nonnested definition, have the same number of components. An argument structure then is a collection of data having a fixed structure whose values are altered in the process of evaluation. The evaluation in turn is carried out by an implementing algorithm, in which the argument structure is represented by a *data structure*. Design of argument structures for efficiency is a similar enterprise to design of data-structures.

Definition 2.3.2

$$\left[\begin{array}{ll} f(v, n, S) = \{v\} & :n = 0 \\ f(v, n, S) = \bigcup_{i=1}^{|S|} f(v \| \langle S_i \rangle, n-1, S) & :n \in N \\ \text{Initially } S \text{ is a finite set of letters, } v = \langle\ \rangle, n \in N. & \end{array}\right.$$

The individual components or arguments of the argument structure (v, n, S) each serve different ends in this definition.

v, the first component of the argument structure, appears on the right of the terminal line, directly determining the terminal value of $f(v, n, S)$. Even when a component does not affect any terminal value, it may still affect the value of a quantity which appears outside the argument of any f call in a recursive line. For example, in the line

$$f(n, s) = \langle s \rangle | * | f(n-1, 0) \| \langle 1 - s \rangle | * | f(n-1, 1) \qquad :n \in N$$

the component s appears outside the argument of any f call in $\langle s \rangle$ and $\langle 1 - s \rangle$. These affected quantities will in turn affect the value of f. A component which affects the terminal value in a nonrecursive line, or values outside the argument structures in a recursive line, is an *info component*.

n, the second component, serves to distinguish among the subdomains associated with the different lines of the definition, and thus helps determine which line is applicable to a given argument structure. Such a component is called a *(definition) control component*. In Chapter 4 we will encounter components which are used only to simplify programming of the definition. These are also referred to as *(program) control components*.

S, the third component, appears unchanged on the left and right of each recursive line in the definition. We call such a component *invariant*. Despite its inactivity, S is necessary because S_i is used in constructing the first argument, and $|S|$ determines the upper limit of the $\cup$ in the definition. S is invariant, but also a control component because $|S|$ gives the number of appearances of f on the right of the recursive line. Because the first component, v, is an info component and S enters into the computation of the first component of the right, S is also an info component.

From here on invariant components will ordinarily be excluded from the argument of recursive definitions. Instead we will list such invariants in the "Initially..." line at the end of the definition. So in Definition 2.3.2 the third argument of f would be removed throughout and the "Initially..." line is changed to:

"Initially $v = \langle\ \rangle$, $n \in N$, S is a finite set of letters and invariant."

Another example of different component types is provided by Definition 2.3.3 for binary n-tuples.

Definition 2.3.3

$$\left[\begin{array}{ll} f(v) = \{v\} & :|v| = n \\ f(v) = f(v\|\langle 0\rangle) \cup f(v\|\langle 1\rangle) & :|v| < n \\ \text{Initially } v = \langle\ \rangle,\ n \in N, \text{ and } n \text{ is invariant.} & \end{array}\right.$$

In this definition,

1. n is invariant and control, and consequently is eliminated in the argument structure.
2. v appears here both in defining different subdomains and in the terminal line outside any argument of f. Such a component is an *information–control component*.

A definition for a given function may be written in many different ways. Note the following for example,

For any definition with information–control components there is an equivalent one which has only pure information and pure control components. For example, in Definition 2.3.3 v has control aspects because the *size of* v is needed to distinguish the terminal and nonterminal domain. Let k denote the size of v (instead of $|v|$). Then v becomes pure information (not a control component); k becomes pure control, with $k = n$ and $k < n$, respectively, denoting the terminal and nonterminal domains; and n remains control (and invariant). This would result in an altered definition in which the recursive line for example becomes

$$f(v, k) = f(v\|\langle 0\rangle, k + 1) \cup f(v\|\langle 1\rangle, k + 1) \qquad :k < n$$

The conditions $k < n$ and $k = n$ could alternatively be written $0 < n - k$ and $0 = n - k$, respectively. Thus the distinction in domains depends on the quantity $n - k$. Since this is the only way that n or k enter into the definition, let $n - k = p$. Replacing components $n - k$ with p, we get the following definition.

Definition 2.3.4

$$\left[\begin{array}{ll} f(v, p) = \{v\} & :p = 0 \\ f(v, p) = f(v\|\langle 0\rangle, p - 1) \cup f(v\|\langle 1\rangle, p - 1) & :p > 0 \\ \text{Initially } v = \langle\ \rangle,\ p \in N. & \end{array}\right.$$

Although removal of information-control components is always possible, it may require duplication of a component. Then, even though both copies of the component have the same value, one copy is used for information and the other for control. (See Problem 10, Section 2.12.)

2.3.3.1 SIMPLIFIED PURE INFORMATION ARGUMENT HANDLING

A definition with a pure information component can be evaluated by including only a stripped down version of that component in the definition. After evaluation, using the history of that component, we can recreate complete information. In practice this approach is often adopted with consequent improvement in efficiency. Consider a definition of the following form.

Definition 2.3.5

$$\left[\begin{array}{ll} f(X, Y) = \langle t_1(X), t_2(Y)\rangle & :T(X) \\ f(X, Y) = \mathrm{SEL}(f(o_1(X), v_1(X, Y)), \ldots, & \\ \qquad f(o_{m(X)}(X), v_{m(X)}(X, Y)) & :\sim T(X) \\ \text{Initially } X \in D1 = d1,\ Y \in D2 = d2. & \end{array}\right.$$

with $\mathrm{SEL}(\langle a_1, b_1\rangle, \langle a_2, b_2\rangle, \ldots, \langle a_n, b_n\rangle) = \langle a_j, b_j\rangle$ where the selection is based on a criterion independent of the b_i's.

Because Y is a pure information component in Definition 2.3.5, the terminal and nonterminal domains (T, $\sim T$) and the number of f calls on the right of the definition (m) are independent of Y, but Y has the additional property that it has no effect on the pairs selected by SEL. Only the final value of $f(X, Y)$ depends on Y; the process of evaluation does not require Y at all.

The solution to the following definition will contain enough information to rapidly reconstruct the solution to Definition 2.3.5.

Definition 2.3.6

$$\left[\begin{array}{ll} f(X, Z) = \langle t_1(X), Z\rangle & :T(X) \\ f(X, Z) = \text{7SEL}(f(o_1(X), Z\|\langle 1\rangle) \cup \cdots & \\ \qquad \cup f(o_{m(X)}(X), Z\|\langle m(X)\rangle)) & :\sim T(X) \\ \text{Initially } X \in D1 = d1,\ Z = \langle\ \rangle. & \end{array}\right.$$

In this definition the pure information component has been replaced by a queue of f call indexes or child #'s, and thus directly records the history of an evaluation. This queue is called the *solution structure* because SEL, $f(X, Z)$, like $f(X, Y)$, selects a number of pairs. Because the first components of these pairs are identically generated in both definitions there is a one-to-one correspondence between the pairs selected by both. Let $\langle a, z\rangle$ be such a pair from $f(X, Z)$; then the corresponding pair of $f(X, Y)$ has an identical first component a, and the second component is computed as follows.

We first define the function $Q(X, Z, Y)$, which uses the queue Z and the initial values of X and Y to reconstruct the second component that would have been produced by $f(X, Y)$.

Definition 2.3.7

$$\left[\begin{array}{ll} Q(X, Z, Y) = Y & :Z = \langle\ \rangle \\ Q(X, Z, Y) = Q(o_i(X), v_i(X, Y), Z - Z_1) & :Z \neq \langle\ \rangle \\ \qquad \text{where } i \text{ is the contents of } Z_1 & \\ \text{Initially } X \in D_1 = d_1,\ Y \in D_2 = d_2,\ Z \text{ is a queue formed as a second} & \\ \text{component generated by Definition 2.3.6.} & \end{array}\right.$$

Corresponding to $\langle a, z\rangle \in f(X, Z)$ then is $\langle a, t_2(Q(d_1, z, d_2))\rangle$ of $f(X, Y)$.

This way of determining the pairs of $f(X, Y)$ is valuable if SEL selects only a few pairs. Then $Q(X, Z, Y)$ need only be applied to a few queues—those associated with solution pairs, and not for computations that produce argument structures which will not be selected.

Examples of this way of approaching pure information components are scattered throughout the text.

Sometimes it is profitable to maintain the values of X, rather than recomputing them when obtaining Q. In that case we could implement Definition 2.3.8 rather than 2.3.7.

Definition 2.3.8

$$\begin{bmatrix} f(X,Z) = \langle t_1(X), Z\rangle & :T(X) \\ f(X,Z) = w(f(o_1(X), \langle X,1\rangle \| Z), \ldots, & \\ \qquad f(o_{m(X)}(X), \langle X, m(X)\rangle \| Z)) & :\sim T(X) \\ \text{Initially } X \in D_1 = d_1,\ Z \in \langle\ \rangle. & \end{bmatrix}$$

If Definition 2.3.8 contains the pair $\langle a, z\rangle$, the corresponding pair of $f(X, Y)$ of Definition 2.3.5 is computed using the following definition of $Q(Y, Z)$.

Definition 2.3.9

$$\begin{bmatrix} Q(Y,Z) = Y & :Z = \langle\ \rangle \\ Q(Y,Z) = Q(v_i(X,Y), Z - Z_1) & :Z \neq \langle\ \rangle \\ \qquad \text{where } X \text{ is the first and } i \text{ the second component of } Z[1] & \\ \text{Initially } Y \in D_2 = d_2,\ Z \text{ is a queue formed as a second component} & \\ \text{generated by Definition 2.3.6.} & \end{bmatrix}$$

In the one-to-one correspondence $\langle a, t_2(Q(d_2, z))\rangle$ of $f(X, Y)$ corresponds to $\langle a, z\rangle \in f(X, Z)$.

Even when a definition does not have a pure information component, we will still speak about its solution structure. We then mean the effect of an evaluation on a queue component, like that in Definition 2.3.8 or 2.3.9, if such a component were added to the argument structure.

2.3.4 Dependence — Graphical Representation

If d is a member of the starting domain of DEF.f, then the members of $\Delta_f(d)$, have a familylike relation to each other.

Let $A \in \Delta_f(d)$. The members of $\Delta_f(A)$, other than A itself, are *descendants* of A, while A is an *ancestor* or *predecessor* of its descendants. Assume that the composition line in whose domain A occurs has the form $f(X) = E(P, f, X)$. Suppose $E(P, f, A)$, that is, the right side of this line with A replacing X, is of the form "$\ldots f(C) \ldots$." C is a primitive argument structure if $E(P, f, A)$ is not nested; if it is nested, C is an expression involving the nonterminal f which, after subsequent substitutions, becomes a primitive argument structure. In either case call the resultant primitive argument structure C'. We say A *directly depends* on the primitive argument structure C'. The set of all primitive argument structures on which A is directly dependent is designated ddep(A), and the members of ddep(A) are called the *children* of A. Note that we consider only definitions in which ddep(A), the number of children of A, is finite. The members of ddep(A) are called *1st descendants* of A. C is a *jth descendant* of A if and only if B is the $(j - 1)$th descendant of A

and C is in ddep(B). A is the *parent* of the members of ddep(A). *A depends* on C or C is in dep(A), and A is the ancestor of C, if and only if C is a jth descendant of A for some j.

The dependence relations associated with DEF.f evaluated at d can be represented by a digraph $G_f(d)$. Associated with each number of $\Delta_f(d)$ there is a vertex of $G_f(d)$, and conversely. Reference to a vertex of G_f will often be used to refer to its associated $\Delta_f(d)$ member and vice versa; the appropriate interpretation will be clear from the context. For A and B in $\Delta_f(d)$ there is an edge from A to B in $G_f(d)$ if and only if A directly depends on B.

Many of the concepts important in the study of recursive definitions have a graphic interpretation. A sampling of these follows. *A depends* on C if and only if there is a *path* from A to C in $G_f(d)$. DEF.f *terminates* at A if and only if *no path containing A has a cycle* or is of infinite length. $|\Delta_f(A)|$ is the *number of different vertices* that can be reached by a path containing at least one branch from A in $G_f(d)$.

2.4 PROOF OF TERMINATION [Manna 80]

Although a definition need not terminate to be useful, the presence of termination facilitates understanding, manipulation, and proof. Inductive proofs become particularly natural. Constructing a proof of termination is aided by assigning integers to the members of Δ_f. The integers serve to partially order Δ_f and provide the framework for induction. Consider the Fibonacci number definition.

Definition 2.4.1: Fibonacci Number

$$\left[\begin{array}{ll} f(x) = 1 & : x \in \{0,1\} \\ f(x) = f(x-1) + f(x-2) & : x \in N_2 \\ \text{Initially } x = d \in N_0 & \end{array}\right.$$

For this definition $\Delta_f(d)$ for any $d \in N_0$ consists of the non-negative integers and nothing else, so the integer function $s(x) = x$ maps each member of Δ_f to an integer. Consider the recursive line of Definition 2.4.1. To evaluate $f(x)$ if the integer $s(x) = x$, we have to first evaluate $f(x)$ at $x - 1$ and $x - 2$, corresponding to the smaller integers, namely, $s(x - 1) = x - 1$ and $s(x - 2) = x - 2$. Then to evaluate f at $x - 1$ and $x - 2$ we must evaluate f at a member of its domain corresponding to still smaller integers, and so on. The evaluation of f at any x then will depend *in a finite number of steps* on its evaluation at domain member y with $s(y)$ equal to 0 or 1. This of course assumes that one of the definition lines is applicable at each member of the domain encountered in this evaluation, that is, it assumes that the function is *complete*.

In general, proof of termination for a complete definition is assured if there exists an integer-valued function s, with the following properties.

Definition 2.4.2: Termination Function s

1. With $X \in \Delta_f$, $s(X)$ is a positive integer or 0, that is, s: $\Delta_f \rightarrow N_0$.
2. For any nonterminal $X \in \Delta_f$ and for all $Y \in \text{ddep}(X)$ (all children of X), $s(X) > s(Y)$.

(The number of children of X must of course be finite, which it will be if the definition conforms to the requirements we have placed on recursive definitions.)

Theorem 2.4.1: The existence of the function s which satisfies Definition 2.4.2 is necessary and sufficient to ensure that a complete definition DEF.f terminates.

PROOF: We demonstrate sufficiency by induction on the integers $s(X)$, $X \in \Delta_f$.

If $X \in \Delta_f$ and $s(X) = 0$, then X must be terminal and $f(X)$ must terminate. If X is not terminal but $s(X) = 0$, then by Definition 2.4.2(2) its children (which must be in Δ_f) have an s-value less than 0, which is impossible by Definition 2.4.2(1).

Now assume $X \in \Delta_f$ and DEF.f terminates at all argument structures $Z \in \Delta_f$ for which $s(Z) < s(X)$. Then:

1. If X is terminal then DEF.f terminates at X.
2. If X is not terminal then a right side in DEF.f of $f(X)$ contains a finite number of f-calls, each of whose arguments is a child, Y, of X. But $s(Y) < s(X)$ for each Y and so the evaluation of DEF.f at Y terminates and so does the evaluation of the right side. Thus $f(X)$ terminates.

The existence of an integer function satisfying Definition 2.4.2 then is sufficient to show termination.

Necessity is demonstrated by showing that given the properties of function s, it is possible to construct such a function for any DEF.f which terminates. Let $X \in \Delta_f$. The function $s(X) = |\Delta_f(X)|$ (called the *dependence number of* X) satisfies Definition 2.4.2. Because DEF.f terminates, $|\Delta_f(X)|$ is finite. Certainly if X is terminal, it has no descendants, that is, $|\Delta_f(X)| = 0$. Furthermore if X depends on Y, then $\Delta_f(X) \supset \Delta_f(Y)$, so $|\Delta_f(X)| > |\Delta_f(Y)|$.

■

2.4.1 Examples of Termination Proofs

We give some simple examples of definitions and the s functions which establish their termination.

In the greatest common divisor definition there are two components in the argument, but only one needs to be considered in establishing termination.

Definition 2.4.3: Greatest Common Divisor

$$\left[\begin{array}{ll} f(x, y) = x & : y = 0 \\ f(x, y) = f(y, \text{rem}(x, y)) & : y \in N \\ \quad\quad \text{where rem}(x, y) = \text{the remainder of } x/y & \\ \text{Initially } x, y \in N \text{ and } y \leq x. & \end{array}\right.$$

Let $s(x, y) = y$. Note $\Delta_f \subseteq N_0 \times N_0$, so $s\colon N_0 \times N_0 \to N_0$. Thus Definition 2.4.2(1) is satisfied. Clearly $s(x, y) = y > \text{rem}(x, y) = s(y, \text{rem}(x, y))$ as long as $y > 0$, and y must be > 0 for (x, y) to have children in Δ_f. So Definition 2.4.2(2) is satisfied. Therefore if Definition 2.4.3 is complete, it terminates. But there is a line for each argument structure in Δ_f, so the definition is complete. In general, once Δ_f is determined, completeness is usually easily shown. In subsequent examples, establishing Δ_f and completeness are left as exercises.

Here is a nested definition for which the s function is easily established.

$$\left[\begin{array}{ll} f(x, y) = y & : x \in \{0, 1, 2\} \\ f(x, y) = f(x - 3, f(x - 1, y^2)) & : x \in N_3 \\ \text{Initially } x \in N \text{ and } y \in Z \text{ (set of reals)}. & \end{array}\right.$$

The function $s(x, y) = x$ works here. $\Delta_f \subseteq N_0 \times Z$, therefore $x \in N_0$. For any (x, y) in the substitutional domain of f, clearly $x \in N_0$, whence Definition 2.4.2(1) is satisfied. Definition 2.4.2(2) is also satisfied. There are two children to consider. The first of these, say c_1, is $\langle x - 1, y^2 \rangle$; the second, say c_2, is $\langle x - 3, f(c1) \rangle$; and therefore $s(c_1) = x - 1$ and $s(c_2) = x - 3$. So both are less than $s(x, y) = x$; therefore the definition terminates.

Even though it is difficult to formulate an s function for some definitions, the definition can be transformed into an equivalent one for which an s function is easily established. For example, consider the Ackermann–Peter function definition.

Definition 2.4.4: Ackermann–Peter Function

$$\begin{array}{ll} 1. \\ 2. \\ 3. \\ \end{array} \left[\begin{array}{ll} f(x, y) = y + 1 & : x = 0 \\ f(x, y) = f(x - 1, 1) & : y = 0 \\ f(x, y) = f(x - 1, f(x, y - 1)) & : x, y \in N \\ \text{Initially } x, y \in N_0,\ x + y \in N. & \end{array}\right.$$

As it stands, it is difficult to establish an s function which must depend on both x and y since neither x nor y decreases consistently in going from the left (parent) to the right (children) of a definition line. But the definition can be transformed into an equivalent one by repeated substitutions.

Consider line 3. If $x \geq y = 1$, then this line becomes

$$f(x, y) = f(x-1, f(x,0)) = \underset{1}{f}(x-1, \underset{2}{f}(x-1,1)) \quad \leftarrow \text{count of nested } f\text{'s}$$

If $x \geq 1,\ y = 2$, then

$$\begin{aligned} f(x, y) &= f(x-1, f(x,1)) && \text{by line (3)} \\ &= f(x-1, f(x-1, f(x,0))) && \text{substituting for } f(x,1) \\ &= f(x-1, f(x-1,1)) = \underset{1}{f}(x-1, \underset{2}{f}(x-1, \underset{3}{f}(x-1,1))) && \text{by line 2} \end{aligned}$$

In general if $x \geq 1,\ y \geq 1$,

$$f(x, y) = \underset{1}{f}(x-1, \underset{2}{f}(x-1, \underset{3}{f}(x-1, \underset{,\ldots,}{\ldots,} \underset{y+1}{f}(x-1,1)\ldots)))$$

Now this line may replace line 3 in Definition 2.4.4 without changing the function defined. With this replacement the s function $s(x, y) = x$ will satisfy Definition 2.4.2, as is easily verified. Thus the revised definition terminates. Since any substitution using line 3 of this revised definition is the same as a finite number, $y + 1$, of substitutions of line 3 of the original definition, the original definition also terminates.

In all the examples given thus far, even when a child requires evaluation of an f call, we have avoided having an f call as an argument of an s function. This has been done by making the s function depend on only those components of the child that were not f calls. This dodge is not always available. For example, consider the following definition.

Definition 2.4.5

$$\begin{array}{ll} 1. & \left[\begin{array}{ll} f(i) = i & :i \leq 0 \\ f(i) = f(f(i-1)-1) & :i > 0 \\ \text{Initially } i \in N. & \end{array}\right. \\ 2. & \end{array}$$

We propose as a function having properties of Definition 2.4.2, $s(i) = i$ if $i > 0$ and $= 0$ otherwise. It follows that $s(i) \geq 0$ for $i \in \Delta_f$. $s(i)$ is also integer valued (this is implicit in the remainder of this argument), so Definition 2.4.2(1) is satisfied.

For Definition 2.4.2(2) we have that $s(i) = i > i - 1 = s(i-1)$ because for all nonterminal values of $i \in \Delta_f$, $i > 0$. We also require that for $i > 0$, $s(i) = i > s(f(i-1)-1) = f(i-1) - 1$, which is not obvious and remains to be shown.

Everything depends on determining the value of $f(i-1)-1$. In fact $f(x) = -x$ for x a nonterminal, that is, $x > 0$, or even for $x = 0$. (For $x = 0$ this follows immediately from the terminal line for the remaining cases we give an argument in the next paragraph.) If i is nonterminal, that is, $i > 1$, then $i - 1 \geq 0$, so $f(i-1) = -i+1$ and $f(i-1)-1 = -i$, and so $s(f(i-1) - 1) = s(-i) = 0$. Therefore $s(i) = i > 0 = s(f(i-1)-1)$. Now we show, as promised, that $f(x) = -x$ for $x \geq 0$.

The proof is by induction on the positive integers. We will show that whether or not the above definition terminates, any function $f(x)$ that satisfies the conditions of that definition must have the property that $f(x) = -x$ for all $x > 0$. This property holds if $i = 1$, as shown by direct evaluation. Now assume (hypothesis) the property for all $i \ni n > i \geq 1$. For $i = n$ line 2 is applicable. Then by hypothesis, $f(n-1) = -(n-1)$, and it follows that

$$\begin{aligned} f(n) &= f(f(n-1)-1) && \text{by line 2} \\ &= f(-(n-1)-1) && \text{by hypothesis} \\ &= f(-n) = -n && \text{by line 1 since } -n < 1 \end{aligned}$$

This completes the proof that $f(i) = -i$ and thus that the given function satisfies all the remaining conditions of Definition 2.4.2. Note that the fact that a definition is uniquely satisfied does not guarantee that it terminates. If line 2 of Definition 2.4.5 is replaced by $f(i) = f(f(i+1)+1)$, the definition will still be satisfied by $f(i) = -i$, but it does not terminate.

Finally we give a definition which has not been proven to terminate, although all the evidence indicates that it does.

Definition 2.4.6: Hailstorm Function

$$\left[\begin{array}{ll} f(n) = 1 & :n = 1 \\ f(n) = f(3n+1) & :n \text{ odd} \\ f(n) = f(n/2) & :n \text{ even} \\ \text{Initially } i \in N. & \end{array}\right.$$

2.4.2 Uniqueness of Terminating Definition

Since a recursive definition defines all the functions that satisfy its equations, it may define any number, or even an infinity of functions. In general, determination of the number, let alone the identity, of these defined functions is difficult. However

Theorem 2.4.2: If a recursive definition DEF.f terminates, then it is satisfied by exactly one function.

PROOF: Since DEF.f terminates, $f(X)$ can be substitutionally evaluated for every X in its starting domain, in fact for every $X \in \Delta_f$. This evaluation determines a function that satisfies DEF.f. Call this function $f'(X)$. Now we need to show that no other functions also satisfy DEF.f.

Let $f''(X)$ be any function that satisfies DEF.f. It will be shown that $f''(X) = f'(X)$ for each $X \in \Delta_f$. Suppose on the contrary that $f''(X) \neq f'(X)$ for some $X \in \Delta_f$. Of all the members X of Δ_f for which $f''(X) \neq f'(X)$ let X_a be one having the smallest dependence number $|\Delta_f(X)|$. (Because DEF.f terminates, the dependence number $|\Delta_f(X)|$ first defined in Theorem 2.4.1 must exist.) But there is one line in DEF.f whose domain contains X_a. If that line is terminal, that is, has the form: $f(X) = t(X)$, then $f'(X_a)$ and $f''(X_a)$ must both equal $t(X_a)$ and therefore must equal each other. This is a contradiction. So that line, if it exists, must be recursive. Say it has the form

$$f(X) = E_k(P, f, X)$$

So

$$f(X_a) = E_k(P, f, X_a)$$

therefore

$$f'(X_a) = E_k(P, f', X_a) \qquad \text{and} \qquad f''(X_a) = E_k(P, f'', X_a)$$

Since the line is recursive, there must be children of X_a in $E_k(P, f', X_a)$. But each child of $X_a \in E_k(P, f', X_a)$ and $E_k(P, f'', X_a)$ is of lower dependence number than X_a by the argument in the proof in Theorem 2.4.1. Thus the children of X_a appearing in the same positions in $E_k(P, f', X_a)$ and $E_k(P, f'', X_a)$ must give equal values, X_a having by hypothesis the smallest dependence number at which they could be unequal. Therefore

$$f'(X_a) = E_k(P, f', X_a) = E_k(P, f', X_a)$$

With this contradiction the theorem is established. ■

2.5 INDUCTIVE PROOFS OF PROPERTIES OF RECURSIVE DEFINITIONS

There are powerful techniques for proving assertions about terminating recursive definitions [Purdom 85, pp. 44–57; Rosser 53, pp. 397–405]. By Theorem 2.4.1 an integer $s(X)$ satisfying Definition 2.4.2 can be associated with each $X \in \Delta_f$. This provides an ordering of Δ_f upon which induction can be carried out. Types of inductive proofs of properties of recursive definitions will be considered in this section. Assume the goal is to show that the predicate $Q(X)$ is true for all $X \in \Delta_f$. The form of the proof could be

PROOFFORM 2.5.1: Induction on a Terminating Recursive Definition

1. Establish termination by finding a function $s(X)$ satisfying Definition 2.4.2. (The function s is also used in step 3.)
2. Show that $Q(X)$ is true for all X which are terminal.
3. For each nonterminal Y in a prescribed subset of Δ_f, with $s(Y) < n$, assume that $Q(Y)$ holds. Under this assumption show that $Q(X)$ is true for all X in Δ_f with $s(X) = n$.

ProofForm 2.5.1 is usually applied independently to each line of a recursive definition. The property $Q(X)$ is shown to hold for Δ_f by showing separately that it holds for each subset of Δ_f associated with each line of the definition.

If a definition contains only nonnested recursive lines, a simplified version of step 3 is applicable. Assume that each recursive line is of the standard nonnested form

$$f(X) = w\big(f(o_1(X)),\ldots, f\big(o_{m(X)}(X)\big)\big) \qquad :P(X)$$

From step 1 we know that $s(o_i(X)) < s(X)$, so we can replace step 3 with

3′. For each $X \in \Delta_f$ for which $P(X)$ holds assume that $Q(o_i(X))$ holds. Under this assumption show that $Q(X)$ is true. [In other words, show that the assumption of the truth of the predicate Q on the right (at the children of X) of the equation implies its truth on the left (at X).]

That step 3′ is an adaptation of step 3 is justified as follows. The assumption that $Q(o_i(X))$ holds is short for the weaker inductive assumption in step 3 that "$Q(X)$ holds for a prescribed subset of Δ_f for each of whose members Y, we have $s(Y) < n$." Having chosen the function s to satisfy Definition 2.4.2 in step 1 of the proofform, we know for any child of X [which by definition is $o_i(X)$] that $s(o_i(X)) < s(X)$. The set of children is the subset in question. If it is shown that $Q(X)$ follows from this inductive assumption, it follows that $Q(X)$ holds for all $X \in \Delta_f$ that satisfy $P(X)$. This gives us

PROOFFORM 2.5.2: Induction on a Nonnested Terminating Recursive Definition

1. Establish termination—by finding a function $s(X)$ satisfying Definition 2.4.2, for example.
2. Show that $Q(X)$ is true for all X which are terminal.
3. For each $X \in \Delta_f$ for which $P(X)$ holds, assume that $Q(o_i(X))$ holds. Under this assumption show that $Q(X)$ is true.

Unlike ProofForm 2.5.1, ProofForm 2.5.2 does not explicitly use the function $s(X)$ in step 3. This simplifies the explicit inductive argument. (ProofForm 2.5.2 does require termination, from which the existence of $s(X)$ can be inferred, so step 1 is essentially the same in both proofforms.)

As an example of the use of ProofForm 2.5.2 consider Definition 2.2.1 for the set of all binary n-tuples.

Definition 2.5.1: Binary N-tuple (reprint of Definition 2.2.1)

$$\left[\begin{array}{ll} f(x, n) = \{x\} & : n = 0 \\ f(x, n) = f(x \| \langle 0 \rangle, n - 1) \cup f(x \| \langle 1 \rangle, n - 1) & : n \in N \\ \text{Initially } x = \langle\ \rangle,\ n \in N_0. & \end{array}\right.$$

As is easy to show inductively, $f(x, n)$ is a set. Now using ProofForm 2.5.2, we will show that $\forall \langle x, n \rangle \in \Delta_f$, $|f(x, n)| = 2^n$.

1. Using the integer function $s(x, n) = n$, this definition terminates.
2. When $n = 0$, $f(x, n) = \{x\}$, thus $|f(x, n)| = 2^n = 2^0 = 1$.
3. Assume $|f(x \| \langle 0 \rangle, n - 1)| = |f(x \| \langle 1 \rangle, n - 1)| = 2^{n-1}$. By Definition 2.5.1, $f(x, n)$ is the union of two sets, each of cardinality 2^{n-1} when $n > 0$. If these sets are disjoint, then $|f(x, n)| = 2^n$ for $n > 0$.

It remains only to show that these sets are disjoint. That is (hypothesis), if $x \neq y$, then for all n, $f(x, n) \cap f(y, n) = \{\ \}$. This time we use ProofForm 2.5.1.

1. Termination is already established.
2. For $n = 0$ the conclusion is obvious.
3. Assume the truth of the hypothesis for pairs of members of Δ_f, $\langle x, k \rangle$ and $\langle y, k \rangle$, in which $k < n$ and $x \neq y$. So, for example, if $x \neq y$, the intersection of $f(x \| \langle 0 \rangle, n - 1)$ or $f(x \| \langle 1 \rangle, n - 1)$ with either of the sets $f(y \| \langle 0 \rangle, n - 1)$ or $f(y \| \langle 1 \rangle, n - 1)$ is empty. Now note that for $n > 0$,

$$\begin{aligned} f(x, n) \cap f(y, n) = {} & [f(x \| \langle 0 \rangle, n - 1) \cup f(x \| \langle 1 \rangle, n - 1)] \\ & \cap [f(y \| \langle 0 \rangle, n - 1) \cup f(y \| \langle 1 \rangle, n - 1)] \end{aligned}$$

By hypothesis the intersection of the sets

$$[f(x \| \langle 0 \rangle, n - 1) \cup f(x \| \langle 1 \rangle, n - 1)] \quad \text{and}$$

$$[f(y \| \langle 0 \rangle, n - 1) \cup f(y \| \langle 1 \rangle, n - 1)]$$

is empty. It follows immediately that the intersection $f(x, n) \cap f(y, n)$ is empty.

In the disjointness argument the inductive approach has been applied in a somewhat different situation than that for which it is prescribed. The property to be demonstrated here involves the relation of pairs, rather than single members of Δ_f, that is, $Q(X, Y)$ is true if and only if $f(X) \cap f(Y)$ is empty. Nevertheless the truth of $Q(X, Y)$ depends on the truth of $Q(Z, W)$, where Z and W are such that both $s(X)$ and $s(Y)$ are greater than both $s(Z)$ and $s(W)$. This makes the inductive argument used analogous to the one prescribed and valid.

Descent induction (use of the minimal principle) is a variant of the standard form which is convenient for showing that a property $Q(X)$ does *not* hold for any X in Δ_f. It takes the following form.

PROOFFORM 2.5.3: Descent Induction on a Terminating Recursive Definition

1. Find the function $s(X)$ satisfying Definition 2.4.2.
2. Show that $Q(X)$ does not hold when X is terminal.
3. Show that if X is nonterminal, $s(X) = n$, and $Q(X)$ is true, then, for at least one $Y \in \Delta_f$ with $s(Y) < n$, $Q(Y)$ is true.

If the recursive lines in the definition are all nonnested in the same form as Eq. (2.2.1), step 3 may be replaced by

3′. Show that if X is nonterminal and $Q(X)$ is true, then $Q(o_i(X))$ is true for at least one value of i.

An example of the use of descent induction is given in the following alternative proof of the statement proven above. In this case we are trying to show that the following proposition ($Q(X)$) is *false*. Referring to Definition 2.5.1:

1. (Proposition) If $x \neq y$, then for all n, $f(x, n) \cap f(y, n) \neq \{\ \}$.
2. When $\langle x, n\rangle$ and $\langle y, n\rangle$ are terminal, if $x \neq y$, then $f(x, n) \cap f(y, n) = \{x\} \cap \{y\} = \{\ \}$ ($Q(X)$ is false).

3′. For $f(x, n) \cap f(y, n) \neq \{\ \}$ to be true, $[f(x\|\langle 0\rangle, n-1) \cup f(x\|\langle 1\rangle, n-1)] \cap [f(y\|\langle 0\rangle, n-1) \cup f(y\|\langle 1\rangle, n-1)] \neq \{\ \}$ must be true, which in turn means at least one of the following $\neq \{\ \}$: $f(x\|\langle 0\rangle, n-1) \cap f(y\|\langle 0\rangle, n-1)$, or $f(x\|\langle 0\rangle, n-1) \cap f(y\|\langle 1\rangle, n-1)$, or $f(x\|\langle 1\rangle, n-1) \cap f(y\|\langle 0\rangle, n-1)$, or $f(x\|\langle 1\rangle, n-1) \cap f(y\|\langle 1\rangle, n-1)$.

The proof of Theorem 2.4.2 is another example of the use of descent induction.

2.5.1 Equivalence of Recursive Definitions

Suppose we wish to show that $f(X) = c(X, g(X))$, where $f(X)$ and $g(X)$ are given, respectively, by DEF.f and DEF.g and c is a domain-transformation function. DEF.f is recursive, DEF.g may be either recursive or nonrecursive, and c is a primitive function. The following proofform is a specialization of ProofForm 2.5.2 for showing that such equalities are true. Like ProofForm 2.5.2, it requires that all nonrecursive lines have the nonnested form of Eq. (2.2.1).

PROOFFORM 2.5.4: Equivalence of Nonnested Recursive Definitions

1. Show termination (to ensure uniqueness—Theorem 2.4.2).
2. For each terminal line in DEF.f of the form

$$f(X) = t_i(X) \qquad :T_i(X)$$

show that

$$c(X, g(X)) = t_i(X) \qquad :T_i(X)$$

If DEF.g is nonrecursive over the domain $\{T_i(X)\}$, whether recursive elsewhere or not, this reduces to showing that primitive expressions are equal.

3′. For each nonterminal line in DEF.f of the form of Eq. (2.2.1),

$$f(X) = w\big(f(o_1(X)), \ldots, f(o_{m(X)}(X))\big) \qquad :P(X)$$

show that

$$c(X, g(X)) = w\big(c(o_1(X), g(o_1(X))), \ldots,$$
$$c\big(o_{m(X)}(X), g\big(o_{m(X)}(X)\big)\big)\big) \qquad :P(X)$$

where the o_i and m functions are those of DEF.f. Here again if DEF.g is nonrecursive over the domain $\{P(X)\}$, then both sides of the above expression are primitive and we are required to show the equality of two primitive expressions. If $g(X)$ is not primitive but is given by a recursive line over the domain $\{P(X)\}$, then $c(x, U)$ where U is the right side of that defining line in DEF.g needs to be shown equivalent to

$$w\big(c(o_1(X), g(o_1(X))), \ldots, c\big(o_{m(X)}(X), g\big(o_{m(X)}(X)\big)\big)\big) \qquad :P(X)$$

Note: If DEF.f terminates, a separate verification that $g(X)$ terminates is not necessary.

As an example of this approach we will show that Definition 2.5.1 does in fact define the set of all binary n-tuples. The meaning of "all binary n-tuples" must be specified more precisely. The assertion to be proven is

$$g(x, n) = \{x\|\langle y_1, \ldots, y_n\rangle \mid y_i \in \{0,1\}\} \text{ satisfies Definition 2.5.1}$$

To prove this by ProofForm 2.5.4 we must show that:

1. Definition 2.5.1 terminates. This has been done.

2. $$\{x\|\langle y_1, \ldots, y_n\rangle \mid y_i \in \{0,1\}\} = \{x\} \qquad : n = 0$$

   which must be true if $\{x\|\langle\ \rangle\} = \{x\}$ when $n = 0$, which is obviously true.

We must also show that the following equality holds. It results from substituting the definition of $g(x, n)$ for $f(x, n)$ throughout the recursive line in Definition 2.5.1.

3. $$\{x\|\langle y_1, \ldots, y_n\rangle \mid y_i \in \{0,1\}\}$$
   $$= \{x\|\langle 0\rangle\|\langle y_1, \ldots, y_{n-1}\rangle \mid y_i \in \{0,1\}\}$$
   $$\cup \{x\|\langle 1\rangle\|\langle y_1, \ldots, y_{n-1}\rangle \mid y_i \in \{0,1\}\}$$

   Since y_1 must be 0 or 1, the set on the left of the equal sign above can be written as follows:

   $$\{x\|\langle 0\rangle\|\langle y_2, \ldots, y_n\rangle \mid y_i \in \{0,1\}\} \cup \{x\|\langle 1\rangle\|\langle y_2, \ldots, y_n\rangle \mid y_i \in \{0,1\}\}$$

Even though the subscripts are not identical, the expressions to the left and right of the equal sign represent the same set of vectors. The identity of the subscripts of y in these expressions is unimportant because in any case each subscripted y takes the values 0 and 1 in every possible combination with the other subscripted y's. Only the number of subscripts is relevant, and this is the same on the right and left. In summary:

Proposition 2.5.1: Proving Equivalence of Recursive Definitions
If DEF.f terminates, then to show that $f(X) = g(X)$ first substitute $g(X)$ for all occurrences of $f(X)$ in every line of DEF.f and then demonstrate that the resultant equalities are valid.

The core of ProofForm 2.5.4 is the demonstration of equality in step 3. Throughout the book we will be proving such equalities in this way. It is worthwhile to formulate it generally. Suppose we wish to determine whether

$x_0 = y_0$. This quest is expressed by writing

$$x_0 =?\ y_0 \qquad (=?\text{ indicates that equality is yet to be established})$$

We then find another equality, $x_1 = y_1$, whose truth implies that $x_0 = y_0$. The quest to establish this equality is expressed as

$$x_1 =?\ y_1$$

In general the $(i + 1)$th sought-for equality $x_{i+1} =? y_{i+1}$ if established must imply that $x_i = y_i$. This continues until an equality expressed by

$$x_n =!\ y_n$$

is produced whose truth is obvious (thus $=!$ instead of $=?$). When this obvious equality is found, the original equality $x_0 = y_0$ is established.

More can be said about how the successive equalities are obtained. Usually $x_{i+1} =?\ y_{i+1}$ is obtained by substituting an expression q' for a subexpression q in either x_i or y_i, where $q = q'$ is an identity. If that is done, not only does

$$x_{i+1} = y_{i+1} \text{ imply } x_i = y_i$$

but also

$$x_i = y_i \text{ implies } x_{i+1} = y_{i+1}$$

The two tentative equalities are equivalent.

MORE EXAMPLES

To show that Definition 2.4.3 really gives the greatest common divisor (gcd) using ProofForm 2.5.4 and this notation, first change Definition 2.4.3 to the obviously equivalent form.

Definition 2.5.2: Greatest Common Divisor

$$\left[\begin{array}{ll} f(x, y) = y & :y \text{ divides } x \\ f(x, y) = f(y, \text{rem}(x, y)) & :\text{otherwise} \\ \text{Initially } x, y \in N \text{ and } x \geq y. & \end{array}\right.$$

We wish to show that $\text{gcd}(x, y) = f(x, y)$.

1. Termination has been demonstrated.
2. If y divides x, then $f(x, y) = y = \text{gcd}(x, y)$.
3. Substitute $\text{gcd}(x, y)$ for $f(x, y)$ throughout the recursive line:

$$\text{gcd}(x, y) =?\ \text{gcd}(y, \text{rem}(x, y))$$

Because $x = k * y + \text{rem}(x, y)$, this is equivalent to

$$\gcd(k * y + \text{rem}(x, y), y) =! \gcd(y, \text{rem}(x, y))$$

The symbol $=!$ is justified because any number that divides both arguments of the gcd on the right of $=!$ must divide both arguments of the gcd on its left, and vice versa.

We can also use the same proofform in establishing an upper bound on the number of f calls that are needed to evaluate this recursive definition for $f(x, y)$, which is the complexity of a recursive implementation of that definition.

Define the function $q(x, y) = x$ on the substitutional domain of Definition 2.5.2. It will be used to keep track of the magnitude of x as the o-function is applied to the argument structure.

1. If $y \leq x/2$, then applying q to the child of (x, y), we get

$$q(y, \text{rem}(x, y)) = y \leq s(x, y)/2 = x/2$$

2. If $y > x/2$, then applying q to the grandchild of (x, y), we get

$$q(\text{rem}(x, y), \text{rem}(y, \text{rem}(x, y))) = \text{rem}(x, y) \leq s(x, y)/2 = x/2$$

   Thus after at most two applications of the o function the first component x of the argument has been decreased to half its value or less.

Therefore if $c(x) \geq$ the number of calls needed to evaluate $f(x, y)$;

$$c(x) = 1 \qquad :x = 1$$

$$c(x) = 2 + c(x/2) \qquad :x > 1$$

It is now easy to prove that $c(x) = 2\log x + 1$. If $x = 1$,

$$2\log 1 + 1 =! 1$$

If $x > 1$,

$$\begin{aligned} 2\log x + 1 &=? 2 + 2\log(x/2) + 1 \\ &=? 3 + 2(\log(x) - 1) \\ &=? 3 + 2\log(x) - 2 \end{aligned}$$

So $2\log(x) + 1 \geq$ the number of calls needed to evaluate $f(x, y)$. (In fact, $\log_\phi x$ is a sharp upper bound, where $\phi =$ the golden ratio.)

As an example of the use of ProofForm 2.5.4 to show that two recursive definitions define the same function, consider a new definition of all binary n-tuples which uses the primitive function $|*|$, cross concatenation.

Definition 2.5.3: Binary n-tuples

$$\left[\begin{array}{ll} g(n) = \{\langle\ \rangle\} & :n = 0 \\ g(n) = \{\langle 0\rangle\}|*|g(n-1) \cup \{\langle 1\rangle\}|*|g(n-1) & :n \in N \\ \text{Initially } n \in N_0. & \end{array}\right.$$

Example: Use of ProofForm 2.5.4

We use ProofForm 2.5.4 to show that $\{x\}|*|g(n)$ with $g(n)$ as defined above is equivalent to $f(x, n)$ of Definition 2.5.1.

1. Termination has been established.
2. For $n = 0$, $f(x, n) = \{x\} = \{x\}|*|\{\langle\ \rangle\} = \{x\}|*|g(n) = \{x\}$ because in Definition 2.5.3 $g(n) = \{\langle\ \rangle\}$.
3. For $n \in N$, the result of substituting $\{x\}|*|g(n)$ for $f(x, n)$ throughout the recursive line in Definition 2.5.1 must be shown to produce a true equality. This is expressed by

$$\{x\}|*|g(n) =?\ \{x\|\langle 0\rangle\}|*|g(n-1) \cup \{x\|\langle 1\rangle\}|*|g(n-1)$$

where $|*|$ takes precedence over $\cup$. Because of their meaning we can replace $\|$ with $|*|$, provided their arguments are each embedded in a set,

$$\{x\}|*|g(n) =?\ (\{x\}|*|\{\langle 0\rangle\})|*|g(n-1) \cup (\{x\}|*|\{\langle 1\rangle\})|*|g(n-1)$$

Because $|*|$ is associative,

$$\{x\}|*|g(n) =?\ \{x\}|*|(\{\langle 0\rangle\}|*|g(n-1)) \cup \{x\}|*|(\{\langle 1\rangle\}|*|g(n-1))$$

Since $|*|$ distributes through $\cup$,

$$\{x\}|*|g(n) =?\ \{x\}|*|\text{“}\{\langle 0\rangle\}|*|g(n-1) \cup \{\langle 1\rangle\}|*|g(n-1)\text{”}$$

The right side of the recursive line in Definition 2.5.3 appears in quotes here. Replace it with the left side of the recursive line to get

$$\{x\}|*|g(n) =!\ \{x\}|*|g(n)$$

This tentative equality is clearly valid. Therefore the original tentative equality is itself valid and the proof is completed.

In the technique called *recursion induction* one shows that two primitive expressions, $g(X)$ and $h(X)$, are equal by finding a terminating recursive definition, DEF.f, which both satisfy. ProofForm 2.5.4 can often be used to show that $g(X)$ and $h(X)$ satisfy DEF.f. Note that DEF.f must terminate for the conclusion to be valid because *termination ensures that $f(X)$ is unique.*

2.5.1.1 Equivalence of Recursive Definitions which Do Not Terminate

The demonstration of steps 2 and 3 in ProofForm 2.5.4 is sufficient to establish that any function that satisfies DEF.g will satisfy DEF.f. Then, depending on whether DEF.g and DEF.f are satisfied by one or more functions, additional conclusions follow.

1. If DEF.f is satisfied by only one function, then DEF.g is also uniquely satisfied by that function.
2. If DEF.f is satisfied by several functions, then the functions that satisfy DEF.g form a subset of those that satisfy DEF.f. In this latter case we might wish to prove steps 2 and 3 with the roles of DEF.f and DEF.g interchanged. If this could be done, then the two definitions would be shown to define the same set of functions.

2.6 SOME GENERAL EQUIVALENCES OF RECURSIVE DEFINITIONS

Some general relations between recursive definitions can be demonstrated using the proofforms considered here. Equivalence of a broad class of nonnested and nested definitions will be shown, but first we prove an equivalence between two general forms of nonnested recursive definitions, DEF.f and DEF.h.

DEF.f is in a standard nonnested form, as discussed in Section 2.2.1, with recursive and terminal lines, respectively, the same as Eqs. (2.2.1) and (2.2.2). The conclusions that are drawn about DEF.f in the theorem hold for any w, o_i, m, t, D, and T that have the properties explicitly stated in the theorem.

The form of DEF.h is similar to that of DEF.f. In fact it represents a subset of the definitions represented by DEF.f. Where DEF.f has the function w, which may be any primitive function, DEF.h has concatenation; otherwise the same primitives are used in both definitions. DEF.h may be thought of as a standard nonnested form for defining sequences or enumerations.

$f(X)$ is shown to be equivalent to $w'(h(X))$, where w' is a function on a sequence defined in terms of the primitive function w from DEF.f, that is, $w'(\langle s_1, \ldots, s_n \rangle) = \langle w(s_1), \ldots, w(s_n) \rangle$. To be verified, this requires only that the primitive function w be associative (such as minimum, maximum, or addition).

In the following proof only steps 2 and 3 of ProofForm 2.5.4 are used. Thus the exact interpretation of the result depends on whether or not DEF.f terminates.

Theorem 2.6.1: Given the following two definitions.
DEF.f:

$$\begin{array}{lll} 1. & \left[f(X) = t(X) \right. & :T(X) \\ 2. & f(X) = w\big(f(o_1(X)),\ldots, f(o_{m(X)}(X))\big) & :\sim T(X) \\ & \text{Initially } X \in D. & \end{array}$$

and DEF.h:

$$\begin{array}{lll} 1. & \left[h(X) = \langle t(X) \rangle \right. & :T(X) \\ 2. & h(X) = h(o_1(X)) \| \cdots \| h\big(o_{m(X)}(X)\big) & :\sim T(X) \\ & \text{Initially } X \in D. & \end{array}$$

and given the function w' defined recursively in terms of w as follows. Let $X = \langle x_1, \ldots, x_n \rangle$,

$$\begin{array}{lll} 1. & \left[w'(X) = w(x_1, \ldots, x_n) \right. & :n > 1 \\ 2. & w'(X) = x_1 & :n = 1 \end{array}$$

and finally, given that w is associative, *then*

$$w'(h(X)) = f(X)$$

PROOF:

1. When $T(X)$, substitute $w'(h(X))$ for $f(X)$ throughout line 1 of DEF.f,

$$w'(h(X)) = ?\, t(X)$$

   By line 1 of DEF.h,

$$w'(\langle t(X) \rangle) = ?\, t(X)$$

   By line 2 of the definition of w',

$$t(X) = !\, t(X)$$

2. When $\sim T(X)$, substitute $w'(h(X))$ for $f(X)$ throughout line 2 of DEF.f,

$$w'(h(X)) = ?\, w\big(w'(h(o_1(X))),\ldots, w'\big(h\big(o_{m(X)}(X)\big)\big)\big) \qquad (*)$$

$h(o_i(X))$ is a vector (as is easily shown using induction), so define $\text{cmps}(h(o_i(X)))$ to be the list of components of $h(o_i(X))$. Then substituting in Eq. (*) with definition 1 of w' and using the cmps notation,

$$w'(h(X)) = ?\, w\Big(w(\text{cmps}(h(o_1(X)))),\ldots, w\Big(\text{cmps}\Big(h\Big(o_{m(X)}(X)\Big)\Big)\Big)\Big)$$

By associativity of w,

$$w'(h(X)) = ?\, w\Big(\text{cmps}(h(o_1(X))),\ldots, \text{cmps}\Big(h\Big(o_{m(X)}(X)\Big)\Big)\Big)$$

By definition of cmps and $\|$,

$$w'(h(X)) = ?\, w\Big(\text{cmps}(h(o_1(X)))\| \cdots \|\text{cmps}\Big(h\Big(o_{m(X)}(X)\Big)\Big)\Big)$$

By definition 1 of w',

$$w'(h(X)) = ?\, w'\Big(h(o_1(X))\| \cdots \|h\Big(o_{m(X)}(X)\Big)\Big)$$

From line 2 of DEF.h,

$$w'(h(X)) = !\, w'(h(X)) \quad \blacksquare$$

The theorem demonstrates that any function that can be defined by a standard nonnested definition (DEF.f) with an associative w can be viewed as equivalent to w operating on the members of a sequence or enumeration—where the enumeration itself is defined by a standard nonnested definition (DEF.h). Furthermore DEF.f differs from DEF.h only in having the general w function in place of the $\|$ in DEF.h. This view is useful in designing recursive definitions.

Another transformation worth noting allows imposition of evaluation order by nesting. Given a nonnested recursive definition, it is always possible to transform it to an equivalent nested one. Furthermore we can get either a nested definition in which only the initial f call is evaluated, as in Theorem 2.6.2, or one in which the evaluation of all intermediate f calls are recoverable, as in Theorem 2.6.3. Where the nonnested definition is neutral as to evaluation order, the equivalent nested definition imposes a depth-first ordering.

2.6.1 Equivalence when *w* Is Associative and the Function Is to Be Computed for only Its Initial Argument

Theorem 2.6.2 below is based on the fact that a standard nonnested recursive line in which w is associative, namely,

$$f(X) = w\Big(f(o_1(X)), f(o_2(X)),\ldots, f\Big(o_{m(X)}(X)\Big), p(X)\Big) \qquad :\sim T(X)$$

can be combined with the variable V as follows:

$$w(V, f(X)) = w\big(V, w\big(f(o_1(X)), f(o_2(X)), \ldots, f(o_{m(X)}(X)), p(X)\big)\big) \qquad :\sim T(X)$$

and because of w's associativity, can be expressed as

$$w(V, f(X)) = w\big(w\big(\ldots w(w(V, f(o_1(X))), f(o_2(X))), \ldots, f(o_{m(X)}(X))\big), p(X)\big) \qquad :\sim T(X)$$

and finally, giving $w(V, f(X))$ the name $g(V, X)$, can be reexpressed by

$$g(V, X) = w\big(g\big(\ldots g(g(V, o_1(X)), o_2(X)), \ldots, o_{m(X)}(X)\big), p(X)\big) \qquad :\sim T(X)$$

A more formal statement and justification of this "nesting procedure" is given in Theorem 2.6.2. In this theorem, w must have an identity 0_w.

Theorem 2.6.2: Given the two definitions

Definition 2.6.1

$$\left[\begin{array}{ll} f(X) = t(X) & :T(X) \\ f(X) = w\big(f(o_1(X)), f(o_2(X)), \ldots, f(o_{m(X)}(X)), p(X)\big) & :\sim T(X) \\ \text{Initially } X = d,\ d \in D. & \end{array}\right.$$

and

Definition 2.6.2

$$\begin{array}{l} 1. \\ \\ \text{or} \\ \\ 2. \end{array} \left[\begin{array}{ll} g(V, X) = w(V, t(X)) & :T(X) \\ g(V, X) = w\big(g\big(\ldots g(g(V, o_1(X)), o_2(X)), \ldots, o_{m(X)}(X)\big), p(X)\big) & :\sim T(X) \\ \\ g(V, X) = g\big(\ldots g(g(w(V, p(X)), o_1(X)), o_2(X)), \ldots, o_{m(X)}(X)\big) & :\sim T(X) \\ \text{Initially } V = 0_w,\ X = d,\ d \in D. & \end{array}\right.$$

then $w(V, f(X))$ is satisfied by $g(V, X)$ of Definition 2.6.2(1) if w is associative, and also by Definition 2.6.2(2) if w is also commutative. So in both these cases $w(0_w, f(X)) = f(X)$ is satisfied by $g(0_w, X)$.

PROOF:

1. For Definition 2.6.2(1) and (2) if $T(X)$, substitute $w(V, f(X))$ for $g(V, X)$ throughout the terminal line of Definition 2.6.2(1) or (2),

$$w(V, f(X)) = ?\, w(V, t(X))$$

and since $f(X) = t(X)$ if $T(X)$,

$$w(V, t(X)) = !\, w(V, t(X))$$

2. For Definition 2.6.2(1) if $\sim T(X)$, substitute $w(V, f(X))$ for $g(V, X)$ throughout the recursive line of Definition 2.6.2(1) in a number of steps:

$$w(V, f(X)) = ?\, w\big(g(\ldots g(w(V, f(o_1(X))),$$

$$o_2(X)), \ldots, o_{m(X)}(X)), p(X)\big)$$

$$w(V, f(X)) = ?\, w\big(g(\ldots w(w(V, f(o_1(X))),$$

$$f(o_2(X))), \ldots, o_{m(X)}(X)), p(X)\big)$$

$$w(V, f(X)) = ?\, w\big(w(\ldots w(w(V, f(o_1(X))),$$

$$f(o_2(X))), \ldots, f\big(o_{m(X)}(X)\big)\big), p(X)\big)$$

which by associativity of w becomes

$$w(V, f(X)) = ?\, w\big(V, f(o_1(X)), f(o_2(X)), \ldots, f\big(o_{m(X)}(X)\big), p(X)\big)$$

which by the recursive line of Definition 2.6.1 becomes

$$w(V, f(X)) = !\, w(V, f(X))$$

3. For Definition 2.6.2(2) if $\sim T(X)$, substitute $w(V, f(X))$ for $g(V, X)$ throughout the recursive line of Definition 2.6.2(2) in a number of steps:

$$w(V, f(X)) = ?\, g\big(\ldots g\big(w\big(w(V, p(X)), f(o_1(X))\big),$$
$$o_2(X)\big), \ldots, o_{m(X)}(X)\big)$$
$$w(V, f(X)) = ?\, g\big(\ldots w\big(w\big(w(V, p(X)), f(o_1(X))\big),$$
$$f(o_2(X))\big), \ldots, o_{m(X)}(X)\big)$$
$$w(V, f(X)) = ?\, w\big(\ldots w\big(w\big(w(V, p(X)), f(o_1(X))\big),$$
$$f(o_2(X))\big), \ldots, f\big(o_{m(X)}(X)\big)\big)$$

which by associativity and commutativity of w becomes

$$w(V, f(X)) = ?\, w\big(V, f(o_1(X)), f(o_2(X)), \ldots, f\big(o_{m(X)}(X)\big), p(X)\big)$$

which by the recursive line of Definition 2.6.1 becomes

$$w(V, f(X)) = !\, w(V, f(X))$$ ■

The nested Definitions 2.6.2(1) and (2) are equivalent to the nonnested Definition 2.6.1, and Definitions 2.6.2(1) and (2) imply a depth-first evaluation. Therefore the depth-first implementations which will be developed for Definition 2.6.1 in Chapter 4 will work for Definitions 2.6.2(1) and (2).

2.6.2 Transformation when *w* is Locally Associative and / or the Function Is to Be Evaluated at All Arguments

A variation on Theorem 2.6.2 for the more general case in which the w function has the property of local associativity rather than associativity is given next. Furthermore the definition of the equivalent nested function $g(V, d)$ in the new theorem makes it easy to obtain the value of $f(X)$ for *all* the argument structures X in $\Delta_f(d)$ while evaluating $g(v, d)$.

$w(x_1, \ldots, x_n)$ is *locally associative* if it can be expressed as $n(v(x_1, \ldots, x_n))$, where v is associative. It is expressed that way in Theorem 2.6.3.

Any function which is associative is necessarily locally associative. To interpret the theorem for a function which is simply associative, let n be the identity function, that is, remove all occurrences of n from the theorem.

Theorem 2.6.3: Given the two definitions

Definition 2.6.3

$$\begin{bmatrix} f(X) = t(X) & :T(X) \\ f(X) = n\big(v\big(f(o_1(X)), f(o_2(X)), \ldots, f\big(o_{m(x)}(X)\big), p(X)\big)\big) & :\sim T(X) \\ \text{Initially } X = d,\ d \in D. & \end{bmatrix}$$

and

Definition 2.6.4

$$\begin{bmatrix} g(V, X) = v(V, t(X)) & :T(X) \\ g(V, X) = v\big(V, n\big(v\big(g\big(\ldots g\big(g(0_v, o_1(X)), o_2(X)\big), \ldots, & \\ \qquad o_{m(X)}(X)\big), p(X)\big)\big)\big) & :\sim T(X) \\ \text{Initially } V = 0_v,\ X = d,\ d \in D. & \end{bmatrix}$$

Then $v(V, f(X))$ is satisfied by $g(V, X)$ of Definition 2.6.4 if v is associative so that $n \circ v$ is locally associative. Thus, $v(0_v, f(X)) = f(X)$ is satisfied by $g(0_v, X)$.

PROOF

1. For Definition 2.6.4 if $T(X)$, substitute $v(V, f(X))$ for $g(V, X)$ throughout the terminal line of Definition 2.6.2(1) or (2),

   $$v(V, f(X)) = ?\, v(V, t(X))$$

   and since $f(X) = t(X)$ if $T(X)$,

   $$v(V, t(X)) = !\, v(V, t(X))$$

2. For Definition 2.6.2(a) if $\sim T(X)$, substitute $v(V, f(X))$ for $g(V, X)$ throughout the recursive line of Definition 2.6.2(1) in a number of steps:

   $$v(V, f(X)) = ?\, v\Big(V, \Big(n\Big(v\Big(g\Big(\ldots g\big(v(0_v, f(o_1(X)))\big), o_2(X)\Big), \ldots, o_{m(X)}(X)\Big)\Big), p(X)\Big)\Big)\Big)$$

   By definition of 0_v,

   $$v(V, f(X)) = ?\, v\Big(V, \Big(n\Big(v\Big(g\Big(\ldots g\big(f(o_1(X)), o_2(X)\big), \ldots, o_{m(X)}(X)\Big)\Big), p(X)\Big)\Big)\Big)$$

and continuing to substitute,

$$v(V, f(X)) = ?\, v\Big(V, \Big(N\Big(v\Big(g\Big(\ldots v(f(o_1(X)), f(o_2(X))), \ldots,$$

$$o_{m(X)}(X)\Big)\Big), p(X)\Big)\Big)\Big)$$

$$v(V, f(X)) = ?\, v\Big(V, \Big(n\Big(v\Big(v\Big(\ldots v(f(o_1(X)), f(o_2(X))), \ldots,$$

$$f\big(o_{m(X)}(X)\big)\Big), p(X)\Big)\Big)\Big)\Big)$$

which by associativity of v becomes

$$v(V, f(X)) = ?\, v\Big(V, n\Big(v\Big(f(o_1(X)), f(o_2(X)), \ldots,$$

$$f\big(o_{m(X)}(X)\big), p(X)\Big)\Big)\Big)$$

which by the recursive line of Definition 2.6.3 becomes

$$v(V, f(X)) = !\, v(V, f(X))$$

Also along the way we have shown that

$$f(X) = n\Big(v\Big(g\Big(\ldots g(g(0_v, o_1(X)), o_2(X)), \ldots,$$

$$o_{m(X)}(X)\Big), p(X)\Big)\Big) \qquad :\sim T(X) \quad \blacksquare$$

2.6.3. Transformation for case of no restriction on the *w* Function

Any w function on the arguments $(x_1, \ldots, x_n)$ can be viewed as locally associative on the arguments $(\langle x_1 \rangle, \ldots, \langle x_n \rangle)$ by the following reasoning. Let $w'(\langle x_1, \ldots, x_n \rangle) = w(x_1, \ldots, x_n)$. Therefore,

$$w'(\langle x_1, \ldots, x_n \rangle) = w'(\langle x_1 \rangle \| \ldots \| \langle x_n \rangle)$$

w' is a unary function with one argument which is a vector; $\|$ is an associative function. Therefore $w'(\langle x_1 \rangle \| \ldots \| \langle x_n \rangle)$ is locally associative.

Using this view of the w function, we can transform any nonnested definition into a nested one using a slight variation of Theorem 2.6.3. Therefore nested definitions have at least equal defining power. In fact they have more. Not only can functions be defined with nested definitions which cannot be defined with nonnested ones, but even while defining the same function, the nested definition can, in fact must, specify more details about *how to evaluate* the definition. It is for the latter reason that we are interested in nested definitions here.

2.7 GENERAL PROCEDURES FOR EVALUATING RECURSIVE DEFINITIONS

In Process 2.3.1, in defining termination, we developed the SE procedure for evaluating recursive definitions. Such a procedure is also useful because it is the raw material from which algorithms for evaluating recursive definitions are built. In fact, process 2.3.1 is a special case of a more general process whose description follows. (To follow our description of this process, it may help to look at Example 2.7.1.)

This process progressively changes a set of equations by adding and removing equations as well as by altering those in the set. Let Q_i be the set of equations as it exists on entering the ith cycle of the process. The jth equation in Q_i, E_j, has the form $f(x_j) = R_j$. If d is the initial argument, and $f(d) = E(P, f, d)$ is the form of the line in DEF.f whose domain contains d, then Q_1 consists of the single initial equation $f(d) = E(P, f, d)$. As the process proceeds and the set of equations changes, the right side of this initial equation may be altered, but the left side will remain unchanged. When the right side of the initial equation in Q_i becomes a primitive argument structure, say equal to z, then $f(d) = z$ and the process may terminate.

Process 2.7.1: General Recursive Definition Evaluation
[To produce Q_{i+1} from Q_i, either of the two procedures—*new equation formation* or *back substitution*—can be employed (if applicable) in any order.]

```
WHILE (there is no answer and either procedure is still applicable) DO
    new equation formation or
    back substitution
  END
```

```
PROC: new equation formation
IF [there is an equation E_a whose right side contains an
   occurrence of f(x_b), where x_b is a primitive
   argument structure and there is no equation in Q_i
   with f(x_b) on the left of its equal sign]
   THEN let f(X) = E(P,f,X) be the form of the composition
            line in DEF.f whose domain contains x_b and
            ADD the new equation: f(x_b) = E(P,f,x_b)
                  to Q_1 to form Q_i+1.
END new equation formation
```

PROC: *back substitution*
IF [there is a pair of equations E_a and E_b in Q_i
with the right side of the E_a containing an
occurrence of the left side of the E_b as follows:

$$E_a: f(x_a) = R_a = \ldots f(x_b)\ldots$$

$$E_b: f(x_b) = R_b]$$

THEN REPLACE the occurrence of the $f(x_b)$ in
R_a with R_b and simplify the result.
So E_a becomes E_a: $f(x_a) = R_a$
= simplification of $(\ldots R_b \ldots)$.
The revised equation set is Q_{i+1}.
END *back substitution*

(If R_b is a primitive argument structure, then this step is called *constant back substitution*.)

Note. The two procedures above are the basic ones for obtaining the evaluation. However, the following addition may be useful in simplifying bookkeeping. Any equation other than the initial one may be removed at any time. Such a removal is pointless unless it is known that the left side of the removed equation will not again appear on the right of another equation in the set.

2.7.1 Substitutional Evaluation (SE)

Process 2.7.1 is nondeterministic, that is, permissive as to the order in which its basic steps are to be carried out. It is worthwhile considering some specializations in which the order is constrained. In fact we have already done so; Process 2.3.1 is a constrained version of Process 2.7.1. If PROC: *new equation formation*, *PROC*: *back substitution* and the modification involving removal of an equation in the note to Process 2.7.1 are alternated, the equation set will start with the initial equation, an equation will be added, its right side will be substituted in the initial equation, and then that equation will be removed, leaving only an altered initial equation. If DEF.f terminates, this sequence of events will eventually terminate with the answer. This specialization of Process 2.7.1 results in the repeated expansion of a single equation. It is essentially the same as SE as given in Process 2.3.1.

2.7.2 Equation Set Evaluation (ESE)

Another useful class of constrained orderings of Process 2.7.1 which emphasize the formation and solution of a set of equations leads to *equation set evaluation*

(ESE). Two members of this class follow. Both refer to the procedures PROC: *new equation formation* and PROC: *back substitution* defined in Process 2.7.1.

Process 2.7.2: Equation Set Evaluation (ESE) 1

```
WHILE (The answer has not been obtained and the
       condition for either procedure is satisfied) DO
       (a) REPEAT PROC: new equation formation as long as
           the conditions of its initial IF are satisfied.
       (b) REPEAT PROC: back substitution in its
           constant back substitution form as long as
           the conditions of its initial IF are satisfied.
END
```

Process 2.7.3: Equation Set Evaluation (ESE) 2

```
WHILE (The answer has not been obtained and the
       condition for either procedure is satisfied) DO
       (a) REPEAT PROC:back substitution in its
           constant-back-substitution form as long as
           the conditions of its initial IF are satisfied.
       (b) DO PROC:new equation formation to form a new
           equation if possible.
END
```

Notes. The description of both forms of ESE can be further refined by more explicitly specifying which new equations PROC: *new equation formation* should form when called. In general there will be a choice. Let equations be ordered according to their generation by PROC: *new equation formation*. To determine which equation to form next, search in order through the right sides of equations already formed until the first occurrence of an f for which there is no equation. Within the right side of each equation, search left to right.

Examples of ESE Processes 2.7.2 and 2.7.3 are now given. First an example of Process 2.7.2 is given. Consider the Fibonacci number definition of Definition 2.4.1 rewritten here,

$$\left[\begin{array}{ll} f(x) = 1 & :x \in \{0,1\} \\ f(x) = f(x-1) + f(x-2) & :x \in N_2 \\ \text{Initially } x \in N_0 & \end{array}\right.$$

Process 2.7.2 is illustrated by finding $f(4)$ in Example 2.7.1.

Example 2.7.1

First moving *downward* we repeat
PROC: *new equation formation.*

$\Vert$

Q_1:	$\vert f(4) = f(3) + f(2)$	
	$\Downarrow$	
Q_2:	$\vert f(4) = f(3) + f(2)$	$f(4) = 3 + 2 = 5 \Leftarrow$ *Answer*
	$\vert f(3) = f(2) + f(1)$	$f(3) = 3$
	$\Downarrow$	$f(2) = 2$
		$f(1) = 1$
		$f(0) = 1$
		$\Uparrow$
	$\vert f(4) = f(3) + f(2)$	$f(4) = f(3) + 2$
Q_3:	$\vert f(3) = f(2) + f(1)$	$f(3) = 2 + 1 = 3$
	$\vert f(2) = f(1) + f(0)$	$f(2) = 2$
	$\Downarrow$	$f(1) = 1$
		$f(0) = 1$
		$\Uparrow$
	$\vert f(4) = f(3) + f(2)$	$f(4) = f(3) + f(2)$
Q_4:	$\vert f(3) = f(2) + f(1)$	$f(3) = f(2) + 1$
	$\vert f(2) = f(1) + f(0)$	$f(2) = 1 + 1 = 2$
	$\vert f(1) = 1$	$f(1) = 1$
	$\Downarrow$	$f(0) = 1$
		$\Uparrow$
	$\vert f(4) = f(3) + f(2)$	$f(4) = f(3) + f(2)$
	$\vert f(3) = f(2) + f(1)$	$f(3) = f(2) + f(1)$
Q_5:	$\vert f(2) = f(1) + f(0)$	$f(2) = f(1) + 1$
	$\vert f(1) = 1$	$f(1) = 1$
	$\vert f(0) = 1$	$f(0) = 1$
		$\Uparrow$
	$\Downarrow$	Now moving *upward* we repeat
	$\Rightarrow$	PROC: *back substitution* in its
		constant back substitution form

For a second example of ESE, this time of Process 2.7.3, consider the Ackermann–Peter function (Definition 2.4.4) repeated here.

$$\left[\begin{array}{ll} f(x, y) = y + 1 & : x = 0 \\ f(x, y) = f(x - 1, 1) & : y = 0 \\ f(x, y) = f(x - 1, f(x, y - 1)) & : x, y \in N \\ \text{Initially } x, y \in N_0,\ x + y \in N. & \end{array}\right.$$

The evaluation of $f(2,1)$ by Process 2.7.2 is traced below. Most of the steps shown represent one new equation formation or back substitution, but a few represent more than one. The equation sets are numbered in the order in which they are formed.

Example 2.7.2

New equation formations

0
$$f(2,1) = f(1, f(2,0))$$

1
$$\begin{aligned} f(2,1) &= f(1, f(2,0)) \\ f(2,0) &= f(1,1) \end{aligned}$$

2
$$\begin{aligned} f(2,1) &= f(1, f(2,0)) \\ f(2,0) &= f(1,1) \\ f(1,1) &= f(0, f(1,0)) \end{aligned}$$

3 and 4
$$\begin{aligned} f(2,1) &= f(1, f(2,0)) \\ f(2,0) &= f(1,1) \\ f(1,1) &= f(0, f(1,0)) \\ f(1,0) &= f(0,1) \\ f(0,1) &= 2 \end{aligned}$$

6
$$\begin{aligned} f(2,1) &= f(1, f(2,0)) \\ f(2,0) &= f(1,1) \\ f(1,1) &= f(0,2)) \\ f(1,0) &= 2 \\ f(0,1) &= 2 \\ f(0,2) &= 3 \end{aligned}$$

8
$$\begin{aligned} f(2,1) &= f(1,3) \\ f(2,0) &= 3 \\ f(1,1) &= 3 \\ f(1,0) &= 2 \\ f(0,1) &= 2 \\ f(0,2) &= 3 \\ f(1,3) &= f(0, f(1,2)) \\ f(1,2) &= f(0, f(1,1)) \end{aligned}$$

10
$$\begin{aligned} f(2,1) &= f(1,3) \\ f(2,0) &= 3 \\ f(1,1) &= 3 \\ f(1,0) &= 2 \\ f(0,1) &= 2 \end{aligned}$$

Constant back substitutions

5
$$\begin{aligned} f(2,1) &= f(1, f(2,0)) \\ f(2,0) &= f(1,1) \\ f(1,1) &= f(0,2) \\ f(1,0) &= 2 \\ f(0,1) &= 2 \end{aligned}$$

7
$$\begin{aligned} f(2,1) &= f(1,3) \\ f(2,0) &= 3 \\ f(1,1) &= 3 \\ f(1,0) &= 2 \\ f(0,1) &= 2 \\ f(0,2) &= 3 \end{aligned}$$

9
$$\begin{aligned} f(2,1) &= f(1,3) \\ f(2,0) &= 3 \\ f(1,1) &= 3 \\ f(1,0) &= 2 \\ f(0,1) &= 2 \\ f(0,2) &= 3 \\ f(1,3) &= f(0, f(1,2)) \\ f(1,2) &= f(0,3) \end{aligned}$$

11
$$\begin{aligned} f(2,1) &= f(1,3) \\ f(2,0) &= 3 \\ f(1,1) &= 3 \\ f(1,0) &= 2 \\ f(0,1) &= 2 \end{aligned}$$

$$
\begin{aligned}
&f(0,2) = 3\\
&f(1,3) = f(0, f(1,2))\\
&f(1,2) = f(0,3)\\
&f(0,3) = 4
\end{aligned}
\qquad
\begin{aligned}
&f(0,2) = 3\\
&f(1,3) = f(0,4)\\
&f(1,2) = 4\\
&f(0,3) = 4
\end{aligned}
$$

12

$$
\begin{aligned}
&f(2,1) = f(1,3)\\
&f(2,0) = 3\\
&f(1,1) = 3\\
&f(1,0) = 2\\
&f(0,1) = 2\\
&f(0,2) = 3\\
&f(1,3) = f(0,4)\\
&f(1,2) = 4\\
&f(0,3) = 4\\
&f(0,4) = 5
\end{aligned}
$$

13

$$
\begin{aligned}
&f(2,1) = 5 \Leftarrow \textit{Answer}\\
&f(2,0) = 3\\
&f(1,1) = 3\\
&f(1,0) = 2\\
&f(0,1) = 2\\
&f(0,2) = 3\\
&f(1,3) = 5\\
&f(1,2) = 4\\
&f(0,3) = 4\\
&f(0,4) = 5
\end{aligned}
$$

These examples illustrate the general proposition that

Lemma 2.7.1: If DEF.f is nonnested, then part (a) of ESE Process 2.7.2, new equation formation, after once being completed, will never be needed again. The same is true for part (b).

PROOF: This is so because in a nonnested definition the execution of part (b) cannot introduce any occurrences of f on the right on an equation and so cannot generate a requirement to form any new equations. ■

On the other hand when DEF.f is nested, it may be necessary to cycle through parts (a) and (b) many times, as in the evaluation of the Ackermann–Peter function $f(2,1)$.

Although in Example 2.7.1 the execution of part (a) ended with a finite set of equations, and it was possible by making substitutions in part (b) to get the answer, either of these events could fail to occur in *a* nonnested definition. If for a definition DEF.f part (a) ends with a finite set of equations for every member of its initial domain, then we say DEF.f is *equation complete*. It is easy to show that a definition that terminates is also equation complete. However, even if it does not terminate, it may still be equation complete. Subsequent back substitution may or may not produce an answer. It can however, be shown that

Lemma 2.7.2: If DEF.f terminates, then not only will DEF.f be equation complete, but back substitution will produce an answer.

On the other hand, DEF.f may be equation complete but not solvable with simple back substitution alone. Such cases are nevertheless often solvable, as we will see in Chapter 5.

The evaluation procedures described in this chapter form the basis for our development of efficient algorithms for implementing recursive definitions. In

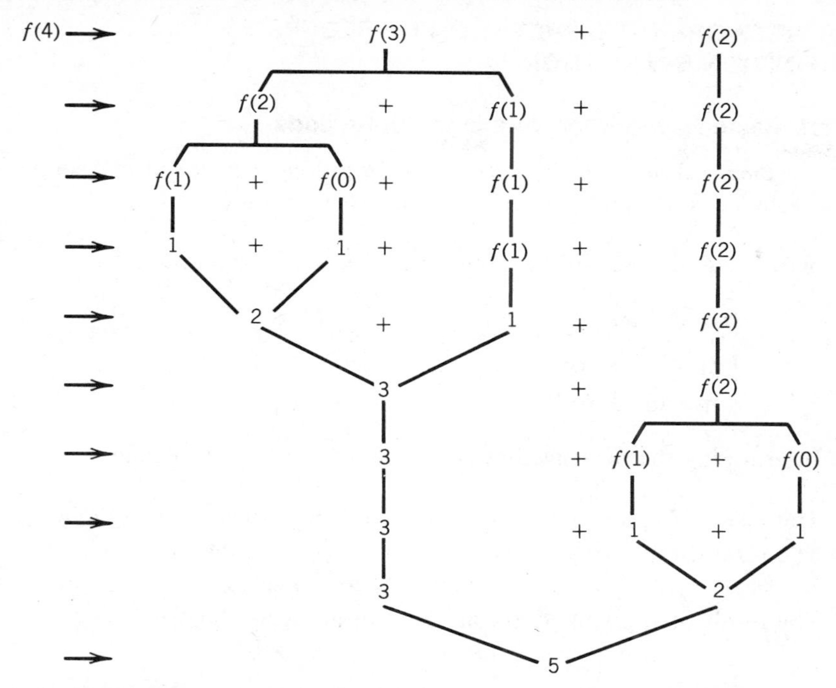

FIGURE 2-2 Network for SE evaluation of Fibonacci number definition (Example 2.71).

fact these algorithms may be viewed as computer simulations of these evaluation procedures.

The ESE procedures (Processes 2.7.2 and 2.7.3) are attractive alternatives to the SE of Process 2.3.1. They eliminate much redundant handling of an argument structure when that argument structure would be generated several times in an SE. In ESE there is one equation $f(X) = R(x)$ for each different argument X which would appear in an SE. Once $R(X)$ is evaluated, its value is substituted for all occurrences of $f(X)$ in other equations in the set formed in ESE; $f(X)$ need not be reevaluated before each such substitution. In SE, by contrast, each occurrence of $f(X)$ requires, in effect, that $f(X)$ be reevaluated. If in Example 2.7.1, the evaluation of the Fibonacci number function $f(4)$, proceeds by SE, we obtain Figure 2-2.

In this SE of $f(4)$, $f(2)$ is evaluated completely at both of its occurrences, whereas in ESE, only one equation was formulated for $f(2)$, the value found once, and that value substituted in the two different equation occurrences of $f(2)$.

ESE will generally require a data structure for efficient storage of and access to the intermediate results. This may obviate its time complexity advantage over SE.

2.8 NETWORK INTERPRETATION OF RECURSIVE DEFINITION EVALUATION

2.8.1 Basic Network for Nonnested Definitions

The *standard form* lines for nonnested definitions, developed in Section 2.2, are included in a standard form definition in Definition 2.8.1.

Definition 2.8.1: Standard for Nonnested Recursive Definition

$$\left[\begin{array}{ll} f(X) = w\big(f(o_1(X)),\ldots, f\big(o_{m(x)}(X)\big)\big) & :T(X) \\ f(X) = t(X) & :\sim T(X) \\ \text{Initially } X \in D. & \end{array}\right.$$

Interpreting the recursive line of Definition 2.8.1, $f(X)$ equals:

The result of applying the primitive function w to each of the following:
The result of applying f to (the result of applying the primitive function o_1 to X itself).
The result of applying f to (the result of applying the primitive function o_2 to X itself).
⋮
The result of applying f to (the result of applying the primitive function $o_{m(x)}$ to X itself).

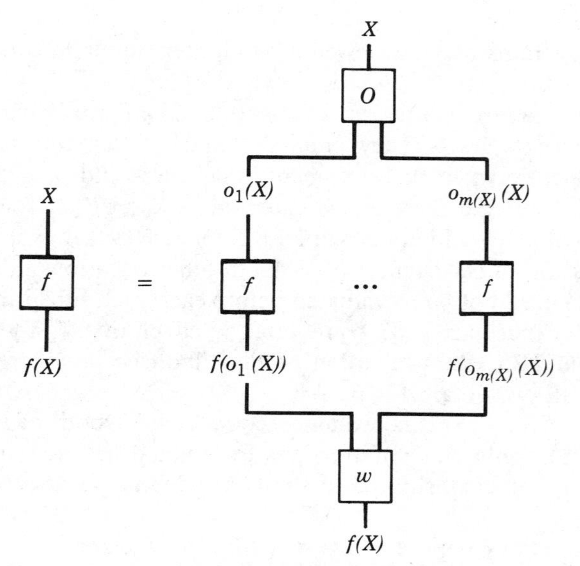

FIGURE 2-3 One-stage network for nonnested recursive line $f(X) = w(f(o_1(X)), \ldots, f(o_{m(x)}(X)))$ $:\sim T(X)$.

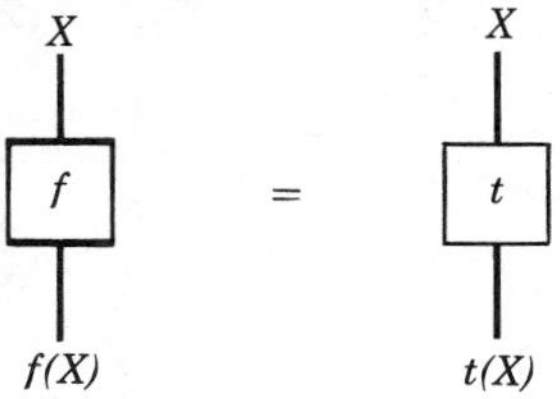

X X

f = t

f(X) t(X)

FIGURE 2-4 One-stage network for nonrecursive line $f(X) = t(X)$.

The idea behind this weave of words and symbols is pictured in the *network* in Figure 2-3. In representing such networks we use an O block whose outputs are $o_1(X)$ through $o_{m(x)}(X)$ when its input is X, as well as w, t, and f *blocks*, each representing the corresponding function. Assume that signals pass into the top of a block and out the bottom, unless stated otherwise.

The analogous network corresponding to the terminal, nonrecursive line in Definition 2.8.1 is given in Figure 2-4. There is a network consisting of interconnected subnetworks, each of the form given in Figures 2-3 and 2-4, for every evaluation of a standard form nonnested definition. Assume that Definition 2.8.1 is to be evaluated with initial argument $d \in D$ and that $\sim T(d)$; then the network would look like that in Figure 2-3 with every X replaced with d. Then assuming that each $o_i(d)$, like d, is in the domain of a nonterminal line, each f block with such an input is replaced by a network like Figure 2-3, yielding Figure 2-5. Further substitution for the f blocks, assuming some inputs are in the terminal and some in the nonterminal domains, yields Figure 2-6.

In general evaluation of a nonnested recursive definition DEF.f with initial argument d can be interpreted as construction of a network according to DEF.f and d, followed by a determination of the network output when its input is d. This *O–w–t Network*, Figure 2-7, always has the form of two trees, the O network and the w network, which are mirror images of each other, separated by a line of t blocks called the *t line*. Figure 2-7 illustrates the form of the network.

As an example of the construction evaluation process consider the following definition for the number of combinations of the first n integers taken m at a time.

Definition 2.8.2: Number of Combinations of m of the First n Integers

$$\begin{bmatrix} f(n,m) = 1 & :n = m \text{ or } m = 0 \\ f(n,m) = f(n-1, m-1) + f(n-1, m) & :n > m > 0 \\ \text{Initially } n, m \in N, n > m & \end{bmatrix}$$

The developed network for the evaluation of $f(4,2)$ is given in Figure 2-8.

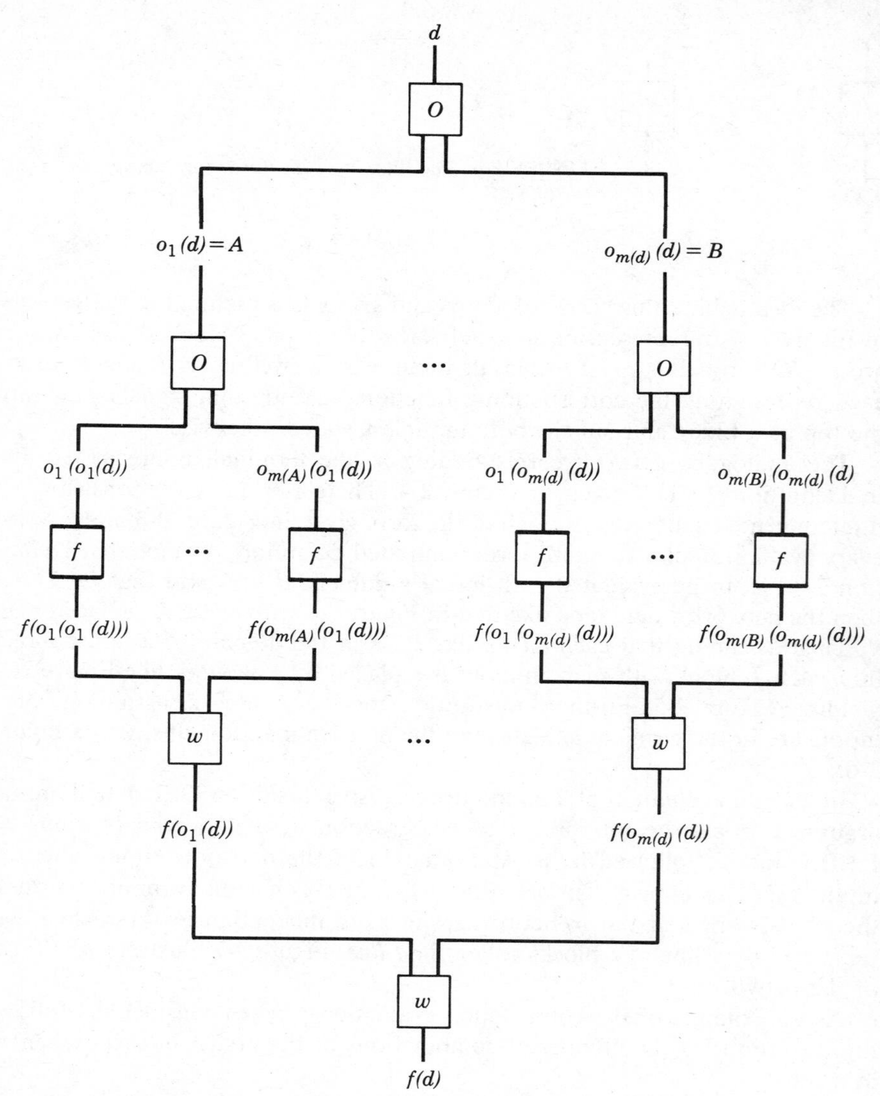

FIGURE 2-5 Two-stage network for nonnested recursive definition.

Example 2.8.1: The number of outputs from O blocks, and accordingly of inputs to w blocks, as well as the depth and uniformity of the O and w networks in an O–w–t network depend on both the function definition and the initial argument of the function which the network simulates. Nevertheless there will invariably be an O network, a tree opening downward, and its mirror image tree, the w network, opening upward, with the two separated and connected through a t line. These structural features exist in the O–w–t

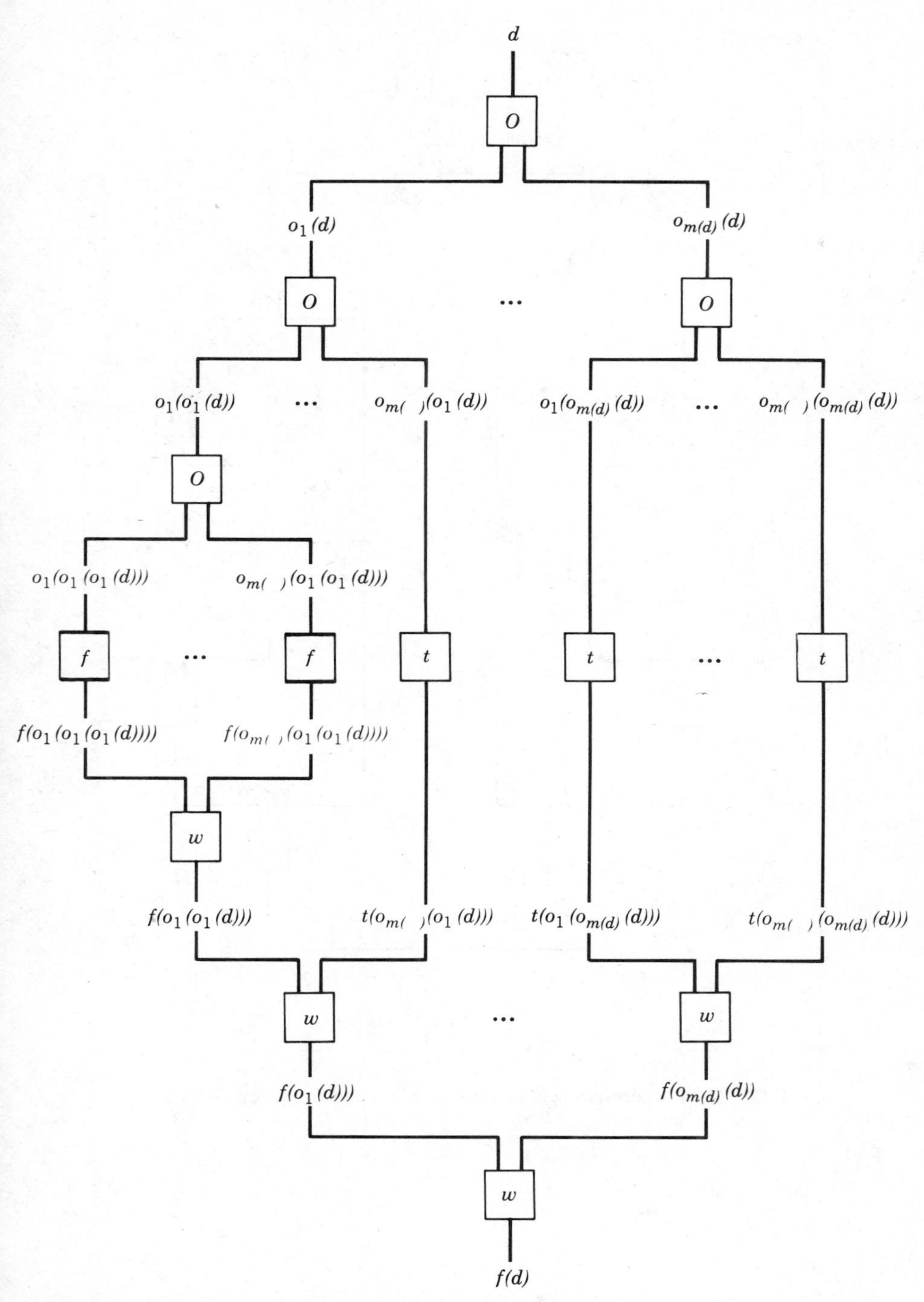

FIGURE 2-6 Multiple-stage network for Nonnested recursive definition. *Note*: The argument of m has been left blank in most places to maintain readability.

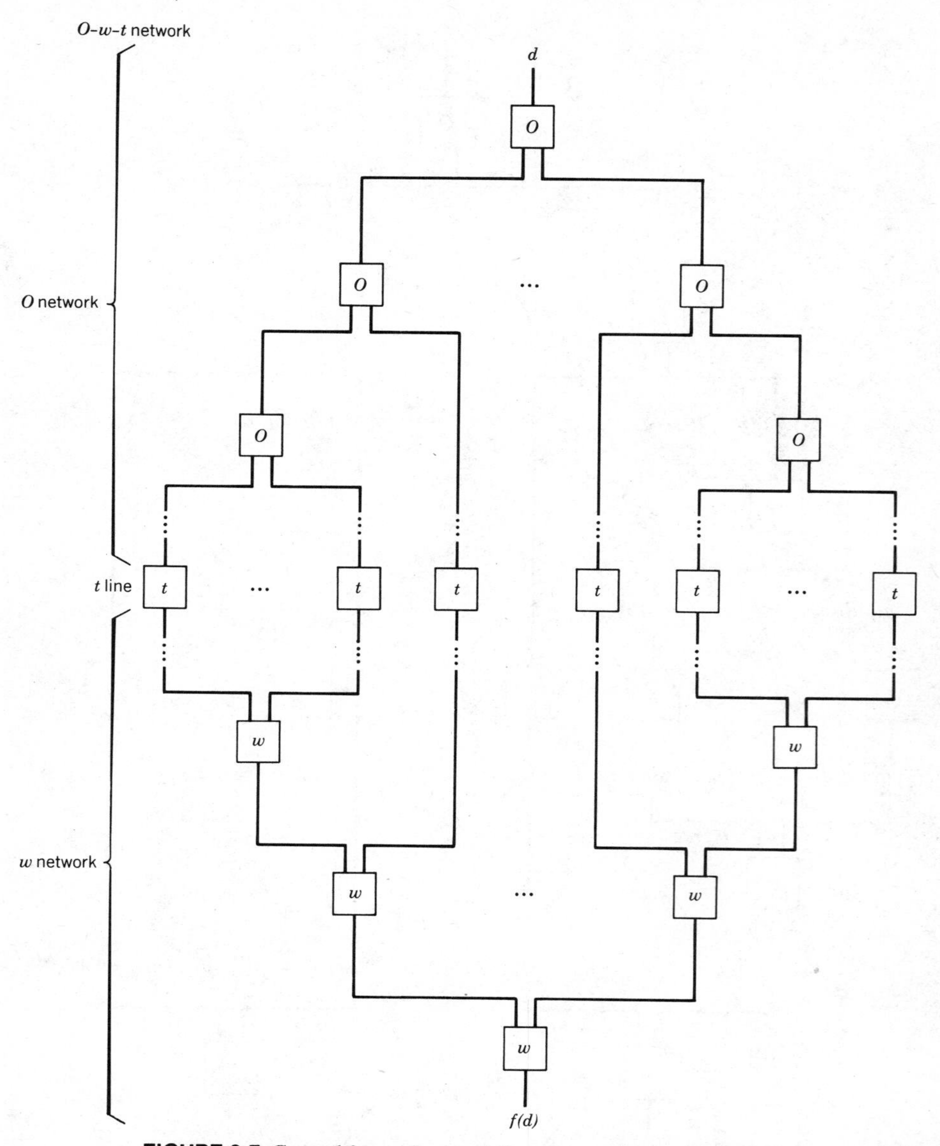

FIGURE 2-7 General form of network for nonnested recursive definition.

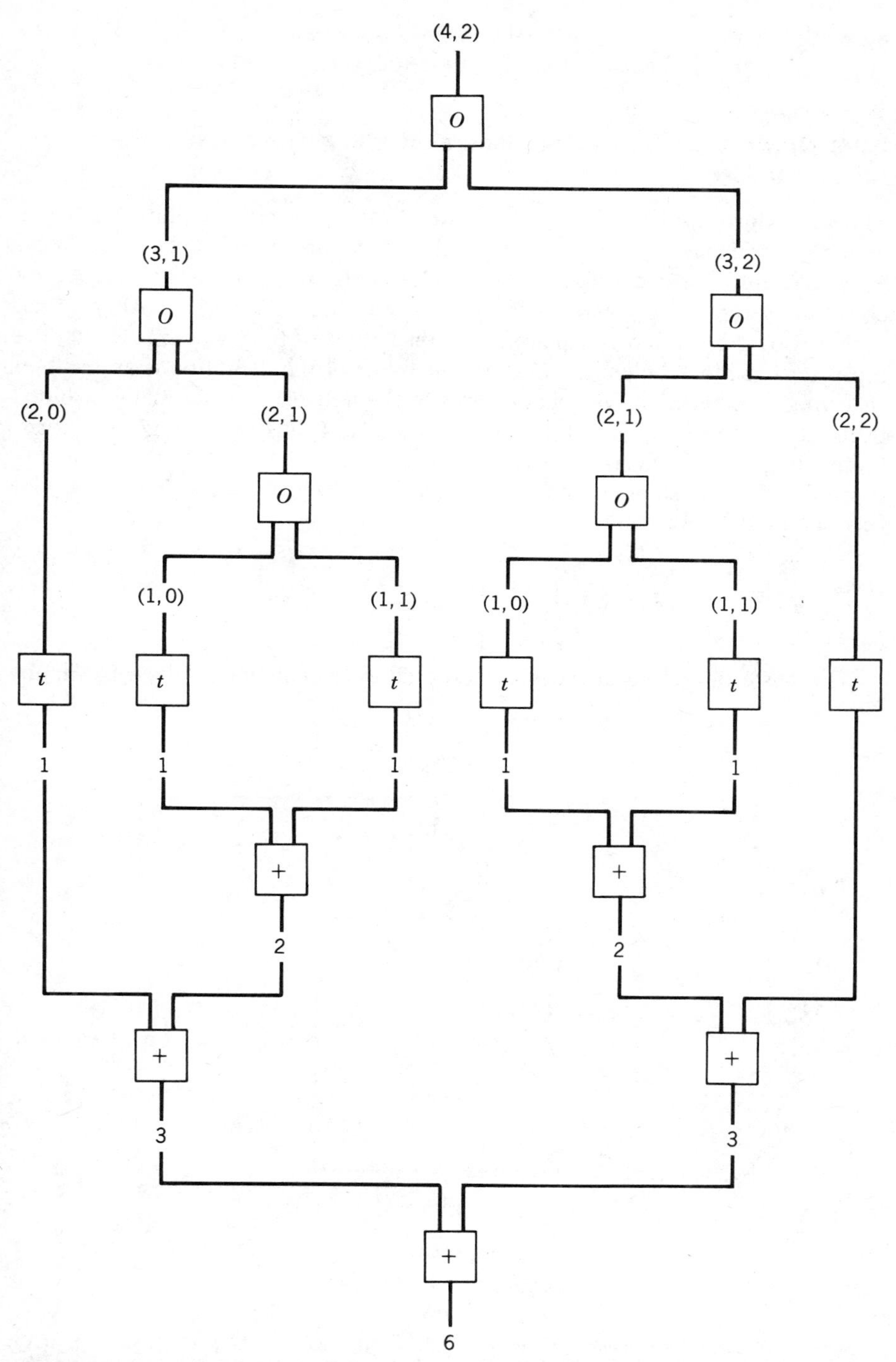

FIGURE 2-8 Network for nonnested recursive definition for the number of combinations of 4 things taken 2 at a time (Definition 2.8.2).

network for any nonnested recursive definition evaluation. They provide a key in planning the implementation of such definitions.

2.8.2 Network for Nonnested Recursion with *w* Dependent on the Argument *X*

The procedure used to obtain the network for a recursive definition is quite general. For each line of the definition there is a corresponding basic network. The problem is to expand an initial f block into a network. This is done by continually replacing f block occurrences with basic networks, each f block replaced by a basic block depending on the subset of the domain to which the input to that block belongs, and thus which line of the definition is applicable. In trying to determine the general form of the network we can make assumptions about the domain of the f block input and illustrate the general form under different circumstances, as we did for Figure 2.5. As a further illustration consider the standard nonnested definition whose recursive line has the form of Eq. (2.2.3), namely,

$$f(X) = w\big(f(o_1(X)),\ldots, f\big(o_{m(x)}(X)\big), p(X)\big) \qquad :\sim T(X)$$

The basic one-stage network corresponding to that line is given in Figure 2-9.

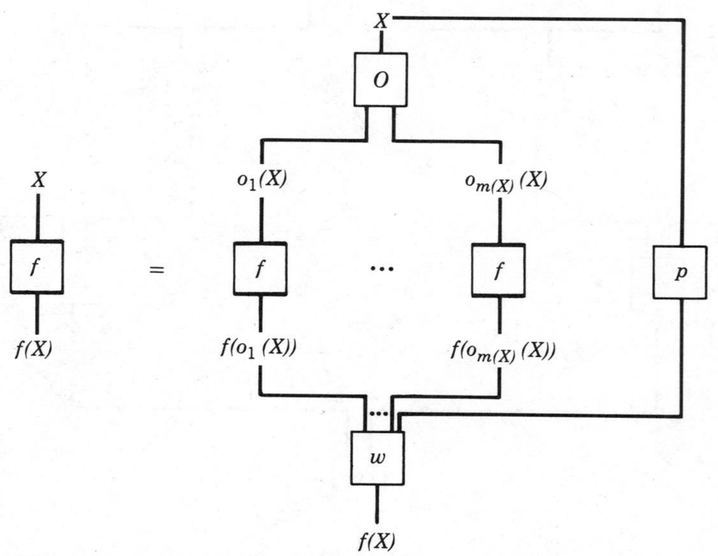

FIGURE 2-9 One-stage network for recursive line $f(X) = w(f(o_1(X)),\ldots, f(o_{m(x)}X)), p(X))$ $:\sim T(X)$.

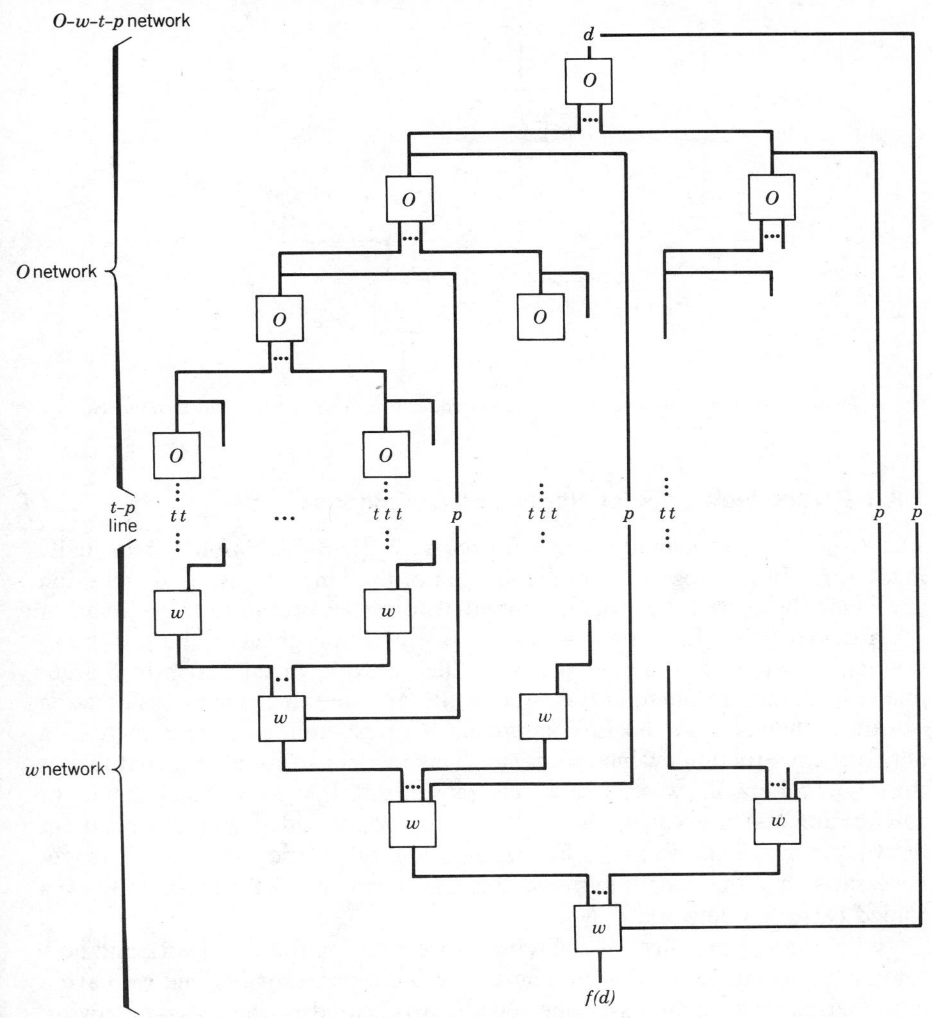

FIGURE 2-10 General form of network for recursive line $f(X) = w(f(o_1(X)), \ldots, f(o_{m(x)}(X)), \ldots, f(o_{m(x)}(X)), p(X)) \quad :\sim T(X)$.

The general form of the expanded network for a definition using the recursive line of the basic network in Figure 2-9, and the terminal line of Figure 2-4, is sketched in Figure 2-10. In this representation we have followed only a few of the lines which go through *p* blocks—from their origin in the *O* network to becoming *w* block inputs. This is sufficient to give the general form of the network. The *p* and *t* blocks are represented, respectively, by the letters *p* and *t*, without any surrounding box. Solid lines present actual lines, dotted lines indicate a part of the network not shown. The *t* blocks at the end of the dotted lines are not placed precisely, but the *p* blocks in the solid lines are.

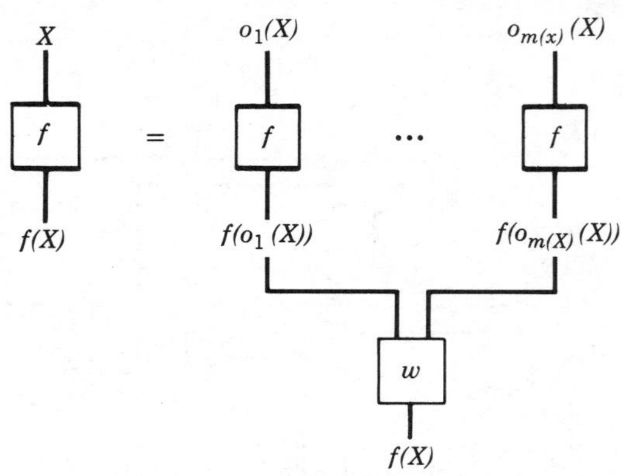

FIGURE 2.11 One-stage network for nonnested recursive definition, O blocks removed.

2.8.3 Folded Networks for Nonnested Definitions

The O–w–t network models the SE process 2.3.1. Substitution is repeatedly made for f blocks according to the domain of their inputs, just as substitution is repeatedly made for f calls in sentential forms according to the domain of f's arguments in SE. In neither case is any attention given to the possibility that block inputs at different places in the network, or equivalently f arguments in different sentential forms of an SE, are the same. That consideration is central however in the ESE approach. Corresponding to ESE there is a network construction evaluation model. This model can be constructed just as the O–w–t network was, after which, recognizing that some block inputs (f call arguments) are equal, these inputs can be identified and the resultant network simplified. With the inputs thus identified, the network no longer necessarily has the simple reflected tree pair form. We call such a network a folded (O–w–t) network.

In illustrating the effect of equal inputs we focus on the w network and how folding occurs there. Corresponding to the basic nonrecursive line we have a modified version of the basic one-stage network used in the O–w–t network development. The O blocks are removed as shown in Figure 2-11.

Now assume that the initial input is d, and (for simplicity) that $m(d) = m(o_1(X)) = m(o_2(X)) = 2$; also assume that $o_1(X)$ and $o_2(X)$ are both in the domain of a recursive line. Replacing the relevant f blocks using the basic network in Figure 2-11, gives the network in Figure 2-12. If $o_1(o_2(d)) = o_2(o_1(d))$, the two f blocks with these inputs can be made one with the resultant output going two different places, just as in ESE the value of a single equation would be substituted in all right sides in which the corresponding left side appeared (Figure 2-13).

In Definition 2.8.2 the condition $o_1(o_2(d)) = o_2(o_1(d))$ holds. This allows repeated folding of pairs of input lines throughout the substitutional domain.

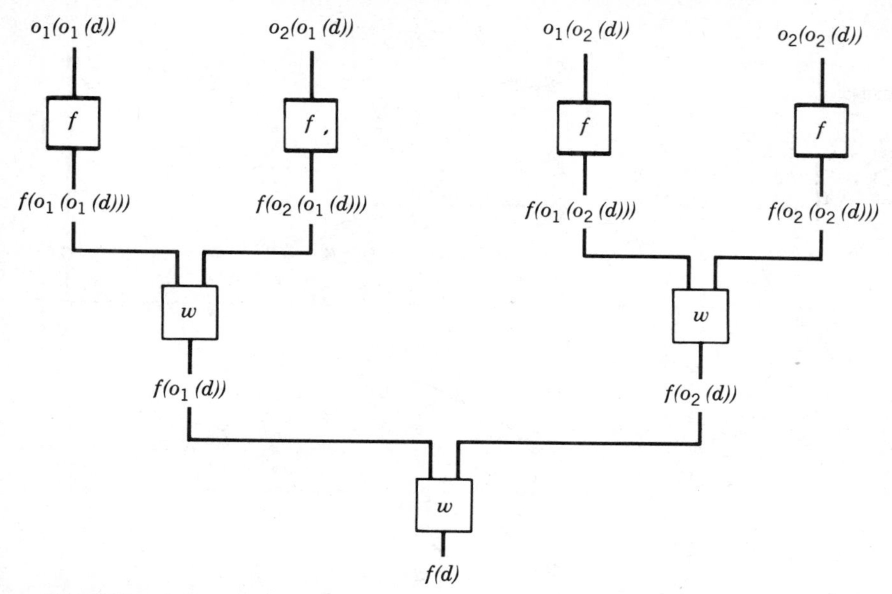

FIGURE 2-12 Two-stage network for nonnested recursive definition, O blocks removed.

The result when this is done for $f(4,2)$ is the network model of an ESE evaluation given in Figure 2-14.

Relations among the argument structures on the right of the recursive line, as in the above example, often lead to patterns of folding, which in turn lead to efficient implementations. We next consider a consequence of such relations, this time in quite general terms.

2.8.3.1 Consequence of Some Relations Among Argument Structures

Recall the following definitions. If $I = \langle i_1, i_2, \ldots, i_n \rangle$, then

$$o_I(X) = o_{i_n}\left(\ldots o_{i_2}\left(o_{i_1}(X)\right)\ldots\right)$$

$$x^m = \underbrace{\langle x, x, \ldots, x \rangle}_{m}$$

Our first two examples assume that the recursive line of the definition in question is of the form

$$f(X) = w(f(o_1), f(o_2))$$

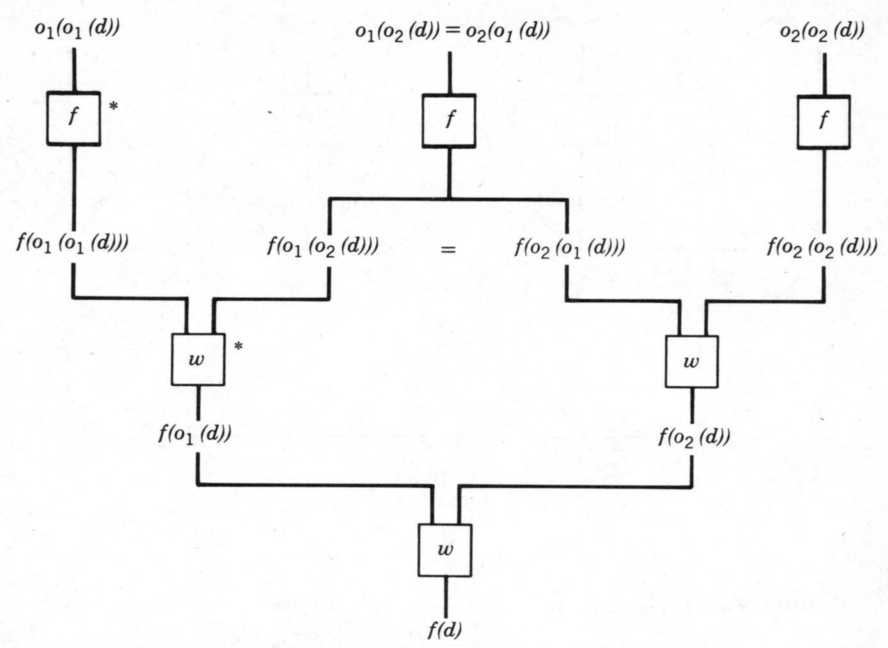

FIGURE 2-13 Two-stage folded network for nonnested recursive definition with $o_1(o_2(d)) = o_2(o_1(d))$.

Suppose $o_{11}(X) = o_2(X)$. Let $I = \langle i_1, i_2, \ldots, i_n \rangle$, where $i_j \in \{1, 2\}$. Then $o_I = o_{1^{M'}}$, where $M = i_1 + i_2 + \cdots + i_n$. This so because: each 2 in the vector I can be replaced by two 1's because $o_{11} = o_2$. It follows that any argument structure that can be generated by applications of o_1 and o_2, which can be generated from an initial argument structure d in any mix, can be represented by one of the following:

$$X, o_1(X), o_{11}(X), o_{111}(X), \ldots, o_{11\ldots 1}(X)\ldots$$

or, more compactly,

$$X, o_{1^1}(X), o_{1^2}(X), o_{1^3}(X), \ldots, o_{1^j}(X), \ldots$$

If the definition terminates, this line of argument structures will terminate with $j =$ an integer, say P. So the line would end with $o_{1^P}(X)$.

Further for any j, the value of f at $o_{1^j}(X)$ depends only on f at $o_{1^{j+1}}(X)$ and $o_{1^{j+2}}(X)$. Thus we have a one-dimensional line of argument structures. We can implement this definition by computing the value of the function at each argument structures in the one-dimensional line one at a time, moving from right to left. If the computation of f at each argument structure on this

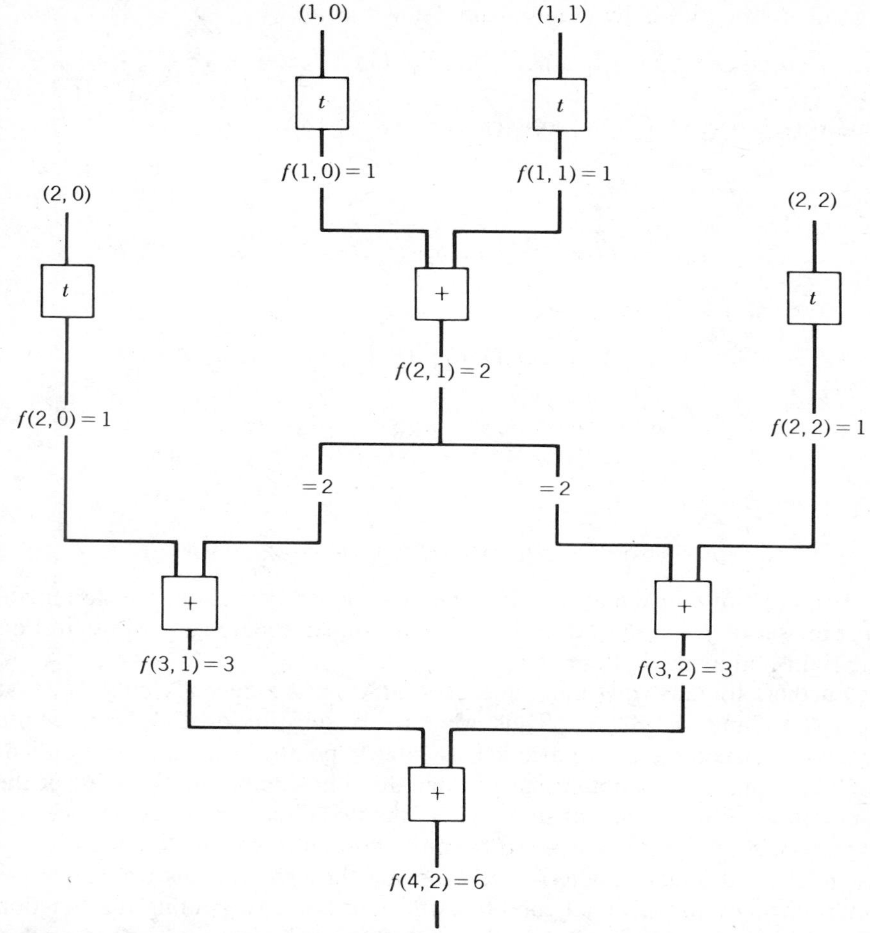

FIGURE 2-14 Folded network for nonnested recursive definition for the number of combinations of 4 things taken 2 at a time (Definition 2.8.2).

line depends only on the value of f at a fixed number of other places on the line, the computation will often be $O(1)$, and the entire computation will take time proportional to the number of argument structures in the line.

Consider a second example.

Suppose $o_{12} = o_{21}$. Let $I = \langle i_1, i_2, \ldots, i_n \rangle$, where $i_j \in \{1, 2\}$. Then $o_I = o_{1^A 2^B}$, where A is the number of 1 components in I, B is the number of 2 components in I, because 2's can be moved right past the 1's in I.

It follows that we can represent all the argument structures that can be generated by applications of o_1 and o_2 in any mix, generated from an initial argument structure d by the two-dimensional array

$$d,\ o_1(d),\ o_{11}(d),\ o_{111}(d), \ldots,\ o_{11\ldots1}(d), \ldots, \ldots$$

$$o_2(d),\ o_{12}(d),\ o_{112}(d),\ o_{1112}(d), \ldots,\ o_{11\ldots12}(d), \ldots$$

$$\vdots$$

$$o_{2\ldots2}(d),\ o_{12\ldots2}(d),\ o_{112\ldots2}(d),\ o_{1112\ldots2}(d), \ldots,\ o_{11\ldots12\ldots2}(d), \ldots$$

or, more compactly,

$$d,\ o_{1^1}(d),\ o_{1^2}(d),\ o_{1^3}(d), \ldots,\ o_{1^j}(d), \ldots$$

$$o_{2^1}(d),\ o_{1^12^1}(d),\ o_{1^22^1}(d),\ o_{1^32^1}(d), \ldots,\ o_{1^j2^1}(d), \ldots$$

$$\vdots$$

$$o_{2^k}(d),\ o_{1^12^k}(d),\ o_{1^22^k}(d),\ o_{1^32^k}(d), \ldots,\ o_{1^j2^k}(d), \ldots$$

If the definition terminates, this line of argument structures will terminate at some set of j, k values truncating the two-dimensional region of interest on the right and at the bottom.

Further, for any j, k the value of f at $o_{1^j2^k}(X)$ depends only on f at $o_{1^{j+1}2^k}(X)$ and $o_{1^j2^{k+1}}(X)$. Thus we have a substitutional domain whose argument structures are embedded as integer points in a finite two-dimensional region. We can implement this definition by computing the value of the function at all the argument structures in the two-dimensional region, one at a time, moving row by row upward from the bottom row (as determined by the terminal conditions), or alternatively moving through columns from right (as determined by the terminal conditions) to left. Since computing the function at each point involves evaluating a w function on the value of f at two other points, it will usually be a constant time computation, so overall complexity will be proportional to the number of integer points in the two-dimensional region.

Much of this can be generalized. For each $k \in N$, a k-dimensional array of argument structures suffices for a considerably larger class of relations than produce the one- and two-dimensional arrays above. The example relations above apply to recursive lines with $m(X) = 2$. There are n-dimensional arrays for any $m(X)$, including cases in which its value really is a function of X. We consider many examples of such definitions in Section 2.10 and in subsequent chapters under the heading "Dynamic Programming."

Consider such a generalization. Assume the recursive line of the definition to be studied is of the form

$$f(X) = w\big(f(o_1(X)), f(o_2(X)), \ldots, f\big(o_{m(x)}(X)\big)$$

Suppose there is a function

$$r_1(X) \ni o_i(X) = \underbrace{r_{11\ldots1}}_{k_i}(X) = r_{1^{k_i}}(X)$$

Then the following one-dimensional array represents every possible argument structure that can be generated from an initial argument structure d,

$$d, r_1(d), r_{11}(d), r_{111}(d), \ldots, r_{11\ldots1}(d), \ldots$$

or, more compactly,

$$d, r_{1^1}(d), r_{1^2}(d), r_{1^3}(d), \ldots, r_{1^j}(d), \ldots$$

For any a, the value of f at $r_{1^a}(d)$ $(= Y)$ is w of the value of f at $r_{1^{a+k_1}}(d), r_{1^{a+k_2}}(d), \ldots, r_{1^{a+k_{m(Y)}}}(d), \ldots$. Assume the cost of this computation is cost (Y). Since f is to be computed at each argument structure Y on the one-dimensional array, the complexity will be proportional to the sum of the costs, cost(Y), over all these argument structures.

The folding in the networks we have considered so far is generated by relations between the o_i's in the recursive line. These relations are independent of the initial argument structure d. Thus the shape or dimensionality of the networks is independent of d. For any nonnested definition which is equation complete, we can form for each initial argument structure d a set of equations involving all the argument structures on which the value of f at d depends. For each equation in this set we can form subnetworks like Figure 2-11, the output corresponding to the f call on the left of the equation, and the inputs corresponding to the f calls on the right of the equation. By identifying any two lines labeled with the same argument structure, we can draw the subnetworks together in such a way that no argument structure will label more than one line in the resultant network. If the definition from which the network is designed does not have any such relations among the o_i's, the shape of the network will largely be determined by the initial argument structure, and will differ with the initial argument structure even within the same definition. When we consider algorithms that simulate ESE in Chapter 6, we will see many examples of such networks dependent on the initial argument structure or inputs to the algorithms.

2.9 NETWORKS FOR NESTED RECURSIVE DEFINITIONS

The same network construction techniques applicable to nonnested definitions apply to nested definitions. Consider Definition 2.9.1, which uses the nested recursive line of Example 2.1.14.

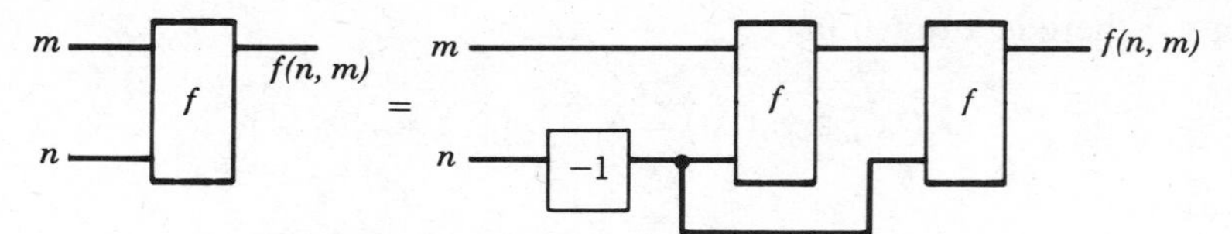

FIGURE 2-15 One stage of network for Definition 2.9.1.

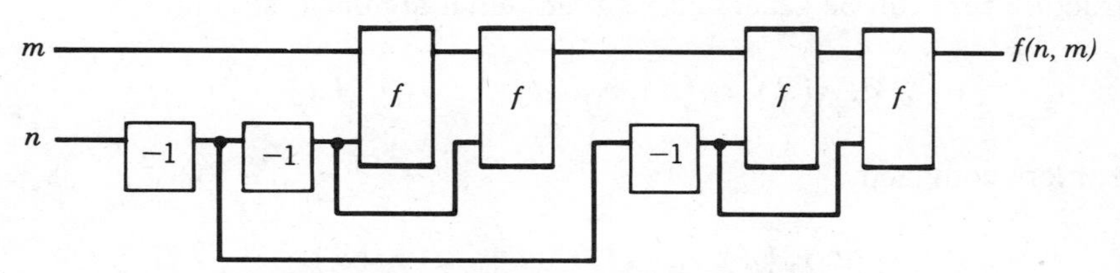

FIGURE 2-16 Two stages of network for Definition 2.9.1.

Definition 2.9.1: Nested Definition

$$\begin{cases} f(n, m) = t(m) & : n = 0 \\ f(n, m) = f(n-1, f(n-1, m)) & : n, m \in N \\ \text{Initially } n, m \in N. & \end{cases}$$

The recursive line can be rewritten in network terms, as shown in Figure 2-15. (Unlike the O–w–t networks, the flow in this one is viewed as horizontal, with inputs coming into a block from the left and outputs emerging from the right.)

Substituting for each occurrence of an f block on the right, using the equality of Figure 2-15, we get the network in Figure 2-16 for $f(n, m)$. This network (Figure 2-16) can be simplified to give 2^2 blocks, each of which has one of its inputs $= n - 1 - 1 = n - 2$ (see Figure 2-17).

Again substituting for all f blocks, we get 2^3 blocks, each of which has its lower input $= n - 3$ (see Figure 2-18).

The general case can be inferred. The final network will have 2^n blocks, each of which will have $n - n$, or 0, as one of its inputs; the other input of the first block will be m. All other blocks will take their input from the output of the previous block. Because of the 0 input and the terminal line of Definition

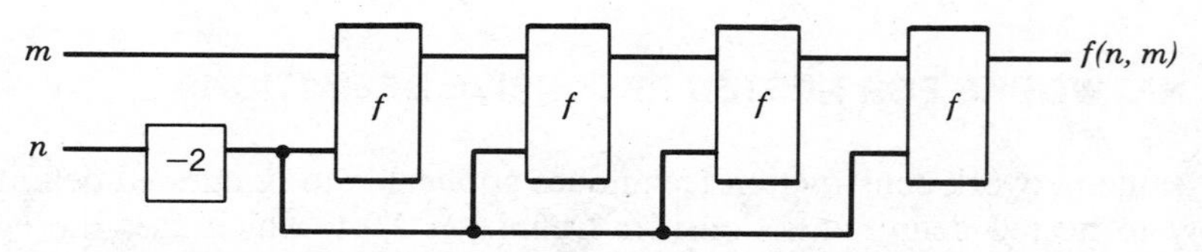

FIGURE 2-17 Two stages of simplified network for Definition 2.9.1.

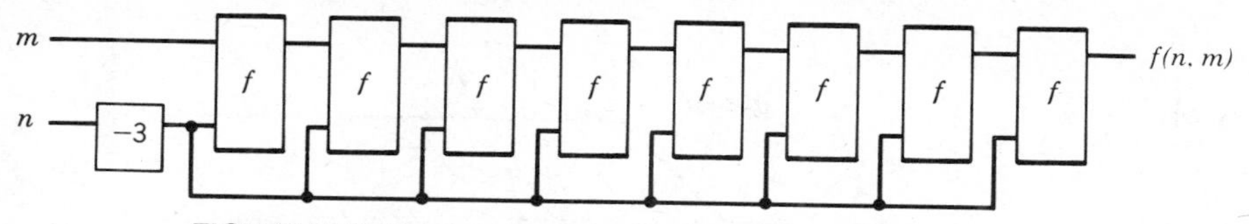

FIGURE 2-18 Three stages of simplified network for Definition 2.9.1.

2.9.1, each block will perform the terminal function t. Thus the function $f(n, m)$, the output of the entire network, is $t^{2^n}(m)$.

Unlike the nonnested case, we cannot say that the networks for all nested definitions share any significant common characteristics, though there certainly are significant subclasses of nested definitions which do.

2.10 PATTERNED FOLDED NETWORKS AND THEIR IMPLEMENTATION

In Section 2.8.3 we saw how relations among the o_i's on the right of the recursive line produce patterned folded networks. We continue that development by considering examples to see how the relations can be viewed more concretely, to develop the resultant folded network completely, including its boundaries, and to explore the characteristics of algorithms for implementing such networks. The definition for counting the number of combinations of n things taken m at a time produces a folded network and provides an example.

2.10.1 Combinations Example

Consider Definition 2.8.2. It is immediately clear that

$$o_1(o_2(X)) = \langle (n-1) - 1, (m) - 1 \rangle = \langle (n-1) - 1, m-1 \rangle = o_2(o_1(X))$$

thus the argument structures of Δ_i of Definition 2.8.2 fold into a two-dimensional structure. Let us approach this conclusion by a different, more concrete route.

Each member of Δ_f of Definition 2.8.2 is a pair of integers $\langle n, m \rangle$, with n greater than m, and both greater than 0. The members of Δ_f thus correspond to integer points in a triangular part of the positive plane in which n is greater than m and a folded network can be laid out with the block output $f(i, j)$ placed at the point (i, j). According to the recursive line, $f(n, m)$ depends on $f(n-1, m-1)$ and $f(n-1, m)$, so the output at point (n, m) depends on that one place diagonally below and to the left, and that one place to the left. To get Figure 2-19 we have taken the example network of Figure 2-3, and reoriented it so that output $f(4, 2)$ is at the point (4, 2). (This is closely related to the Pascal's triangle [Tucker]). The rectangular axes along which i and j vary, labeled, respectively, A and B (A–B system), are shown. In addition an

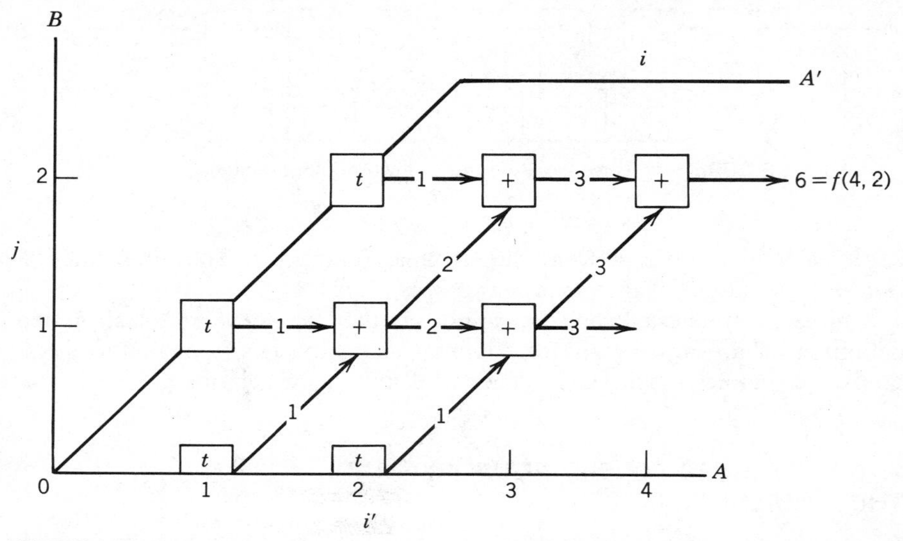

FIGURE 2-19 Example of two-dimensional network positioned in space (Definition 2.8.2, combinations, $f(4.2)$). *Note*: Inputs to blocks are on the left and below, outputs on the right and above.

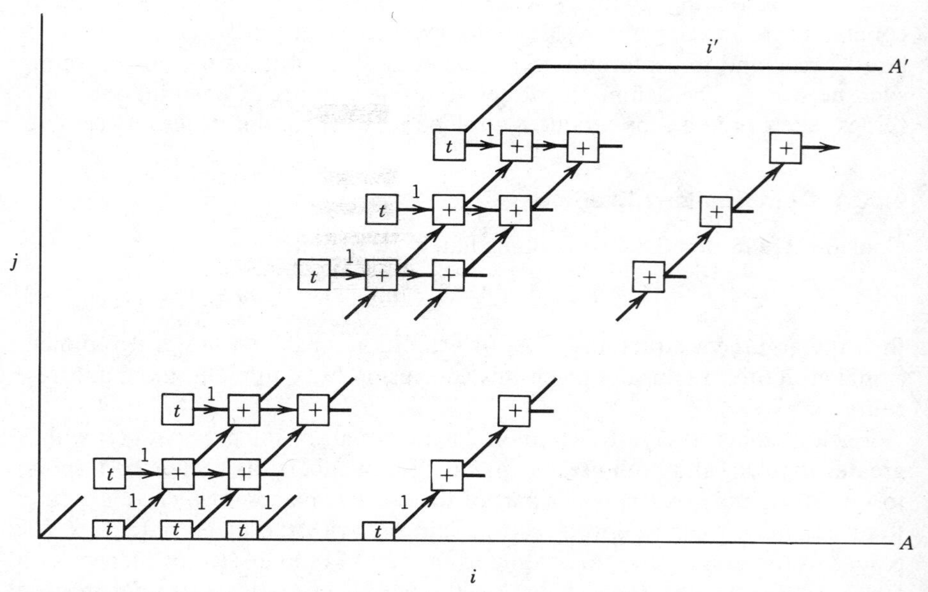

FIGURE 2-20 Two-dimensional network positioned in space (Definition 2.8.2, combinations).

axis labeled A', from which the variable i' varies, is shown. The position of each block might just as well be given in relation to the A', B axis (in the A'–B system)—as we will see there is good reason to do so. The values of the block outputs are also shown.

In Figure 2-20 the network of Figure 2-19 is generalized to compute $f(n, m)$ as given in Definition 2.8.2 for any n and m in its starting domain.

The t blocks correspond to the terminal lines of Definition 2.8.2, and thus their outputs, which appear at $m = 0$ (the abscissa) and at $m = n$ (the diagonal), are all 1. Notice that the structure of the upper right-hand corner is the same as that which was developed step by step for the $f(4, 2)$ example.

The pattern formed by the network suggests that it can be simulated by a pair of nested FOR loops, with indices I and J, respectively, stepping through the row and column positions of the block outputs. A simple simulation uses an $(n + 1) \times (m + 1)$ matrix F to hold the output of each block, including those at zero-valued coordinates. These outputs are initialized to 1. Then the activity of each row (J value) of the matrix is simulated starting with 1 and ending on row m. In each row the simulation starts at the column position (I value) one place to the right of the diagonal and proceeds to the last column on that row whose output can affect the final result. Rather than rowwise, simulation can proceed columnwise. As we will see, the choice may make a difference.

Algorithm 2.10.1

```
DCL F(n + 1, m + 1) array; INIT(1)
  FOR J = 1 to m DO
    FOR I = J + 1 to J + n − m DO
      F(I, J) ← F(I − 1, J − 1) + F(I − 1, J)
    END
  END
```

This implementation can be improved by changing from orthogonal axes labeled A and B along which I and J vary, to the axes A' and B along which I' and J vary. A point $I = a$, $J = b$ on the original axes becomes $I' = a - b$, $J = b$. This transformation is illustrated in Figure 2-21. From this figure we can see that the point represented by (I, J) in the A–B system is represented by $(I - J, J)$ in the A'–B system, that is, $I' = I - J$.

The difference in representation in the two coordinate systems is further illustrated in Figure 2-22, where a typical block in the network is shown with the coordinates of its inputs and outputs given in both systems.

The representation is simpler in the A'–B system, so we propose to transform Algorithm 2.10.1 to this system. The arguments of the F's represent the same points, but now in the A'–B system. So the line inside the DOs becomes

$$F(I - J, J) = F(I - J, J - 1) + F(I - J - 1, J)$$

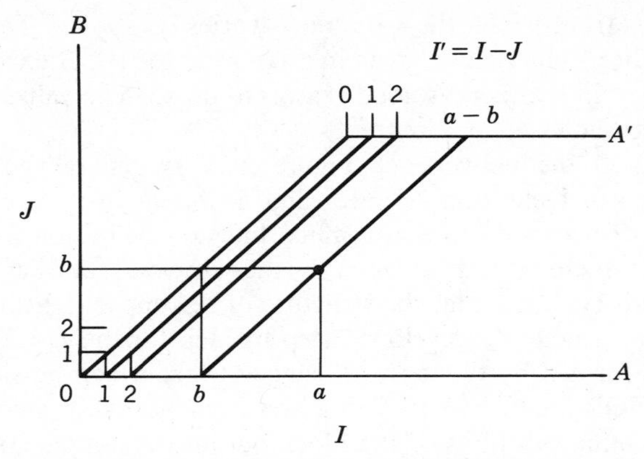

FIGURE 2-21 Linear transformation of two-dimensional argument (Definition 2.8.2, combinations).

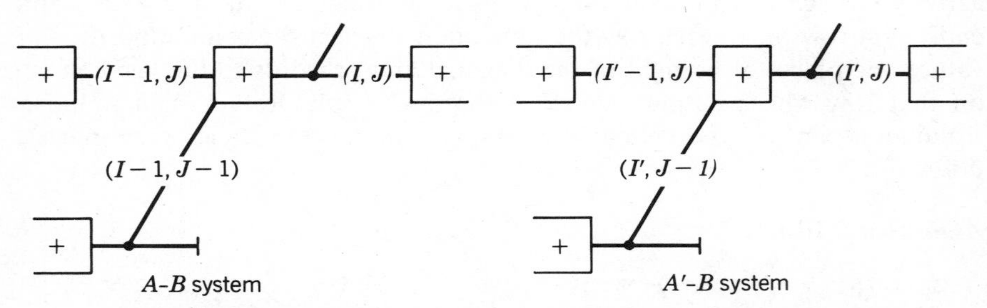

FIGURE 2-22 Effect of linear transformation on two-dimensional network.

Now replacing $I - J$ with I' in this line and noting that as I goes from $J + 1$ to $J + n - m$, $J - I = I'$ goes from 1 to $n - m$, we get

Algorithm 2.10.2

```
DCL F(n − m + 1, m + 1) array; INIT(1)
  FOR J = 1 to m DO
    FOR I' = 1 to n − m DO
      F(I', J) ← F(I', J − 1) + F(I' − 1, J)
    END
  END
```

Algorithm 2.10.2 simulates the network rowwise; an interchange of the DOs would give a columnwise simulation and produce the same output.

A final improvement takes advantage of the fact that in computing the output of a block b in row r, storage is required only for the outputs of blocks in $r - 1$, plus the output of the one block before b in r. The output of a block b in row r is needed in computing the output of the next block in r, and for computing the output of a block in $r + 1$. Therefore as block outputs are computed row by row, it is necessary to save a block output until it is needed, one row's worth of block outputs later, but no longer. So only one row of block outputs need be saved at any time in Algorithm 2.10.3.

Algorithm 2.10.3

```
DCL F(n − m + 1) vector; INIT(1)
  FOR J = 1 to m DO
    FOR I' = 1 to n − m DO
      F(I') ← F(I') + F(I' − 1)
    END
  END
```

If we had interchanged the FOR loops, the line inside the FOR loops would be $F(J) = F(J - 1) + F(J)$; the declaration would be DCL $F(m + 1)$ array; INIT(1). So the amount of storage necessary to save intermediate results would be $m + 1$ instead of $n - m + 1$. This difference in temporary storage requirements is the major consequence of the alternative order of simulation caused by an interchange of DOs. This is an example of the general proposition, developed in Chapter 4, that the major consequence of alternative sequencings of blocks in a parallel network is differences in the use of temporary storage.

Although the transformations and refinements were made at the program development stage of the design, the transformations could have been carried out at the earlier recursive definition stage. Definition 2.10.1 following is equivalent to Definition 2.8.2 in that $f(n, m)$ of Definition 2.8.2 equals $g(n - m, m)$ of Definition 2.10.1.

Definition 2.10.1: Number of Combinations of m' of $n' + m'$ Items

$$\left[\begin{array}{ll} g(n', m') = 1 & : n' \text{ or } m' = 0 \\ g(n', m') = g(n', m' - 1) + g(n' - 1, m') & : n', m' \in N \\ \text{Initially } n', m' \in N. & \end{array}\right.$$

The arguments in Δ_f of Definition 2.8.2 are transformed one to one into those of Definition 2.10.1; that is, each point (i, j) in the area of the plane in which $i > j$ is transformed to (i', j) with the following transformation equations:

$$i' = i - j$$

$$j' = j$$

Since this is a one-to-one transformation, the old coordinates may be expressed in terms of the new ones by solving the transformation equations.

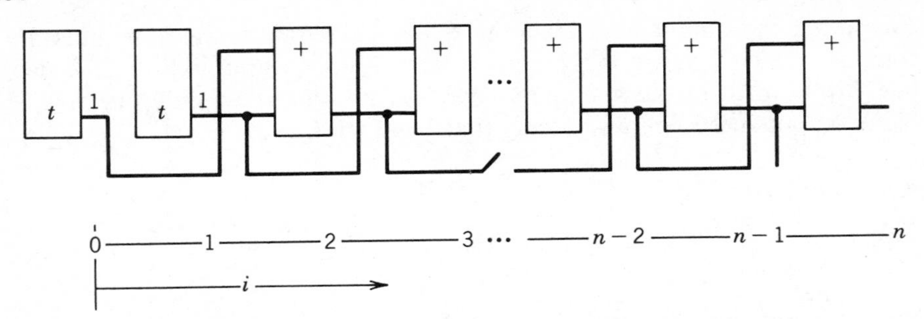

FIGURE 2-23 Folded network for Fibonacci definition (Definition 2.4.1).

Thus the predicates in Definition 2.10.1 can be expressed in terms of the new coordinates. Analogous transformations of the members of the substitutional domain Δ_f often result in refinements in the implementing algorithm. In Chapter 3 other such transformations will be studied.

2.10.2 Fibonacci Example

Another example of a definition which gives a folded network of uniform structure is Definition 2.4.1 for the Fibonacci numbers. Since $o_{11}(X) = o_2(X)$, the network is one-dimensional. The network can also be constructed in a way analogous to that used for the previous example, Definition 2.8.2 for combinations. We recognize immediately that the argument structures in Δ_f are simply integers and naturally form a one-dimensional space. The resultant network is shown in Figure 2-23, laid out along its single coordinate so that $f(i)$ is the block output at position i on that axis.

A single FOR loop, Algorithm 2.10.4, is adequate to implement such a one-dimensional network.

Algorithm 2.10.4

```
DCL F(n) vector; INIT(1)
  FOR I = 2 to m DO
    F(I) ← F(I − 1) + F(I − 2)
  END
```

As in the previous example, temporary memory can be greatly reduced, here because only two previous outputs are necessary in computing the output of a given block. In Algorithm 2.10.5 these two values are kept in X_1 and X_2. The variable T is used for temporary storage.

Algorithm 2.10.5: Fibonacci Numbers

```
X[1] ← 1; X[2] ← 1
  FOR I = 2 to m DO
    T ← X[1]
    X[1] ← X[1] + X[2]
    X[2] ← T
  END
```

2.10.2.1 A Note on Proof of the Fibonacci Implementation

We have justified the correctness of a network or program in implementing the evaluation of a recursive definition. These justifications, though we hope convincing, are not proofs. An actual proof can often be constructed using the same approach applied to showing equivalence between pairs of recursive definitions. We show that Algorithm 2.10.5 implements Definition 2.4.1, Fibonacci numbers, effectively by substituting that algorithm f throughout the definition, and showing that the resultant (proposed) equality is true. Actually it is the output of Algorithm 2.10.5 that is to be substituted for f. That output resides in X_1 when the program stops. The result found in X_1 at any time depends on the number of times the program has passed through the FOR loop. The values in X_1 and X_2 after m passes through the FOR loop are, respectively, denoted by X_1^m and X_2^m. To verify that Algorithm 2.10.5 satisfies the recursive line of Definition 2.4.1, we need to show that

$$X_1^n = ?\ X_1^{n-1} + X_1^{n-2}$$

X_1^n is formed in the last pass through Algorithm 2.10.5 with the following final steps. (The value is actually formed in the assignment statement followed by $*$.)

$$T \leftarrow X_1^{n-1}$$

$$X_1^n \leftarrow X_1^{n-1} + X_2^{n-1} \qquad (*)$$

$$X_2^n \leftarrow T$$

In the next to last pass through the DO loop the following occurs:

$$T \leftarrow X_1^{n-2}$$

$$X_1^{n-1} \leftarrow X_1^{n-2} + X_2^{n-2}$$

$$X_2^{n-1} \leftarrow T$$

from which it is clear that

$$X_2^{n-1} = X_1^{n-2}$$

Substituting in the (*) statement above for X_2^{n-1}, yields

$$X_1^n =!\ X_1^{n-1} + X_1^{n-2} \qquad (*)$$

2.11 IMPLEMENTATION OF k-DIMENSIONAL DEFINITIONS BY SIMULATION OF THEIR FOLDED NETWORKS

The combinations and Fibonacci examples above, with their k-dimensional folded networks and nested FOR loop implementations, are members of a large class of definitions with these same characteristics. To illustrate how the design of algorithms might be, at least partially, automated for a class of problems, we will explore this generalized class in some detail. As the class is defined, it is completely independent of the w function. If properties of the w function are considered, better implementations can be found for some subclasses; this could be handled as a refinement. Nonetheless, the design of algorithms for many members of this class can be largely automated, and provides a good illustration of what we would like to do for all recursive definitions.

The synthesis procedures developed here for implementing k-dimensional definitions are based on elementary concepts of linear algebra as found, for example, in [Edelen, Kydoniefs 72].

The class will be described using vector variables. To motivate the generalization, notice that for the vector variable $V = \langle n', m' \rangle$, the vector constants $R_1 = \langle 1,0 \rangle$, $R_2 = \langle 0,1 \rangle$, $A_1 = \langle 1,0 \rangle$, and $A_2 = \langle 0,1 \rangle$, and $\cdot$ the vector dot product, Definition 2.10.1 for combinations can be rewritten as follows.

Definition 2.11.1: Number of Combinations of m' of $n' + m'$ Items

$$\left[\begin{array}{ll} g(V) = 1 & :\exists i \in \{1,2\} : (A_i \cdot V < 1) \\ g(V) = g(V - R_1) + g(V - R_2) & :\forall i \in \{1,2\} : (A_i \cdot V \geq 1) \\ \text{Initially } V = \langle n', m' \rangle,\ n', m' \in N. & \end{array}\right.$$

This definition can be implemented with nested FOR loops because each argument of f on the right of the recursive line is a vector variable minus a vector constant, and because of the nature of the terminal and nonterminal conditions. We next define a class of definitions with these key properties, thus having nested FOR loop implementations.

In the following definition W is any associative function, $V = \langle v_1, \ldots, v_N \rangle$ is a vector variable, $R_i = \langle r_{1i}, \ldots, r_{Ni} \rangle$ and $A_j = \langle a_{1j}, \ldots, a_{Nj} \rangle$ are vector constants in which each r_{ki} and a_{kj} is a *positive* number, and b_j is an integer constant. (The "positive" restriction can be relaxed.) The terminal and nonterminal conditions in the definition are linear inequalities in which for each j: $\mathrm{rel}_j(x, y)$ is a predicate variable whose value can be one of two relations, either $x < y$ or $x \geq y$. The terminal domain is the space on, and to one side of, hyperplanes in the positive part of k-dimensional space; the nonterminal domain is that portion of the positive part of space that is not terminal.

Definition 2.11.2: k-Dimensional Definition

$$\left[\begin{array}{ll} f(V) = t(V) & :\exists j \in N^p : \sim \mathrm{rel}_j(A_j \cdot V, b_j) \\ f(V) = W_{i=1}^{i=M}[f(V - R_i)] & :\forall j \in N^p : \mathrm{rel}_j(A_j \cdot V, b_j) \\ \text{Initially } V = L = \langle L1, \ldots, LN \rangle,\ L1, \ldots, LN \text{ are each} \in N_0. & \end{array}\right.$$

Now consider the design of an algorithm to implement Definition 2.11.2. The folded network for this definition will be a k-dimensional network with w blocks located at the integer points in a k-dimensional space. The M inputs to a block located at the point V come from blocks at the points $V - R_1$ or "R_1 away," $V - R_2$ or "R_2 away," ..., $V - R_M$ or "R_M away." This is shown in Figure 2-24. The output of the block at V similarly goes to blocks "R_1 away," "R_2 away," ..., "R_M away," provided that in each case the point "R_j away" is not outside the region to which the computation is confined.

As illustrated in our combination and Fibonacci examples, optimization of a k-dimensional definition's FOR loop implementation has three parts. The first is reduction of the R's on the right side of the recursive line to "lowest terms," the second is setting the range of the FOR loops, and the last is minimization of temporary storage for the intermediate result. We consider each of these in turn. As noted in this development we will not attempt to take advantage of any special properties of w or other primitive functions in Definition 2.11.2. Much better optimizations are sometimes possible when this is done, as the well-known formulas for Fibonacci and combinations (which depend on the fact that w is addition) attest.

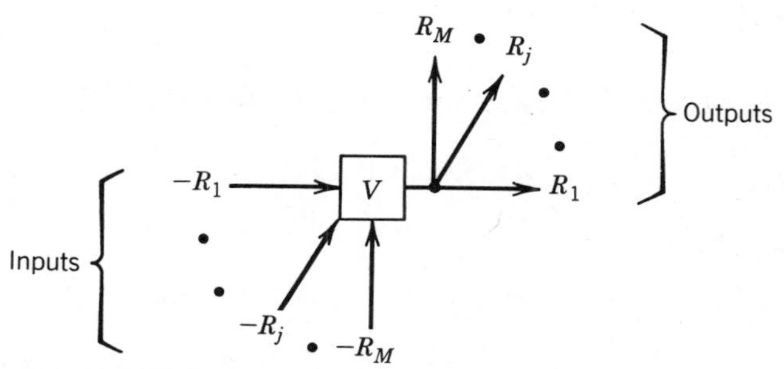

FIGURE 2-24 Typical block in network for k-dimensional definition.

2.11.1 Reduction of *R*'s on the Right Side of the Recursive Line

The goal of simplification of the recursive definition is to minimize the number of R_i's and to maximize the number of zero components in the surviving R_i's.

The subsequent development includes a quick review of some concepts of elementary linear algebra. We make extensive use of *linear column transformations* on a set of column vectors. Assume the R_i vectors are arranged as rows of the matrix X. The columns of X are vectors: the ith column vector is designated C_i. The linear column transformation works on the columns C_i of X.

Definition 2.11.3: Linear Column Transformation. A linear column transformation of matrix X is any sequence of applications of the following basic transformations in which a and b are numbers unrestricted except that $a \neq 0$:

1. Replace C_i of X with $a * C_i$ (multiply a column by a constant).
2. Replace C_i of X with $C_i + C_j$ (add two columns).
3. Remove an all-0 column.

Notice that the first two basic transformations can be combined into:

$$\text{Replace } C_i \text{ of } X \text{ with } a * C_i + b * C_j,\ a \neq 0.$$

For any definition D of the form of Definition 2.11.2 we can form a matrix R_D, or just R, whose rows are the vectors R_j of that definition. R then has M rows—the number of right-side f calls in D—and N columns—the dimensionality of the vector variable of D. As an example consider the definition for combinations repeated here as Definition 2.11.4.

Definition 2.11.4: Number of Combinations of m of the First n Integers

$$\left[\begin{array}{ll} f(n, m) = 1 & : n = m \text{ or } m = 0 \\ f(n, m) = f(n-1, m-1) + f(n-1, m) & : n > m > 0 \\ \text{Initially } n, m \in N,\ n > m. & \end{array}\right.$$

The various vectors associated with this definition are:

Variable vector: $V = \langle n, m \rangle$.
Row vectors: $R_1 = \langle 1, 1 \rangle$, $R_2 = \langle 1, 0 \rangle$.

The recursive line rewritten in the vector notation would be

$$f(\langle n, m \rangle) = f(\langle n, m \rangle - \langle 1, 1 \rangle) + f(\langle n, m \rangle - \langle 1, 0 \rangle) \qquad : n > m > 0$$

The matrix **R** is

$$\mathbf{R} = \begin{Vmatrix} n & m \\ 1 & 1 \\ 1 & 0 \end{Vmatrix}$$

Column vectors: $C_1 = \langle 1,1 \rangle$, $C_2 = \langle 1,0 \rangle$.

In order to simplify the definition D, we will want to perform linear column transformations on $\mathbf{R}_D$ to obtain a matrix $\mathbf{R}'_D$ in which as many of the *row* vectors as possible are unit vectors. (A unit vector has a 1 in one position and 0's elsewhere, for example, $\langle 1,0,\ldots,0 \rangle$, $\langle 0,1,0,\ldots,0 \rangle$.)

There are systematic ways to do this, which are studied in linear algebra. One procedure follows:

1. Let R_a be a row with a nonzero entry in column 1 at C_{1a}. If there is no such row (that is, if column 1 is all 0's), remove column 1 and repeat step 1. Divide C_1 by scalar C_{1a} to get C_1'. If $C_{ia} = d \neq 0$, $i > 1$, then subtract $d * C_1'$ from column C_i, making $C_{ia} = 0$. The result is a unit vector in row a, having a 1 in column 1.
2. Let R_b be a row with a nonzero entry in column 2. If there is no such row (that is, if column 2 is all 0's), remove column 2 and repeat step 2. Divide C_2 by C_{b2} to get C_2'. If $C_{ib} = e \neq 0$, $i = 1$ or $i > 2$, then subtract $e * C_2'$ from column C_i, making $C_{i2} = 0$. The result is a unit vector in row b having a 1 in column 2.
3. The procedure is applied in a similar way to succeeding columns.

All-zero columns occur in the procedure as a consequence of linear dependence among the original row vectors or the existence of more rows than columns. A slightly different procedure is needed if division ever gives a non-integer here.

If the number of linearly independent columns is greater than or equal to the number of rows, then the procedure guarantees that all the rows are unit vectors. Otherwise the number of unit vectors equals the number of linearly independent columns, and some row vectors are not unit vectors. For the Fibonacci definition we would get a matrix with one column and two row vectors, one of which is a unit vector, and no linear column transformation could make them both unit vectors.

The column transformations also transform the row vectors. This transformation has a geometric interpretation. A linear column transformation is equivalent to rotating the axis and changing the scale of the row vectors.

For our example,

$$\mathbf{R} = \begin{Vmatrix} n & m \\ 1 & 1 \\ 1 & 0 \end{Vmatrix} \xrightarrow{C_1 = C_1 - 1 * C_2} \begin{Vmatrix} n' & m' \\ 0 & 1 \\ 1 & 0 \end{Vmatrix} = \mathbf{R}'$$

The row vectors R_i's of the transformed matrix $\mathbf{R}'_D$ now become the R'_i's of another definition, say D', of the form of Definition 2.11.2. So $\mathbf{R}_{D'} = \mathbf{R}'_D$. The rows of $\mathbf{R}_{D'}$ are unit vectors, so D' is a simplification of D. The relation between $\mathbf{R}_D$ and $\mathbf{R}_{D'}$ can be expressed in a straightforward way by recording the linear transformations on the column vectors that transformed the former to the latter. For our running example, initially $n = C_1$, $m = C_2$. After the transformation, $n' = C_1$, $m' = C_2$ and

$$n' = n - m$$
$$m' = m$$

Here n and m are vector variables. The given relations of these and of n' and m' are linear vector equations.

Now we wish to make definition D', with R_i given by $\mathbf{R}_{D'}$, equivalent to D, with R_i given by $\mathbf{R}_D$. For this purpose we need to express the elements of $\mathbf{R}_D$ in terms of $\mathbf{R}_{D'}$. We already know how to get linear equations giving the column vectors (C'_j) of $\mathbf{R}_{D'}$ in terms of the column vectors C_j of $\mathbf{R}_D$. We need only solve these equations. In our running example, solving

$$n' = n - m$$
$$m' = m$$

we get

$$n = n' + m = n' + m'$$
$$m = m'$$

For any k-dimensional definition we get a similar set of linear equations giving the column vectors of D in terms of those in D'. Calling the linear function involved $\text{lin}_{i'}$, and remembering that the vector variable v_i has as its value the column vector $C_{i'}$, we represent the set of linear equations as

$$\{ v_i = \text{lin}_i(v'_1, \ldots, v'_N) | i \in N^N \} \tag{2.11.1}$$

These equations are all that is required to produce a simplified definition with the maximum number of R_j's equal to unit vectors, and the minimum number of columns. Now the description of how these are used to construct the equivalent simplified definition D' is given.

1. The matrix $\mathbf{R}'$ results from the transformation. In forming $\mathbf{R}'$ we may have removed some all-zero columns. The components of the variable vector V corresponding to deleted columns are deleted. V' is the new variable vector corresponding to $\mathbf{R}'$. If, for example, V originally was $\langle v_1, v_2, v_3, v_4, v_5 \rangle$ and rows 2 and 4 of $\mathbf{R}$ become all-zero, then v_2 and v_4 would not be in V'. We also rename the components of V'. So $V' = \langle v'_1, v'_3, v'_5 \rangle$ becomes the new argument of f.

2. The R_j's in D' are the rows of $\mathbf{R}'$.
3. The nonterminal line of D', excluding the predicates, is obtained by replacing all occurrences of $v_1, \ldots, v_n$ in this line in D with $v'_1, \ldots, v'_n$.
4. The nonterminal and terminal predicates, which in D are given as linear inequalities in the v_i's, are given their D' form by substituting $\text{lin}_i(v'_1, \ldots, v'_N)$ for each v_i throughout these inequalities.
5. The terminal value in D is a primitive function in the v_i's. It is given in its D' form by substituting $\text{lin}_i(v'_1, \ldots, v'_N)$ for each of those v_i's.
6. The initial value of the vector V in D has the form

$$v_i = Li \text{ for all } i \in N^N$$

In D' it becomes

$$\text{lin}_i(v'_1, \ldots, v'_N) = Li \text{ for all } i \in N^N$$

This completes the construction of the new equivalent definition. A variable, v'_j's appearing in the predicates and initial values of this new definition, may not appear as an argument of f. Such a variable will be assigned a value in initialization, and be invariant, though useful, throughout evaluation of the definition. We do not require invariants to appear as arguments in the definition.

Applying the transformation to our running example involving Definition 2.8.2, we get the following.

$$\mathbf{R}' = \begin{array}{c} \begin{array}{cc} n' & m' \end{array} \\ \left\| \begin{array}{cc} 0 & 1 \\ 1 & 0 \end{array} \right\| \end{array}$$

and

$$n = n' + m', \ m = m'$$

Since there are no all-zero columns, the vector variables for the new definition are n' and m'.

Substituting these results according to the transformation above gives

$$\left[\begin{array}{ll} g(n', m') = 1 & : n' + m' \text{ or } m' = 0 \\ g(n', m') = g(n', m' - 1) + g(n' - 1, m') & : n' + m' > 0 \\ \text{Initially } n' + m' \text{ and } m' \in N,\ n' + m' \geq m' & \end{array}\right.$$

After obvious simplifications, this definition is identical to Definition 2.10.1.

2.11.2 The Range of the FOR Loops

Assume we have a k-dimensional definition simplified by the transformation of the previous paragraphs. The implementation is constructed by simulating the corresponding network. In simulating a network NET with no cycles, the blocks may be simulated in any order satisfying the constraint:

1. If the output of block A is an input to block B, then A must be simulated before B.
2. If each block in NET is to be simulated only *once*, then temporary storage requirements for the simulation of NET are determined as follows. If S is the set of all blocks that have been simulated at time t, then all outputs from blocks in S which are inputs to a block outside of S must be in storage at time t.

As we have argued, a k-dimensional definition will generate a *patterned* folded network whose typical block is given in Figure 2-24. An algorithm having n nested FOR loops, with storage to hold the result of *every* block whose output is computed, is adequate for simulating the network. Although this is usually more storage than is necessary, we describe such a "full" implementation first and refine it later. Algorithm 2.11.1 gives this algorithm under the assumption that the argument structure is kept in the variable $V = \langle V1, V2, \ldots, Vn \rangle$, with $Vi = Li \geq 0$ initially, and each intermediate function $f(V_1, \ldots, V_n) = f(V)$ is kept in a k-dimensional matrix F at $F(V)$.

Algorithm 2.11.1: Algorithm for k-dimensional Definition, Definition 2.11.2

```
DCL F(L1, L2,..., Ln) array; INIT(c)
  FOR V1 = start_1 to end_1 DO
    FOR V2 = start_2 to end_2 DO
     .

      .
      FOR V_n = start_n to end_n DO
        F(V) ← w(F(V − R_1), F(V − R_2),..., F(V − R_M))
      END
      .

     .
    END
  END(R_1, R_2,..., R_M are n-vectors and constant)
```

The values of start_i's and end_i's can be found by examining the region of k-dimensional space containing all points at which f must be evaluated to eventually determine $f(L1, L2, \ldots, Ln) = f(L)$. This "relevant" region is bounded by a set of hyperplanes. Figure 2-25 displays such a region, which in linear algebra terminology forms a *convex hull*.

The relevant region is only partially specified by the linear equations, which are nonterminal conditions. In the figure these constraints specify the region

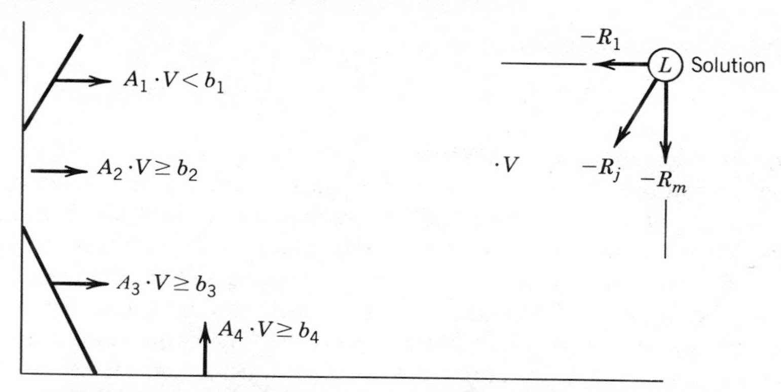

FIGURE 2-25 Boundary conditions in network for k-dimensional definition.

indicated by the arrows pointing and extending indefinitely to the right. Additional constraints which confine V to a completely bounded relevant region must be found. The value of the function at any point in the region depends on its values at neighboring points whose relative positions are given by the vectors R_j. The additional constraints are a consequence of this relation dependence. Any point V at which the function is to be computed must be reachable by traversing some number, say x_1, of the R_1 vectors, plus x_2 of the R_2 vector, plus ..., plus x_m of the R_m vectors, with all the x_j's ≥ 0. This is expressed as follows:

$$V = L - (x_1 \cdot R_1 + \cdots + x_m \cdot R_m)$$

Let $X = \langle x_1, \ldots, x_m \rangle$ and R be the matrix with row vectors R_1 to R_m,

$$V = L - (X \cdot \mathbf{R})$$

$$L - V = X \cdot \mathbf{R}$$

and if $\mathbf{R}$ has an inverse, then

$$(L - V) \cdot \mathbf{R}^{-1} = X$$

But we have said that each component of X, $x_i \geq 0$. So remembering that 0^m is an all-zero vector with m components,

$$(L - V) \cdot \mathbf{R}^{-1} \geq 0^m$$

which, together with the inequalities $X \geq 0^m$, is the required constraint if $\mathbf{R}$ has an inverse. If $\mathbf{R}$ does not have an inverse, the equation

$$L - V = X \cdot \mathbf{R}$$

is equivalent to the inequalities

$$L - V \geq X \cdot \mathbf{R} \qquad \text{and} \qquad L - V \leq X \cdot \mathbf{R}$$

This provides additional linear inequalities which, together with the nonterminal conditions, completely delineate the relevant region. However, not every integer point in the region will necessarily occur in the substitutional domain. The extreme values of vector components in the relevant region are given by solving these inequalities. These extreme values are the start_i's and end_i's of the nested FOR loops. The inequalities may be solved by a linear programming or a Gaussian elimination procedure. The approach is applicable whether or not the R_j's are unit vectors. As an illustration, we now find the FOR loop ranges for implementing Definition 2.8.2.

$$L = \langle N, M \rangle$$

$$V = \langle I, J \rangle$$

$$R_1 = \langle 1,1 \rangle$$

$$R_2 = \langle 1,0 \rangle$$

$$\mathbf{R} = \begin{Vmatrix} 1 & 1 \\ 1 & 0 \end{Vmatrix}, \qquad \mathbf{R}^{-1} = \begin{Vmatrix} 0 & 1 \\ 1 & -1 \end{Vmatrix}$$

From the procedure for developing the range of FOR loops,

$$\langle N - I, M - J \rangle \cdot \begin{Vmatrix} 0 & 1 \\ 1 & -1 \end{Vmatrix} \geq 0^2$$

Multiplying out we get the linear constraints

$$M - J \geq 0$$

$$N - I - M + J \geq 0$$

Those from the nonterminal conditions are

$$I - J > 0 \text{ (or } \geq 1)$$

$$J > 0 \text{ (or } \geq 1)$$

Solving for J we get the following equivalent set of inequalities, in which $[A, B] \geq [C, D]$ is shorthand for the four inequalities $A \geq C$, $A \geq D$, $B \geq C$, and $B \geq D$,

$$[M, I - 1] \geq J \geq [I - (N - M), 1] \qquad \text{range } J$$

These relations give the range of J in terms of the constants N and M and the variable I. We next eliminate J from the equation, obtaining the range of I in terms of constants N and M only.

These relations can be satisfied for some J if and only if the following, obtained by dropping J, can be satisfied:

$$[M, I-1] \geq [I-(N-M), 1]$$

This is shorthand for

$$\begin{aligned} N \geq I &\geq 2 \qquad \text{range } I \\ M &\geq 1 \\ (N-M) &\geq 1 \end{aligned}$$

The line labeled "range I" gives the allowable range of I in terms of constants only. This is the range of I for the outer FOR loop. The line "range J" gives the range of the inner FOR loop. The range of J is given in terms of constants and the variable I, whose value is set when control passes to the inner FOR. The range specification for J is awkward. There is an alternative. Solve the original set of inequalities for I, rather than J, first. I then is given in terms of J and constants. The set of inequality equations becomes

$$\begin{aligned} (N-M)+J \geq I &\geq J+1 \qquad \text{range } I \\ M \geq J &\geq 1 \end{aligned}$$

Now I can be dropped,

$$\begin{aligned} (N-M)+J &\geq J+1 \\ M \geq J &\geq 1 \end{aligned}$$

and J separated from the others to get

$$\begin{aligned} M \geq J &\geq 1 \qquad \text{range } J \\ (N-M) &\geq 1 \end{aligned}$$

In this case J is set in the outer FOR loop according to the line labeled "range J," and I in the inner FOR loop according to "range I." This gives a simpler and preferable way to set the ranges.

2.11.2.1 Storage Saving

In Algorithm 2.11.1 storage is maintained for every block output, the output of the block at $\langle V1, V2, \ldots, Vn \rangle$ being saved in $F[V1, V2, \ldots, Vn]$. This is more storage than necessary. $F[V1, \ldots, Vn]$ needs to be saved only until its last use; that is, at least until $F[V1 - H_1, V2, \ldots, Vn]$, with $H_1 = \text{MAX}_{j=1}^{M}[r_{1j}]$ being the largest first component of any R_j, is computed. Certainly $F[V1, \ldots, Vn]$ will not be needed any longer after $F[V1 + H_1 + 1, 0, \ldots, 0]$, but it may still be needed after $F[V1 + H_1, V2, \ldots, Vn]$ is computed, to compute f further along in the hyperplane whose first component is $V1 + H_1$. So in general $F[V1 + H_1 + 1, \ldots, Vn]$ cannot necessarily be stored in the same location as $F[V1, \ldots, Vn]$. In the remainder of this section, without giving a completely general method for minimizing the storage for k-dimensional algorithms, we give methods for doing so for some subclasses.

In the one-dimensional case $F[V1 + H_1]$ can overwrite $F[V1]$. This can be accomplished by making $F[V1 + H_1 + i]$ and $F[V1 + i]$ with $0 \le i \le H_1 - 1$ the same storage location. This in turn can be accomplished by making $F(X) = F(X_{\text{mod } H_1})$.

Algorithm 2.11.2: Algorithm for One-Dimensional Definition

$$H_1 = \max(r_{11}, r_{12}, \ldots, r_{1M})$$

```
DCL F(H₁) array; INIT(c)
  FOR V1 = start₁ to end₁ DO
    F(V1mod H₁) ← w(F([V1 − R₁]mod H₁), . . . , F([V1 − R_M]mod H₁))
  END
```

The storage requirement for this one-dimensional case then is $O(H_1)$.

For a two-dimensional definition it is no longer certain that $F[V1, V2]$ can be overwritten by $F[V1 + H_1, V2]$. If, for example,

$$F(V1, V2) = F(V1 - 1, V2) + F(V1 - 1, V2 - 1) + F(V1, V2 - 1)$$

then $H_1 = 1$. The typical block is shown in Figure 2-26.

$F(V1 + 1, V2)$ cannot overwrite $F(V1, V2)$ because it will be necessary to keep $F(V1, V2)$ for the subsequent computation of $F(V1 + 1, V2 + 1)$. For

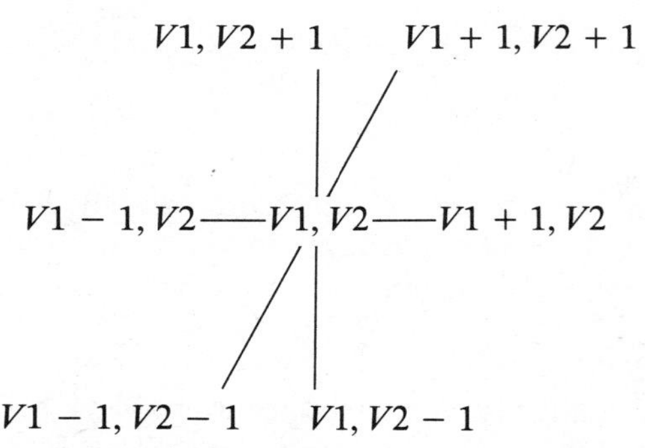

FIGURE 2-26 Typical block when $F(V1, V2) = F(V1 - 1, V2) + F(V1 - 1, V2 - 1) + F(V1, V2 - 1)$.

this case it is necessary to find among all R_j with $r_{1j} = H_1$, $H_2 = \text{MAX}_{j=1}^{M}[r_{2j}]$. Now $F(V1 + H_1, V2 + H_2)$ [that is, $F(V1 + 1, V2 + 1)$ in our example] can reuse the storage for $F(V1, V2)$. This can be done fairly gracefully by linearizing the two-dimensional space. To do this let u be the maximum value of $V2$, and r the maximum range of values of $V2$ for any value of $V1$. Then we can transform each point $[V1, V2]$ in the two-dimensional space into $uV1 + V2$ in the corresponding one-dimensional space. Furthermore we need to make storage available for $Z = rH_1 + H_2$ different values so we can let $F(uV1 + V2) = F([uV1 + V2]_{\text{mod } Z})$. This leads to Algorithm 2.11.3.

Algorithm 2.11.3: Algorithm for Two-Dimensional Definition

$$H_1 = \text{MAX}(r_{11}, r_{12}, \ldots, r_{1M})$$

$$H_2 = \text{MAX}_{r_{1j}=H_1}(r_{2j})$$

$$u = \text{end}_2 = \text{maximum value of } V2$$

$$r = \text{maximum range of values in } V2$$

$$Z = rH_1 + H_2$$

```
DCL F(Z) array; INIT(c)
  FOR V1 = start1 to end1 DO
    FOR V2 = start2 to end2 DO
      F([uV1 + V2]mod z) ← w(F([u(V1 − r11) + (V2 − r21)],..., F([u(V1 − r1M) +
                                                              (V2 − r2M) mod Z
    END
  END
```

Because of possible nonuniformity in the two-dimensional space in which the computation must take place, this approach will not necessarily use the minimal amount of storage, but it will cost very little in time. The storage requirement is $O(Z)$.

In general, even though we cannot be sure that $f[V1, V2, \ldots, Vn]$ can be overwritten by $F[V1 + H_1, V2, \ldots, Vn]$, we can be quite certain that it can be safely overwritten by $F[V1 + H_1 + 1, V2, \ldots, Vn]$. This in turn implies that $F[V1, V2, \ldots, Vn]$ and $F[V1 + H_1 + 1, V2, \ldots, Vn\}$ can be stored in the same location. We can accomplish this reuse of storage easily by transforming index 1 in such a way that $F(V1) = F(V1 + H_1 + 1)$. This is accomplished by making $F(X) = F(X_{\text{mod}(H_1+1)})$.

Algorithm 2.11.4 makes use of this storage reduction.

Algorithm 2.11.4: Algorithm for k-Dimensional Definition

$$Z = H_1 + 1 = \max(r_{11}, r_{12}, \ldots, r_{1M}) + 1$$

$$\text{Let } V_z = \langle V1_{\text{mod}(H_1+1)}, V2, \ldots, Vn \rangle$$

$$\text{Let } [V - R_j]_z = \langle (V1 - R_{j1})_{\bmod Z}, V2 - R_{j2}, \ldots, Vn - R_{jn} \rangle$$

```
DCL F(Z, L2, ..., Ln) array; INIT (c)
  FOR V1 = start_1 to end_1 DO
    FOR V2 = start_2 to end_2 DO
      .
        .
          FOR Vn = start_n to end_n DO
            F(V_z) <- w(F([V - R_1]_z), ..., F([V - R_M]_z))
          END
          .
        .
    END
  END
```

This algorithm requires

$$O((H_1 + 1) * (\text{maximum of } V2) * (\text{maximum of } V3) * \cdots * (\text{maximum of } Vn))$$

This is simple and not overly time consuming, but it may use more space than necessary.

2.12 PROBLEMS

1. *Recursive Definition Evaluation–Exercise.* Try to evaluate Definitions 2.12.1 and 2.12.2 by hand, using the SE procedure, Process 2.3.1, for $f(7)$. Definition 2.12.2's evaluation does not terminate.

Definition 2.12.1

$$\left[\begin{array}{ll} f(n) = -1 & :n = 0 \\ f(n) = 1 & :n = 1 \text{ or } 2 \\ f(n) = -(\text{minimum}(f(n-1), f(n-2), f(n-3))) & :n \geq 3 \\ \text{Initially } n \in N_0. & \end{array}\right.$$

Definition 2.12.2

$$\left[\begin{array}{ll} f(n) = n & :n = 1 \text{ or } > 7 \\ f(n) = f(n+1) + f(n-1) & :n \leq 7 \\ \text{Initially } n \in \{1, \ldots, 7\}. & \end{array}\right.$$

2. *Recursive Definition Evaluation ESE–Exercise.* Try to evaluate $f(7)$ for Definition 2.12.2 by hand using the ESE procedure (Process 2.7.1, 2.7.2, or 2.7.3.) That procedure will leave you with a set of equations—evaluate that set by any means you can.

3. *Recursive Definition Termination—Exercise.* Prove that Definitions 1.3.1 and 2.12.1 terminate. (Display a positive integer valued function for each with the properties given under Definition 2.4.2.) Prove that Definition 2.12.2 does not terminate.
4. *Recursive Definition Evaluation Network—Exercise.* Draw a picture of the general form of the w–t folded network for Definition 2.12.1.
5. *Recursive Definition Equivalent to Closed Form—Proof.* Given the definition

$$\left[\begin{array}{ll} f(p,m,n) = p & :n = 0 \\ f(p,m,n) = f(p+m,m,n-1) & :n \in N \\ \text{Initially } m,n,p \in N. & \end{array}\right.$$

Prove that $f(p,m,n) = p + (m * n)$.

6. *Recursive Definition Equivalent to Closed Form—Proof.* Given the definition min(S) is the smallest component of S,

$$\left[\begin{array}{ll} f(S) = \langle\,\rangle & :S = \langle\,\rangle \\ f(S) = \langle \mathrm{MIN}(S)\rangle \| f(S - \mathrm{MIN}(S)) & :S \neq \langle\,\rangle \\ \text{Initially } S = \langle S_1, S_2, \ldots, S_n\rangle, S_i \text{ is a number.} & \end{array}\right.$$

Prove that $f(S) = \langle S_{i_1}, S_{i_2}, \ldots, S_{i_n}\rangle$ where $S_{i_1} \le S_{i_2} \le \cdots \le S_{i_n}$.

7. *Recursive Definition Equivalent to Closed Form—Proof.* Given the definition

$$\left[\begin{array}{ll} f(v,n,s,S) = \{v\} & :n = 0,\ s = S \\ f(v,n,s,S) = \{\,\} & :n = 0,\ s \neq S \\ f(v,n,s,S) = f(v\|\langle 0\rangle, n-1, s, S) & \\ \qquad\qquad \cup f(v\|\langle 1\rangle, n-1, 1-s, S) & :n \in N \\ \text{Initially } v = \langle\,\rangle,\ n \in N,\ s = 0,\ S = 0 \text{ or } 1. & \end{array}\right.$$

Prove that $f(v,n,s,S) = \{v\|\langle y_1,\ldots,y_n\rangle | (s + y_1 + \cdots + y_n)_{\mathrm{mod}\,2} = S\}$ where $y_i \in \{0,1\}$, from which it would follow that

$$f(\langle\,\rangle, n, 0, S) = \{\langle y_1,\ldots,y_n\rangle | (y_1 + \cdots + y_n)_{mod\,2} = S\}.$$

8. *Recursive Definitions Equivalent to Each Other.* Prove that definitions 1 for $f(X)$ and 2 for $g(Y, X)$ below are equivalent Definition 2 is the same as definition 1 except that a *program control component* Y has been added. Since these components do not affect the value of the function, they may seem useless. However, they can be of use in making the implementation more efficient.

Definition 1

$$\left[\begin{array}{ll} f(X) = t(X) & :T(X) \\ f(X) = w\big(f(o_1(X)), \ldots, f\big(o_{m(x)}(X)\big), p(X)\big) & :\sim T(X) \\ \text{Initially } X = d \in D & \end{array}\right.$$

Definition 2

$$\left[\begin{array}{ll} g(X, Y) = t(X) & :T(X) \\ g(X, Y) = w(g(o_1(X), o_1(X)), \ldots, & \\ \qquad\qquad \big(g\big(o_{m(x)}(X), o_1(X)\big), p(X)\big) & :\sim T(X) \\ \text{Initially } X = d \in D,\ Y \in Q & \end{array}\right.$$

9. *Removing Function References in the (Non-)Terminal Conditions.* Consider a nonnested recursive definition with f the *condition* of one or more nonterminal lines. For example,

Definition 2.12.3

$$\left[\begin{array}{ll} f(n) = 1 & :n \in 0, 1 \\ f(n) = f(n-1) + 2 * f(n-2) & :f(n-1) \geq 5 \\ f(n) = f(n-1) + f(n-2) & :f(n-1) < 5 \\ \text{Initially } n \in N_0. & \end{array}\right.$$

Is there always an equivalent nonnested recursive definition with no occurrence of the nonterminal in any line condition?

10. *Recursive Definition Transformations.* The next definition gives the cost of the shortest path in a directed graph G from the vertex "begin" to the vertex "goal." The recursive line is based on the observation that the cost of the minimum cost path from vertex x to "goal" equals the minimum cost of each neighbor's minimum cost path to the "goal" plus the edge cost to that neighbor from x. If v is a vertex of G, then

Definition 2.12.4: Cost of Shortest Path

Remember that:

$N(v) =$ the outward degree or number of outward neighbors of v

$n_i(v) =$ the ith outward neighbor of v

$c_i(v) =$ the cost of the ith outward edge of v

$$\begin{cases} g(v) = 0 & :v = \text{"goal"} \\ g(v) = \text{MIN}_{i=1}^{N(v)}[c_i(v) + g(n_i(v))] & :v \neq \text{"goal"} \\ \text{Initially } v = \text{"begin."} \end{cases}$$

v is an information–control component.

Obtain an equivalent definition in which there is no information–control component.

11. *Recursive Definition—Termination.* Here is another cost of shortest path definition,

$$\begin{cases} f(c, v) = c & :v = \text{"goal"} \\ f(c, v) = \text{Min}[f(c + c_1(v), n_1(v)), f(c + c_2(v), n_2(v)), \ldots, \\ \qquad f(c + c_{N(c)}(v), n_{N(v)}(v))] & :v \neq \text{"goal"} \\ \text{Initially } c = 0,\ v = \text{"begin" vertex.} \end{cases}$$

a. Does this definition terminate?

b. Obtain an equivalent nested definition.

12. *General Transformation of Recursive Definition—Making m(X) = 2.* Consider any standard nonnested definition,

$$\begin{cases} f(X) = t(X) & :T(X) \\ f(X) = w(f(o_1(X)), \ldots, f(o_{m(x)}(X))) & :\sim T(X) \\ \text{Initially } X \in D. \end{cases}$$

and the following "binarized" $[m(X) = 2]$ definitions

$$\begin{cases} g(X, j) = t(X) & :T(X) \\ g(X, j) = w(g(o_j(X), 1), g(X, j + 1)) & :\sim T(X) \text{ and } j < m(X) \\ g(X, j) = w(g(o_{m(X)}(X), 1)) & :\sim T(X) \text{ and } j = m(X) \\ \text{Initially } X \in D,\ j = 1. \end{cases}$$

Assume that w is associative and that w(a single argument) = that argument. Show that $f(X)$ is satisfied by $g(X, 1)$).

13. *Recursive Definition—Nested for Loop Implementation.* Consider the definition

$$\begin{cases} f(m, n, p) = 1 & :n = 0 \text{ or } m = p \\ f(m, n, p) = f(m, n - 1, p) + f(m + 1, n, p) & :n > 0 \text{ and } m < p \\ \text{Initially } m < p,\ m, n, p \in N. \end{cases}$$

Write (not for running) a nested FOR loop algorithm to implement this definition. Prove that it terminates.

14. *k-Dimensional Definitions—Generalization.* In Definition 2.11.2 we require every component of every R_j to be positive. If this requirement is violated it is possible to produce definitions that do not terminate. How? On the other hand the requirement can be eased and still provide assurance that the definition will terminate. Discuss.

15. *k-Dimensional Definitions—Storage Use in Implementation.* Describe the "linearization" approach to storage saving for a three-dimensional recursive definition.

CHAPTER THREE

DESIGN OF RECURSIVE DEFINITIONS

Designing recursive definitions entails discovery and is thus an example of heuristic problem solving as considered in [Polya 45] as well as in most books on artificial intelligence, such as [Nilsson 80], and, as applied to algorithms and recursive definitions, in [Aho, Hopcroft, Ullman 75, chap. 2; Roberts 86; Goodman, Hedetniemi 79]. The design of functional and recursive programs is covered in [Henderson 83; Burge 82].

The analysis of recursive definitions, their evaluation by SE and ESE, and their interpretation as networks, as developed in Chapter 2, provide the basis for our study of their design. First we consider designing recursive definitions looking just for correctness. Later we will study ways to improve such initial feasible designs. Two broad approaches are considered. The first is: "think recursively," that is, ask how $f(X)$, the function's value at point X is related to its values at neighboring points, $o_i(X)$, in its domain. This approach is aided by constructing a w network for a sample output. The definition is a generalization of the functional relations in the final stage of that network. The second approach is: "think iteratively," that is, develop an O network stage by stage for sample inputs and, generalizing from the network, determine the O and other primitive functions for the associated recursive definition. We do not have powerful theorems to offer for definition building, but we do have many examples and advice, which becomes progressively more specific and useful as the range of functions to be defined is narrowed. The discussion is focused on the development of nonnested definitions expressed in *standard (nonnested) form*, with recursive lines of the form

$$f(X) = w\big(f(o_1(X)), \ldots, f\big(o_{m(X)}(X)\big)\big) \qquad :\sim T(X)$$

or

$$f(X) = w\big(f(o_1(X)), \ldots, f\big(o_{m(X)}(X)\big), p(X)\big) \qquad :\sim T(X)$$

and nonrecursive lines of the form

$$f(X) = t(X) \qquad :T(X)$$

Theorem 2.6.1 gives a result useful in developing recursive definitions of this form. It can be interpreted as follows. Suppose the w functions, W, of such a definition, Def.f, is associative, such as addition or minimum. Rather than developing Def.f directly, we can first develop another definition, Def.f', to enumerate the set to which W is applied by Def.f. Def.f' uses the w function W' = union if W is commutative, and concatenation otherwise. A recursive definition equivalent to Def.f is $W \circ f'$ (W applied to the result of f'.) Thus,

Any function design problem that can be solved by a standard form definition with an associative w can be formulated as a problem of enumerating a set and applying a desired w to that set.

Under this characterization the recursive definition to be designed is always an enumeration of an (ordered) set using concatenation or union as *its* w function. For example, finding a minimum path from a to b in a graph is equivalent to enumerating all paths from a to b and finding the minimum of these—minimum is the w function. The validity of this "enumeration-w" characterization is also apparent from the form of the O–w–t network. The outputs of the t line, ordered from left to right, form the (ordered) set S in question. Let $W(S)$ be the output of the w network. Clearly if w is associative, $W(S)$ is equivalent to applying w to arguments consisting of members of S. In any case this characterization shows the importance of set enumeration and explains the emphasis on recursive definitions for enumerations in the following sections.

Even if the w function is associative and the enumeration approach applicable, it may be a somewhat awkward approach. For example, consider a function $f(T)$ whose argument T is a binary tree and whose value is the number of nodes in that tree. The recursive line results from the realization that the number of nodes in T is 1 plus the number of nodes in its left subtree [LS(T)] plus the number of nodes in its right subtree [RS(T)],

$$f(T) = 1 + f(\mathrm{LS}(T)) + f(\mathrm{RS}(T))$$

Alternatively and more awkwardly, using the enumeration approach, we could think of this as developing in two steps. First obtain a definition which enumerates a sequence of 1's, one for every node in the tree whose definition is

$$g(T) = \langle 1 \rangle \| g(\mathrm{LS}(T)) \| g(\mathrm{RS}(T))$$

and then obtain the first recursive line by changing $\|$ to $+$ and g to f.

3.1 RECURSIVE APPROACH TO BUILDING RECURSIVE DEFINITIONS

A recursive line relates the value of a function at a point in its domain to its value at *nearby* points. If the definition is to terminate, there must be an ordering of the points of the domain (see Theorem 2.4.1). For a terminating definition it is with respect to this ordering that we speak of nearby points. Whether or not the definition terminates, "nearby points" will usually be implicit in the problem description; for example, neighbors in a graph and children in a tree are usually appropriate nearby points. Along with its recursive line or lines, a terminating recursive definition has at least one nonrecursive, or terminal line. The terminal line is applicable at the lowest domain points in the ordering or, in the nonterminating case, at any point X at which the function value is generally expressible as a primitive function of X. To construct a recursive definition DEF.f "recursively," the extent and ordering of f's domain must be established. Also, and this is central, we must find relations between f's value at a point X other than the lowest, $f(X)$ and its value at the children of X, $f(o_i(X))$; and perhaps a primitive function of X, $p(X)$. That relation is expressed with primitive functions. A way to approach the formulation of this central relation is to assume that f's value is known at some points and express its value at a higher neighboring point in terms of these known f's. This is related to the reduction approach to problem solving in artificial intelligence. In addition to the recursive relation we must define f with a primitive function of those points at the lowest points in the domain. It cannot depend on its own value at children since it has no children.

3.1.1 Examples of the Recursive Approach

Several examples will show that even such general advice as given thus far is of value in formulating recursive definitions.

3.1.1.1 Tower of Hanoi

In the Tower of Hanoi puzzle there are three pegs and n disks, each disk having a different size. Initially the disks are mounted in size order, with the smallest disk on top, on one of the pegs, say i. The problem is to generate a sequence of moves by which all n disks can be transferred from i to another peg, say j, under the following constraints. The disks are to be moved one at a time, always from the top of one pile on a peg to the top of the pile on another peg. A disk may be placed only on one larger than itself. A move from peg i to j is designated $i \rightarrow j$. In summary, we wish to design a function $h(n, i \rightarrow j)$ whose value is the sequence of moves necessary to transfer the $n \in N$ disks from peg i to j, where i and $j \in \{1,2,3\}$. The domain then includes all pairs $\langle n, i \rightarrow j\rangle$ with

$$n \in N \text{ and } i \rightarrow j \in \{1 \rightarrow 2, 1 \rightarrow 3, 2 \rightarrow 1, 2 \rightarrow 3, 3 \rightarrow 1, 3 \rightarrow 2\}.$$

Ordering domain pairs according to the magnitude of their first component n seems reasonable. There does not appear to be any way in which the second component of the pairs $i \to j$ can naturally enter into the ordering. Based on this hypothesized domain, the lowest members of the domain would be those with $n = 1$, and for such the terminal line is

$$h(n, i \to j) = \langle i \to j \rangle \qquad :n = 1$$

To construct the recursive line, assume that the value of the sequence of moves $h(n-1, i \to j)$ is known for all possible $i \to j$. Can we then construct the function $h(n, i \to j)$ in terms of those assumed known? That is, if we know the sequence of moves necessary for legitimately moving $n-1$ disks initially on peg i to peg j for any i and j, can we construct from those the sequence of moves necessary to move n disks from any i to any j? We can. Assume i, j, and k are each in $\{1, 2, 3\}$ and different from each other. With n disks initially on i we can move the top $n-1$ of them to disk k with the sequence $h(n-1, i \to k)$, then move the remaining disk on i to peg j, and then move the $n-1$ disks from k to j with the sequence $h(n-1, k \to j)$. By concatenating these sequences we produce a sequence that moves all n disks from i to j. This gives us the recursive line, which together with the nonrecursive line and the initial domain description gives

Definition 3.1.1: Tower of Hanoi

$$\left[\begin{array}{ll} h(n, i \to j) = \langle i \to j \rangle & :n = 1 \\ h(n, i \to j) = h(n-1, i \to k) \| h(1, i \to j) \| h(n-1, k \to j) & :n > 1 \\ \text{Initially } n \in N;\ i, j, k \text{ are each different members of } \{1, 2, 3\}. & \end{array}\right.$$

The recursive approach is often aided by looking at examples, using the w network as a model, and generalizing. Consider the output of a definition of $h(n, i \to j)$ with $j = 3$ and $i \to j$. The necessary moves to get three disks from peg 1 to 2 are $\langle 1 \to 2, 1 \to 3, 2 \to 3, 1 \to 2, 3 \to 1, 3 \to 2, 1 \to 2 \rangle$. So this sequence is the output of the final w block of the O–w–t network for $h(3, 1 \to 2)$. This might have been arrived at by trial and error. What are the inputs to this final w block? Since w is concatenation, the vector for $h(3, 1 \to 2)$ should be partitionable into subvectors expressible in the form $h(x, i \to j)$, with x probably less than 3; $x = 2$ is the simplest assumption. We know, again perhaps by trial and error, that

$$h(2, 1 \to 2) = \langle 1 \to 3, 1 \to 2, 3 \to 2 \rangle \qquad \text{and} \qquad h(1, 1 \to 2) = \langle 1 \to 2 \rangle$$

Now it would be necessary to recognize that the moves $h(3, 1 \to 2)$ can be partitioned into the first three: $\langle 1 \to 2, 1 \to 3, 2 \to 3 \rangle$, the middle one $\langle 1 \to 2 \rangle$, and the last three: $\langle 3 \to 1, 3 \to 2, 1 \to 2 \rangle$, and each expressed as a h of some domain point. The first three are seen to be $h(2, 1 \to 3)$ by analogy with $h(2, 1 \to 2)$ $(= \langle 1 \to 3, 1 \to 2, 3 \to 2 \rangle)$, the middle is $h(1, 1 \to 2)$ $(= \langle 1 \to 2 \rangle)$, and the last three are $h(2,3 \to 2)$. Thus,

$$h(3,1 \to 2) = h(2,1 \to 3) \| h(1,1 \to 2) \| h(2,3 \to 2)$$

This result would then be generalized to obtain the recursive line in Definition 3.1.1.

3.1.1.2 Gray Code

The problem of generating a Gray code is closely related to that of generating the moves for the Tower of Hanoi puzzle. An n-bit Gray code is a sequence of binary n-tuples ordered so that successive n-tuples differ in only one bit position. For example, $\langle 000, 001, 011, 010, 110, 111, 101, 100 \rangle$ is a 3-bit Gray code (with 000 short for $\langle 0,0,0 \rangle$). Assuming that we know how to produce an $(n-1)$-bit Gray code designated $g(n-1)$, can we construct $g(n)$? Concatenating 0 onto the front of each vector in $g(n-1) = \langle\langle 0 \rangle\rangle |*| g(n-1)$ gives a sequence of n-tuples which still have the Gray code property, but it gives only half of the n-tuples needed. If we reverse the sequence $g(n-1)$, we still have a Gray code. Designate this reversed $g(n-1)$ by $g'(n-1)$ and note that the last n-tuple in $g(n-1)$ equals the first n-tuple in $g'(n-1)$. Recalling that $|*|$ is cross concatenation, we see that $\langle\langle 0 \rangle\rangle |*| g(n-1) \| \langle\langle 1 \rangle\rangle |*| g'(n-1)$ gives the complete sequence $g(n)$. But now we need a definition for $g'(n)$. This is easily seen to be $\langle\langle 1 \rangle\rangle |*| g(n-1) \| \langle\langle 0 \rangle\rangle |*| g'(n-1)$. These two lines may be reduced to one if $g(n, s)$ is used to represent the Gray code sequence and the reverse Gray code sequence, respectively, as $s = 0$ or 1. (This scheme of adding an argument can always be used to reduce pairs of mutually recursive definitions to one.) Then including the terminal line and the initial domain, we get

Definition 3.1.2: Gray Code

$$\left[\begin{array}{ll} g(n,s) = \langle\langle\ \rangle\rangle & : n = 0 \\ g(n,s) = \langle\langle s \rangle\rangle |*| g(n-1,0) \| \langle\langle 1-s \rangle\rangle |*| g(n-1,1) & : n \in N \\ \text{Initially } n \in N,\ s = 0. & \end{array}\right.$$

This definition represents the Gray code with the sequence of n-tuples that constitute it. Alternatively it can be represented by a sequence of integers, each $\leq n$, indicating which n-tuple position is to be changed to produce the next n-tuple in the Gray code, assuming the initial n-tuple consists of all 0's. This choice of representing a sequence of changes to a system by the sequence of states through which it passes, or the sequence of state changes which produce the sequence, is often available. In the Tower of Hanoi example it was the sequence of changes to the state of the disks on their three pegs which was defined, instead of the sequence of actual disk–peg configurations. There is clearly a close relation between these choices. In fact, under general conditions encompassing the two definitions above, we can mechanically transform one of these forms of definition into the other. This transformation is given in Section 3.8.

3.1.1.3 Permutations

Let $f(n)$ be the set of all permutations of the first n integers, each permutation being represented as a vector. Then assuming $f(n-1)$ is known, we can construct $f(n)$ by

> "Inserting" n between the 1st and 2nd component of each vector in $f(n-1)$.
>
> Then between the 2nd and 3rd component of each vector in $f(n-1)$.
>
> ⋮
>
> Then between the $(n-2)$th and $(n-1)$th component of each vector in $f(n-1)$.
>
> Finally after the $(n-1)$th component of each vector in $f(n-1)$, and then taking the union of the result.

To write the recursive line compactly, an "insert" primitive is useful. With Y being a vector $= \langle y_1, \ldots, y_n \rangle$, j an integer $\leq n$, and c any allowable component value of the vector Y, the function written $y_{j-} \leftarrow c$ is equal to the vector with $|Y| + 1$ components obtained by inserting c between y_{j-1} and y_j in Y. By convention $y_{n+1-} \leftarrow c = Y \| \langle c \rangle$. With Y being a vector $\in Z$, $Z_j \leftarrow * c$ is defined as $\{Y_{j-} \leftarrow c \,|\, Y \in Z\}$.

The definition for permutations incorporating the design of the recursive line as described above, including a terminal line for permutations of length 0, is

Definition 3.1.3: Permutations by Insertion

$$\left[\begin{array}{ll} f(n) = \{\langle\ \rangle\} & :n = 0 \\ f(n) = [f(n-1)_{1-} \leftarrow * \, n] \cup \cdots \cup [f(n-1)_{n-} \leftarrow * \, n] & :n \in N \\ \text{Initially } n \in N. & \end{array}\right.$$

3.1.1.4 Partitions

Consider the set of all (unordered) combinations of integers $\in N$ which add up to a given number, r. These are the *partitions* of r. The partitions of 5 are $\{14, 122, 113, 1112, 11111, 23, 5\}$. In general the set of all partitions of n is

> The set of all such combinations whose smallest member is 1
> unioned with all such combinations whose smallest member is 2
>
> ⋮
>
> unioned with all such combinations whose smallest member is j
>
> ⋮
>
> unioned with all such combinations whose smallest member is $\lfloor n/2 \rfloor$
> together with n itself

The smallest member of a nontrivial partition of n cannot be greater than $\lfloor n/2 \rfloor$. The relation as stated above is not quite recursive. It gives the function equal to the set of *partitions of r* in terms of the sets of all *partitions of r whose smallest member is j*, with j ranging from 1 to $\lfloor r/2 \rfloor$. Thus it gives partitions of r in terms of partitions of r. To express the partitions of r in terms of partitions of numbers less than r, interpret $f(j, r)$ to be all partitions of r whose smallest member is greater than or equal to j, and note that all partitions of r whose smallest member is k may then be expressed as $\{\langle k \rangle\}|*|f(k, r-k)$. We can now give the required recursive description. For example, to express *all partitions of 7 whose smallest member is at least 1*, we would get

$$f(1,7) = \{\langle 1 \rangle\}|*|f(1, 6[=7-1])$$

$$\cup \{\langle 2 \rangle\}|*|f(2, 5[=7-2]) \cup \{\langle 3 \rangle\}|*|f(3, 4[=7-3]) \cup \{\langle 7 \rangle\}$$

A recursive line of this kind is not applicable to $f(3,4)$, for example, because the smallest number in the partition is to be 3, so any other number must be at least 3, with a total of at least 6, which is too large. On the other hand $\{\langle 4 \rangle\}$ itself is a partition of 4 whose smallest member is greater than or equal to 3, so $f(3,4) = \{\langle 4 \rangle\}$. This value will be assigned to $f(3,4)$ in a terminal line. In general the terminal line is applicable to $f(j, r)$ if and only if $\lfloor r/2 \rfloor - j < 0$, because when that happens, no partition with more than one member can have a smallest member $= j$.

A generalization of the recursive and terminal line described above gives

Definition 3.1.4: Partitions of an Integer

$$\left[\begin{array}{ll} f(j, r) = \{\langle r \rangle\} & :\lfloor r/2 \rfloor - j < 0 \\ f(j, r) = \displaystyle\bigcup_{i=0}^{\lfloor r/2 \rfloor - j} [\{\langle j+i \rangle\}|*|f(j+i, r-(j+i))] & \\ \qquad \cup \{\langle r \rangle\} & :\lfloor r/2 \rfloor - j \geq 0 \\ \text{Initially } j = 1,\ r \in N. & \end{array}\right.$$

(For future use note that although the recursive line above is not applicable if $\lfloor r/2 \rfloor - j < 0$, if it is interpreted to be equal to $\{\langle r \rangle\}$, for example, by interpreting $\bigcup_{i=0}^{-1} = \{\ \}$, we can evaluate f by the nonterminal line alone.)

3.1.1.5 Combinations

The set of all combinations of m of the first n integers, $C(n, m)$, can be defined in terms of smaller sets. The combinations $C(n, m)$ can be partitioned into

1. Those containing n, which is the set of all $m - 1$ combinations of the first $n - 1$ integers with n attached to each combination, $\{\langle n\rangle\}|*|C(n-1, m-1)$, and
2. Those which do not contain n, which is the combinations of m of the first $n - 1$ integers, $C(n - 1, m)$.

This gives the recursive line

$$C(n, m) = C(n-1, m) \cup \{\langle n\rangle\}|*|C(n-1, m-1)$$

Notice the similarity of this definition, for enumerating the set of combinations, with that for counting the number of such combinations given in Definition 2.8.2. *This relation between the count of the number of members of a set and the enumeration of that set occurs frequently and is sometimes helpful in obtaining the definition of one, given that of the other.*

3.1.1.6 Graph Coloring

Consider a partition of the vertices of a graph G into c sets such that no two vertices in a set are connected by an edge. If a different color is chosen for each of the c sets and assigned to all the vertices in the corresponding set, G would be colored so that no two vertices connected by an edge have the same color. A coloring with this property is called a *vertex coloring* of G by c colors. Given a graph, we wish to design a recursive definition that produces all vertex colorings of G.

Assume G contains the pair of vertices x and y, between which there is no edge. Then G can be colored with a set of colorings

1. C_1, in which x and y have the same color, union
2. C_2, in which x and y have different colors.

Furthermore C_1 is equivalent to all ways of coloring the graph G_1 obtained by identifying x and y in G, and C_2 is all ways of coloring the graph G_2 obtained from G by adding an edge from x to y into G. Thus if G has any pair of vertices with no edge between them, then all ways of coloring G equals all ways of coloring G_1 together with all ways of coloring G_2. G_1 has fewer vertices than G, whereas G_2 has more edges than G. Thus as we move right to left in the recursive statement, either the number of vertices decreases or the number of edges increases. Eventually then we must color a graph with one vertex, or a completely connected graph. A single vertex requires a single color. A completely connected graph with n vertices requires n color, and a graph is completely connected only if there is no pair of vertices not connected by an edge. Both a recursive and a terminal line of the definition can be developed from this information. The recursive line would be

$$f(G) = f(G_1) \cup f(G_2)$$

To express the definition formally, it will be necessary to rewrite the above line in terms of primitives, such as "identify $G(x, y)$," whose output is a graph G_1 built from G by replacing vertices x and y of G by one vertex with edges to every vertex formerly adjacent to either x or y. Furthermore the terminal line will also have to be expressed formally.

3.1.1.7 Divide and Conquer Schema [Aho, Hopcroft, Ullman 75, sec. 2.6; Purdom 85]

As long as the w function for a recursive definition is associative, its recursive line can be viewed as the design of a set enumeration. In fact almost all our examples involved the development of a statement of how a set whose nature depends only on the integer n, say $S(n)$, is the union or concatenation of smaller sets, part$(k, S(n))$, or, equivalently, how a set $S(n)$ can be partitioned into subsets, part$(k, S(n))$. That is,

$$S(n) = \bigcup_{k \in K(n)} \text{part}(k, S(n))$$

For example, when S is a set of vectors, the partition is often into blocks characterized by the same kth component or, perhaps, by a component equal to k.

Furthermore part$(k, S(n))$ is defined in such a way that it depends on S at an argument less than n, such as part$(k, S(n)) = r(k, S(n - u(k)))$, where r and u are primitive functions, so that we have

$$S(n) = \bigcup_{k \in K(n)} r(k, S(n - u(k)))$$

For example, when S is a set of vectors partitioned by the presence of a particular component, the smaller set would be the one with that component removed.

This recursive scheme can be further generalized to arguments other than integers. If X is any argument structure and $o_i(X)$ is "lower" in the domain of definition, the schema we use for guidance in constructing a recursive definition "recursively" is

$$S(X) = \bigcup_{k \in K(X)} r(k, S(o_k(X)))$$

This is the *divide and conquer schema*,

Frequently we find that set $K(X)$ has only two members, that is, $S(n) = r(k_1, S(o_1(X)) \cup r(k_2, S(o_2(X))$. Often there is a key element present in one of these parts and absent from the other. Combinations and colorings are each examples of this case.

In fact, however, virtually any nonnested recursive definition can be given in a binary form (see exercise 18).

As a final example of divide and conquer note that

The set of all binary n-tuples $S(n)$ is:
The set of all binary $(n - 1)$-tuples $S(n - 1)$ with $\langle 0 \rangle$ concatenated on each $= \langle 0 \rangle \| S(n - 1)$ $[r(0, S(n - 1))]$, unioned with
The set of all binary $(n - 1)$-tuples $S(n - 1)$ with $\langle 1 \rangle$ concatenated with each $= \langle 1 \rangle \| S(n - 1)$ $[r(1, S(n - 1))]$.

Concatenation r can be to the front or back of each vector. (The set of n-tuples produced will be the same, though the order in which the set members are generated in a leftmost SE evaluation will differ.) Such concatenation, to the front of each vector, forms the heart of Definition 1.11.1, which generates the n-tuples in the usual numerical order.

As described here, the divide and conquer schema is applicable when the w function is associative. Examples of designs for nonassociative w's are given in Section 3.3.

3.2 ITERATIVE OR GROWTH APPROACH TO FUNCTION DEFINITIONS

The structure of the O–w–t network provides a guide to the design of recursive definitions. For example, assume that we experience difficulty in designing a definition to perform the function w on the members of a set S, although we have one for producing the set S' which includes S. Consider the O network for S'. To get S we need only suppress those t block outputs which give outputs in S' but not in S. If we can formulate the conditions, say C, under which the t block output is to be suppressed, then this change is easily incorporated in the definition. Add "& $\sim C$" to each existing terminal line, and add the line "$f(X) = 0_w \quad :C$," where 0_w is defined so that $w(0_w, X) = X$. For an associative w function an argument with this property can be easily included if it is not already present. For example, if w is addition, then $0_w = 0$; if w is multiplication, then $0_w = 1$. This technique for producing a subset S of S' when we know how to produce S' is inefficient and to be avoided if possible. However, sometimes it is unavoidable.

Like "thinking recursively," the technique we have called "think iteratively" is based on the structure of the O–w–t network.

Assume that w is associative, so the O network can be thought of as enumerating a set or sequence upon whose members w will finally operate. The input–output functioning of an O block is related to the syntax of the standard form recursive definition. The outputs of an O block with input X are the $o_1(X), \ldots, o_{m(X)}(X)$, terms that go into the composition of that recursive line. The principle by which the O block generates its outputs as a function of its input is called the *generating principle*. Therefore:

Given the generating principle, we can immediately write down the recursive line of the corresponding definition.

There is an O network for each input of a given definition. Each O network pictures the growth of the enumerated set of terminal values from the initial argument structure. Stage by stage, repeated application of the generating principle grows larger and larger sets. The developing sets progressively approach the form and cardinality they will have in the final enumerated set. The argument structures which appear in the evaluated O network, in their treelike relation to one another, will be called a *growth tree*. No matter which member of the starting domain seeds the growth tree, the same generating principle produces the growth. Our approach based on these considerations is to somehow cultivate sample growth trees from members of the starting domain and from these induce the generating principle. How such sample growth trees may be obtained is illustrated in the following examples.

3.2.1 Permutations

As a first example, consider the growth tree for all permutations of the first three integers. These could be grown in stages starting with all permutations of no integers, then of the first integer, then of the first two integers, and so on. At each stage we propose growing the permutations of the first n integers from those at the first $n - 1$ integers. This is shown in Example 3.2.1, where permutations are represented by vectors.

Example 3.2.1

$$\langle\ \rangle$$

$$\langle 1 \rangle$$

$$\langle 1,2 \rangle \qquad\qquad\qquad \langle 2,1 \rangle$$

$$\langle 1,2,3 \rangle \langle 1,3,2 \rangle \langle 3,1,2 \rangle \langle 2,1,3 \rangle \langle 2,3,1 \rangle \langle 3,2,1 \rangle$$

To make this a growth tree, we must connect groups of stage n vectors to individual vectors at stage $n - 1$. We can either try to see how individual vectors at stage $n - 1$ "produce" groups of vectors at level n, or we can try to partition the entire group of vectors at level n into blocks whose members share some common relation to an individual at stage $n - 1$. (The ordering of the permutations in each stage could be considered a bias in our presentation, but it is a natural and consistent ordering.) In Example 3.2.2 we show the result of making such connections in a completed growth tree, or O network, with the unboxed letter O representing an O block.

Example 3.2.2

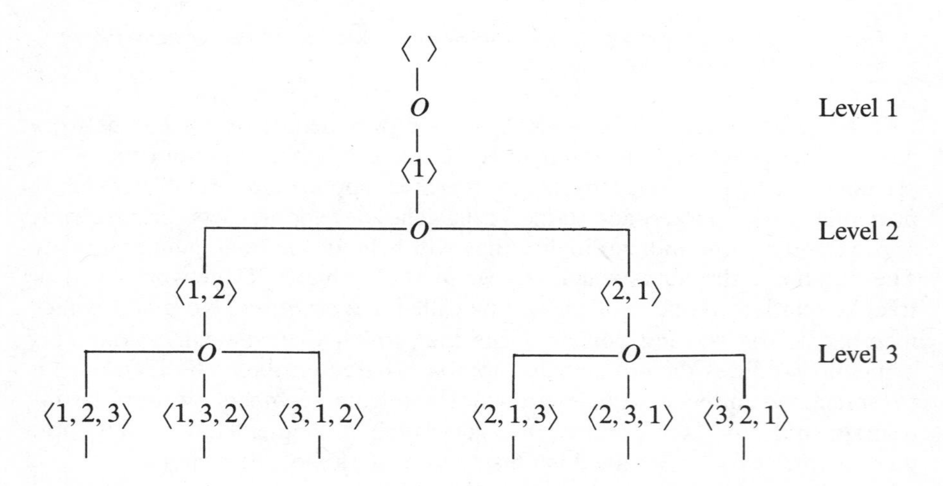

At the ith level the output vectors of each O are produced from the input vector by inserting i successively after the last, after the next to last, . . . , after the first position, and finally after the 0th (that is, before the first) position of that vector. The part of the generating principle, or O function, which specifies where the insertions are to be made, is thus determined. But how, given the input vector only, can we determine what to insert? We will insert the integer one greater than the largest component in that vector. (Alternatively we could, but will not, add an integer as a second component, starting at 1 and increased by 1 in each O block, explicitly giving the number to be inserted.) Still a question remains: How can we tell from the input to an O block that termination should occur at level 3? Information must be added to the input. If 3 was the initial value of a second component of the O block input, and this component was decreased by 1 at each level, then termination would be signaled when that value reaches 0. Such a component is included in the following modification of the O network of Example 3.2.2. (In representing vectors in the next figure, all brackets and commas in those brackets have been dropped.)

Example 3.2.3

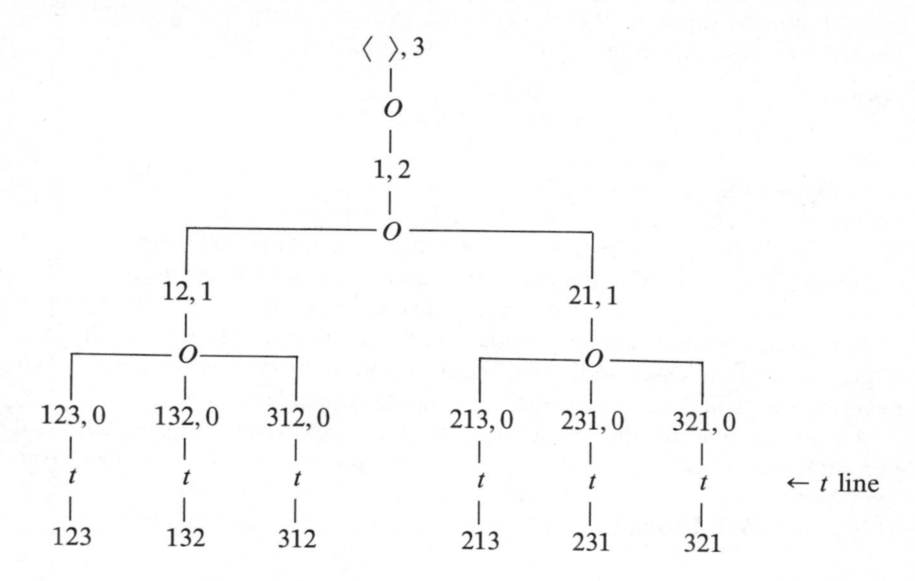

In Example 3.2.3 we have constructed a sample O network with a well-defined generating principle that seems to satisfy all requirements. If we can generalize the generating principle, a recursive definition is easily constructed from this network. First, the inputs and outputs of O blocks have two components, the vector x and the integer n. Further, the ith output from an O block with input (x, n) is the result of inserting an integer 1 greater than the largest member of x, MAX(x), after the ("last" + $1 - i$)th position. Since the position "last" equals the size of x = MAX(x), this becomes the (MAX(x) + $1 - i$)th position. Thus the first component of the ith output is

$$x_{(\mathrm{MAX}(x)+1-i)+} \leftarrow (\mathrm{MAX}(x) + 1)$$

The second component is $n - 1$. This gives $o_i(x, n)$ as it appears in the standard form recursive line. There are MAX(x) + 1 outputs from each O block, and therefore MAX(x) + 1 = $m(x, n)$ of the standard form recursive line. The union of the first component vectors, or more precisely of the sets whose only members are those vectors, will be the final output. Therefore the w of the recursive line is union ($\cup$). The line is

$$\begin{aligned} f(x, n) = f\big(x_{\mathrm{MAX}(x)+} &\leftarrow (\mathrm{MAX}(x) + 1), n - 1\big) \\ &\cup \cdots \cup f\big(x_{1+} \leftarrow (\mathrm{MAX}(x) + 1), n - 1\big) \\ &\cup f\big(x_{0+} \leftarrow (\mathrm{MAX}(x) + 1), n - 1\big) \qquad : n \in N \end{aligned}$$

The network in Example 3.2.3 includes a t line. Each t block has a two-component input of the form $\langle v, 0\rangle$, with a corresponding output $\{v\}$. Thus the corresponding terminal line in the standard form is

$$f(x, n) = \{x\} \qquad : n = 0$$

Note that MAX$(x) = |x|$ in the O network and could be replaced by it in the definition. Whether it should be, depends on which is easier to compute. In fact there are better alternatives, as we will see later in this chapter.

Generalizations of this type are often quite easy once a small growth tree, or O network, for a sample input is completely specified. The inventive step is in obtaining a proper sample growth tree. Some other examples of such trees, requiring in some cases additional input–output components for control, but nevertheless clearly exhibiting their generating principle, follow.

An alternative to the previous "insertion" approach to producing all permutations of the first n integers obtains output vectors in an O block from the input vector v by concatenating each integer $\leq n$ not already in v onto v. This is shown in Example 3.2.4.

Example 3.2.4

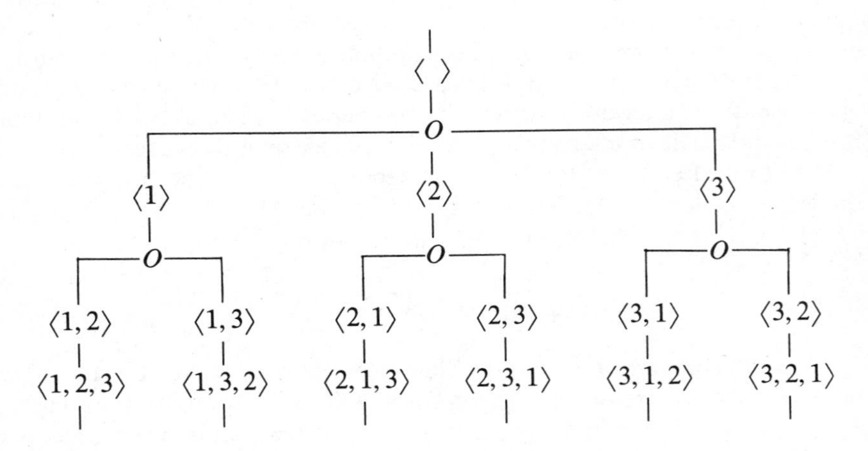

Explicit information about n must be included in the argument structure, both to complete the definition of the O function and to establish the terminal condition. If n is added as a component, then the terminal condition is $|v| = n$. For $|v| < n$ the generating principle is incorporated in the following recursive line.

$$f(v, n) = \bigcup_{j=1}^{n,\, j \notin v} [f(v \| \langle j \rangle, n)]$$

With a slight modification we can get a recursive line for the set of all those permutations of the first n integers in which the integer j does not appear in position j. These are called *derangements*. The following recursive line could be used in their definition:

$$f(v, n) = \bigcup_{j=1}^{n,\, j \notin v,\, j \neq |v|+1} [f(v \| \langle j \rangle, n)]$$

Another simple modification is also adequate for a recursive line to define the set of all *selections*, permutations of symbols when the same symbols can appear more than once. Let the symbols be integers in N^n. Let integer i appear m_i times. The recursive line for this case would become

$$f(v, n) = \bigcup_{j=1}^{n \,\&\, \text{number of } j\text{'s in } v < m_j} [f(v \| \langle j \rangle, n)]$$

It is interesting that the last two recursive lines are easy modifications of the last "concatenation" approach to permutation enumeration, but it is difficult to see how they can be developed as a modification of the previous "insertion" approach.

3.2.2 Graph Coloring

Refer to the definition given in Section 3.1.1.6 for the coloring problem in terms of partitions of the vertices of the graph. There are a number of ways to approach the design of the definition to enumerate the colorings, although the idea of representing the colorings by partitions is already a significant step. We develop one of these approaches—from the growth point of view. Another approach is related to packing and is mentioned when packing is considered as the next topic.

Assume that the graph G to be colored has n vertices, labeled $\{1, 2, \ldots, n\}$. Then we start the growth of the coloring with a completely connected graph of n vertices, say $G0$. The coloring of $G0$ is given by the partition $\{\{1\}, \{2\}, \ldots, \{n\}\}$. Now we peel away edges of $G0$ which are not in G, one at a time, and see how this gives new graphs and therefore new colorings. This approach is tried in Figure 3-1 on a sample graph with four vertices. The last line is incomplete, giving only four of the set of five colorings.

In Figure 3-1, G is the graph to be colored. We need to determine the generating principle. First we see that there are two or fewer outputs of an O block. This is because when an edge (x, y) is removed, the coloring represented by the input to that block is still viable, so it is one of the outputs.

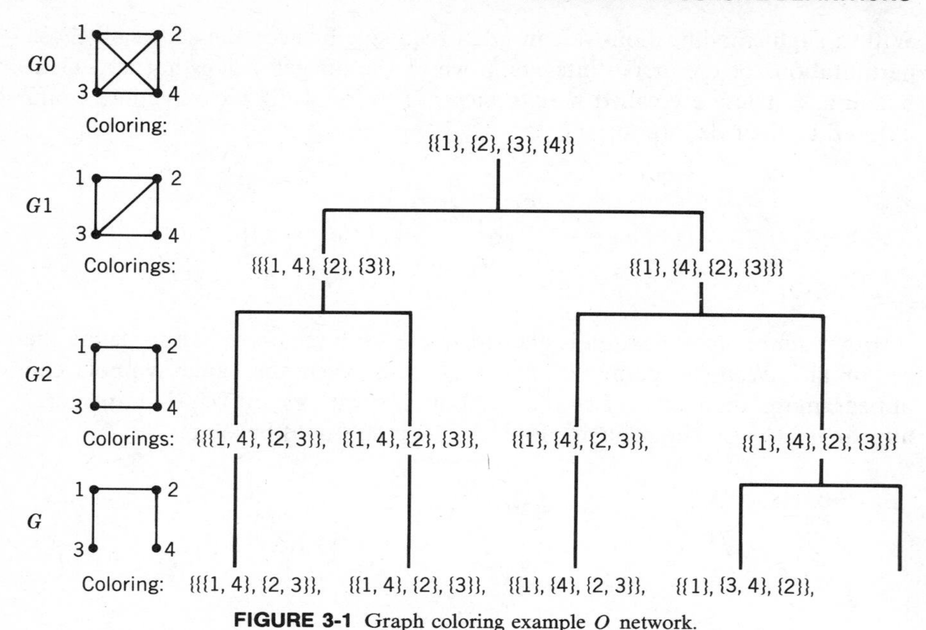

FIGURE 3-1 Graph coloring example O network.

There is another output if the only edge connecting the set containing x, say X, and the set containing y, say Y, is (x, y). In that case the sets X and Y can be replaced by $X \cup Y$, which is now completely disconnected.

3.2.3 Packing

Consider a set of n boards, with board i having length L_i. We wish to find all ways to partition the set into groups so that when the boards in a group are placed end to end, their cumulative length is less than or equal to a given constant TOT. A partially developed growth tree for a set of boards in which $\langle L_1, \ldots, L_n \rangle = \langle 6, 3, 3, 2, 2, 2, 2 \rangle$ and TOT $= 10$ is shown in Example 3.2.5. Each group of boards is represented by a *group vector*. The input to, as well as the output of, an O block is a set of group vectors. The kth output of an O block at level j results from concatenating the jth board to the kth group vector in the input set of group vectors whose sum will not thereby exceed TOT. In addition there is a final output from each O block, which consists of a copy of its input group vectors together with a group vector containing the jth component of L as its only entry. In the example the commas and the outside brackets delineating the partitions or vectors of group vectors have been dropped, and outputs involving more than two group vectors are shown in only a few cases. The starred argument structures are duplicated elsewhere in the O network.

Example 3.2.5

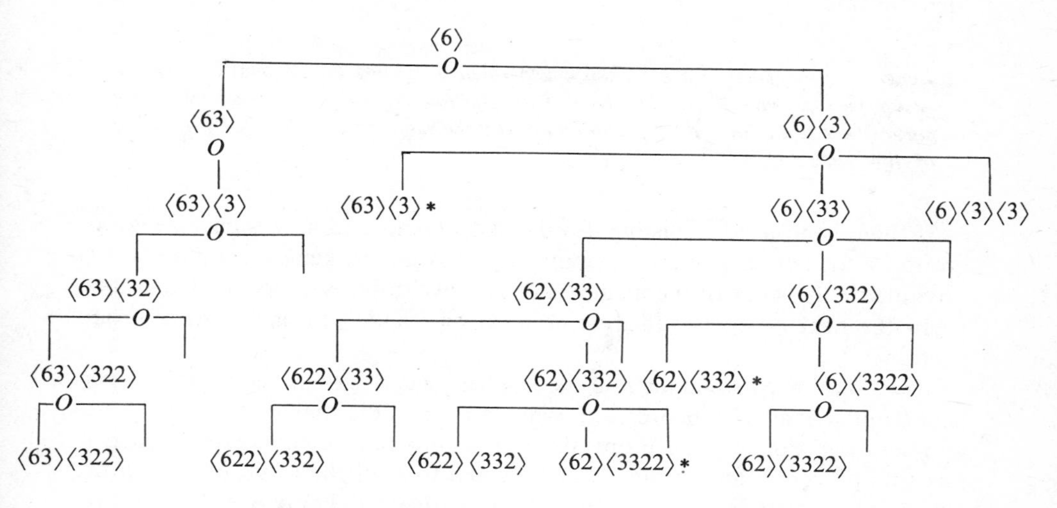

The generating principle, and thus the O function, should be clear from Example 3.2.5, even though some additional information is necessary to complete its detailed description.

Generating Principle: The kth child is generated by adding the jth board to the kth group vector, with k ranging from 1 to (number of group vectors in parent) + 1.

Notice that an algorithm to enumerate all ways of coloring a graph can be based on this approach. If the length list is replaced by the vertices of a given graph G, and a vertex v is added to a group vector of vertices V only if v is not connected by an edge to any vertex in V, then each group vector will represent a set of vertices assigned the same color. The sets of vectors formed then represent all ways of assigning colors to vertices—the same color if in the same vector, and different colors if in different vectors.

It is also important to note that the O network based on this approach will sometimes contain identical argument structures at two different sites, for example, $\langle 6, 3\rangle\langle 3\rangle$ in Example 3.2.5. In a straightforward SE procedure, such duplicates require inefficient duplicate processing. Even though the enumeration does not duplicate any board placements, different sets of boards can have the same total length. This is certain to happen if more than one board has the same length. Many of the duplicates can be avoided by the design of the recursive definition.

Assume boards of the same length need not be distinguished. Then the generating principle can be revised so as not to reproduce configurations which only differ in the identity of two equal-length boards. The lengths should be arranged so that equal-length boards are grouped, and each argument structure X should be made to include an integer ℓ, indicating into which group vector the last board was placed. The revised generating principle then becomes

Generating Principle: The kth child is generated by adding the jth board to the kth group vector, with k ranging from i = ℓ to (number of group vectors in the parent) + 1 if the jth and (j − 1)th boards are the same size, while i = 1 if they are of different sizes.

If there is no need to distinguish the exact composition of equal-size vector groups, a further decrease in the number of board arrangements enumerated is possible. If the sums of the members of any two group vectors are equal, then when the next board is added, it need only be added to one of these group vectors.

If we only need to enumerate the smallest sets of group vectors which can accommodate all the boards, we can add an additional constraint on the enumeration. For this constraint, the L_i's should be in nonincreasing order. If any group vector sum becomes exactly equal to L as the result of adding the jth board to a configuration, then that alternative is the only one that need be pursued. This constraint is certainly not enough to eliminate all abortive enumeration paths; it will eliminate some. It is illustrated in Example 3.2.6.

Example 3.2.6

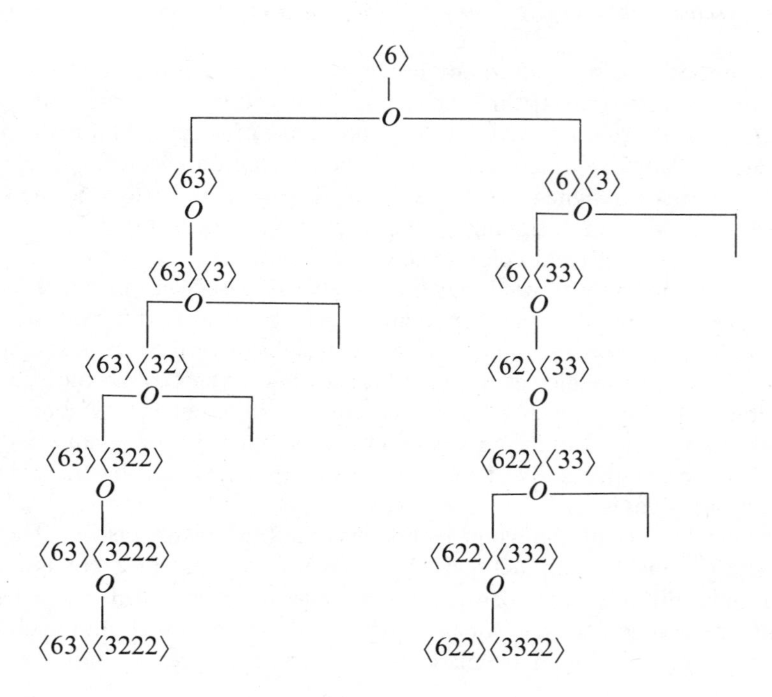

3.2.4 Partitions

As another example, consider the growth of partitions of a given number. The partitions of 6 are shown in Example 3.2.7.

Example 3.2.7

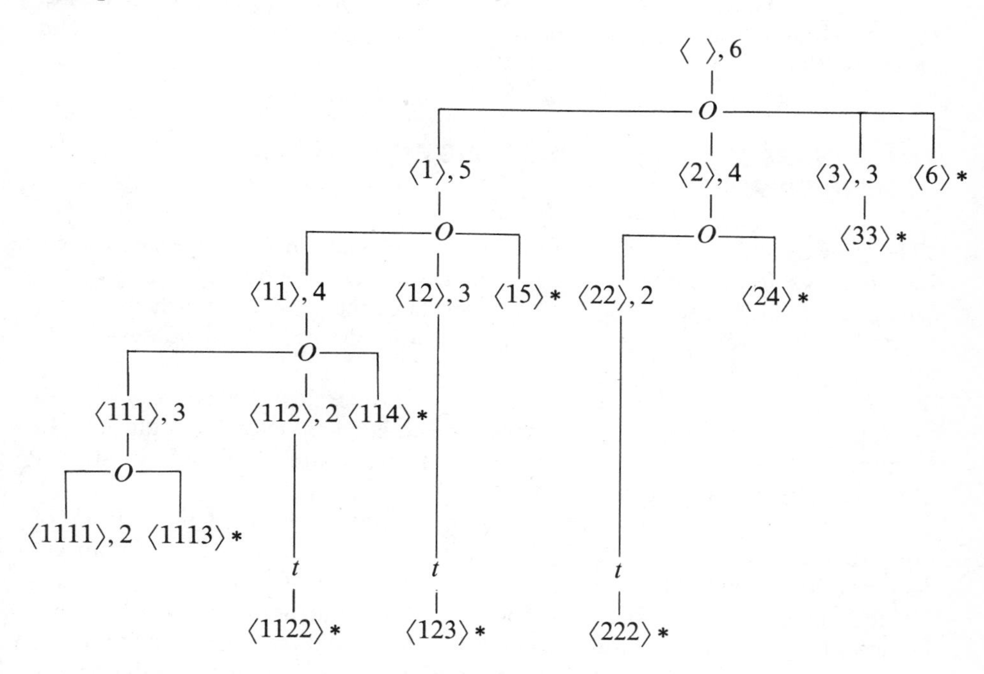

Each O block input in the network in Example 3.2.7 has two components. The first is the vector v, the second is a number to be partitioned r. In the generalization of this example to obtain the generating principle, a third part j, equal to the last component of v, is added. The starred argument structures are all terminal values.

The generating principle exemplified by Example 3.2.7 is incorporated in the recursive line of the following definition in which $v = \langle v_1, \ldots, v_k \rangle$ and $j = v_k$.

Definition 3.2.1: Partitions of an Integer

$$\left[\begin{array}{ll} f(v, j, r) = \{v \| \langle r \rangle\} & :\lfloor r/2 \rfloor - j < 0 \\ f(v, j, r) = \bigcup_{i=0}^{\lfloor r/2 \rfloor - j} [f(v \| (j+i), j+i, r-(j+i))] & \\ \qquad \cup \{v \| \langle r \rangle\} & :\lfloor r/2 \rfloor - j \geq 0 \\ \text{Initially } v = \langle\ \rangle,\ j = 1,\ r \in N. & \end{array}\right.$$

The upper limit on the $\bigcup$ is determined by the need to continue i until $j + i = (v_k + i) = \lfloor r/2 \rfloor$, which means until $i = \lfloor r/2 \rfloor - j$.

Example 3.2.7 shows terminal values at outputs of O blocks, as well as in the usual place at the output of t blocks. In general this can happen in an O–w–t network for a nonnested definition with a recursive line of the form of Eq. (2.2.3). These terminal values arise as arguments of w [$p(X)$ in Eq. (2.2.3)], which are not f calls. Definition 3.2.1 has a recursive line with such an argument.

3.3 NONASSOCIATIVE *w* FUNCTIONS AND COMBINED APPROACHES

The growth approach is appropriate when the function to be designed has a set as its value or can be viewed as some associative function operating on the members of such a set. In such a case, design of the generating principle, or O function, is virtually the entire problem. Little attention need be devoted to the w network design.

Although it also has been illustrated almost exclusively on problems in which w is associative, the recursive approach is more generally applicable. In contrast to the growth approach, it seems to center attention on the w function and its arguments.

Notice that associative functions differ in ways that affect the efficacy of different design approaches. Functions such as minimum and maximum are *selection functions*. Their value is always equal to one of their arguments. This is not true of addition, multiplication, and union. Furthermore, even though a function is associative and the same evaluation is obtained no matter how its arguments are parenthesized, some ways of parenthesizing may produce more efficient programs than others. In such a case, as well as in those in which the w function is not associative, the structure and function of the w network become more important in the design of the recursive definition. The increased complexity of the w network is often accompanied by a decrease in O network complexity. Examples of such problems will now be considered. We will use the recursive approach to get the basic structure of the w network, but the relation between the O and w networks, and the growth approach will also play a role.

3.3.1.1 Merge–Sort

Let x and y be vectors with numbers as components. In each of these vectors the component values are nondecreasing, that is, each is sorted. The value of the function merge(x, y) is a vector containing all the components from x and y and sorted in the same nondecreasing order. Merge is associative and therefore

$$\text{merge}(\text{merge}(\text{merge}(x, y), z), w) = \text{merge}(\text{merge}(x, y), \text{merge}(z, w))$$

However, assuming the merge is implemented in the simplest straightforward way and that all the vectors x, y, z, w are the same size n, the association of merging on the right of the equality will be more efficient in the worst case. In the worst case merging n and m items costs $n + m - 1$ comparisons, so the worst case for the first order of association is $9n - 3$ and for the second, $8n - 3$. An algorithm to sort the originally unsorted components of the vector $v = \langle v_1, \ldots, v_p \rangle$ can be constructed by implementing the function merge($\langle v_1 \rangle, \ldots, \langle v_p \rangle$). For efficiency we want to design a recursive definition for this function which always merges pairs of sorted vectors whose sizes are as nearly equal as possible. (That this is the most efficient choice of relative size is demonstrated in Chapter 4.) Focusing on the w part of the O–w–t network, we see that its shape determines a way of combining or associating arguments which emerge from the t line, the function $w \equiv$ merge. Now if the resultant merge network is to combine pairs of equal-size vectors, the final w, or merge, block in that network must be as shown in Example 3.3.1.

Example 3.3.1

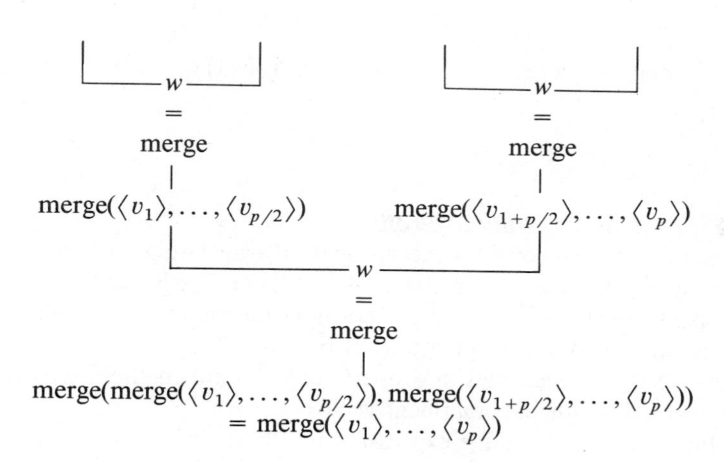

It is a characteristic of every O–w–t network that

The O and w networks are mirror images of each other, and the input to an O block is the argument of the output of the mirror image w block.

Thus the structure of the top of the O network must be the mirror image of the bottom of the merge network we have shown. That O block's input then is $\langle v_1 \rangle, \ldots, \langle v_p \rangle$, and its two outputs are $\langle v_1 \rangle, \ldots, \langle v_{p/2} \rangle$ and $\langle v_{1+p/2} \rangle, \ldots, \langle v_p \rangle$. A reasonable name for the O function then is "split," since it splits a vector in as close to two equal pieces as possible.

Example 3.3.2

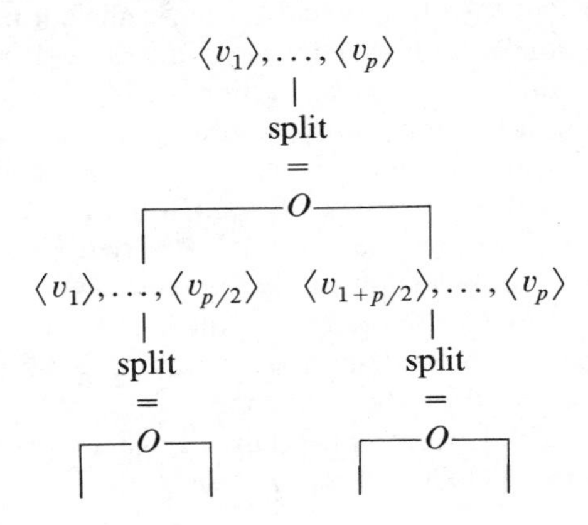

Another stage of the network could be similarly constructed, but already the generating principle is clear and the recursive line can be expressed as

$$f(v) = \text{merge}(f(\text{split}_1(v)), f(\text{split}_2(v))) \qquad :|v| > 1$$

3.3.0.2 *Translation – Infix to Prefix*

As a second example consider the problem of translating from infix to prefix notation, given as input a parse tree of an algebraic expression consisting of binary operations only. For example, consider the parse tree for the expression $a * b + c * d$, shown in the upper part of Example 3.3.3. The terminal symbols, or leaves, represent variables and operations. To aid in the discussion, these are represented as single component vectors.

The parse tree analyzes the expression into basic infix parts, consisting of a variable or result of a previous computation, followed by an operation, followed by a variable or result of a previous computation, $\langle a_1\rangle\langle \text{op}\rangle\langle a_2\rangle$. These basic units appear as sequences of three leaves of the tree with a common parent, and at higher levels as groups of three nodes with a common parent.

We would like to do the translation by applying a simple infix-to-prefix translation to the basic units, composed of leaves of the parse tree, first, and then to units appearing higher in the tree. The basic unit is $\langle a_1\rangle\langle \text{op}\rangle\langle a_2\rangle$, to which we apply the function pre to get is prefix form, $\text{pre}(\langle a_1\rangle, \langle \text{op}\rangle, \langle a_2\rangle) = \langle \text{op}, a_1, a_2\rangle$. This is shown on a network attached to the bottom of the parse tree in Example 3.3.3.

Example 3.3.3

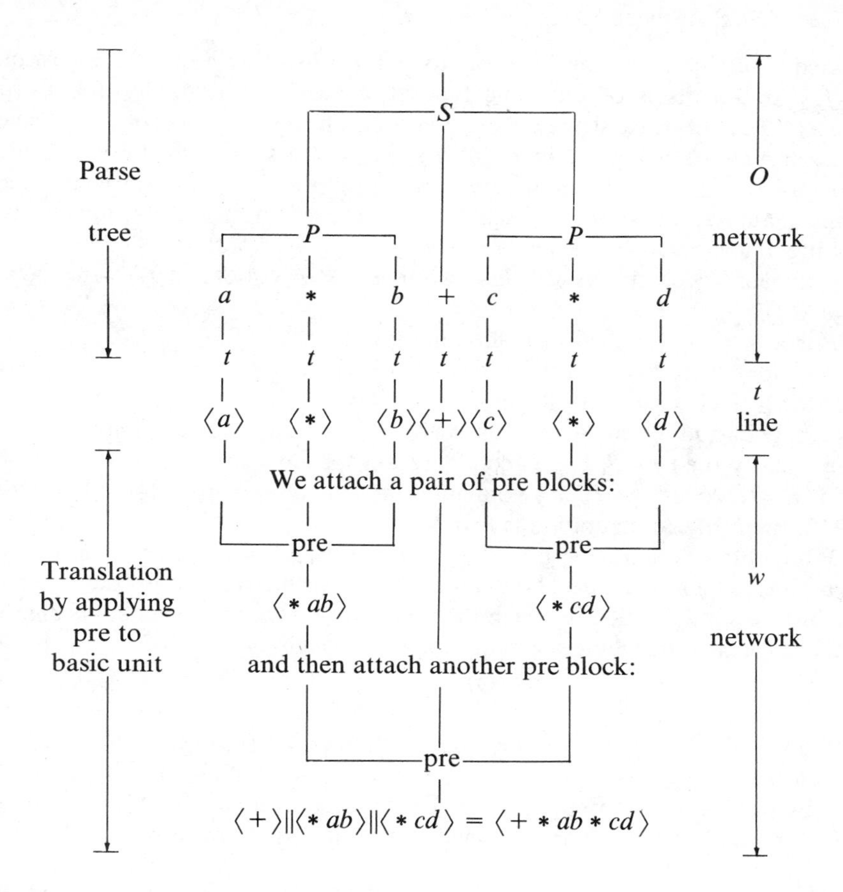

Notice that Example 3.3.3 is in fact a small model of an O–w–t network. The parts of the network are identified there. They are generalizable by identifying the pre function with the w function of that network. Let q be a node in the parse tree T, $d_i(q)$ the ith daughter of q in the parse tree, and info(q) the information at node q. (In the example parse tree $\langle a \rangle$, $\langle + \rangle$ are equal to info(q) at some of the nodes q.) Now the recursive definition is

Definition 3.3.1: Infix-to-Prefix Translation

$$\begin{cases} f(q,T) = \langle \text{info}(q) \rangle & :q \text{ a terminal in } T \\ f(q,T) = \text{pre}(f(d_2(q),T), \\ \qquad\qquad f(d_1(q),T), f(d_3(q),T)) & :q \text{ a nonterminal in } T \\ \text{Initially } T \text{ is the given parse tree, } q \text{ is its root node.} \end{cases}$$

3.3.1 Nested Recursive Definitions — String Matching Example [Baase 78, sec. 4.2; Sedgewick 83, chap. 19]

Nested recursion is feasible as a way to define a function f only if at least one of f's arguments is of the same type as f itself, for example, if both are integers. Functions on strings of symbols often have this property, with one or more arguments being the index of a symbol in a string and the result also being such an index. We now develop a nested definition to answer a simple string matching question; in Section 6.1, we give a nested definition, Early's parsing algorithm.

Consider two strings of m symbols, $P = \langle p_1, \ldots, p_n \rangle$ and $Q = \langle q_1, \ldots, q_m \rangle$.

For $k \leq m$, $\langle q_1, \ldots, q_k \rangle$ is called the *kth prefix of* Q.

Consider these two strings lined up, with Q below P, so that p_j is above q_k where $j \geq k$. Thus p_{j-i} is above q_{k-i} for $0 \leq i \leq k-1$ (see Figure 3-2). We say that P and Q lined up this way are in the (j, k) *configuration* (and equivalently the $(j-i, k-i)$ configuration for any $0 \leq i \leq k-1$).

If P, Q are in the (j, k) configuration and $p_{j-i} = q_{k-i}$ for all $0 \leq i \leq k-1$, then that configuration is *matched*.

We define a function $f(j, k, P, Q)$, or $f(j, k)$ for short since P and Q are invariant, which indicates how much Q must be shifted to the right to establish a configuration in which the prefix of Q whose last element is aligned with p_j forms a matching configuration at j. Precisely:

$$f(j, k) = r$$

where r is the largest integer such that configuration (j, r) is matched.

Another way to say this is that r is the largest integer such that sliding Q to the right at least 1 position (j, r) is the *first* matched configuration obtained. For example here is a P and Q lined up so that p_5 is above q_4

$$\begin{array}{ll} P = & 1\,2\,1\,2\,1\,1 \\ Q = & \;\;1\,2\,1\,2\,1\,1. \end{array}$$

Now if we move Q to the right at least one but only the fewest position necessary to get another match we have

$$\begin{array}{ll} P = & 1\,2\,1\,2\,1\,1 \\ Q = & \;\;\;\;1\,2\,1\,2\,1\,1. \end{array}$$

Now p_5 is above q_3 so $f(5, 4) = 3$.

$$\begin{array}{ll} P = & \langle p_1, \ldots, p_{j-k+1}, \ldots, p_j, \ldots, p_m \rangle \\ & \qquad\qquad \| \qquad\qquad\quad \| \\ Q = & \qquad\quad\; \langle q_1, \ldots, \qquad\; q_k, \ldots, q_m \rangle \end{array}$$

Figure 3-2 Matched configuration (j, k). If any of the equalities is invalid, then (j, k) is still a configuration, but is not matched.

For

$$f(j, k) = r$$

we must have the property that $(j - 1, r - 1)$ is a matched configuration. But $f(j - 1, k - 1)$ gives an upper bound on $r - 1$. Tentatively we assume that $r - 1$ is $f(j - 1, k - 1)$, and if $p_j = q_{f(j-1,k-1)+1}$, conclude that we are correct and that $r = f(j, k) = f(j - 1, k - 1) + 1$. This explains the first recursive line in Definition 3.3.2 in which $P = Q$.

If on the other hand

$$p_j \neq q_{f(j-1,k-1)+1}$$

then the upper bound on $r - 1$ of $f(j - 1, k - 1)$ is not its actual value, the next possible value will be determined by applying the reasoning of the above paragraph to finding $f(j, f(j - 1, k - 1) + 1)$. This accounts for the second recursive line in Definition 3.3.2.

The nonrecursive lines of Definition 3.3.2 are designed to insure that the definition of $f(j, k)$ works when $j \leq 2$. Let p_0 match every symbol. Together with the necessity for $f(2, x)$ to fit the definition, this determines the terminal lines.

Definition 3.3.2: String Self-Matching

$$\left[\begin{array}{ll} f(j,k) = -1 & :k = 0 \\ f(j,k) = 0 & :k = 1 \\ f(j,k) = f(j-1,k-1)+1 & :k > 1 \text{ and } p_j = p_{f(j-1,k-1)+1} \\ f(j,k) = f(j, f(j-1,k-1)+1) & :k > 1 \text{ and } p_j \neq p_{f(j-1,k-1)+1} \\ \text{Initially } j, k \in N,\ j \geq k. & \end{array}\right.$$

Notice that the predicates associated with the recursive lines of the definition include a reference to f. Though this is a usage not allowed in any of our standard forms, an equivalent definition can be given in standard form as follows:

$$\begin{aligned} \text{choice}(x, y, z, w) &= x \quad \text{if } z = w \\ &= y \quad \text{if } z \neq w \end{aligned}$$

and replace the two recursive lines with

$$f(j, k) = \text{choice}\big(f(j - 1, k - 1) + 1, f(j, f(j - 1, k - 1) + 1),$$

$$p_j, p_{f(j-1,k-1)+1}\big) \qquad :k > 1$$

$P = Q$	1	2	1	2	1	1
j	1	2	3	4	5	6

k						
6						1
5					3	
4				2	3	1
3			1	2	1	
2		0	1	1	1	1
1	0	0	0	0	0	0
j	1	2	3	4	5	6

FIGURE 3-3 Example of $f(j, k), 1 \leq j \geq k \leq 6$.

Consider the result of computing $f(j, k)$ for the example shown in Figure 3-3 with $1 \leq j, k \leq 6$. The details of the computation of $f(5, k)$, $1 \leq k \leq 5$, are shown below, assuming that $f(j, k)$, $1 \leq j, k \leq 4$ are as given. Notice that $f(5, k)$ ultimately depends only on $f(5, k - x)$ and $f(4, k - y)$ for $1 \leq x$, $y \leq k$. In general $f(j, k)$ depends only on $f(j, k - x)$ and $f(j - 1, k - y)$ for $1 \leq x$, $y \leq k$. Given these values $f(j, k)$ can be computed by the application of the appropriate line of the recursive definition, as determined by a simple test. Thus we need no more than k, j such evaluations, starting with $f(0, 0)$, to determine $f(j, k)$.

$$
\begin{aligned}
f(5,1) &= 0 \\
f(5,2) &= f(4,1) + 1 = 1 \quad \text{if} \quad p_5 = p_{1(=f(4,2)+1)} && \text{yes!} \\
f(5,3) &= f(4,2) + 1 = 2 \quad \text{if} \quad p_5 = p_2 && \text{no!} \\
&= f(5,2) \qquad = 1 && \text{yes!} \\
f(5,4) &= f(4,3) + 1 = 3 \quad \text{if} \quad p_5 = p_3 && \text{yes!} \\
f(5,5) &= f(4,4) + 1 = 3 \quad \text{if} \quad p_5 = p_3 && \text{yes!}
\end{aligned}
$$

As a further example we compute $f(6, 6)$, showing and computing all the $f(6, x)$ required; the required $f(5, y)$ have already been computed.

$$
\begin{aligned}
f(6,6) &= f(5,5) + 1 = 4 && \text{if } p_6 = p_4 \quad \text{no!} \\
&= \quad f(6,4) = f(5,3) + 1 = 2 && \text{if } p_6 = p_2 \quad \text{no!} \\
&\qquad\qquad = f(6,2) = f(5,1) + 1 = 1 && \text{if } p_6 = p_1 \quad \text{yes!}
\end{aligned}
$$

3.3.1.1 The Use of Matching Function f(j, k) in Pattern Matching

The function $f(j, k, P, Q)$ can be used in searching one string, say s, for the occurrence of a second (called a pattern), say p. At each instance a symbol of

$$
\begin{array}{llll}
s = & \langle s_1, \ldots, s_{j-k+1}, \ldots, s_{j-1}, s_j, \ldots, s_n \rangle \\
 & \qquad\qquad \| \qquad\qquad \| \\
p = & \qquad \langle p_1, \ldots, \qquad p_{k-1}, p_k, \ldots, p_m \rangle \\
 & \qquad\qquad\qquad\qquad \| \quad \|? \\
\text{slide } p \rightarrow & \qquad\qquad \langle p_1, \ldots, p_{r-1}, p_r, \ldots, p_m \rangle
\end{array}
$$

FIGURE 3-4 Pattern matching.

p is matched against a symbol of s. Assume the kth symbol of p is being matched against the jth symbol of s. This implies that $p_1, \ldots, p_{k-1}$ matches $s_{j-k+1}, \ldots, s_{j-1}$. so the section of s from $j - k + 1$ to $j - 1$ is the same as p from 1 to $k - 1$. We can thus think of the matched configuration $(j - 1, k - 1)$ existing between p and itself (Figure 3-4). If the match of p_k and s_j is successful, p is in the matched configuration (j, k) with itself, and we would next match the $(k + 1)$th symbol of p against the $(j + 1)$th symbol of s. If the match fails, then we want to find the next matched configuration (after $(j - 1, k - 1)$) of p with itself, $(j - 1, r - 1)$, with $r - 1 < k - 1$. But note that $r - 1$ is simply $f(j - 1, k - 1, p, p)$, so we know how to compute it. As long as $f(j - 1, k - 1, p, p) > 0$, we next test for equality of p_r and s_j. If, however, $f(j - 1, k - 1, p, p) = 0$, we need next to compare p_1 and s_{j+1} (the next input symbol). Finally note that $f(j, k, p, p)$ for all $j, k \leq m$, can be computed and tabled before any string matching is started, and this table can be used for matching the pattern p against any string.

3.3.2 Procedures, Functions Defining Action Sequences

So far we have assumed that defined functions and their arguments always have a value: This call-by-value assumption, although adequate for most of our needs here, results in very awkward expressions for some recursive definitions. We will briefly consider ways of removing the call-by-value constraint while still maintaining its spirit.

We can think of the *value* of a function as being an "action" or a sequence (vector) of "actions." An action can be an assignment, thus for example, the action $[x \leftarrow y]$ represents the assignment of y to variable x, similarly it could be an output statement like [output ← data(n)] which outputs the data at node n of a linked structure. Definition 3.3.3 gives an example of the use of such actions in formulating the familiar in order traversal of a binary tree T.

Definition 3.3.3: Recursive Definition Generating Actions—In-order Traversal

$$f(n, T) = \text{noop} \qquad :T = \text{nil}$$

$$f(n, T) = f(c.1(n), T) \| \langle [\text{output} \leftarrow \text{data}(n)] \rangle \| f(c.2(n), T) \qquad :T \neq \text{nil}$$

Initially T is a binary tree, each node x of which has a first (left) and second (right) child pointer, $c.1(x)$ and $c.2(x)$, and also a data field, data(x).

We use brackets to identify an action.

The simplest kind of an *action* definition is one in which a sequence of actions and nothing else is generated, and in which none of the actions affect the arguments of subsequent occurrences of f. A class of such definition is given in Definition 3.3.4.

Definition 3.3.4: Recursive Definition Generating Actions $\beta(X)$ and $\alpha_i(X)$ are actions performed on X

$$\left[\begin{array}{ll} f(X) = \beta(X) & :T(X) \\ f(X) = \|_{i=1 \text{ to } m(X)} \langle \alpha_i(X) \rangle \| f(o_i(X)) & :\sim T(X) \\ \text{Initially } X = d \in D. & \end{array}\right.$$

These definitions can be handled and interpreted just as we have handled other recursive (that is, call-by-value) definitions. However, an action can generate a result which may then affect the arguments of subsequent f calls. Quicksort provides an example of a recursive procedure which performs a sequence of actions on subarrays of an array of numbers A, with the arguments of the quicksort function specifying the range. These ranges, however, are themselves determined by the result of the actions performed by the algorithm. A quicksort definition using an extended action notation is given in definition 3.3.5.

Definition 3.3.5: Recursive Definition Generating Actions–Quicksort I

$$\left[\begin{array}{ll} q(A, i, j) = \langle [\text{noop}] \rangle & :i = j \\ q(A, i, j) = \langle [k \leftarrow \text{split}(A, i, j)] \rangle \| q(A, i, k-1) \| q(A, k+1, j) & :i < j \\ \text{Initially } A \text{ is a length } n \text{ array, } i = 1,\ j = n. & \end{array}\right.$$

The action $[k \leftarrow \text{split}(A, i, j)]$ results in rearranging the numbers in A in positions i through j so that there is a k, $i \le k \le j$, and $A(k - u) \le A(k) < A(k + v)$, $u, v > 0$, $k - u \ge i$, $k + v \le j$.

This example has properties that differ from those of the nonnested definitions considered previously. It provides a way of expressing an ordering on the computation of argument structures which do not otherwise exist for nonnested definitions. The value given to k by the primitive (with respect to this definition) function is used as an argument in the q calls on the right of the definition. This can alternatively be done in a form closer to the standard nonnested form.

Let the combined calculation of k and operation on the array A given by split(A, i, j) be separated so [split1(A, i, j)] rearranges A, and split2(A, i, j) is "compute and return the position k." Then we can express the quicksort in a standard nonnested form which generates a sequence of actions.

Definition 3.3.6: Recursive Definition Generating Actions—Quicksort II

$$\left[\begin{array}{ll} q(A, i, j) = \langle[\text{noop}]\rangle & :i = j \\ q(A, i, j) = \langle[\text{split1}(A, i, j)]\rangle \| q(A, i, \text{split2}(A, i, j)) & \\ \qquad\qquad \| q(A, \text{split2}(A, i, j), j) & \\ \text{Initially } A \text{ is a length } n \text{ array, } i = 1,\ j = n \in N & \end{array}\right.$$

3.4 DEFINITIONS OF SETS OF VECTORS — BACKTRACKING

The approaches to recursive function design studied so far balance their wide applicability with a lack of concreteness. For application to a smaller class of design problems more concrete advice is available. We consider such a smaller class now, the class of *backtrack set definitions* [Reingold, Nievergelt, Deo 77; Horowitz, Sahni 78].

3.4.1 Backtrack Set Definitions

To help describe what we mean by a backtrack set we give examples in which a set is defined in different ways, one of which is a backtrack set definition, while the others are near misses.

In the following pair of definitions of sets of permutations the first is not a backtrack set definition, the second is:

1. $\text{perm}(n) = \{\langle v_1, \ldots, v_n\rangle | v_j \in N^n \text{ and } v_j \neq v_i,\ i \neq j\}$
2. $\text{perm}(n) = \{\langle v_1, \ldots, v_k\rangle \mid \overbrace{v_j \in N^n \text{ and } v_j \neq v_{j-a}, 0 < a < j}^{P(v,\ j)} \text{ and }$ $k \text{ is smallest index} \ni \text{there is no } v_{k+1} \ni P(v, k+1)\}$

Both definitions yield all the permutations of the first n integers. They differ in two key properties:

1. In the first definition the length of the vector is specified as n in the expression $\langle v_1, \ldots, v_n\rangle$ on the *left* of $|$. In the second definition the vector length is a variable k whose length is determined by the constraint "k is the smallest index $\ni$ there is no $v_{k+1} \ni P(v, k+1)$" on the *right* of $|$.
2. In the first definition the constraint is given as a restriction on the value of each vector component (they must be drawn from N^n) and on each pair of vector components (they must be unequal). In the second definition the constraint on each component is given in terms of components of lower index.

Intuitively these two properties of the second definition will characterize backtrack set definitions.

Next consider three definitions for sets of combinations.

1. comb(n, m) = any maximal subset of the set

$$\{\langle v_1, \ldots, v_m\rangle | v_j \in N^n \text{ and } v_j \neq v_i,\ i \neq j \text{ and } j, i \in N^m\}$$

which does not contain both a vector and its permutation.

2. comb$(n, m) = \{\langle v_1, \ldots, v_m\rangle | v_j \in N^n$ and $v_j > v_{j-a}, 0 < a < j$ and $j, a \in N^m\}$

3. comb$(n, m) =$

$$\{\langle v_1, \ldots, v_k\rangle | \overbrace{v_j \in N^n \text{ and } n - m + j \geq v_j > v_{j-a}, 0 < a < j}^{P(v,\, j)} \text{ and } k \text{ is smallest index} \ni \text{there is no } v_{k+1} \ni P(v, k+1)\}$$

The first has neither of the properties required of a backtrack set definition. In the second the constraint on each component is given in terms of components of lower index, but it does not have the property that the vector length is a variable k whose length is determined by a constraint of the form: "k is the smallest index $\ni$ there is no $v_{k+1} \ni P(v, k+1)$." The third has both of these properties, and is a backtrack set definition.

The difference between the second and third combination definitions is not just notational. Using the third definition, we can construct a vector in comb(n, m) by using the constraints to the right of | to choose a v_1; then given v_1 to choose a v_2; and so on. Finally after computing $v_1, \ldots, v_k$ we will be unable to chose a v_{k+1} to satisfy the constraint. The vector $\langle v_1, \ldots, v_k\rangle$ thus constructed *is guaranteed* to be in comb(n, m). This is to be contrasted with a vector $v_1, \ldots, v_j$ similarly built using the constraints to the right of | in the second definition until there is no v_{j+1}. In this case the vector we produce *is not necessarily* in comb(n, m). Its length may be less than m and it will then be discarded. For example, using the second definition of comb(7, 4) in this way, we begin a vector construction with $< 2, 5, 7$ and another with $< 6, 7$, and in both instances have to discard these constructs.

The difference between the first and second definitions is also of interest. The requirement of no permutations in the first definition is avoided in the second by requiring that $v_j > v_{j-a}$, $0 < a < j$. Often we wish to exclude various reorderings of a vector in the defined set; generally this is difficult, but in this and similar cases it is nicely accomplished by ordering the vector components.

We now state the properties of a backtrack set definition more generally.

Assume we are defining the backtrack set $f(X)$, $X \in D$. To construct $f(X)$ we need a sequence of predicates, one for each integer $\in N$. The jth predicate in the sequence depends on X and gives the relation which must be satisfied by the j variables $v_1, \ldots, v_j$. It is designated $P[X]_j(v_1, \ldots, v_j)$ or, when less detail is adequate, $P[X]_j$, or even P_j. (This notation is used instead of $P_j(X, v_1, \ldots, v_j)$ because its dependence on X is usually different from and independent of its dependence on the v_i's.) Examples of predicates which could appear in such expressions are

1. For $X = (n, m)$, n and m are integers:

$$P[n, m]_j \equiv \text{“}v_j \in N^n \text{ and } v_j > v_{j-a}, 0 < a < j \text{ and } j, a \in N^m\text{”}$$

This one was used in defining combinations above.

2. For $X = B = \langle b_1, \ldots, b_n \rangle$, $b_j \in N$ and $b_i \neq b_j$ for $i \neq j$,

$$P[B]_j \equiv \text{“}v_j = b_{k_j} \text{ and } k_j > k_{j-1},\ j, k_j \in N^n\text{”}$$

This insures that if v_j is the kth member of B, all subsequent components, v_{j+a}'s of v, will be members of B with index greater than k. Thus two vectors which satisfy this constraint cannot be permutations of each other, and the constraint can be used in defining all combinations of the components of the vector B as follows:

$$\text{comb}(B) = \left\{ \langle v_1, \ldots, v_k \rangle \mid v_j = b_{k_j} \text{ and } k_j > k_{j-1},\ j, k_j \in N^n \right\}$$

To determine v_j it is necessary to first infer k_{j-1} from B and v_{j-1}.

A significant class of vector sets can be described in terms of such predicates. A backtrack set $g(X)$ is the set of all vectors $\langle v_1, \ldots, v_k \rangle$, each of which satisfies all the predicates $P[X]_1(v_1)$ through $P[X]_k(v_1, \ldots, v_k)$ and for which $P_{k+1}[X](v_1, \ldots, v_{k+1})$ cannot be satisfied with any $\langle v_1, \ldots, v_{k+1} \rangle$. More compactly we have

Definition 3.4.1: Backtrack Set Definition

$$g(X) = \left\{ \langle v_1, \ldots, v_k \rangle \mid P_1 \,\&\, P_2 \,\&\, \cdots \,\&\, P_k \text{ and } k \ni \text{for } j \leq k \{P_j\} \neq \{\ \} \right.$$
$$\left. \text{and is finite and } \{P_{k+1}\} = \{\ \} \right\}$$

The constraints on k starting with $k \ni$ are present in all definitions of this form. From now on when giving backtrack set definitions we will abbreviate this set of conditions with "end k." So in place of Definition 3.4.1 we will write

$$g(X) = \{ \langle v_1, \ldots, v_k \rangle \mid P_1 \,\&\, P_2 \,\&\, \cdots \,\&\, P_k \text{ and end } k \}$$

When constraints are given in this way, it is possible to select the vector components in the defined set one at a time. First v_1 may have any value that satisfies $P[X]_1(v_1)$, say a_1; then v_2 may have any value that satisfies the predicate $P[X]_2(a_1, v_2)$ provided such a value exists. If there is none, that is, if $\{P[X]_2(a_1, v_2)\}$ is empty, then the vector $\langle a_1 \rangle$ would be part of the defined set $g(X)$. In general if v_1 through v_j were chosen to be a_1 through a_j and $\{P[X]_{j+1}(a_1, \ldots, a_{j+1})\}$ is empty, then $\langle a_1, \ldots, a_j \rangle$ is in $g(X)$. The fact that the members of a backtrack set $g(X)$ can be generated "componentwise" implies that there is a simple recursive definition which enumerates that set.

Sometimes, although a set cannot be specified with a backtrack set definition, it can be formulated as a subset S' of a backtrack set S, where S' can be selected from S by application of *local selection function*. Such a set definition is called a *backtrack selection definition*. We next define a local selection function.

A *selection function* is one whose value is a collection of its arguments. If the arguments are integers, "all arguments greater than 5," or "the minimum of the arguments" is a selection function; if the arguments are sets, "sets containing 3" is, but "union" is *not*, a selection function. Selection function $\text{sel}(x_1, \ldots, x_n)$ is a *local selection function* if the determination of whether or not some argument x_i is selected depends only on its value, and not on the values of any other x_j. More precisely, there is a property Q of elements in the definition domain, (such that

$$x_i \in \text{sel}(x_1, \ldots, x_n) \text{ if } Q(x_i).$$

Thus "the minimum of the arguments" is not local while "all arguments greater than 5" and "all strings with a letter appearing more than once" are.

It follows that a selection function, sel, is local if and only if

$$\text{sel}(x_1, \ldots, x_n) = \text{sel}(x_1) \cup \cdots \cup \text{sel}(x_n)$$

So there are two parts to a backtrack selection definition.

1. The backtrack set definition, Definition 3.4.1.
2. The local selection function.

The use of the local selection function should be avoided if possible because it implies the generation of a larger set than actually needed, some of whose members are to be eliminated. The problem of designing a backtrack set definition which exactly matches a given set is considered in Section 3.4.4.

We now give some examples of backtrack selection definitions.

Example 3.4.1: Permutations of the Members of a Set $S = \{s_1, \ldots, s_n\}$

1. The set is

$$\text{perm}(S) = \{\langle v_1, \ldots, v_k \rangle | v_1 \in S \ \& \ v_2 \in S - \{v_1\} \ \& \cdots$$
$$\& \ v_k \in S - \{v_1, \ldots, v_{k-1}\} \text{ and end } k\}$$

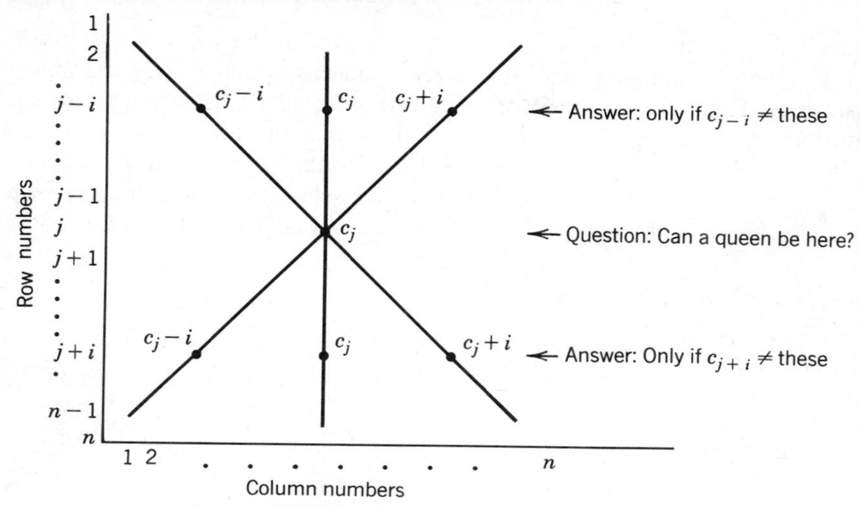

FIGURE 3-5 Constraints for the n-queens problem.

2. The selection function sel is

$$\text{sel}(\text{perm}(n, S)) = \text{perm}(n, S)$$

(In this case no selection function is necessary or, equivalently, the identity function can be used.)

As a second example consider the problem of determining all ways in which n queens can be placed on an $n \times n$ chessboard with no queen threatening another, that is, no two queens are on the same row, the same column, or the same diagonal. The constraints are illustrated in Figure 3-5, in which c_j refers to the column that the queen in row j occupies. There can be a queen at (j, c_j) provided there is none at $(j \pm i, c_j - i)$, $(j \pm i, c_j + i)$, and $(j \pm i, c_j)$ for all $1 \leq j \pm i \leq n$. It follows that

$$c_{j-i} \neq c_j - i \text{ nor } c_j \text{ nor } c_j + i$$

$$c_{j+i} \neq c_j - i \text{ nor } c_j \text{ nor } c_j + i$$

Solving each for c_j gives

$$c_j \neq c_{j \mp i} \mp i \text{ nor } c_{j \mp i} \text{ (where the } \mp \text{ are independent of each other)}$$

A formal definition of the configurations we want can now be given:

$$\text{queens}(n) = \{\langle c_1, \ldots, c_n \rangle \mid \text{for } c_j,\ j, i \in N^n;\ j - i > 0,\ j + i \leq n;$$

$$c_j \neq (c_{j \mp i} \text{ nor } c_{j \mp i} \mp i)\}$$

Ideally we would like a backtrack set definition in the form of Definition 3.4.1 to give all configurations of exactly n queens which satisfy the above constraints. There does not appear to be one. Instead we use a backtrack set definition of all configurations of queens in which the first k rows contain legitimately placed queens, but the $(k + 1')$th cannot be legitimately occupied, for any $k \leq n$. This set is then refined by a local selection of the members with n queens (there may be none). That definition, given as Example 3.4.2, follows.

Example 3.4.2: *n* Queens

1. The set

$$\text{queens}(n) = \{\langle c_1, \ldots, c_n \rangle | c_j,\ j, i \in N^n;\ j - i > 0,\ j + i \leq n;$$

$$c_j \neq (c_{j-i} \text{ nor } c_{j-i} \mp i)\}$$

2. The selection function sel

$$x \text{ is in sel(queens } (n)) \text{ if } x \text{ is in queens}(n) \text{ and } |x| = n$$

A somewhat smaller set than that defined by Example 3.4.2 could be enumerated by taking advantage of symmetries of the chess board, thus requiring the selection function to discard fewer members of the backtrack set. If c_1 is limited to values in $N^{n/2}$, for example, then for each pair of configurations of n queens, symmetric about a vertical line through the center of the board, only one will be produced.

3.4.2 Backtracking — Recursive Definition

From a backtrack selection set definition D of set S, a recursive definition DEF.f for generating S is easily constructed. The local selection function of D determines the terminal line, and the backtrack set definition of D the recursive line of DEF.f. The recursive line is given in terms of the primitives and predicates in Definition 3.4.2.

In the following, as in Theorem 2.6.1, we use cmps(v) to mean the list of components of the vector v when these are to constitute arguments of a function. For example, $g(u, \text{cmps}(v), z)$ is shorthand for $g(u, v_1, \ldots, v_n, z)$.

Theorem 3.4.1: If to the set of vectors $g(X)$ given by Definition 3.4.1 we apply the local selection function sel, then the result is equal to the set defined by $f(\langle\ \rangle, 1, X)$, where $f(v, j, X)$ is defined as follows.

Definition 3.4.2: Assume $|v| = j = 1$ and recall that with P a predicate, $\{P\}$ is the set that satisfies P

$$\left[\begin{array}{ll} f(v, j) = \text{sel}(v) & :\{P[X]_j(\text{cmps}(v), u)\} = \{\ \} \\ f(v, j) = \bigcup_{u \in \{P[X]_j(\text{cmps}(v), u)\}} [f(v \| \langle u \rangle, j+1)] & \\ & :\{P[X]_j(\text{cmps}(v), u)\} \neq \{\ \} \\ \text{Initially } v = \langle\ \rangle,\ j = 1,\ X \in \text{domain of } g(X) \text{ and invariant.} & \end{array}\right.$$

PROOF: This equivalence could be proven by the technique of ProofForm 2.5.4. Instead we sketch an alternative. A growth principle which generates a tree whose leaves are clearly the set $g(X)$ defined in Definition 3.4.1 will be developed. From the correspondence between a growth tree and a standard form definition it follows that Definition 3.4.2 gives that tree. The argument structure of the constructed tree contains a vector v and an index j. Initially, at level 1, j is 1 and v is empty. One child is generated for each member of $\{P[X]_1(u)\}$, with a different member of the set $\{P[X]_1(u)\}$ concatenated to v at each child. In general assume that at each child of level j a different member, u, of $\{P[X]_j$(cmps(v of the parent), $u\}$ is concatenated to the parent vector v. Assume also that j increases by 1 in going from parent to child. Such a tree would clearly produce the members of the set $g(X)$ defined by the backtrack set definition at its terminals. Termination is to occur when $j = k + 1$. Now the t line consists of sel blocks. With w being union, $g(X)$ is also produced by the entire network. We know the correspondence between an O–w–t network and a standard form recursive definition. That correspondence will produce the recursive definition of the theorem. ■

The theorem can be extended to cover a selection function sel which is not local. Instead of sel forming the terminal function as in Definition 3.4.2, make the terminal function equal to $\{v\}$ and replace the $\cup$ in the recursive line with sel.

Definition 3.4.2 of Theorem 3.4.1 can be simplified. Add a new argument C to f. Initially set C to $\{P[X]_1(u)\}$ and define the function ud (we will call it an *update* function) so that

$$\text{ud}(X, v, j, C) = \{P[X]_{j+1}(v_1, \ldots, v_j, u)\}$$

where $C = \{P[X]_j(v_1, \ldots, v_{j-1}, u)\}$. By straight substitution in Definition 3.4.2 we get the more compact equivalent form

Definition 3.4.3: Backtrack Selection with Update

$$\left[\begin{array}{ll} f(v, j, C) = \text{sel}(v) & :C = \{\ \} \\ f(v, j, C) = \bigcup_{u \in C} [f(v \| \langle u \rangle, j+1, \text{ud}(X, v, j, C))] & :C \neq \{\ \} \end{array}\right.$$

Initially $v = \langle\ \rangle$, $j = 1$, $C = \{P[X]_1(u)\}$, $X \in$ domain of $g(X)$ and invariant.

The use of an updating function ud does not always simplify the evaluation, but it will if updating C is less complex than computing $\{P[X]_j(v_1, \ldots, v_{j-1}, u)\}$.

As an example let us apply Theorem 3.4.1, using Definition 3.4.3, to the definition for permutations in Example 3.4.1. The recursive line of the resultant definition is:

$$f(v, j, C) = \bigcup_{u \in C} [f(v \| \langle u \rangle, j+1, \text{ud}(S, v, j, C))] \qquad :C \neq \{\ \}$$

where $\text{ud}(S, v, j, C) = S - v_1 - \cdots - v_j - u$.

The update function $\text{ud}(X, v, j, C)$ can be simplified. Initially $C = S$, so in the children of the initial argument structure can be given as follows:

$$C \text{ at child} = \text{ud}(S, v, j, C) = S - u = C \text{ at parent } - u$$

and in general by induction,

$$\text{ud}(S, v, j, C) = C - u$$

The result is given in Definition 3.4.4.

Definition 3.4.4: Permutations with Update Improvement

$$\text{perm}(S) = f(\langle\ \rangle, 1, S)$$

where $f(v, j, C)$ is given by

$$\left[\begin{array}{ll} f(v, j, C) = \{v\} & :C = \{\ \} \\ f(v, j, C) = \bigcup_{u \in C} [f(v \| \langle u \rangle, j+1, C - u)] & :C \neq \{\ \} \end{array}\right.$$

Initially $v = \langle\ \rangle$, $j = 1$, $C = S$, S is a finite set and invariant.

Such simplified updating functions are often possible. The idea of updating arguments as an alternative to computations in a definition is widely applicable. It will be considered in a more general setting in section 3.6.

As another example, we apply Definition 3.4.3 to the n-queen backtrack selection definition and get

Definition 3.4.5: n Queens

$$\text{queens}(n) = f(\langle\ \rangle, 1, N^n)$$

where $f(c, j, C)$ has the following definition. (The constraint c_j, i, $j \in N^n$ and $i < j$ holds throughout the definition.)

$$\left[\begin{array}{ll} f(c, j, C) = \{c\} \text{ if } |c| = n & :C = \{\ \} \\ \quad\quad\quad\ \ = \{\ \} \text{ otherwise} & \\ f(c, j, C) = \bigcup_{u \in C} \Big[f\big(c \| \langle u \rangle, j+1, & \\ \quad \big\{ v \in N^n | v \neq \big(c_{j+1-i} \text{ nor } c_{j+1-i} \mp i\big), i < j+1 \big\}\big)\Big] & :C \neq \{\ \} \\ \text{Initially } c = \langle\ \rangle,\ j = 1,\ C = N^n,\ n \in N \text{ and invariant.} & \end{array}\right.$$

3.4.3 Unordered Backtrack Set Definitions [Wadel 86]

In the backtrack set definition for a set of vectors S, the components of the members of S are chosen in a fixed order: v_1 is chosen, then v_2 is chosen constrained by the chosen v_1, then v_3, and so on. In the *unordered backtrack set definition* the order in which these components are selected is not fixed in the definition; that is, the definition allows the choice of any vector component, for example, v_3 first, then constrained by that choice any vector component next, for example, v_1 constrained by the choice of v_3.

Suppose we have a set of vectors, S, for which every backtrack set definition overproduces vectors, and it is therefore necessary to incorporate a selection function. This holds for n-queens, for example. For such a set S it may be possible to take a different approach to limiting overproduction of vectors by adusting the order in which components are added to the growing vectors in S. To do this, however we first need a form of definition similar to a backtrack set definition, but differing in that it is permissive as to the order in which vector components are to be added. In implementing such an "unordered" backtrack set definition, D, we need to chose a definite order to add components. This order can be expressed in the recursive definition that implements D, and further can be chosen to depend on information gathered during generation of vectors and to minimize their overproduction.

In general, in an unordered backtrack set definition, the length of vectors is fixed at n. Component values are chosen one at a time in any order. The value chosen is constrained by the values of the previously chosen components. For each of the n chosen components, each in the range 1 to n, the set of legitimate values, based on already chosen components, must not be empty.

The predicate $Q[X]_j$ is the condition that vector component v_j must satisfy in relation to other components. It must do so in such a way that no matter in what order $v_1, v_2, \ldots, v_n$ are chosen, each obeying the corresponding predicate $Q[X]_1, Q[X]_2, \ldots, Q[X]_n$, when the selection function is applied to the result, the set produced is the same.

Definition 3.4.7: Unordered Backtrack Set Definition

$$g(X, n) = \{\langle v_1, \ldots, v_n \rangle | Q[X]_1 \,\&\, Q[X]_2 \,\&\, \cdots$$
$$\&\, Q[X]_n,\ i_j \in N,\ i_j \neq i_k,\ \{Q[X]_j\} \neq \{\ \} \text{ and is finite}\}$$

In analogy to the ordered case, an unordered backtrack set definition D, together with a local selection function, gives an unordered backtrack selection definition.

Recursive Definition 3.4.8 enumerates an unordered backtrack selection set. It differs from the one for the ordered case only in that it must "insert" rather than concatenate new components, and it incorporates a way of ordering the choice of vector components. The index of the next component to be chosen depends on the components in the vector v already selected and is given by $h(v)$.

Definition 3.4.8: Unordered Backtrack Selection Set—Recursive Definition
0 here represents a value different from any legitimate value of a vector component in the defined set. $h(v)$ is a function from a vector with n components (including 0) to an integer $\leq n$.

$$f(v, j) = \text{sel}(v) \qquad : j = n + 1$$
$$f(v, j) = \bigcup_{z \in \{Q[X]_{h(v)}\}} [f(v_{i_j} \leftarrow z, j + 1] \qquad : \{Q[X]_{h(v)}\} \neq \{\ \}$$

Initially $v = 0^n$, $j = 1$, $C = \{Q[X]_{i_1}\}$, $X \in$ domain of $g(X, n)$ and n are invariant.

n queens can be defined using the form of Definition 3.4.7, together with a selection function. Assume the components v_{i_k}, $1 \leq k < j$ have been selected then the predicate $Q[X]_{i_j}$ which expressed the constraints on choosing v_{i_j} is given as follows:

v_{i_j} is chosen as an integer $\leq n$ such that the position $\langle j, v_{i_j} \rangle$ is not in the same column, nor on the same diagonal as $\langle k, v_{i_k} \rangle$, $l \leq k < j$. If however there is no such integer then v_{i_j} = "#", where # is a special mark indicating this impossibility. Note that also v_{i_j} = # whenever for $1 \leq k < j$, v_{i_k} = #.

No matter what the order of component choice, the vectors of length n with no # in them will be the same using this predicate in Definition 3.4.7. If the selection function

$$\text{sel}(v) = \{v\} \text{ if and only if } v \text{ contains no } \#, \{\ \} \text{ otherwise}$$

is added to get a backtrack selection definition, the result will give the same set, independent of the order of component selection. But we can do better. By adding the line

$$f(v, j) = \{\ \} \qquad :\{Q[X]_{h(v)}\} = \{\#\} \tag{3.4.1}$$

we will avoid developing a vector beyond the point at which it is first determined that the desired conditions cannot be satisfied.

The function $h(v)$, necessary for the corresponding recursive Definition 3.4.8, can now be designed for an efficient enumeration of potential board arrangement vectors. This can be done by designing $h(v)$ so that Eq. (3.4.1) added to Definition 3.4.8 applies after a few components as possible have been chosen. That will limit the number of vectors produced which will eventually fail. A heuristic for $h(v)$ is: "choose from all unassigned components the one for which the set $\{Q[X]_{h(v)}\}$ has the smallest size." The resultant recursive definition is given in more detail using an appropriate update argument structure in Definition 3.6.9.

3.4.4 Formulating Constraints in Backtracking Order

The backtrack set definition, Definition 3.4.1, is a special case of the following, in which V_j is a subset of the vector $\langle v_1, \ldots, v_k \rangle$.

Definition 3.4.9

$$g(X) = \{\langle v_1, \ldots, v_k \rangle | P[X]_1(k, V_1) \,\&\, P[X]_2(k, V_2) \,\&\, \cdots$$
$$\&\, P[X]_p(k, V_p)\}$$

Here the constraints on the components as well as on the length of the vectors in $g(X)$ are given by p different predicates. A definition D in the form of Definition 3.4.9 may very well not be a backtrack set definition. Thus it may be difficult to find a recursive definition for D which does not overproduce vectors only to later reject them. If a component v_j in D is constrained by a number of different predicates, a value consistent with one predicate may violate another. In a backtrack set definition this cannot happen as long as the predicates are applied in order—P_1, P_2, and so on. If the sets defined by the predicates are finite, it is always possible to determine component values that will not be rejected later. Nevertheless, since the predicates are functions of a variable X, and the time to determine permanent, incontrovertible component values is an important consideration, useful algorithms for choosing acceptable component values may be difficult to find. We will therefore consider the problem of transforming a definition of the form of Definition 3.4.9 into a backtrack set definition.

Definition 3.4.10, is a subcase of Definition 3.4.9, for which we can often obtain an equivalent backtrack set definition. The constraints, or predicates, are linear inequalities, and the components are positive integers. (Note that although vector length is fixed in this definition, it is still possible to effectively define sets containing vectors of more than one length. This is done by discarding marked components or, as illustrated at the end of this section, by combining a number of backtrack set definitions, each designed for a different value of k.)

Definition 3.4.10: $\{v_{i1}, \ldots, v_{in_i}\}$ represents a subset of V. Let $\mathrm{Lin}[X]_i \equiv a_{i1}(X)v_{i1} + \cdots + a_{in_i}(X)v_{in_i} \geq b_i(X)$ and $b_i(X) \in N$,

$$g(X) = \{\langle v_1, \ldots, v_k \rangle \mid \mathrm{Lin}[X]_1 \ \& \ \mathrm{Lin}[X]_2 \ \& \cdots \& \ \mathrm{Lin}[X]_p \text{ and } v_i \in N_0 \text{ and } k = r(X)\}$$

Such a definition can still be quite far from a backtrack set definition. Given such a definition, one could choose v_1 through v_{j-1} so as not to violate the linear constraint, assuming the remaining v_j's are 0, and then find that with any choice of v_j no choice is possible for v_{j+1}. If $j + 1 < n$, then all that effort was wasted. This can be prevented if the constraints are combined and reformulated as a backtrack set definition, which when implemented by recursive Definition 3.4.2, will not overgenerate vectors.

For a definition of the form of Definition 3.4.10 we need a way to find an equivalent in the form of Definition 3.4.1. This can be done by "solving" the constraints given in the first form, where the solution is a set of the constraints in the second form. The solution technique is modeled on the Gaussian elimination approach for a set of inequalities. It will be explained by an example. Consider the following definition, in the form of Definition 3.4.10, for all partitions of the integer n into p parts.

Definition 3.4.11

$$\mathrm{part}(p, n) = \{\langle v_1, \ldots, v_p \rangle \mid n = v_1 + \cdots + v_p \text{ and } n \in N \text{ and } v_i \in N \text{ for } i \in N^p\}$$

The example, given in detail in Figure 3-6, illustrates how linear inequality constraints can be "solved" by Gaussian elimination. Although the solution technique is the same as the one developed to solve any set of linear inequalities, it is here given a somewhat more difficult job. We need to obtain a *general integer* solution to a set of linear equalities. The process solves for one variable at a time, in two steps. First the variable to be solved for is *separated*, then it is *eliminated*. A general solution is obtained by doing these

j	Constraints			
	$n = v_1 + \cdots + v_p$			and $v_i \geq 1,\ i \in N^p$
p	$n - (v_1 + \cdots + v_{p-1})$	$= v_p$	≥ 1	and $v_i \geq 1,\ i \in N^{p-1}$
	$n - (v_1 + \cdots + v_{p-1})$		≥ 1	and $v_i \geq 1,\ i \in N^{p-1}$
$p - 1$	$n - 1 - (v_1 + \cdots + v_{p-2})$	$\geq v_{p-1}$	≥ 1	and $v_i \geq 1,\ i \in N^{p-2}$
	$n - 1 - (v_1 + \cdots + v_{p-2})$		≥ 1	and $v_i \geq 1,\ i \in N^{p-2}$
$p - 2$	$n - 2 - (v_1 + \cdots + v_{p-3})$	$\geq v_{p-2}$	≥ 1	and $v_i \geq 1,\ i \in N^{p-3}$
		$\vdots$		
2	$n - (p - 2) - v_1$	$\geq v_2$	≥ 1	and $v_i \geq 1,\ i \in N^1$
	$n - p + 2 - v_1$		≥ 1	
1	$n - p + 1$	$\geq v_1$	≥ 1	
	$n - p + 1$		≥ 1	
	n	$\geq p$		

FIGURE 3-6 Solution of inequalities for permutations of partitions.

steps symbolically. Linearity keeps this process simple. The integer constraint is assured because whenever a separated variable lies at or between two values, at least one of the values is an integer. The reasoning used in the separation and elimination steps is illustrated by giving details on their use in eliminating the first variable of Definition 3.4.11.

One of the starting constraints is

$$n = v_1 + \cdots + v_p$$

which is true if and only if $n - (v_1 + \cdots + v_{p-1}) = v_p$. Thus the first variable to be removed, v_p, has been separated from the other variables in this constraint. The remainder of the initial constraint is

$$v_i \geq 1,\ i \in N^p$$

which is true if and only if $v_p \geq 1$ and $v_i \geq 1,\ i \in N^{p-1}$. The separation step for v_p is complete. In summary we have

$$n - (v_1 + \cdots + v_{p-1}) = v_p \geq 1 \text{ and } v_i \geq 1,\ i \in N^{p-1}$$

The separation step is followed by the elimination step.

We can certainly find a v_p that satisfies this constraint if the following constraint, from which v_p has been eliminated, can be satisfied:

$$n - (v_1 + \cdots + v_{p-1}) \geq 1 \text{ and } v_i \geq 1,\ i \in N^{p-1}$$

because if this is satisfied, v_p need only be set equal to

$$n - (v_1 + \cdots + v_{p-1})$$

The separation–elimination process applied to all variables is shown in Figure 3-6. The lines displaying the separated variables are numbered by the index of that variable in the vector of which it is a component. Each numbered line is followed by a line giving the result of eliminating that variable.

The first line is simply the constraint from the original set definition. The entire sequence of lines is constructed to satisfy the following logical connections.

1. The constraints given on an unnumbered line are equivalent to those on the following line.
2. The constraints on a numbered line can be satisfied if the unnumbered line following it can be satisfied.

If the numbered lines in Figure 3-6, excluding in each everything to the right of "and," are read from the bottom up, they give the predicates P_1, P_2 necessary for a backtrack set definition (Definition 3.4.1). This definition is equivalent to the original constraint in Definition 3.4.11. Thus Theorem 3.4.1 applies and the following recursive definition enumerates the set given by Definition 3.4.11.

Definition 3.4.12: Permutations of Partitions of n in p Parts

$$\text{part}(p, n) = f(\langle\ \rangle, 1)$$

where $f(v, j)$ is given by

$$\begin{array}{ll} 1. & f(v, j) = \{v\} \qquad :j > p \\ 2. & f(v, j) = f(v \| \langle n - (v_1 + \cdots + v_{j-1})\rangle, j + 1) \qquad :j = p \\ 3. & f(v, j) = \bigcup_{n-p+j-(v_1+\cdots+v_{j-1}) \geq u \geq 1} [f(v\|\langle u\rangle, j+1)] \qquad :j < p \\ & \text{Initially } v = \langle\ \rangle,\ j = 1,\ n,\ p \in N. \end{array}$$

Lines 2 and 3 are equivalent to the recursive line of Definition 3.4.2. Lines 1 and 2 can be combined and replaced with the terminal line

$$f(v, j) = \{v \| \langle n - (v_1 + \cdots + v_{j-1})\rangle\} \qquad :j = p$$

We can alternatively base the definition on a slightly modified version of Definition 3.4.3. For this purpose let $C = \{1, \ldots, h\}$, where $h = n$ initially, and by convention $C = \{\ \}$ means $h = 0$. Instead of using C as the update argument, we use h; that is, C is a set of integers whose highest value is h, so h is used as the argument from which C can be obtained easily. Terminal and nonterminal conditions which depend on whether or not C is empty are replaced by conditions on j and p because

$$C \neq \{\ \} \quad \text{if } j + 1 < p$$
$$C = \{\ \} \quad \text{if } j + 1 = p$$

Now an update function $\text{ud}(p, n, v, j, h)$, to be applied to h rather than $C = \{1, \ldots, h\}$ is required. We want $h = n - p + j - (v_1 + \cdots + v_{j-1})$. Thus,

1. If $j < p$, since we want $\text{ud}(p, n, v, j, h)$ to give the updated value of h, that is,

$$\text{ud}(p, n, v, j, h) = n - p + j + 1 - (v_1 + \cdots + v_{j-1} + v_j)$$

therefore

$$\text{ud}(p, n, v, j, h) = h + 1 - v_j$$

2. If $j = p$,

$$\text{ud}(p, n, v, j, h) = 0$$

because the call generated on the right must have $C = \{\ \}$. Thus we get

Definition 3.4.13: Permutations of Partitions of n in p Parts

$$\text{part}(p, n) = f(\langle\ \rangle, 1)$$

where $f(v, j)$ is given by

$$\begin{array}{ll} 1. & f(v, j, h) = \{v\} \qquad :j > p \\ 2. & f(v, j, h) = \bigcup_{u=h} [f(v \| \langle u \rangle, j+1, h+1-v_j)] \qquad :j = p \\ 3. & f(v, j, h) = \bigcup_{h \geq u \geq 1} [f(v \| \langle u \rangle, j+1, h+1-v_j)] \qquad :j < p \\ & \text{Initially } v = \langle\ \rangle,\ j = 1,\ h = n,\ n, p \in N \text{ and invariant.} \end{array}$$

Also, as previously, lines 1 and 2 can be replaced with

$$f(v, j, h) = \{v \| \langle h + 1 - v_j \rangle\} \qquad :j = p$$

This definition gives every permutation of every set of p integers which sum to n, rather than all combinations (in nondecreasing size order) of any number of integers that add to n as given by Definition 3.2.1. Our Gaussian elimination technique for getting backtrack definitions can also derive such a definition of partitions.

This is done in two steps. First get all combinations of p integers which sum to n, each represented by a vector of length p with members in nondecreasing order. This set must satisfy the constraint

$$n = v_1 + \cdots + v_p \text{ and } v_i \geq v_{i-1} \qquad , i \in N_2^p \text{ and } v_1 \geq 1$$

j	Constraints			
	$n = v_1 + \cdots + v_p$			and $v_i \geq v_{i-1}: N_2^p$ and $v_1 \geq 1$
p	$n - (v_1 + \cdots + v_{p-1})$	$= v_p$	$\geq v_{p-1}$	and $v_i \geq v_{i-1}: N_2^{p-1}$ and $v_1 \geq 1$
	$n - (v_1 + \cdots + v_{p-1})$		$\geq v_{p-1}$	and $v_i \geq v_{i-1}: N_2^{p-1}$ and $v_1 \geq 1$
	$n - (v_1 + \cdots + v_{p-2})$	$\geq 2v_{p-1}$		and $v_i \geq v_{i-1}: N_2^{p-1}$ and $v_1 \geq 1$
$p-1$	$n/2 - (v_1 + \cdots + v_{p-2})/2$	$\geq v_{p-1}$	$\geq v_{p-2}$	and $v_i \geq v_{i-1}: N_2^{p-2}$ and $v_1 \geq 1$
	$n/2 - (v_1 + \cdots + v_{p-2})/2$		$\geq v_{p-2}$	and $v_i \geq v_{i-1}: N_2^{p-2}$ and $v_1 \geq 1$
	$n/2 - (v_1 + \cdots + v_{p-3})/2$	$\geq \frac{3}{2}v_{p-2}$		and $v_i \geq v_{i-1}: N_2^{p-2}$ and $v_1 \geq 1$
$p-2$	$n/3 - (v_1 + \cdots + v_{p-3})/3$	$\geq v_{p-2}$	$\geq v_{p-3}$	and $v_i \geq v_{i-1}: N_2^{p-3}$ and $v_1 \geq 1$
	$n/3 - (v_1 + \cdots + v_{p-3})/3$		$\geq v_{p-3}$	and $v_i \geq v_{i-1}: N_2^{p-3}$ and $v_1 \geq 1$
			$\vdots$	
2	$n/(p-1) - v_1/(p-1)$	$\geq v_2$	$\geq v_1$	and $v_1 \geq v_{i-1}, v_1 \geq 1$
	$n/(p-1) - v_1/(p-1)$		$\geq v_1$	≥ 1
1	n/p	$\geq v_1$	≥ 1	
	n/p		≥ 1	
	n		$\geq p$	

FIGURE 3-7 Solution of inequalities for partitions as combinations.

This is now converted to backtrack set form by a Gaussian elimination technique, Figure 3-7.

The recursive function definition corresponding to the constraints in Figure 3-7 is obtained by Theorem 3.4.1. It is

Definition 3.4.14: Combinations of Partitions of n in p Parts

$$\text{part}(p, n) = f(\langle\ \rangle, 1)$$

where $f(v, j)$ is given by

$$\left[\begin{array}{ll} f(v, j) = \{v \| \langle n - (v_1 + \cdots + v_{j-1})\rangle\} & : j = p \\ f(v, j) = \bigcup_{\frac{n-(v_1+\cdots+v_{j-1})}{p-j+1} \geq v_j \geq (1, v_{j-1})} [f(v \| \langle v_j \rangle, j+1)] & : j < p \\ \text{Initially } v = \langle\ \rangle,\ j = 1,\ n, p \in N \text{ and invariant.} & \end{array}\right.$$

An update version of this definition can be developed similarly to our development of Definition 3.4.13.

3.4.4.1 Different Vector Lengths

The solution for partitioning n into p parts can be used to get the recursive definition for all partitions of n into *any number of parts, from* 1 *to* n, with members in nondecreasing order. We need only let p take all values in its range between 1 and n at each constraint, that is, all values which allow the predicate to be satisfied at all.

In general the constraints specify an upper value that v_j can take. That constraint is

$$\frac{n-(v_1+\cdots+v_{j-1})}{p-j+1} \geq v_j \geq 1 \qquad \text{when } p > j$$

When $p = j$, the first $\geq$ becomes $=$.

Now we have that, given a value of j, p can take any value from j to n. For $p = j$ the constraint to be satisfied is

$$n-(v_1+\cdots+v_{j-1}) = v_j \equiv w(j-1)$$

When this constraint is satisfied, v_j becomes the last component of a terminal vector. The vector $v \| w(j-1)$ thus produced when $p = j$, is given as a separate term in the definition.

For $p > j$, the constraint is

$$\frac{n-(v_1+\cdots+v_{j-1})}{p-j+1} \geq v_j \geq 1 \qquad \text{for all } p > j$$

Note that $p = j+1$ will allow all the v_j's that any other value of p will allow. Therefore the constraint becomes

$$\frac{n-(v_1+\cdots+v_{j-1})}{2} \geq v_j \geq 1 \qquad \text{for all } p \geq j$$

The result of these changes is a definition which no longer depends on p and incorporates the new constraints.

Definition 3.4.15: Combinations of All Partitions of n

$$\text{part}(n) = f(\langle\ \rangle, 1)$$

where $f(v, j)$ is given by

$$f(v, j) = \bigcup_{\frac{n-(v_1+\cdots+v_{j-1})}{2} \geq u \geq (1, v_{j-1})} [f(v\|\langle u\rangle, j+1)]$$

$$\cup \{v\|\langle w(j-1)\rangle\} \qquad : j < n$$

Initially $v = \langle\ \rangle$, $j = 1$, $n \in N$ and invariant.

As for the other partition definitions, an update version of this definition is easily developed.

3.5 STRONGLY EQUIVALENT TRANSFORMATIONS: ARGUMENT STRUCTURE CHANGES WHICH LEAVE THE DEFINITION UNCHANGED

Having obtained a feasible recursive definition, we will want to improve it, that is, convert it into one which, though *equivalent* that is, defines the same function, has a more efficient implementing algorithm. Many approaches to improvement are considered in this book. The next two sections survey some of these.

Before looking at changes which improve a definition, we first consider changes in the argument structure of a terminating standard form nonnested definition, which are guaranteed to leave the defined function unchanged. They may or may not improve the definition, but are necessarily the kind we must consider for improvement.

For convenience the standard nonnested form is given once again.

Definition 3.5.1

$$\left[\begin{array}{ll} f(X) = t(X) & :T(X) \\ f(X) = w\big(f(o_1(X)),\ldots, f(o_{m(X)}(X))\big) & :\sim T(X) \\ \text{Initially } X \in D. & \end{array}\right.$$

In general changing the domain of function DEF.f to any set which has a one-to-one correspondence with Δ_f will give an *equivalent* definition. Let the equivalent definition be DEF.h, with substitutional domain Δ_h having a one-to-one correspondence with Δ_f. Let the correspondence be given by the function

$$r\colon \Delta_f \to \Delta_h$$

Because of the one-to-one correspondence, r has an inverse, r^{-1}.

Our goal is to have $h(r(X)) = f(X)$. We therefore define the predicate T', the terminal value t', the primitive function giving the number of h calls, m', and the O functions o_i' for DEF.h so that $T'(r(X)) = T(X)$, $t'(r(X)) = t(X)$, and $m'(r(X)) = m(X)$. Also $o_i'(r(X))$ should correspond under r to $o_i(X)$, and therefore $o_i'(r(X)) = r(o_i(X))$. In general the primed functions of DEF.h are each given in terms of r and the corresponding unprimed functions.

Definition 3.5.2: Strongly Equivalent Transformation. If $Y \in \Delta_h$,

$$T'(Y) = T\big(r^{-1}(Y)\big)$$

$$t'(Y) = t\big(r^{-1}(Y)\big)$$

$$m'(Y) = m\big(r^{-1}(Y)\big)$$

$$o_i'(Y) = r\big(o_i\big(r^{-1}(Y)\big)\big)$$

and finally if X is in the starting domain of DEF. f, then $r(X)$ is the starting domain D' of DEF. h.

All the parts of DEF. h have been assembled. Here is DEF. h.

Definition 3.5.3

$$\left[\begin{array}{ll} h(Y) = t'(Y) & :T'(Y) \\ h(Y) = w\big(h(o'_1(Y)), \ldots, h\big(o'_{m'(Y)}(Y)\big)\big) & :\sim T'(Y) \\ \text{Initially } Y \in D'. & \end{array}\right.$$

With all these conditions met, we can conclude that

Theorem 3.5.1: With reference to Definitions 3.5.1 and 3.5.3, for any X in D, if $Y = r(X)$, then $f(X) = h(Y) = h(r(X))$.

PROOF: By satisfaction:

1. If $T(X)$,

$f(X) = t(X)$ from Definition 3.5.1

$h(r(X)) =?\, t(X)$ by substitution of $h(r(X))$ for $f(X)$ throughout

$h(r(X)) =?\, t'(r(X))$ by definition of t'

$h(Y) =?\, t'(Y)$ by substituting Y for $r(X)$

$h(Y) =!\, t'(Y)$ from Definition 3.5.3

Also if $T(X)$, then $T'(r(X)) = T'(Y)$ by definition of T'.

2. If $\sim T(X)$,

$f(x) = w\big(f(o_1(X)), \ldots, f\big(o_{m(X)}(X)\big)\big)$ from Definition 3.5.1

$h(r(X)) =?\, w\big(h(r(o_1(X))), \ldots, h\big(r\big(o_{m(X)}(X)\big)\big)\big)$

by substitution of $h(r(X))$ for $f(X)$ throughout

$h(r(X)) =?\, w\big(h(o'_1(r(X))), \ldots, h\big(o'_{m'(r(X))}(r(X))\big)\big)$

by definition of o'_1 & m'

$h(Y) =?\, w\big(h(o'_1(Y)), \ldots, h\big(o'_{m'(Y)}(Y)\big)\big)$

by substitution of Y for $r(X)$

$h(Y) =!\, w\big(h(o'_1(Y)), \ldots, h\big(o'_{m(X)}(Y)\big)\big)$ by Definition 3.5.3

Also if $\sim T(X)$, then $\sim T'(r(X)) = \sim T'(Y)$ by definition of $\sim T'$.

■

Not only does $f(X) = h(r(X))$, but also the structures of DEF.f and DEF.h are similar. More precisely, they are related by the properties in Definition 3.5.2. Two definitions related in this way are called *strongly equivalent*.

The change from Definition 2.33 to Definition 2.3.4 is an example of a transformation based on a one-to-one correspondence of substitutional domains. Remember that, though it is not shown because it is invariant, the argument structure of f in Definition 2.3.3 technically has a second argument n. The mapping function for this transformation is

$$r(v, n) = (v, n - k)$$

and the inverse is

$$r^{-1}(v, p) = (v, p + |v|)$$

From Definition 3.5.2 we compute the primitives of Definition 2.3.4 from those of Definition 2.3.3.

$$T'(v, p) = T(r^{-1}(v, p)) = T(v, p + |v|) = (|v| = p + |v|) = p(= 0)$$

$$t'(v, p) = t(r^{-1}(v, p)) = t(v, p + |v|) = \{v\}$$

$$m'(v, p) = m(r^{-1}(v, p)) = m(v, p + |v|) = 2$$

$$\begin{aligned} o_1'(v, p) &= r(o_1(r^{-1}(v, p))) \\ &= r(o_1(v, p + |v|)) \\ &= r(v \| \langle 0 \rangle, p + |v|) \\ &= (v \| \langle 0 \rangle, p + |v| - |v \| \langle 0 \rangle|) \\ &= (v \| \langle 0 \rangle, p + |v| - |v| - 1) \\ &= (v \| \langle 0 \rangle, p - 1) \end{aligned}$$

Similarly,

$$o_2'(v, p) = (v \| \langle 1 \rangle, p - 1)$$

Let the argument structure of recursive definition DEF.f be $X = \langle x_1, \ldots, x_n \rangle$. A general and frequently useful transformation in such an argument structure is to add a vector, say, $Y = \langle y_1, \ldots, y_m \rangle$, to X, where Y is a function of X, that is, $Y = c(X)$. The transformation is

$$r(X) = X \| c(X).$$

$r(X)$ is one to one. It has an inverse for $Z \in$ range of r, $r^{-1}(Z)$ is the first n components of Z.

Using this $r(X)$ in conjunction with the transformation of Definition 3.5.2, a definition DEF.f can be transformed into a strongly equivalent one, DEF.g, with an elongated argument structure containing redundant information. This information can result in the primitives m', T', o_i', or t' of DEF.g being simpler than those of DEF.f. A simple example occurs when the argument structure X of DEF.f contains a string whose length, L, is changed by the application of the o_i functions, and the terminal condition $T(X)$ depends on L. L, can be added as a component to X, and thus be available for computing $T'(f(X))$. *Update* functions like those discussed in the backtrack set section, and further considered in the next sections, provide additional examples of redundancy-adding strongly equivalent transformations.

Another commonly useful transformation involves replacing a variable-length string argument with a fixed-length vector. A string or vector s of fixed maximum length M, which grows by concatenation and shrinks by deconcatenation, can be equally well represented by a vector V of length M with component $V_i = s_i$ for $1 \leq i \leq |s|$, together with a pointer $p \leq M$ whose value is $|s|$. Changing the component of V one position beyond its pointer and adding 1 to the pointer corresponds to concatenation, while subtracting 1 from p corresponds to deconcatenation. A function r from an argument structure in which one component is a string like s, to one in which that component is replaced by a pair of components like (V, p), will certainly have an inverse (provided there is no other change). Thus this string-to-vector transformation is easily incorporated into an equivalent definition. The string represented by the vector plus pointer is independent of vector components whose indices are greater than the pointer. This additional space will become significant when the pointer is increased. If it contains the correct entries at that time, no new entry is necessary. In Chapter 4 there is an example of this use of normally meaningless storage in the vector-plus-pointer representation. This transformation, like that above, involves adding redundancy, but more subtly than by simple addition of components to the argument structure.

The initial design of an argument structure may be redundant without any benefit from the redundancy. In particular one of the arguments, x_j, of the argument structure $X = \langle x_1, \ldots, x_n \rangle$ may be a function of the remaining arguments. To show this, an inductive argument is often required. First it is shown for the initial values of X, and then using the definition's recursive lines, and assuming it holds for an arbitrary nonterminal argument structure X, it is shown true for X's children. Thus it will hold over the entire substitutional domain. In the new argument structure the component x_j no longer appears. This argument structure transformation, like the one in the previous paragraph, has an inverse and thus can be used to produce a strongly equivalent definition. Although such a change may save some memory, it is unlikely to save any time.

Conditions under which strongly equivalent transformations, involving additions and deletions of argument structure components, are likely to improve efficiency are explored further in the next section.

3.6 UPDATE TRANSFORMATIONS: FINITE DIFFERENCING AND ARGUMENT STRUCTURE CHANGES WHICH LEAVE THE DEFINITION UNCHANGED AND IMPROVE IMPLEMENTATIONS

Before considering more general issues, we give another example, and introduce some additional notation [Aho, Ullman 72a; Paige 81; Sethi, Aho, Ullman 86].

Consider a definition for producing all permutations of the components of a given vector y. It builds permutations by removing one component from y at each stage of the growth process and concatenating it onto the growing permutation. This is illustrated in Figure 3-8, in which it is assumed that initially $y = \langle a, b, c \rangle$. The generating principle implied by Figure 3-8 leads to Definition 3.6.1.

Definition 3.6.1: Permutations

$$\left[\begin{array}{ll} f(x, y) = \{x\} & : |y| = 0 \\ f(x, y) = f(x \| \langle y_1 \rangle, y - y_1) \cup \cdots & \\ \qquad\qquad \cup f(x \| \langle y_{|y|} \rangle, y - y_{|y|}) & : |y| > 0 \\ \text{Initially } x = \langle\ \rangle,\ y \text{ is a finite vector.} & \end{array}\right.$$

The cardinality of y, $|y|$, is used to determine $m(x, y)$, the number of f calls on the right, and $T(x, y)$, the terminal condition. Can the cardinality computation necessary on every f call be made less costly? In moving from the argument of f on the left of the recursive line to those on the right, that is, from the parent to the children argument structures, the cardinality of the

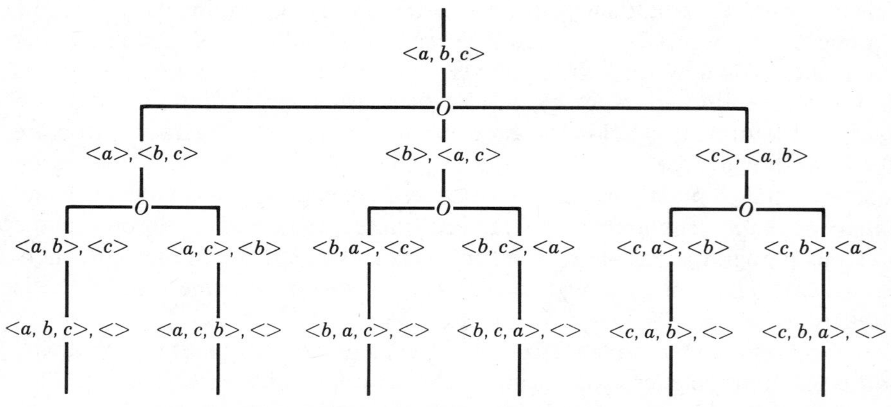

FIGURE 3-8 Vector component permutations, example O tree.

second component decreases by 1, or

$$|y|@\text{parent} - |y|@\text{child}_i = 1$$

$$|y|@\text{child}_i = |y|@\text{parent} - 1$$

So let a third component, z, be added, so that

$$\left[\begin{array}{l} \text{Initially } z = |y| \\ z@\text{child}_i = z@\text{parent} - 1 \end{array}\right.$$

It will follow that $z = |y|$ at every argument structure in Δ_f. This updating is done in Definition 3.6.2.

Definition 3.6.2: Permutations with Update

$$\left[\begin{array}{ll} f(x, y, z) = \{x\} & :z = 0 \\ f(x, y, z) = f(x \| \langle y_1 \rangle, y - y_1, z - 1) \cup \cdots & \\ \qquad\qquad \cup f(x \| \langle y_z \rangle, y - y_z, z - 1) & :z > 0 \\ \text{Initially } x = \langle\ \rangle,\ y \text{ is a finite vector, } z = |y|. & \end{array}\right.$$

At the cost of maintaining an additional integer component z, the computation of the length of a string has been replaced by subtracting 1. This is normally worthwhile (although some implementations of strings keep a length with every string so that the length of a string is available without additional computational cost).

3.6.1 Update Transformations—General

Under the general class of strongly equivalent definitions falls the class of update transformations. We have already seen examples of these in Section 3.4.2.

We can generalize the argument structure transformation illustrated in going from Definition 3.6.1 to Definition 3.6.2. Assume that for a given definition of the standard form, Definition 3.5.1, it is necessary to compute some function op(X) at each $X \in \Delta_f$ in turn, in order to compute $T(X)$ or $m(X)$ or $o_i(X)$. Add a component $z = \text{op}(X)$ to the argument structure. Since d is the initial value of X, the initial value of z is op(d). The simplest way to compute op at the children of X is to make it equal to op($o_i(X)$). This gives us the following definition equivalent to the standard form definition.

$$\left[\begin{array}{ll} f(X, z) = t(X) & :T(X) \\ f(X, z) = w\big(f(o_1(X), \text{op}(o_1(X))), \ldots, & \\ \qquad\qquad f\big(o_{m(X)}(X), \text{op}\big(o_{m(X)}(X)\big)\big)\big) & :\sim T(X) \\ \text{Initially } X = d \in D,\ z = \text{op}(d). & \end{array}\right.$$

This definition gives no advantage over the previous one because $\text{op}(o_i(X))$ must still be computed every time a new argument structure $o_i(X)$ is created. However, we are often able to express $\text{op}(o_i(X))$, for each i, by a simple function on z, that is, on $\text{op}(X)$. Assume there are functions ud_i, called *update functions*, such that

Definition 3.6.3: Update Primitive

$$\text{ud}_i(X', z) = \text{op}(o_i(X))$$

where X' is a subset of the arguments of X.

That is, assume that $\text{op}(o_i(X))$ can be computed from $z = \text{op}(X)$ and perhaps some of the components of $X = X'$. Then the following definition produced by a one-to-one transformation of the argument structure is equivalent to the previous one.

Definition 3.6.4: Transformation Using an Update Function

$$\left[\begin{array}{ll} f(X, z) = t(X) & :T(X) \\ f(X, z) = w(f(o_1(X), \text{ud}_1(X', z)), \ldots, & \\ \qquad f(o_{m(X)}(X), \text{ud}_{m(X)}(X', z))) & :\sim T(X) \\ \text{Initially } X = d \in D,\ z = \text{op}(d). & \end{array}\right.$$

That is, we assume $z = \text{op}(X)$ is a new component in a new argument structure and compute is next values $z@\text{child}_i = \text{op}(o_i(X)) = \text{ud}(X', z)$. As long as z is initially set to $\text{op}(d)$, for d the initial value of X, its subsequent values will be correctly computed by ud. This change, requiring added memory for the added component, is worthwhile if ud is substantially less complex than op.

Let x be a component of the argument of a defined function. From now on we will refer to the initial value, the value on the left, and the value in the ith f call on the right of a recursive line of x, respectively, as x *initially*, $x@$ *parent*, and $x@child_i$.

3.6.2 Example of Update Functions — Strength Reduction, Vector Length

Formulating $\text{op}(o_i(X)) = z@\text{child}_i$ as an update on $\text{op}(X) = z@\text{parent}$ is a general statement of a familiar compiler optimization technique called *strength reduction*. It is usually illustrated by a situation equivalent to that in Example 3.6.1, in which $\text{op}(X)$ requires multiplication of a component of X, x_j, by a constant a. However, because $o_i(X)$ results in that component of X being incremented by a constant b_i, $\text{op}(o_i(X))$ ($z@\text{child}_i$) can be computed by an update which requires only an addition to the previous value of op(X) (z@parent).

Example 3.6.1

$$X = (x_1, \ldots, x_j, \ldots, x_n)$$

where x_j is a number,

$$\text{op}(X) = \text{op}(x_1, \ldots, x_j, \ldots, x_n) = a * x_j$$

$$o_i(x_1, \ldots, x_j, \ldots, x_n) = (x_1, \ldots, x_j + b_i, \ldots, x_n)$$

Therefore

$$\text{op}(o_i(X)) = a * (x_j + b_i) = a * x_j + a * b_i$$

So we have the update function

$$\left[\begin{array}{l} z \text{ initially} = a * (x_j \text{ initially}) \\ z@\text{child}_i = \text{op}(o_i(X)) = z@\text{parent} + a * b_i \end{array}\right.$$

or

$$\text{ud}_i(z) = z + c_i,$$

where $c_i = a * b_i$ only depends on the child# i.

In this case the update approach replaces the multiplication of a possibly different constant vector at each child, with addition of a possibly different constant number at each child.

Earlier in introducing an update function in Definition 3.6.2 for permutations we considered a special case of the following situation.

Example 3.6.2

$$X = (x_1, \ldots, x_j, \ldots, x_n),$$

where x_j is a vector,

$$\text{op}(X) = \text{op}(x_1, \ldots, x_j, \ldots, x_n) = |x_j|$$

$$o_i(x_1, \ldots, x_j, \ldots, x_n) = (x_1, \ldots, x_j \| b_i, \ldots, x_n)$$

where b_i is a vector with $|b_i| = k_i$. So

$$\text{op}(o_i(X)) = |(x_j \| b_i)| = |x_j| + |b_i| = |x_j| + k_i$$

Therefore we have the update function

$$\begin{bmatrix} z \text{ initially} = |x_j \text{ initially}| \\ z@\text{child}_i = \text{op}(o_i(X)) = z@\text{parent} + k_i \end{bmatrix}$$

or

$$\text{ud}_i(z) = z + k_i$$

In this case the update approach replaces concatenation of a possibly different constant vector at each child, with addition of a possibly different constant number at each child. In Definition 3.6.2 this update function is used with $k_i = 1$.

In subsequent examples we give other combinations of op and o_i functions, which lead to efficient update functions. The problem of identifying such functions and finding the update function is called *finite differencing*. It has received extensive study and application, notably in [Paige 81].

It appears that the opportunity for useful update function arises when some part of the argument structure changes incrementally. However, the definition may be formulated using such high-level primitive functions that the possibility for incremental change is obscured. The design of that high-level primitive function itself will usually be in terms of still simpler primitive functions. At that point it is well to be aware of the recursive definition context in which that design must operate. The next example illustrates this situation.

3.6.3 Examples of Update Functions — Graph Cycle Detection

Consider the function $g(A)$ on a graph A, $g(A) = 1$ if A has a cycle and 0 otherwise. To realize this function we will use a recursive definition for a function $f(A, B, i)$, where A and B are graphs and i is an integer. $f(A, B, i)$ will be designed so that given appropriate initial conditions for B and i, $f(A, B, i) = g(A)$.

The edges in A are $e_1, \ldots, e_m$. The idea is to move copies of edges, one at a time, from A to the same position in a new graph B. B initially has the same vertices as A, but no edges. If A has a cycle, then there will eventually be a *first cycle formed in B*. It will be relatively easy to detect when, if ever, an edge generates a first cycle in B. We need only look for a cycle in a graph known to have at most *one* cycle. The approach is illustrated in the upper row of Figure 3-9. The remainder of the figure illustrates the operation of an improved definition. Assume the predicate cycle(G), which is true if and only if G has *one* cycle, is available. By $B + e_i$ we mean the graph that results from adding edge e_i to the current version of B. The definition reflecting this approach is

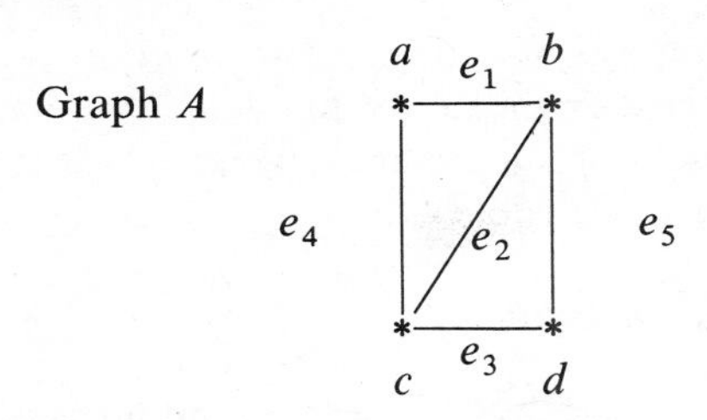

Graph	Stage of algorithm 1	2	3	4	5
B	* * * *	*—* * *	*—* *—*	*—* / *—*	*—* / *—*
Edge added		(a, b)	(c, d)	(b, c)	(a, c)
Blocks	$\{\{a\}\{b\}\{c\}\{d\}\}$	$\{\{a, b\}\{c\}\{d\}\}$	$\{\{a, b\}\{c, d\}\}$	$\{\{a, b, c, d\}\}$	

FIGURE 3-9 Example of the cycle detection procedure.

Definition 3.6.5: Cycle Detection

$$
\begin{array}{ll}
f(A, B, i) = 1 & : i \leq m \text{ and cycle } (B + e_i) \\
f(A, B, i) = 0 & : i = m \text{ and } \sim \text{ cycle } (B + e_i) \\
f(A, B, i) = f(A, B + e_i, i + 1) & : i < m \text{ and } \sim \text{ cycle } (B + e_i)
\end{array}
$$

Initially A is a graph and invariant, e_i is the ith edge in A; B has the same vertices as A but has no edges; $i = 1$.

But "cycle" is not a primitive predicate in many languages. So let us try to develop it from more primitive functions. Assume B has no cycle, and that the vertices of B are partitioned into subsets, called blocks, such that v_1 and v_2 are vertices in the same block if and only if there is a path between them in B. Then the addition of a new edge to B forms a cycle if and only if it is incident on two vertices in the same block. Thus the cycle predicate can be replaced as follows:

1. Let $e_i = \langle u, v \rangle$.
2. Partition B into $\{B_j | j = 1 \text{ to } p\}$.
3. Then: $\exists j \ni u$ and v are both in B_j if and only if cycle $(B + e_i)$ is true.

The recursive definition can now be given in terms of the partition, removing the reference to the cycle predicate. In the definition, part(G) is the partition of the graph G, and in(P, e) is a predicate which is true if both vertices of edge e are in the same block of P.

Definition 3.6.6: Cycle Detection

$$\left[\begin{array}{ll} f(B,i) = 1 & :i \le m \text{ and } \text{in}(\text{part}(B), e_i) \\ f(B,i) = 0 & :i = m \text{ and } \sim \text{in}(\text{part}(B), e_i) \\ f(B,i) = f(B + e_i, i + 1) & :i < m \text{ and } \sim \text{in}(\text{part}(B), e_i) \\ \multicolumn{2}{l}{\text{Initially } A \text{ is a graph and invariant, } e_i \text{ is the } i\text{th edge in } A;\ B \text{ has}} \\ \multicolumn{2}{l}{\text{the same vertices as } A \text{ but has no edges; } i = 1.} \end{array}\right.$$

To implement this definition we need a data structure for B in which part(B) and in(part(B), e) can be computed. Since these operations have to be repeated each time a new argument structure is computed, the complexity of computing cycles will be very sensitive to their complexity. It appears that this is quite expensive. But the partition can be computed incrementally with an update function, as described below, and perhaps this can be done efficiently. We suppress singleton blocks in order to simplify the resulting definition.

If B is a graph, then part($B + e$), the partition that results when the new edge e is added to the existing partition part(B), can be given in terms of part(B). Let part(B) = P, part($B + e$) = inc.part(P, e). With $e = (u, v)$, inc.part(P, e) is defined as follows.

Definition 3.6.7: Primitive Graph Function inc.part

1. If neither u nor v are in any block of P, then inc.part(P, e) = $P \cup \{\{u, v\}\}$.
2. If u (respectively, v) is in block B_j of P and v (u) is not in any block of part(B), then inc.part(P, e) = P, with $v(u)$ added to B_j.
3. If u is in block B_j and v is in block B_k of P, $j \neq k$, then inc.part(P, e) = P, with B_j and B_k removed and replaced with a single block containing all members from B_j and B_k.
4. If u and v are both in the same block of P, then B, and thus A, has a cycle.

Given this way of constructing a new partition from an existing one, we can add the partition P = part (B) as an argument of our definition and get an equivalent definition involving the primitive function described above for updating P or part(B). In summary

$$\left[\begin{array}{l} P \text{ initially} = \{\{a_1\},\ldots,\{a_n\}\}, \text{ where } a_i \text{ is one of } n \text{ vertices in } A \\ P@\text{child} = \text{inc.part}(P@\text{parent}, e_i) \end{array}\right.$$

Note that, unlike in former examples, there is only one child, so the update is not a function of the child # and depends on parts of the argument other than the update component P alone.

Definition 3.6.8: Cycle Detection with Update

$$\begin{array}{ll} f(B, i, P) = 1 & :i \leq m \text{ and } \mathrm{in}(P, e_i) \\ f(B, i, P) = 0 & :i = m \text{ and } \sim \mathrm{in}(P, e_i) \\ f(B, i, P) = f(B + e_i, i + 1, \mathrm{inc.part}(P, e_i)) & :i < m \text{ and } \sim \mathrm{in}(P, e_i) \end{array}$$

Initially A is a graph and invariant, e_i is the ith edge in A; B has the same vertices as A but has no edges; $i = 1$; $P = \{\{a_1\}, \ldots, \{a_n\}\}$.

The lower rows in Figure 3-9 illustrate the evolution of the partitions P as Definition 3.6.8 is evaluated for the graph in that figure. We still need a data structure for the partition P such that inc.part(P, e) and in(P, e) can be computed efficiently. The union–find or compressed tree for disjoint sets structure is such a data structure (see [Tarjan 83]).

3.6.4 *n*-Queen Update for Unordered Backtrack Selection Definition [Nadel 86]

Now we consider an update transformation for the n-queen definition in its *unordered* backtrack selection form, Definition 3.4.8.

An $n \times n$ matrix M and a length n array CNT are maintained. M represents the chess board state, with unavailable squares marked 1 and available ones marked 0. CNT(j) gives the number of available positions in row j of m. Initially CNT(j) is n. When a queen is placed at $\langle i, c_i \rangle$, then all positions in all rows which are in the column or on either diagonal of $\langle i, c_i \rangle$ are marked 1, and whenever a 1 replaces a 0 at $\langle x, y \rangle$, then CNT(x) is decreased by 1. The update functions are given assuming a queen is to be placed in row k; how k is chosen will be explained later.

M initially = all 0's
M@child$_i$ = result of placing at 1 at all positions which are in the column or a diagonal of $\langle k, i \rangle$ of M@parent.
= ud1(M, k, i)

CNT initially = n^n
CNT@child$_i$ = result of subtracting 1 from CNT(x) @parent whenever ud1(M, k, i) changes 0 to 1 in row x
= ud2(CNT, M, k, i)

It is now advisable to place a queen in the column (in row k) in which there is a minimum number of nonconflicting column locations. Once any row has zero such columns available, the enumeration path can be terminated. This scheme for positioning queens seems likely to make this happen as soon as possible, eliminating impossible queen arrangements as soon as possible and thus enumerating all legitimate arrangements with as little extra effort as possible.

This motivates the choice for function $h(v)$ of Definition 3.4.8;

$$h(\text{CNT}) = i \ni \text{CNT}(i) \text{ is minimum}$$

With this choice we can produce an update function for k;

$$\left[\begin{array}{l} k \text{ initially} = 1 \\ k@\text{child}_i = \text{MIN is set to } k\text{. Each time CNT}(x) \text{ is decreased by} \\ \qquad\qquad 1 \text{ in updating CNT, the index of } \min(\text{CNT}(x), \text{CNT}(k)) \\ \qquad\qquad (x \text{ or } k) \text{ replaces MIN. After ud2(CNT, } M, k, i) \text{ is complete:} \\ \qquad\quad = \text{ud3}(\text{CNT}, M, k, i) = \text{MIN} \end{array}\right.$$

Specializing Definition 3.4.7 with the update functions above, and noting that

$$Q[n]_k = i \ni M(k, i) = 0$$

$$(\{Q[n]_k\} = \{\ \}) \equiv (\text{CNT}(k) = 0)$$

we get the following definition.

Definition 3.6.9: *n*-Queen Unordered Backtrack Selection Definition with Update Function

$$\left[\begin{array}{ll} f(v, j, M, \text{CNT}, k) = \{v\} & : j = n + 1 \\ f(v, j, M, \text{CNT}, k) = \{\ \} & :\text{CNT}(k) = 0 \\ f(v, j, M, \text{CNT}, k) = \bigcup_{i \in \{i \ni M(k,i)=0\}} [f(v_k \leftarrow i, j+1, \text{ud1}(M, k, i), & \\ \qquad\qquad\qquad \text{ud2}(\text{CNT}, M, k, i), & \\ \qquad\qquad\qquad \text{ud3}(\text{CNT}, M, k, i)] & \\ & :\text{CNT}(k) \neq 0 \\ \text{Initially } v = 0^n,\ j = 1,\ \text{CNT} = n^n,\ M \text{ is all 0's},\ i = 1,\ n \text{ is invariant.} & \end{array}\right.$$

Note that given the prescribed data structures, the complexity necessary to find all $i \ni M(k, i) = 0$ as required by the definition is proportional to the width of a row in M. By linking together the 0 components in each row of M, this can be cut to the number of 0's in row k.

Without the data structures M, CNT, and k, it would be necessary to count the number of available 0's in each row of the board and find the minimum at every f call. This would require a higher order of complexity than that necessary using M and CNT with their update functions. The update functions were introduced largely to provide information which would permit an intelligent choice for ordering, given by h(CNT). This choice of ordering, though providing speedup on the average, may in some cases require more work than an implementation without it. We call such a nonuniform improvement a *heuristic*.

3.7 EFFICIENCY IMPROVEMENT BY FIXING ORDER

In addition to illustrating update changes in argument structures as a way of improving efficiency, the n-queen example also illustrates how efficiency can be improved by fixing the order of evaluation of the f calls or o_i's on the right. We explore the benefits for the efficiency of implementation of fixing *free* orderings in a more general setting.

An ordering of the primitive function evaluation in a definition DEF.f whose change would not change f is *free*. Free orderings must be fixed in any sequential implementation. We can sometimes find reasons to fix free orderings at the higher, definition, level of design.

First, to get further insight into the characteristics of the n-queen definition, which make the ordering heuristic of the previous section useful, we develop a modest generalization of that definition.

Definition 3.7.1: Vertically Commutative Definition

$$r(Y, y) \subset Y$$

$$f(X, Y) = t(X, Y) \qquad :Y = \{\ \}$$

$$f(X, Y) = \bigcup_{y \in (Y \cap R_k) k \ni (Y \cap R_k \neq \{\ \})} f\big(o_y(X, Y), r(Y, y)\big) \qquad :Y \neq \{\ \}$$

Initially $X \in D$, Y is a finite set, partitioned initially into subsets R_k with $1 \leq k \leq n$, R_k and n are invariant.

This definition is controlled by removing elements from the set Y. It terminates if Y is empty. As defined above, with y *any* member of $Y \cap R_k$, the value of $f(X, Y)$ will not depend on the order in which members of Y are removed, but efficiency of implementation may. Order here refers to what is done at successive levels of the O network— thus vertical in the definition title. This definition is a generalization Definition 3.4.8 for n queens. The set Y corresponds to the board positions which may still be legitimately occupied by a queen, $r(Y, y)$ corresponds to the set that remains legitimate after the queen is placed at position y, R_k is the positions in row k of the board. It is advisable to choose k's which, after as few choices as possible, will lead to an argument structure that *cannot* lead to a solution, thus truncating the generation of parts of the O network that fail to contribute to the output. In general it is difficult to determine which choices of k will do this, but it does seem reasonable to consider choosing k so that $|Y \cap R_k|$ is minimum, as was done for n queens.

Consider another kind of ordering, which does not affect the value of the function defined, and which therefore we are free to fix for efficiency. The standard recursive line is used in defining, not in implementing, a function. Though indices of its O functions run from 1 to $m(X)$, the implementation of the O functions need not follow that order. If the recursive line is commutative, then the order of the O functions is free, but can be fixed for efficiency. The following definition gives a class for which it might be advantageous to order the right side O functions or, in O network terminology, the outputs of a single O block—thus the "horizontal" in the title.

Definition 3.7.2: Horizontally Commutative Definition

$$f(X,C) = \langle X, C\rangle \qquad :T(X,C)$$
$$f(X,C) = \text{MIN-SEL}\big(f(o_1'(X), C + c(r_1(X))), \ldots, f\big(o_{m(x)}'(X), C + c\big(r_{m(x)}(X)\big)\big)\big) \qquad :\sim T(X,C)$$

Initially $X \in D$, C is a number; MIN-SEL selects on the second component of f.

There are many standard implementations for such recursive definitions, like the depth-first algorithms considered in subsequent chapters, which compute lowest indexed O functions as soon as possible.

For such implementations it would be good if the O functions with smaller values of $c(r_j(X))$ had smaller indices, because this would tend to lead rapidly to a MIN-SEL solution. Since MIN-SEL is commutative, we are free to assign this order. As we will see, if we choose a depth-first based implementation (such as branch and bound, Section 5.4.1), it will tend to reach the minimum solution early. This cannot be guaranteed for all definitions of the form of Definition 3.7.2; by itself in this general setting it is a heuristic, worth consideration but guaranteeing no uniform improvement. We also later use this heuristic in going from minimum spanning tree, Definition 7.3.1, to Definition 7.3.3. In that instance we will be able to guarantee that only the first child has to be calculated to find the minimum cost. This provides a large improvement in worst-case performance. On the other hand, if applied to a definition like that for the minimum path, there is no worst-case improvement.

3.8 STATE SEQUENCE ↔ CHANGE SEQUENCE TRANSFORMATION

In Section 3.5 we studied equivalence transformations involving changes in the definition argument structure. Now we sample transformations involving more extensive changes. First we consider a transformation between closely related, but nonequivalent, definitions. (Compare to Section 3.3.2).

Any sequence of vectors, say,

$$S = \langle s_1, \ldots, s_n \rangle$$

can be thought of as a *sequence of states* through which a process passes. There must also be a corresponding *sequence of state changes* which that process experiences. This is expressed by

$$\mathrm{chg}_S = \langle s_2 - s_1, s_3 - s_2, \ldots, s_n - s_{n-1} \rangle$$

For example, Definition 3.1.1 gives the sequence of moves or state changes that solve the Tower of Hanoi puzzle. A state, which is a configuration of n disk, each on one of three pegs numbered 1, 2, and 3, is represented by a vector of n components. Component i identifies by 1, 2, or 3, the peg on which disk i rests. To move three disks from peg 1 to peg 2, the state change chg_S and the corresponding state sequence S are

Example 3.8.1

chg_S		$1 \to 2$		$1 \to 3$		$2 \to 3$		$1 \to 2$		$3 \to 1$		$3 \to 2$		$1 \to 2$	
S:	$\langle 111 \rangle$		$\langle 211 \rangle$		$\langle 231 \rangle$		$\langle 331 \rangle$		$\langle 332 \rangle$		$\langle 132 \rangle$		$\langle 122 \rangle$		$\langle 222 \rangle$

The state and state change sequences are easy to obtain from each other, so we expect a simple transformation to exist between their definitions. The corresponding definitions are likely to differ in the efficiency of their implementations. For example, in a sequence of state changes there are likely to be recurrences which do not exist in the state sequence, so argument structures in the state, but not the state change, form of definition are likely to be unique. Therefore the state, but not the state change, definition, is likely to have a uniform inverse, a property affecting implementation. Neither differences in the representation of sequences nor in the efficiency of their implementation motivate our development of these transformations. Rather the main reason is that, as with other transformations in this book, it serves as a precise statement and concise reminder of the equivalence of concepts embedded in the two forms of the definition. After developing these transformations, we will interpret them in this manner.

For a formal statement of the transformation from state to state change definitions, a general definition of "state" is needed. A state here is a finite set of $\langle$object, place$\rangle$ pairs. Each pair indicates that the given object is at the specified place. Objects are drawn from a finite set, as are the places. No object is in more than 1 place but a place may have more than 1 object. Often a vector is used to represent a state. The vector *component* may represent the *object*, while the *index* of that component represents the *place*, or alternatively the vector *component* may represent the *place* while the *index* of that component represents the object. The latter is used above to represent Tower of Hanoi states.

Formally, a *state element* is a pair $\langle o, p \rangle$, where o and p are drawn respectively from a finite set of objects, O, and a finite set of positions, P, P has a distinguished element, called nil, where the pair $\langle o, \text{nil} \rangle$ indicates that object o has not yet been assigned a position.

For example, if the set of objects consists of four coins, d_1 through d_4, and the set of places corresponds to locations 1 through 9, then $\langle d_3, 6 \rangle$ indicates that coin d_3 is in location 6.

A *partial state* is a set of $n = |O|$ state elements $\langle o, p \rangle$, $o \in O$, $p \in P$, no two of which have the same o component. A partial state Z is a *complete state* if every member of O is in some state element with non-nil p element.

In the example above, $\{\langle d_1, \text{nil} \rangle, \langle d_2, 1 \rangle, \langle d_3, 2 \rangle, \langle d_4, \text{nil} \rangle\}$ and $\{\langle d_1, 3 \rangle, \langle d_2, \text{nil} \rangle, \langle d_3, 4 \rangle, \langle d_4, 1 \rangle\}$ are partial states. We will usually omit state elements with nil position, so that the two partial states above would be represented by $\{\langle d_2, 1 \rangle, \langle d_3, 2 \rangle\}$ and $\{\langle d_1, 3 \rangle, \langle d_3, 4 \rangle, \langle d_4, 1 \rangle\}$, respectively.

In general, a state change is a description of how to change a partial state to a second partial state. There are various representations for such changes, and we will see some of these in the examples. We now give one representation which is then used in the subsequent theorem.

An *elementary state change* is a triple, $[o, x \rightarrow y]$, with $o \in O$, $x, y \in P$ and $y \neq x$, $y \neq \text{nil}$.

A set of elementary state changes, in which no $o \in O$ occurs more than once, is called a *state change*.

In the example, $[d_1, 2 \rightarrow 5]$ indicates that coin d_1 is moved from location 2 to location 5; $[d_3, \text{nil} \rightarrow 7]$ indicates that d_3's position is now known to be location 7.

A *sequence of state changes* is a sequence of nonempty **sets** of elementry changes.

1. If W and U are partial states, $W + U$ is a set of partial *states*

$$W + U = \{\langle o, x \rangle | \langle o, x \rangle \in W \text{ or } (\langle o, \text{nil} \rangle \in W \text{ and } \langle o, x \rangle \in U)\}$$

   (Note $W + U$ may not equal $U + W$)

2. If W and U are partial states, $W - U$ is a set of elementary state *changes*:

$$W - U = \{[o, x \rightarrow y] | \langle o, x \rangle \in U \text{ and } \langle o, y \rangle \in W \text{ and } x \neq y\}.$$

3. If $Q = \langle Q_1, \ldots, Q_m \rangle$, and each Q_i is a partial state and Z is a partial state, then $Z + + Q$ is a *vector* of partial *states*:

$$Z + + Q = \langle Z + Q_1, \ldots, Z + Q_m \rangle$$

4. If $Q = \langle Q_1, \ldots, Q_m \rangle$ and $R = \langle R_1, \ldots, R_m \rangle$, and both Q_i and R_i are partial states, then $Q - - R$ is a *vector of sets* of elementary state *changes*:

$$Q - - R = \langle Q_1 - R_1, \ldots, Q_m - R_m \rangle$$

To illustrate these definitions, consider the sequence of vectors giving the Tower of Hanoi states in Example 3.8.1, in which $\langle xyz \rangle$ is short for $\langle x, y, z \rangle$.

$$S = \langle\langle 111\rangle, \langle 211\rangle, \langle 231\rangle, \langle 331\rangle, \langle 332\rangle, \langle 132\rangle, \langle 122\rangle, \langle 222\rangle\rangle$$

Other vectors can be constructed from S. The following examples will be of use later:

$$Q = S \| \langle\langle 222\rangle\rangle = \langle\langle 111\rangle, \langle 211\rangle, \langle 231\rangle, \langle 331\rangle, \langle 332\rangle, \langle 132\rangle, \langle 122\rangle, \langle 222\rangle, \langle 222\rangle\rangle$$

$$R = \langle\langle 111\rangle\rangle \| S = \langle\langle 111\rangle, \langle 111\rangle, \langle 211\rangle, \langle 231\rangle, \langle 231\rangle, \langle 331\rangle, \langle 332\rangle, \langle 132\rangle, \langle 122\rangle, \langle 222\rangle\rangle$$

Each vector represents a complete state. Object i represented by position i of the vector is disk i, and the place represented by component j of the vector is peg j. So the vector $\langle xyz \rangle$ represents the complete state $\{\langle 1, x\rangle, \langle 2, y\rangle, \langle 3, z\rangle\}$.

The difference $(-)$ operation is designed so that the corresponding change sequence can be expressed as a difference between successive members of the sequence. If $W = \langle x, y, z\rangle$ and $U = \langle a, b, c\rangle$ represent complete states, then

$$\begin{aligned} W - U = \langle x, y, z\rangle - \langle a, b, c\rangle \quad & \text{represents } \{\langle 1, x\rangle, \langle 2, y\rangle, \langle 3, z\rangle\} \\ & - \{\langle 1, a\rangle, \langle 2, b\rangle, \langle 3, c\rangle\} \\ = \{[1, a \to x], [2, b \to y], [3, c \to z]\} \quad & \text{if } x \neq a, y \neq b, z \neq c \\ = \{[1, a \to x], [3, c \to z]\} \quad & \text{if } x \neq a, y = b, z \neq c \\ = \ldots \end{aligned}$$

The difference of successive members of W can be represented with the $--$ operation, using the vectors of state representations Q and R defined above:

$$\begin{aligned} Q -- R = \langle & \langle 111\rangle - \langle 111\rangle, \langle 211\rangle - \langle 111\rangle, \langle 231\rangle - \langle 211\rangle, \\ & \langle 331\rangle - \langle 231\rangle, \langle 332\rangle - \langle 331\rangle, \langle 132\rangle - \langle 332\rangle, \langle 122\rangle - \langle 132\rangle, \\ & \langle 222\rangle - \langle 122\rangle, \langle 222\rangle - \langle 222\rangle\rangle \end{aligned}$$

since, for example, $\langle 211\rangle - \langle 111\rangle$ represents

$$\{\langle 1,2\rangle, \langle 2,1\rangle, \langle 3,1\rangle\} - \{\langle 1,1\rangle, \langle 2,1\rangle, \langle 3,1\rangle\} = \{[1, 1 \to 2]\}$$

These differences can be expressed as a sequence of *elementary state changes*,

$$Q -- R = \langle \{\ \}, [1,1 \to 2], [2,1 \to 3], [1,2 \to 3], [3,1 \to 2],$$
$$[1,3 \to 1], [2,3 \to 2], [1,1 \to 2], \{\ \} \rangle$$

and because an empty set { } is no change at all,

$$Q -- R = \langle [1,1 \to 2], [2,1 \to 3], [1,2 \to 3], [3,1 \to 2],$$
$$[1,3 \to 1], [2,3 \to 2], [1,1 \to 2] \rangle$$

(The set symbols { , } have been dropped when the set contains a single change.)

Notice that these elementary state changes differ from those in Example 3.8.1 in that they contain an initial component identifying the disk to be moved. This information is redundant here because knowing the top disk is always moved, the peg from which it is to be moved identifies it. Notice, however, that the first component values have a pattern, one we will see again in the state change sequence for the Gray code.

The example above can be generalized to give a lemma which will be useful in the theorem that relates state and state change sequences.

Lemma 3.8.1: If $V = \langle V_1, \ldots, V_n \rangle$, then

$$V \| \langle V_n \rangle -- \langle V_1 \rangle \| V = \langle V_2 - V_1, V_3 - V_2, \ldots, V_n - V_{n-1} \rangle$$

PROOF: The proof follows directly from the definition of $\|$, $-$, and $--$.

In the theorem that follows we will be considering a function $f(X)$ whose value is a sequence of vectors with a recursive definition line generalizing the form of the line for Tower of Hanoi and Gray code state sequence definitions. That generalization is

$$f(X) = a(X) \| (r_1(X) ++ f(o_1(X))) \| \cdots$$
$$\| \left(r_{m(x)}(X) ++ f\left(o_{m(x)}(X)\right)\right) \| b(X)$$

where $a(X)$, $b(X)$, and $f(o_i(X))$ are all sequences of complete states. $r_i(X)$ is a state element to be added to ($++$), thus becoming an invariant, of every complete state in the sequence $f(o_i(X))$. Hence each term $(r_i(X) ++ f(o_i(X)))$ is a sequence of complete states, each of which contains the invariant $r_i(X)$. $f(X)$ is thus the concatenation of a number of sequences of complete states, each of which contains an invariant. In the Tower of Hanoi that invariant is the position of the largest disk; in the Gray code it is the high-order bit.

The recursive line in the corresponding state change definition for $\text{chg}_f(X)$ is obtained from that for $f(X)$ by simply expressing the difference between successive complete states in $f(X)$, using the notation we have developed, and letting $\text{chg}_f(o_i(X))$ be the sequence of state changes between successive complete states in $f(o_i(X))$.

In the following theorem we will be using the notation $f(X)_i$ for the ith component and $f(X)_{\text{last}}$ for the last component in the sequence $f(X)$.

Theorem 3.8.1: Given the following two recursive definitions, in which $a(X)$ and $b(X)$ are vectors of complete states, $r_i(X)$ is a partial state, and $f(X)$ is designed to be a vector of complete states.

1. *State sequence*

$$\left[\begin{array}{ll} f(X) = \langle t(X) \rangle & :T(X) \\ f(X) = a(X) \| (r_1(X) + + f(o_1(X))) \| \cdots & \\ \qquad \| \left(r_{m(x)}(X) + + f\left(o_{m(x)}(X)\right)\right) \| b(X) & :\sim T(X) \\ \text{Initially } X = d \in D. & \end{array}\right.$$

2. *State change sequence*

$$\left[\begin{array}{ll} \text{chg}_f(X) = \langle\ \rangle & :T(X) \\ \text{chg}_f(X) = a'(X) \| \text{chg}_f(o_1(X)) & \\ \quad \| [r_2(X) + f(o_2(X))_1] - [r_1(X) + f(o_1(X))_{\text{last}}] & \\ \quad \| \text{chg}_f(o_2(X)) & \\ \quad \vdots & \\ \quad \| \left[r_{m(x)}(X) + f\left(o_{m(x)}(X)\right)_1\right] - \left[r_{m(x)-1}(X) + f\left(o_{m(x)-1}(X)\right)_{\text{last}}\right] & \\ \quad \| \text{chg}_f\left(o_{m(x)}(X)\right) \| b'(X) & :\sim T(X) \\ \text{Initially } X = d \in D. & \end{array}\right.$$

Here $f(o_i(X))_1$, and $f(o_i(X))_{\text{last}}$ must be expressible as primitive functions, or must be known to be identical, in which case they would cancel each other; and where

$$a'(X) = \langle a(X)_2 - a(X)_1, a(X)_3 - a(X)_2, \ldots, a(X)_{\text{last}} - a(X)_{\text{last}-1},$$
$$\left[r_1(X) + f(o_1(X))_1)\right] - a(X)_{\text{last}} >$$

$$b'(X) = \langle b(X)_1 - \left[r_{m(x)}(X) + f\left(o_{m(x)}(X)\right)_{\text{last}}\right]$$
$$b(X)_2 - b(X)_1, \ldots, b(X)_{\text{last}} - b(X)_{\text{last}-1} \rangle$$

Then for any X in Δ_f, $f(X)$ and $\text{chg}_f(X)$ are vectors, which can be represented as:

$$f(X) = \langle f(X)_1, \ldots, f(X)_{\text{last}} \rangle,$$

$$\text{chg}(X) = \langle \text{chg}_f(X)_1, \ldots, \text{chg}_f(X)_{\text{last}} \rangle$$

And, more importantly,

$$\text{chg}_f(X) = \langle f(X)_2 - f(X)_1, f(X)_3 - f(X)_2, \ldots, f(X)_{\text{last}} - f(X)_{\text{last}-1} \rangle$$

PROOF: We prove the result when the vectors $a(X)$, $a'(X)$, $b(X)$, and $b'(X)$ are all $= \langle\ \rangle$. The essence of the proof is unaffected by this simplification.

The fact that $f(X)$ is a vector is easily verified. By Lemma 3.8.1,

$$\text{chg}_f(X) =?\ \langle f(X)_2 - f(X)_1, \ldots, f(X)_{\text{last}} - f(X)_{\text{last}-1} \rangle$$

is equivalent to

$$\text{chg}_f(X) =?\ (f(X) \| \langle f(X)_{\text{last}} \rangle) - -(\langle f(X)_1 \rangle \| f(X))$$

These interchangeable tentative equalities are now proven to be equalities.

When $T(X)$,

$$\text{chg}_f(X) =?\ (f(X) \| \langle f(X)_{\text{last}} \rangle) - -(\langle f(X)_1 \rangle \| f(X))$$

Since from definition 1 $f(X) = \langle t(X) \rangle$, substituting on the right gives

$$\text{chg}_f(X) =?\ (\langle t(X) \rangle \| \langle t(X) \rangle) - -(\langle t(X) \rangle \| \langle t(X) \rangle)$$

$$=!\ \langle\ \rangle \qquad \text{by definition 2}$$

When $\sim T(X)$. Note that the first component of $f(X)$ $(= f(X)_1)$ can be expressed in terms of the right side of the equation for $f(X)$ in definition 1 as follows:

$$f(X)_1 = r_1(X) + f(o_1(X))_1$$

An analogous expression serves for the last component of $f(X)$, $(f(X)_{\text{last}})$:

$$f(X)_{\text{last}} = r_{m(x)}(X) + f\big(o_{m(x)}(X)\big)_{\text{last}}$$

Now in the equation

$$\text{chg}_f(X) =?\ (f(X) \| \langle f(X)_{\text{last}} \rangle) - -(\langle f(X)_1 \rangle \| f(X))$$

substituting for $f(X)$ with the recursive line of definition 1 gives

$$\text{chg}_f(X) = ?\big((r_1(X) + + f(o_1(X)))\| \cdots \|(r_{m(x)}(X) + + f(o_{m(x)}(X)))\big)$$

$$\| \langle r_{m(x)}(X) + f(o_{m(x)}(X))_{\text{last}} \rangle -- \langle r_1(X) + f(o_1(X))_1 \rangle \|$$

$$\big((r_1(X) + + f(o_1(X)))\| \cdots \|(r_{m(x)}(X) + + f(o_{m(x)}(X)))\big)$$

The expression on the right of the equal sign in the previous equation is rewritten with its terms rearranged in accordance with the properties of + and −−,

1. $\langle (r_1(X) + f(o_1(X))_1) - (r_1(X) + f(o_1(X))_1) \rangle$
2. $\left[\begin{array}{l} \|\langle (r_1(X) + f(o_1(X))_2) - (r_1(X) + f(o_1(X))_1) \rangle \| \cdots \\ \|\langle (r_1(X) + f(o_1(X))_{\text{last}}) - (r_1(X) + f(o_1(X))_{\text{last}-1}) \rangle \end{array}\right.$
3. $\|\langle (r_2(X) + f(o_2(X))_1) - (r_1(X) + f(o_1(X))_{\text{last}}) \rangle$
4. $\left[\begin{array}{l} \|\langle (r_2(X) + f(o_2(X))_2) - (r_2(X) + f(o_2(X))_1) \rangle \| \cdots \\ \|\langle (r_2(X) + f(o_2(X))_{\text{last}}) - (r_2(X) + f(o_2(X))_{\text{last}-1}) \rangle \| \end{array}\right.$

   ⋮
5. $\|\langle (r_{m(x)}(X) + f(o_{m(x)}(X))_1) - (r_{m(x)-1}(X) + f(o_{m(x)-1}(X))_{\text{last}}) \rangle$
6. $\left[\begin{array}{l} \|\langle (r_{m(x)}(X) + f(o_{m(x)}(X))_2) - (r_{m(x)}(X) + f(o_{m(x)}(X))_1) \rangle \| \cdots \\ \|\langle (r_{m(x)}(X) + f(o_{m(x)}(X))_{\text{last}}) - (r_{m(x)}(X) + f(o_{m(x)}(X))_{\text{last}-1}) \rangle \end{array}\right.$
7. $\|\langle (r_{m(x)}(X) + f(o_{m(x)}(X))_{\text{last}}) - (r_{m(x)}(X) + f(o_{m(x)}(X))_{\text{last}}) \rangle$

The terms in lines 2, 4, and 6 can be gathered using the properties of + and − to give a more familiar form. In lines 1 and 7 the simplification is another consequence of the property of − and +.

1. $\{\ \}$
2. $\|\langle f(o_1(X))_2 - f(o_1(X))_1, \ldots, f(o_1(X))_{\text{last}} - f(o_1(X))_{\text{last}-1} \rangle$
3. $\|\langle (r_2(X) + f(o_2(X))_1) - (r_1(X) + f(o_1(X))_{\text{last}}) \rangle$
4. $\|\langle f(o_2(X))_2 - f(o_2(X))_1, \ldots, f(o_2(X))_{\text{last}} - f(o_2(X))_{\text{last}-1} \rangle \| \cdots$
5. $\|\langle (r_{m(x)}(X) + f(o_{m(x)}(X))_1) - (r_{m(x)-1}(X) + f(o_{m(x)-1}(X))_{\text{last}}) \rangle$
6. $\|\langle f(o_{m(x)}(X))_2 - f(o_{m(x)}(X))_1, \ldots, f(o_{m(x)}(X))_{\text{last}} - f(o_{m(x)}(X))_{\text{last}-1} \rangle$
7. $\|\{\ \}$

Simplified and written horizontally this becomes

$$\begin{aligned}\mathrm{chg}_f(X) = \mathrm{chg}_f(o_1(X))\|\langle(r_2(X) + f(o_2(X))_1)\\ - (r_1(X) + f(o_1(X))_{\mathrm{last}})\rangle\|\mathrm{chg}_f(o_2(X)) \cdots\\ \|\langle(r_{m(x)}(X) + f(o_{m(x)}(X))_1)\\ - (r_{m(x)-1}(X) + f(o_{m(x)-1}(X))_{\mathrm{last}})\rangle \cdots \|\mathrm{chg}_f(o_{m(x)}(X)) \quad \blacksquare\end{aligned}$$

In order to express chg_f, we need to know $f(o_i(X))_1$, and $f(o_i(X))_{\mathrm{last}}$. These are not usually difficult to determine since f is a sequence of states, through which it is necessary to pass in going from a starting state to a final state. These extreme states are often specified explicitly by the argument of f, and are thus easily determined for a given argument.

Corollary 3.8.1.1: If we add to Theorem 3.8.1 the additional constraints on definition 1, state sequence, that

1. $a(X)_{\mathrm{last}} = f(o_1(X))_1$ and if r_1 contains $[j, x]$ with $x \neq \mathrm{nil}$, then both $f(o_1(X))_1$ and $a(X)_{\mathrm{last}}$ contain $[j, \mathrm{nil}]$,
2. $f(o_{m(x)}(X))_{\mathrm{last}} = b(X)_1$, and if $r_{m(x)}$ contains $[j, x]$ with $x \neq \mathrm{nil}$, then both $f(o_{m(x)}(X))_1$ and $b(X)_1$ contain $[j, \mathrm{nil}]$,
3. $f(o_i(X))_{\mathrm{last}} = f(o_{i+1}(X))_1$, and if r_{i+1} contains $[j, x]$ with $x \neq \mathrm{nil}$, then both $f(o_i(X))_{\mathrm{last}}$ and $f(o_{i+1}(X))_1$ contain $[j, \mathrm{nil}]$,

then the theorem still holds with definition 2, state change sequence, simplified to read

2′. *State Change Sequence*

$$\left[\begin{array}{ll}\mathrm{chg}_f(X) = \langle\,\rangle & :T(X)\\ \mathrm{chg}_f(X) = a'(X)\|\mathrm{chg}_f(o_1(X))\|\langle r_2(X) - r_1(X)\rangle\| \cdots & \\ \qquad \|\langle r_{m(x)}(X) - r_{m(x)}(X)\rangle\|\mathrm{chg}_f(o_{m(x)}(X))\|b'(X) & :\sim T(X)\\ \text{Initially } X = d \in D. & \end{array}\right.$$

where

$$\begin{aligned}a'(X) &= \langle a(X)_2 - a(X)_1, a(X)_3 - a(X)_2, \ldots, a(X)_{\mathrm{last}} - a(X)_{\mathrm{last}-1},\\ &\qquad (r_1(X) + f(o_1(X)_1)) - a(X)_{\mathrm{last}}\rangle\\ b'(X) &= \langle b(X)_1 - (r_{m(x)} + f(o_{m(x)}(X)_{m(x)})),\\ &\qquad b(X)_2 - b(X)_1, \ldots, b(X)_{\mathrm{last}} - b(X)_{\mathrm{last}-1}\rangle\end{aligned}$$

Then for any X in Δ_f, $f(X)$ and $\mathrm{chg}_f(X)$ are vectors:

$$f(X) = \langle f(X)_1, \ldots, f(X)_{\mathrm{last}} \rangle$$

$$\mathrm{chg}(X) = \langle \mathrm{chg}_f(X)_1, \ldots, \mathrm{chg}_f(X)_{\mathrm{last}} \rangle$$

and, more importantly,

$$\mathrm{chg}_f(X) = \langle f(X)_2 - f(X)_1, f(X)_3 - f(X)_2, \ldots, f(X)_{\mathrm{last}} - f(X)_{\mathrm{last}-1} \rangle$$

PROOF: It is sufficient to alter definition 2 in Theorem 3.8.1 according to the added constraint to get definition 2′ above. ■

Alternatively with the removal of the constraints

$$a(X)_{\mathrm{last}} = f(o_1(X))_1, \qquad f\big(o_{m(x)}(X)\big)_{\mathrm{last}} = b(X)_1$$

from definition 1 in Theorem 3.8.1 it is straightforward to obtain still another, more complex, form of definition 2 in that theorem.

In applying Theorem 3.8.1 or its corollary it is well to make sure that

$$\forall i \langle (r_{i+1}(X) + f(o_{i+1}(X))_1) - (r_i(X) + f(o_i(X))_{\mathrm{last}}) \rangle \neq \langle\ \rangle$$

If it did equal $\langle\ \rangle$, then for some i, $r_{i+1}(X) + f(o_{i+1}(X))_1 = r_i(X) + f(o_i(X))_{\mathrm{last}}$. Therefore a pair of consecutive states will be identical, an unnecessary redundancy.

To go from a change to a state sequence is more difficult than the opposite way. We are given $r_2 - r_1$ and must find r_1 and r_2. In general this is not uniquely defined, whereas $r_2 - r_1$ when given r_1 and r_2 is.

This corollary provides a way of uniquely determining a corresponding state-change sequence definition when given a state sequence definition. Given the ($r_i(X)$'s) in the sequence definition the $r_i(X) - r_{i-1}(X)$'s of the state-sequence definition are easily found.

However given a state-change sequence the corresponding state sequence is not uniquely determined, though we can often derive one or more corresponding state sequence definitions.

Assume that $r_i(x) - r_{i-1}(x)$ of a given state change definition equals $\{[o_{i_k}, a_{i_k} \to b_{i_k}] \mid k = 1 \text{ to } n_i\}$. Then letting $r_i(x) = \{\langle o_{i_k}, b_{i_k} \rangle \mid k = 1 \text{ to } n_i\}$ and $r_{i-1}(X) = \{\langle o_{i_k}, a_{i_k} \rangle \mid k = 1 \text{ to } n_i\}$ will provide us with terms that will complete a state sequence definition.

However any other state element could be added to $r_i(x)$ and also to $r_{i-1}(x)$ and the same $r_i(x) - r_{i-1}(x)$ would result. Similar comments apply to Theorem 3.8.1. Both kinds of transformation are illustrated in subsequent examples.

3.8.1 Example of Transformation — Tower of Hanoi

The transformation was designed, in part, to generalize the different ways the solution to the Tower of Hanoi puzzle and the Gray code can be formulated. Let us see how it applies to these examples.

We have already given the solution to the Tower of Hanoi puzzle, as a state change sequence, in Definition 3.1.1. It is repeated with some variation below.

Definition 3.8.1: Tower of Hanoi–Change of State Sequence

$$\left[\begin{array}{lll} 1. & \mathrm{chg}(n, i \to j) = \langle\,\rangle & : n = 0 \\ 2. & \mathrm{chg}(n, i \to j) = \mathrm{chg}(n-1, i \to k) & \\ & \qquad \| \langle \{[n, i \to j]\} \rangle \| \mathrm{chg}(n-1, k \to j) & : n \in N \\ & \text{Initially } n \in N,\ ij \in \{1,2,3\}, \text{ and } i \neq j. & \end{array}\right.$$

The difference in this definition is that $[n, i \to j]$ in line 2 has replaced $\langle i \to j \rangle$ in the original. This is explained as follows. To apply Theorem 3.8.1 we represent the state of the n disks, with a set of pairs $\{\langle i, s_i \rangle | i \in N\}$, s_i being the peg position (1, 2, or 3) of the ith largest disk. To apply the theorem it is also necessary, in specifying a move as a change of state, to indicate which disk is moving. $[n, i \to j]$ in line 2 does this by indicating that the move from i to j is with the nth largest disk. In fact $i \to j$ would be sufficient to represent the change since the disk that is to be moved must be the one on top of peg i and needs no other identification. Alternatively $[n, j]$ would be sufficient to indicate that the disk n is to be moved to peg j. However, the extra information in $[n, i \to j]$ will be helpful in getting the state sequence.

Assume that Corollary 3.8.1.1 is applicable. The function r must be interpreted. Definition 3.8.1 is correlated to definition 2 of the corollary. So $r_2(n, i \to j) - r_1(n, i \to j) = \{[n, i \to j]\}$. Thus $r_1(n, i \to j)$ can consistently be $\langle n, i \rangle$, and $r_2(n, i \to j)$ can be $\langle n, j \rangle$. Application of the corollary then gives the following.

Definition 3.8.2: Tower of Hanoi—State Sequence

$$\left[\begin{array}{ll} f(n, i \to j) = \langle\,\rangle & : n = 0 \\ f(n, i \to j) = (\{\langle n, i \rangle\} ++ f(n-1, i \to k)) & \\ \qquad \| (\{\langle n, j \rangle\} ++ f(n-1, k \to j)) & : n \in N \\ \text{Initially } n \in N,\ i, j \in \{1,2,3\}, \text{ and } i \neq j. & \end{array}\right.$$

This definition does indeed satisfy the requirements that make Corollary 3.8.1.1 applicable; however, other definitions will also satisfy these requirements, so we must independently verify that this one is correct.

A somewhat simpler form is obtained if the set of pairs $\{\langle i, s_i \rangle | i \in N\}$, which constitute a state, are represented by a vector $\langle s_1, \ldots, s_n \rangle$. Taking account of the definition and of properties of + and + +, and recognizing that $f(n-1, j \to k)$ is now a sequence of vectors of length $n - 1$, the + + operation can be replaced by cross concatenation |*|. The recursive line becomes

$$f(n, i \to j) = (\langle i \rangle |*| f(n-1, i \to k)) \| (\langle j \rangle |*| f(n-1, k \to j)) \qquad : n \in N$$

The interpretation of the recursive line of this definition is

> The sequence of states by which n disks are moved from i to j equals
> The sequence of states by which $n - 1$ disks are moved from disk i to disk k with an nth disk being held fixed on peg i
> followed by
> The sequence of states by which $n - 1$ disks are moved from disk k to disk j with an nth disk being held fixed on peg j

3.8.2 Example of Transformation — Gray Code

Assume we are given the definition which produces the Gray code state sequence. In the following variation on Definition 3.1.2 each 0–1 vector is built by inserting 1's and 0's into vectors of fixed length, instead of by successive concatenation of 1's and 0's.

Definition 3.8.3: Gray Code—State Sequence by Insert in Vector

$$\left[\begin{array}{ll} f(n, s) = \langle\langle 0, 0, \ldots, 0 \rangle\rangle & : n = 0 \\ f(n, s) = f(n-1, 0)_n \leftarrow * \, s \| f(n-1, 1)_n \leftarrow * \, (1 - s) & : n \in N \\ \text{Initially } n \in N = p,\ s = 0. & \end{array}\right.$$

The change in state involves the one bit in the 0–1 state vector. To make this definition look more like the state definition in Corollary 3.8.1, interpret $f(n, s)$ to be a sequence of state element pairs rather than a sequence of vectors. The state elements are pairs $\langle n, y \rangle$, where n is the index of the component (bit) in the vector to be changed, and y is the value of that component after the change. Note that ++ and $\leftarrow *$ are related by the property that if X is a vector of vectors then $X_n \leftarrow * \, y$, meaning insert y into the nth component of every vector in X, is equivalent to $\{\langle n, y \rangle\} + + X$ in our current notation. Then the recursive line of Definition 3.8.3 can be written

$$f(n, s) = \{\langle n, s \rangle\} + + f(n-1, 0) \| \{\langle n, 1 - s \rangle\} + + f(n-1, 1) \qquad : n \in N$$

which implies that, in the corollary's terms,

$$r_1(n, s) = \{\langle n, s \rangle\};\ r_2(n, s) = \{\langle n, 1 - s \rangle\}, \text{ so}$$

$$r_2(n, s) - r_1(n, s) = \{[n, s \to 1 - s]\}$$

Definition 3.8.4: Gray Code—Change of State Sequence

$$\left[\begin{array}{ll} \mathrm{chg}_f(n, s) = \langle\ \rangle & :n = 0 \\ \mathrm{chg}_f(n, s) = \mathrm{chg}_f(n-1, 0) & \\ \qquad \| \langle \{ [n, s \rightarrow (1 - s)] \} \rangle \| \mathrm{chg}_f(n-1, 1) & :n \in N \\ \text{Initially } n \in N,\ s = 0. & \end{array}\right.$$

This definition is similar to that for the Tower of Hanoi change sequence. Both produce a sequence of elementary state change pairs. The sequence of the first members of these pairs produced by both definitions is identical if both start with the same first argument n. That is, the sequence of the indices of disks moved in the Tower of Hanoi is the same as the sequence of the indices of bit position changes produced by the Gray code definition.

3.8.3 Example of Transformation— Generalized Tower of Hanoi

Consider an n-disk Tower of Hanoi arrangement with the usual three pegs, this time designated from left to right as $-1, 0, 1$. A disk on -1 can only move right, one on 1 can only move left, a dsk on 0 must move left/right if the move that brought it to 0 was left/right. Furthermore we require all n disks to start on peg -1 and end on peg 1. Each move can only be by one position. First consider the sequence of changes in state necessary, where $a \rightarrow b$ means move the top disk on peg a to the top of peg b.

The idea behind this definition for $\mathrm{chg}_h(n, d)$ is

The sequence of moves to move n discs from peg $-d$ to d equals

The sequence to move $n - 1$ disks from $-d$ to d
(largest disk on $-d$ must be exposed)

followed by

The move of the largest, disk n, from $-d$ to 0
(largest disk moves; it can only move one position)

followed by

The sequence to move $n - 1$ disks from d back to $-d$
(open peg d for largest disk now on 0)

followed by

The move of the largest, disk n, from 0 to d
(move largest disk from 0 to d)

followed by

The sequence to move $n - 1$ from $-d$ to d
(move all other disks from 0 to d)

The only elementary state changes are to and from disk 0, designated $d \rightarrow 0$ and $0 \rightarrow d$.

Definition 3.8.5: Tower of Hanoi Variant—Change of State Sequence

$$\left[\begin{array}{ll} \text{chg}_h(n,d) = \langle\,\rangle & :n = 0 \\ \text{chg}_h(n,d) = \text{chg}_h(n-1,d) & \\ \qquad \Vert\langle\{[n,-d \to 0]\}\rangle\Vert\text{chg}_h(n-1,-d) & \\ \qquad \Vert\langle\{[n,0 \to d]\}\rangle\Vert\text{chg}_h(n-1,d) & :n \in N \\ \text{Initially } n \in N,\ d = 1. & \end{array}\right.$$

Corollary 3.8.1.1 again applies; using vectors to represent states and using $|*|$ to replace $++$, we get for the state sequence

Definition 3.8.6: Tower of Hanoi Variant—State Sequence

$$\left[\begin{array}{ll} f(n,d) = \langle\,\rangle & :n = 0 \\ f(n,d) = (f(n-1,d)|*|\langle -d\rangle)\Vert(f(n-1,-d)|*|\langle 0\rangle) & \\ \qquad \Vert(f(n-1,d)|*|\langle -d\rangle) & :n \in N \\ \text{Initially } n \in N,\ d = 1. & \end{array}\right.$$

This again is not the only state sequence definition that would fit the theorem.

A Gray code sequence for base-n numbers can be obtained by generalizing that for the base-2 Gray code whose definition has been given. The definition can then be presented in either state or state change form by use of Theorem 3.8.1.

3.8.4 Example of Transformation— Chinese Ring Puzzle

A set of n rings are placed on a rod in such a way that a ring at the *rightmost* position $i = 1$ can be moved on or off with no constraint. For position $i > 1$ the ring can be removed if and only if position $i - 1$ has a ring and positions $i - 2$ down to 1 have no ring. A ring can be placed at position i if and only if position $i - 1$ has a ring and positions $i - 2$ down to 1 of the rod have no ring. The object is to obtain the sequence of ring moves, on and off the rod, which will result in the removal of a ring at position n of the rod (see Figure 3-10).

The state of the rings is represented by a vector. The ring positions are numbered *right to left* from 1 to n. A 0 at position i means there is no ring present, a 1 means there is.

The following scheme provides the basic approach to the solution.

Change sequence to go from a single 1 at position n to 0's in positions 1 to n ($\text{chg}_f(n,0)$) =

Sequence to change bit $n-1$ to 1 ($\text{chg}(n-1,1)$) followed by

Change bit n from 1 to 0 $[n,1 \to 0]$ followed by

Ring i may be removed

if there are rings at positions i, $i-1$, and anywhere else except at $i-j, j \geqslant 2$; as, for example,

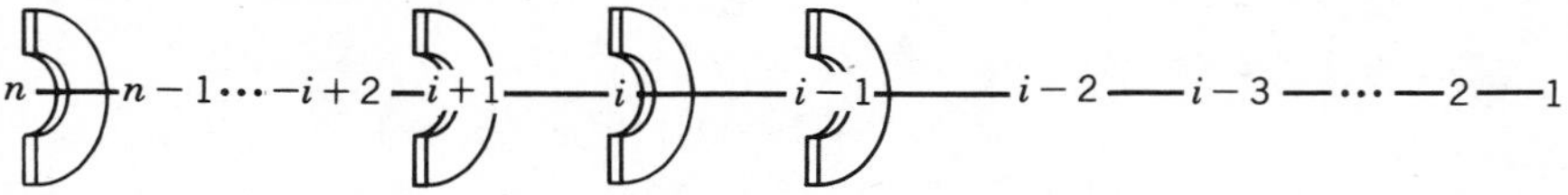

n — $n-1 \cdots -i+2$ — $i+1$ — i — $i-1$ — $i-2$ — $i-3$ — $\cdots$ — 2 — 1

Representation $\langle 10\cdots0011100\cdots000 \rangle$
$n \quad i \quad 1$

A ring may be added at i

if there are rings at positions $i-1$ and anywhere else except at i, and $i-j, j>1$; as, for example,

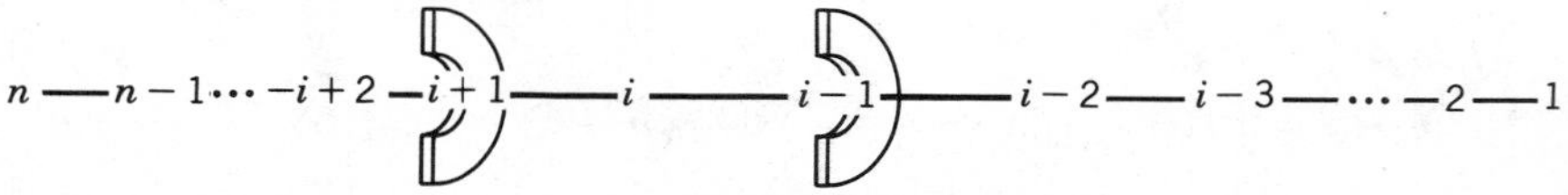

n — $n-1 \cdots -i+2$ — $i+1$ — i — $i-1$ — $i-2$ — $i-3$ — $\cdots$ — 2 — 1

Representation $\langle 00\cdots010100\cdots000 \rangle$
$n \quad i \quad 1$

FIGURE 3-10 Removing and adding rings.

Sequence to change bit $n-1$ to 0 (chg$(n-1, 0)$)

And also:

Change sequence to go from 0's in positions 1 to n to one with a single 1 at position n(chg$(n, 1)$) =

Sequence to change bit $n-1$ to 1 (chg$(n-1, 1)$)
Change bit n from 0 to 1 $[n, 0 \rightarrow 1]$
Sequence to change bit $n-1$ to 0

If we let $s \in \{0, 1\}$, the recursive definition can be expressed compactly as

Definition 3.8.7: Chinese Ring Puzzle—State-Change Sequence

$$\left[\begin{array}{ll} \text{chg}(n, s) = \langle\, \rangle & : n = 0 \\ \text{chg}(n, s) = \text{chg}(n-1, 1) \| [n, (1-s) \rightarrow s] \| \text{chg}(n-1, 0) & : n \in N \\ \text{Initially } n \in N,\ s = 1 & \end{array}\right.$$

This is virtually identical to the Gray code given by Definition 3.8.4, and the state sequence definition, is easily derived.

3.8.5 Example of Transformation — Gasoline Distribution Puzzle

(Jeep problem [Goodman, Hedetniemi 79])

A car with a total gasoline capacity of 1 unit is to travel a distance of $n/3$ units to the right of its starting point. The car uses 1 unit of gas to travel 1 unit of distance. Although there is an unlimited supply of gasoline at the start, the only gasoline at any other location must be carried there by the car. The positions at which gas will be stored are $0, 1/3, 2/3, \ldots, n/3$. The state of the car and all these positions will be represented by the triple

[car position, tank contents, quantity of gas at positions]

where

1. The car position is one of $\{0, 1/3, 2/3, \ldots, n/3\}$.
2. The tank contents are 0, 1/3, 2/3, or 1 unit of gas.
3. The quantity of gas at positions is a vector $\langle q_1, \ldots, q_n \rangle$, with q_i being 0, 1/3, 2/3, or 1, indicating the quantity of gas located at position $i/3$.

A state change will be represented by the pair

[position change, tank change]

where

1. During each state change the car will travel rightward or leftward between positions. This will be indicated by [→] and [←], respectively, in the position change part of the state change;
2. During a position change the car's tank will lose 1/3 unit of gas because of its traveled, and it will acquire or lose some gas as a result of leaving it or picking it up at the position from which it moves. This will be indicated by giving the net change in the contents of the gas tank in the tank change part of the state change.

The object is to find a sequence of these actions that will result in getting the car a distance $n/3$ units to the right of the starting point.

To do this we develop a function $\text{chg}_f(n)$ which gives the sequence of changes needed, starting with the car at position 1/3, with 2/3 units of gas in its tank, to *add* 1/3 unit of gasoline at each of the positions $1/3, 2/3, \ldots, n/3$, and end with the car at position 0 with an empty tank.

$\text{chg}_f(n)$ will be defined recursively using the following approach:

$\text{chg}_\text{f}(\text{n}) =$

1. Do the following sequence of actions:
   $\text{chg}_f(n-1) \| \langle[\rightarrow, 2/3]\rangle \| \text{chg}_f(n-1) \| \langle[\rightarrow, 2/3]\rangle \| \text{chg}_f(n-1)$
   $\langle[\rightarrow, 2/3]\rangle$ is a move to the right from position 0 with an empty tank to position 1/3 with 2/3 of a tank of gas. At 0, 1 unit of gas was picked up, then 1/3 was used up in going to 1/3, leaving 2/3.

Assuming $\text{chg}_f(n-1)$ has the assigned meaning, the total change will finally result in a state in which at each position 1/3, 2/3, 3/3, 4/3, ..., $(n-1)/3$ there is 1 unit of gasoline and the car is at position 0.

2. Starting with 1 unit of gasoline in the car at position 0, move to position 1/3 ($[\rightarrow, 2/3]$).
3. Now move rightward to successive positions taking 1/3 of a unit of gasoline away from the 1 unit stored there and using 1/3 unit for each change in position. 2/3 of a unit will thus remain at each position. The sequence of actions is represented by $\langle[\rightarrow, 0]^{n-1}\rangle$. When the car arrives at position $n/3$, it will have 2/3 unit of gasoline in its tank.

   This will result in a state in which at each position 1/3, 2/3, 3/3, 4/3, ..., $(n-1)/3$, there is 2/3 units of gasoline, 0 at $n/3$, the car is at position $n/3$ with 2/3 unit of gasoline in its tank.
4. Return leftward from position $n/3$, leave 1/3 unit of gasoline there, and return to position $(n-1)/3$, decreasing the tank contents by 2/3 to 0 ($\langle[\leftarrow, -2/3]\rangle$).

   This will result in a state in which at each position 1/3, 2/3, 3/3, 4/3, ..., $(n-1)/3$, there is 2/3 units, while at $n/3$ there is 1/3 unit of gasoline. The car is at position $(n-1)/3$.
5. Move leftward to position 0, removing 1/3 unit of gasoline at each position, with net change of 0 in tank ($\langle[\leftarrow, 0]^{n-1}\rangle$).

   This will result in a state in which at each position 1/3, 2/3, 3/3, 4/3, ..., $n/3$, there is 1/3 unit of gasoline and the car is at 0. This is the final state of $\text{chg}_f(n)$.

The definition that results for the state change sequence is

Definition 3.8.8: Gas Distribution Puzzle—State Change Sequence

$$
\left[\begin{array}{ll}
\text{chg}_f(n) = \langle[\leftarrow, -2/3]\rangle & : n = 1 \\
\text{chg}_f(n) = \text{chg}_f(n-1) \| \langle[\rightarrow, 2/3]\rangle \| & \\
\qquad \text{chg}_f(n-1) \| \langle[\rightarrow, 2/3]\rangle \| \text{chg}_f(n-1) \| & \\
\qquad\qquad \langle[\rightarrow, 2/3)]\rangle \| \langle[\rightarrow, 0]^{n-1}\rangle \| & \\
\qquad\qquad \langle[\leftarrow, -2/3]\rangle \| \langle[\leftarrow, 0]^{n-1}\rangle & : n \in N_2 \\
\text{Initially } n \in N. &
\end{array}\right.
$$

To motivate the definition for $f(n)$, we give the state sequence corresponding to definition 3.8.8, the state change for chg_f and the corresponding state sequence for $f(2)$.

State: [car position, tank contents, quantity of gas at positions]

State change: [position change, tank change]

States:	*States Changes*
$\langle 1/3, 2/3, \langle 0, 0\rangle\rangle$	
	$[\leftarrow, -2/3]$
$\langle 0, 0, \langle 1/3, 0\rangle\rangle$	
	$[\rightarrow, 2/3]$
$\langle 1/3, 2/3, \langle 1/3, 0\rangle\rangle$	
	$[\leftarrow, -2/3]$
$\langle 0, 0, \langle 2/3, 0\rangle\rangle$	
	$[\rightarrow, 2/3]$
$\langle 1/3, 2/3, \langle 2/3, 0\rangle\rangle$	
	$[\leftarrow, -2/3]$
$\langle 0, 0, \langle 1, 0\rangle\rangle$	
	$[\rightarrow, 2/3]$
$\langle 1/3, 2/3, \langle 1, 0\rangle\rangle$	
	$[\rightarrow, 0]$
$\langle 2/3, 2/3, \langle 2/3, 0\rangle\rangle$	
	$[\leftarrow, -2/3]$
$\langle 1/3, 0, \langle 2/3, 1/3\rangle\rangle$	
	$[\leftarrow, 0]$
$\langle 0, 0, \langle 1/3, 1/3\rangle\rangle$	

Referring to Theorem 3.8.1 for the development of $f(n)$'s definition, we see that the recursive line for $f(n)$ will have three occurrences of $f(n-1)$ corresponding to the three occurrences of chg$(n-1)$. Furthermore $r_1\,(n-1)$, $r_2(n-1)$, $f(n-1)_1$, and $f(n-1)_{\text{last}}$ will have to satisfy the following relation in the state sequence we seek.

$$[r_2(n-1) + f(n-1)_1] - [r_1(n-1) + f(n-1)_{\text{last}}] = [\rightarrow, 2/3] \ldots (1)$$

But we have defined the starting state

$$f(n-1)_1 = \langle 1/3, 2/3, \langle 0, \ldots, 0\rangle\rangle(n, 0\text{'s})$$

And the final state of $f(n-1)$

$$f(n-1)_{\text{last}} = \langle 0, 0, \langle 1/3, \ldots, 1/3, 0\rangle\rangle(n-1, 1/3\text{'s})$$

And it is reasonable to expect that

$$r_1(n-1) = \langle 0, 0, \langle 0, \ldots, 0 \rangle\rangle$$

From which, interpreting + and − roughly as vector addition and subtraction we infer that

$$r_2(n-1) = \langle 0, 0, \langle 1/3, \ldots, 1/3, 0 \rangle\rangle$$

because then the right of 1 becomes

$$[\langle 1/3, 2/3, \langle 1/3, \ldots, 1/3, 0 \rangle\rangle - \langle 0, 0, \langle 1/3, \ldots, 1/3, 0 \rangle\rangle]$$

which is a change in the cars position of 1 position right from 0 to 1/3 given by $\rightarrow$, and a change of 2/3 in its tank from 0 to 2/3, and no change in the amount of gas stored at any position, remaining at 1/3. But this is what the left side of (1) says.

In a similar way we can get $r_3(n-1)$ and then using the following notation, the definition for $f(n)$. If $f(n-1)$ is a sequence of states, each of the form $\langle t, p, \langle q_1, \ldots, q_n \rangle\rangle$, then $\langle\langle a, b, \langle c_1, \ldots, c_n \rangle\rangle| + |f(n-1)$ is the sequence of states formed by replacing each state $\langle\langle t, p, \langle q_1, \ldots, q_n \rangle\rangle\rangle$ in $f(n-1)$ with

$$\langle\langle t + a, p + b, \langle q_1 + c_1, \ldots, q_n + c_n \rangle\rangle\rangle$$

Definition 3.8.9: Gas Distribution Puzzle—State Sequence

$$\left[\begin{array}{ll}
f(n) = \langle\langle 2/3, 1/3, \langle 0, \ldots, 0 \rangle\rangle, \langle 0, 0, \langle 1/3, \ldots, 0 \rangle\rangle\rangle & : n = 1 \\
f(n) = f(n-1) \| \langle\langle 0, 0, \langle 1/3, \ldots, 1/3, 0 \rangle\rangle | + | f(n-1) \| & \\
\quad \langle\langle 0, 0, \langle 2/3, \ldots, 2/3, 0 \rangle\rangle\rangle | + | f(n-1) & \\
\quad \langle\langle 2/3, 1/3, \langle 1, 1, \ldots, 1, 0 \rangle\rangle\rangle \| & \\
\quad \langle\langle 2/3, 2/3, \langle 2/3, 1, \ldots, 1, 0 \rangle\rangle\rangle \| \ldots & \\
\quad \| \langle\langle 2/3, n/3, \langle 2/3, 2/3, \ldots, 2/3, 0 \rangle\rangle\rangle \| & \\
\quad \langle\langle 0, n - 1/3, \langle 2/3, 2/3, \ldots, 2/3, 1/3 \rangle\rangle\rangle \| \ldots & \\
\quad \| \langle\langle 0, 1/3, \langle 2/3, 1/3, \ldots, 1/3, 1/3 \rangle\rangle\rangle \| & \\
\quad \langle\langle 0, 0, \langle 1/3, 1/3, \ldots, 1/3 \rangle\rangle\rangle) & : n \in N_2 \\
\text{Initially } n \in N. &
\end{array}\right.$$

In all examples preceding this, the interpretation of what "object" as used in the state element, $\langle$object, place$\rangle$ has been quite natural. Here however the "objects" are: the car, the car's gas tank, and all the gas storage positions. The first two are objects in common usage-the gas storage positions is not. The "object" part of the state element pair need only have properties implied by its definition and not by the usual meaning of the word.

In addition to the examples above Theorem 3.8.1 applies to many common recursive definitions, such as those that generate binary numbers, permutations, and so on. The application to the binary number, and permutation definition however require the interpretation of the vector positions as "objects" of the state element. Another characteristic of the binary number, and permutation state-change sequence definition is that the changes they generate typically affect a number of objects (vector positions). In the preceding examples only one object is typically affected by each change.

3.9 THE CARDINALITY OF A RECURSIVELY DEFINED SET

In Section 3.2 it was noted that the relation between a recursive definition that enumerates a set and one that counts the number of members in that set is often very close. If the w function is either a union of sets known to be disjoint, or a concatenation, and the terminal line values are each a member of the enumerated set, then there is a simple transformation which give the corresponding count definition, namely, replace the w function with addition and replace each terminal line right side with a 1.

For example, if this is done to the Tower of Hanoi definition, Definition 3.1.1, we get

$$\left[\begin{array}{ll} h(n, i-j) = 1 & :n = 1 \\ h(n, i-j) = h(n-1, i-k) + h(1, i-j) + h(n-1, k-j) & :n > 1 \\ \text{Initially } n \in N,\ i,\ j \in \{1,2,3\}. & \end{array}\right.$$

The second component of the argument is irrelevant to the value of this definition, since it does not affect the number of moves required. Removing it and noting that there are then two occurrences of $h(n-1)$ on the right of the recursive line, and that $h(1) = 1$, we obtain

$$\left[\begin{array}{ll} h(n) = 1 & :n = 1 \\ h(n) = 2 * h(n-1) + 1 & :n > 1 \\ \text{Initially } n \in N. & \end{array}\right.$$

As is easily verified, the solution to this definition is $h(n) = 2^n - 1$, which is the number of individual disk moves necessary to move n disk between any two pegs legitimately in the Tower of Hanoi puzzle. There are general methods for solving such recursive definitions or difference equations. For the simpler cases, like the one above, repeated substitution and induction are sufficient to suggest a solution, which may then be easily proven.

The transformation above is between definitions for related but very different functions. The next transformation relates definitions that define the same function.

3.10 TRANSFORMATION OF A NUMBER OF DEFINITIONS INTO ONE — JAMMING [Aho, Hopcroft, Ullman 75; Sethi, Aho, Ullman 86]

Next we consider a "one-for-many" transformation in which two or more different definitions are combined into one.

Jamming is the process of combining two DO loops with similar loop conditions into a single loop. Here this process is applied to recursive definitions.

Given two definitions with the same substitutional domain and terminal conditions, but differing in their w, t, and p functions.

Definition 3.10.1

a)
$$\left[\begin{array}{ll} f(x) = t(X) & :T(X) \\ f(X) = w\big(f(o_1(X)),\ldots, f\big(o_{m(x)}(X)\big), p(X)\big) & :\sim T(X) \\ \text{Initially } X \in D. & \end{array}\right.$$

b)
$$\left[\begin{array}{ll} f'(X) = t'(X) & :T(X) \\ f'(X) = w'\big(f'(o_1(X)),\ldots, f'\big(o_{m(x)}(X)\big), p'(X)\big) & :\sim T(X) \\ \text{Initially } X \in D. & \end{array}\right.$$

By *interspersing* their lines we can rewrite these as a single definition.

Definition 3.10.2

$$\left[\begin{array}{ll} f(X) = t(X) & :T(X) \\ f'(X) = t'(X) & :T(X) \\ f(X) = w\big(f(o_1(X)),\ldots, f\big(o_{m(x)}(X)\big), p(X)\big) & :\sim T(X) \\ f'(X) = w'\big(f'(o_1(X)),\ldots, f'\big(o_{m(x)}(X)\big), p'(X)\big) & :\sim T(X) \\ \text{Initially } X \in D. & \end{array}\right.$$

Evaluating an interspersed pair is simpler than evaluating two completely independent definitions. Assuming both are to be evaluated for the same initial argument structure, only one O Network need be simulated since the same argument structures will be generated in the evaluation of the two definitions. One t line is enough if the output of each t block is a pair, that is, if X is the input, then the output of the combined t block is $\langle t(x), t'(X)\rangle$. Similarly one w network is sufficient, provided each w block has two outputs, that is, if $\langle X, Y\rangle$ is the input to such a combined w block, then $\langle w(X), w'(Y)\rangle$ is the output. This becomes clearer when the two definitions are written together as a single *two-component vector* function.

Definition 3.10.2(1)

$$\left[\begin{array}{ll} \langle f(X), f'(X)\rangle = \langle t(X), t'(X)\rangle & :T(X) \\ \langle f(X), f'(X)\rangle = \langle w\big(f(o_1(X)),\ldots, f\big(o_{m(x)}(X)\big), p(X)\big), & \\ \qquad w'\big(f'(o_1(X)),\ldots, f'\big(o_{m(x)}(X)\big), p'(X)\big)\rangle & :\sim T(X) \\ \text{Initially } X \in D. & \end{array}\right.$$

Using the function multiple $[w - w']$ defined in Section 1.2, this can be rewritten more compactly as

Definition 3.10.2(2)

$$\left[\begin{array}{ll} \langle f(X), f'(X)\rangle = \langle t(X), t'(X)\rangle & :T(X) \\ \langle f(X), f'(X)\rangle = [w - w'](\langle f(o_1(X)), f'(o_1(x))\rangle,\ldots, & \\ \qquad \langle f\big(o_{m(x)}(X)\big), f'\big(o_{m(x)}(X)\big)\rangle, \langle p(X), p'(X)\rangle) & :\sim T(X) \\ \text{Initially } X \in D. & \end{array}\right.$$

Let $h(X) = \langle f(X), f'(X)\rangle$ and $P(X) = \langle p(X), p'(X)\rangle$. Then Definition 3.10.2(2) may be rewritten.

Definition 3.10.3

$$\left[\begin{array}{ll} h(x) = \langle t(X), t'(X)\rangle & :T(X) \\ h(x) = [w - w']\big(h(o_1(X)),\ldots, h\big(o_{m(x)}(X)\big), P(X)\big) & :\sim T(X) \\ \text{Initially } X \in D. & \end{array}\right.$$

In summary, to evaluate two recursive definitions [Definitions 3.10.1(1) and (2)] with the same starting domain and the same o_i functions, it is only necessary to evaluate a single definition (Definition 3.10.3). The interspersed form, Definition 3.10.2(2), in fact already contains the essence of the single definition. Even if some of the o_i's are missing in one or the other of the two initial definitions, or if p or p' is missing in either definition, the result holds.

A similar result also holds for nested definitions.

An example of the jamming of nested definitions will be given in the discussion of biconnected components, Section 5.5.4.1.

3.10.1 Use of Nesting to Express Implementation Details

In Chapter 2, Section 2.6.1, we showed the equivalence of a class of nested and nonnested definitions. Nesting allows us to express implementation details at the definition level, which are otherwise only expressible in the algorithm. As an example of this power of nested definitions, consider a nested definition computing the shortest path.

Definition 3.10.4: Cost of Shortest Path

$$\left[\begin{array}{ll} g(V,c,v)=\min(V,c) & :v=\text{"goal"} \\ g(V,c,v)=V & :v\neq\text{"goal,"}\ c\geq V \\ g(V,c,v)=g(\ldots g(g(V,c+c_1(v),n_1(v)), & \\ \qquad c+c_2(v),n_2(v))\ldots c+c_{N(v)}(v),n_{N(v)}(v)) & :v\neq\text{"goal,"}\ c<V \\ \text{Initially } V=\infty,\ c=0,\ v=\text{"begin."} & \end{array}\right.$$

This definition is obtained by applying the transformation in Section 2.6.1 to the definition given in Problem 11 of Section 2, and adding the second line and the test of whether the developing cost c is greater than V, the minimum path cost computed so far. The second line gives a terminal value for $g(V, c, v)$ if $V \leq c$, because this condition implies that we already have a path smaller than any we can get by continuing to add to the cost c. Thus at the definition level we have managed to express the implementation detail that a path of length greater than or equal to one already found need not be further developed.

Generalizing this example, we get a class of nested definitions, in which implementation details are expressible. In Definition 3.10.5, which is a modification of Definition 2.6.2, a new relation $R(V, X)$ appears in the added line,

$$g(V,X)=V \qquad :R(V,X) \text{ and } \sim T(X),$$

and in the nonterminal condition, where $\sim T(X)$ has been partitioned into,

$$R(V,X) \text{ and } \sim T(X) \qquad \text{and} \qquad \sim R(V,X) \text{ and } \sim T(X)$$

$R(V, X)$ and the added line are generalizations of $c \geq V$, and the second line of the shortest path definition.

Definition 3.10.5: General Nested Definition Form with Implementation

$$\left[\begin{array}{lll} & g(V,X)=w(V,t(X)) & :T(X) \\ & g(V,X)=V & :R(V,X) \text{ and } \sim T(X) \\ 1. & g(V,X)=w(g(\ldots g(g(V,o_1(X)),o_2(X)),\ldots,o_{m(x)}(X)),p(X)) & \\ & & :\sim R(V,X) \text{ and } \sim T(X) \\ & \text{or} & \\ 2. & g(V,X)=g(\ldots g(g(w(V,p(X)),o_1(X)),o_2(X)),\ldots,o_{m(x)}(X)) & \\ & & :\sim R(V,X) \text{ and } \sim T(X) \\ & \text{Initially } V=0_w,\ X=d,\ d\in D. & \end{array}\right.$$

To understand how the relation $R(V, X)$ might arise in a general setting, we first need an interpretation of the significance of V in Definition 3.10.5.

Remember that Definition 3.10.5 without the second line is equivalent to DEF.f, a standard form nonnested definition (Definition 2.6.1), defining $f(X)$. When the first argument structure involving X, say (V, X), is encountered in the evaluation of $g(0_w, d)$, V contains the result of processing all previously encountered argument structures in a depth first evaluation of DEF.f. This interpretation of V is now developed more precisely. Refer to the recursive line in Definition 3.10.5 and Theorem 2.6.2, and let: $X = o_I(d)$ with $I = \langle i_1, \ldots, i_n \rangle$. Then V of (V, X) is given as

$$V = g\Big(g\Big(\cdots g\Big(\cdots g\Big(g\Big(\cdots g(0_w, o_1(d)), \ldots, o_{i_1-1}(d)\Big), \ldots, o_{i_1 1}(d)\Big), \\ o_{i_1 i_2 -1}(d)\Big), \ldots, o_{i_1 i_2 \ldots i_{n-1} 1}(d)\Big), \ldots, o_{i_1 i_2 \ldots i_{n-1}}(d)\Big)$$

which, because we have shown that $w(V, f(X))$ satisfies $g(V, X)$ of Definition 2.6.2(1), becomes

$$V = w\Big(0_w, \Big(w\Big(f(o_1(d)), \ldots, f\big(o_{i_1-1}(d)\big), w\Big(f\big(o_{i_1 1}(d)\big), \ldots, \\ f\big(o_{i_1 i_2 -1}(d)\big), \ldots, w\Big(f\big(o_{i_1 i_2 \ldots i_{n-1} 1}(d)\big), \ldots, f\big(o_{i_1 i_2 \ldots i_n -1}(d)\big), f(X)\Big)\Big)\Big)\Big)\Big)$$

Because w is associative, this becomes

$$V = w\Big(0_w, w\Big(f(o_1(d)), \ldots, f\big(o_{i_1-1}(d)\big), f\big(o_{i_1 1}(d)\big), \ldots, \\ f\big(o_{i_1 i_2 -1}(d)\big), \ldots, f\big(o_{i_1 i_2 \ldots i_{n-1} 1}(d)\big), \ldots, f\big(o_{i_1 i_2 \ldots i_n -1}(d)\big), f(X)\Big)\Big)$$

$$= w\Big(0_w, W_{K \text{ lexically} < I}\big[f(o_K(d))\big]\Big), \text{ where } W_k(y_k) = w(y_1, \ldots, y_k).$$

Now assume that in addition to being associative, w has the property that whenever the predicate $R(o_K(d), X)$ is true, then

$$w(\ldots f(o_K(d)), \ldots, A, f(X), B \ldots) = w(\ldots f(o_K(d)), \ldots, A, B \ldots)$$

Assume, that is, that whenever $R(o_K(d), X)$ is satisfied for an $o_K(d)$ which appeared earlier in the depth-first evaluation, there is no need to pursue X in order to determine $f(X)$. (This would occur if, for example, w is "$\cup$" and R is "$=$," or w is "minimum," R is "$<$," and DEF.f is monotonic). Then clearly it would be true that $g(V, X) = V$ and Definition 3.10.5 would hold.

Definition 3.10.5 was built from Definition 2.6.2, developed in Section 2.6.1, by adding a condition and a line to provide pruning notation at the definition level. In that same section we also developed Definition 2.6.4, a direct transformation of a standard nonnested definition with a *locally associative* w function, a composition of associative function v and unary function n. We now add a relation and a line to this definition to get Definition 3.10.6, useful in expressing efficiencies. We have also added another argument Y, which will be needed in future uses of nesting. Like V, Y usually serves to keep information about X's history, while X serves to determine the value of the defined function. Assume comparability exists among values of argument X, for example, duplicate values of X occur in the O tree, and that it is legitimate to prune argument structures with comparable X's. In implementing pruning, the occurrence of an argument structure comparable to one previously seen can be treated as a terminal condition. Since V and Y carry the history of values of X, it is possible to represent such comparability and thus that terminal condition T'' as a function of V, Y, and the current value of X, say $T''(V, Y, X)$. The resultant terminal value is called $u(V, Y, X)$. So if $T''(V, Y, X)$, then $g(V, Y, X) = u(V, Y, U)$. The new nonterminal condition is designated $\sim T'''(V, Y, X)$. The new terminal line improves efficiency because in the direct implementation of Definition 3.10.6 for those nonterminal argument structures (V, Y, X) for which $T'(V, Y, X)$ is satisfied, a nonrecursive line, rather than a recursive line with its additional calls, will be executed.

Definition 3.10.6

$$
\left[\begin{array}{ll}
g(V,Y,X) = \langle r(V,Y,X), Y\rangle & :T'(V,Y,X) \\
g(V,Y,X) = \langle u(V,Y,X), Y\rangle & :T''(V,Y,X) \\
g(V,Y,X) = v\big(V, n\big(v\big(g\big(\dots g\big(g(0_v, q(Y), o_1(X)), o_2(X)\big), \dots, & \\
\qquad\qquad o_{m(x)}(X)\big), p(X)\big)\big) & :\sim T'''(V,Y,X) \\
\text{Initially } V = 0_{w'},\ X = d,\ d \in D\ \ (T'''(V,Y,X) = T'(V,Y,X) \cup & \\
T''(V,Y,X)). &
\end{array}\right.
$$

The efficiency expressed in this definition will be achieved if the definition is implemented directly as a recursive procedure, as well as a suitably augmented algorithm.

3.11 PROBLEMS

1. *Combinations, Colorings, Packings—Complete Definitions.* In all the following cases, invent reasonable primitives if needed, use English if pressed.
   a. In Section 3.1.1.5 the recursive line of a definition for all combinations of m of the first n integers is given. Complete that definition for combinations.
   b. In Section 3.1.1.6 an approach to designing a definition to enumerate all colorings of a graph is outlined. Write out a complete definition for colorings.

c. Example 3.2.5 gives a sample O network for enumerating all ways of packing a given set of blocks. Give the complete definition for packings.

2. *Parity—Obtain Definition.* The parity of a vector is a function defined as follows: parity$(\langle x_1, \ldots, x_n \rangle) = (x_1 + \ldots + x_n)_{\text{mod } 2}$. Write a simple recursive definition of a function f which for any $n \in N$ produces a sequence of 2^n 0's and 1's which give the parity of the sequence of 2^n binary numbers, each viewed as a vector, in increasing numerical order. For example, if $n = 2$, the binary numbers are $\langle 0, 0 \rangle$, $\langle 0, 1 \rangle$, $\langle 1, 0 \rangle$, $\langle 1, 1 \rangle$, and the parity sequence = the result of f is $\langle 0, 1, 1, 0 \rangle$. (Try the growth approach.) Generalize your result for any modulus.

3. *Multisets—Obtain Definitions.* A multiset is a set which may have many occurrences of the same member. Let $S(n_1, \ldots, n_m)$ be a multiset of integers containing n_i copies of the integer i. Design recursive definitions that give

a. All permutations of a given $S(n_1, \ldots, n_m)$.
b. All combinations of a given $S(n_1, \ldots, n_m)$.
c. All partitions of a given $S(n_1, \ldots, n_m)$ that sum to M.

4. *Minimum Covering Problem—Obtain Definition.* Given m sets of numbers $SS = \{S1, S2, \ldots, Sm\}$, choose a set of numbers Z so that there is at least one number in Z which "represents" each set in SS. A number represents Si if it is in Si. Such a set Z is said to cover SS. If Z covers SS and no other cover of SS has fewer members than Z, then Z is a minimum cover of SS. If $SS = \{\{1, 3\}, \{2, 4\}, \{3, 2, 6\}, \{3\}, \{5, 6\}\}$, then some covers (Z) are: $\{1, 2, 3, 5\}$; $\{1, 2, 3, 6\}$, $\{2, 3, 6\}$. The last of these is a minimum cover.

Give a recursive definition for producing a minimum cover given a collection SS as input. Describe a high-level understandable algorithm for implementing the same, making it as efficient as you can.

5. *Cliques—Obtain Definition.* A clique in a graph is a completely connected subset of vertices. A maximal clique is a clique not properly included in any other clique.

a. Write a recursive definition that enumerates all cliques in a given graph G.

b. Write a recursive definition that enumerates all maximal cliques in a given graph G.

6. *NIM, A Game—Obtain Definition.* The player who moves (mover) in state S of a game is guaranteed a win if the mover in any possible next state of the game is guaranteed a loss. In certain "end" states the outcome for the mover is determined. In one version of the game of NIM the state of the game is the number of counters remaining. Each player alternately removes 1, 2, or 3 counters. The game ends as soon as the number of counters remaining becomes 0 and the mover at that state wins.

Give a definition for a function $f(n)$ which $= 1$ if the outcome of the game for the mover in state n is a guaranteed win, otherwise $f(n) = 0$.

7. *Fitting Triminoes on a Checkerboard—Obtain Definition.* A trimino is a 2-by-2 checkerboard with a corner square removed, as pictured below: A 2^k-by-2^k checkerboard can be covered with nonoverlapping triminoes leaving one noncovered square.

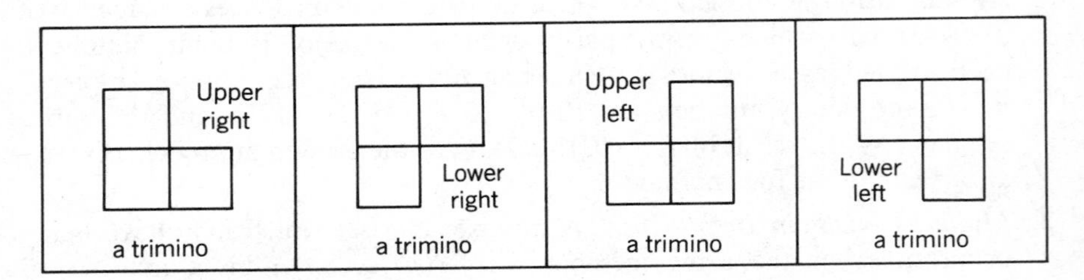

a. Prove inductively that it can be covered in such a way that the non-covered square is anywhere in the checkerboard.

b. Write a recursive definition for a function which enumerates the placement for each trimino in the cover of the 2^k-by-2^k checkerboard. The placement of a trimino is given by giving a position (row, column) of the corner square of the trimino on the checkerboard together with an indication of the orientation. There are four orientations: ur (upper right), lr (lower right), ul (upper left), ll (lower left). They each indicate the position of the concavity of the trimino.

8. *Reachability in a Digraph—Obtain Definition.* Give a recursive definition and an algorithm to find all vertices reachable from a given vertex a in a weighted digraph G.

9. *Depth-First Degree Sequences in Trees—Obtain Definition.* For an oriented tree with n nodes, DFDS(n) is a sequence $\langle d_1, \ldots, d_n \rangle$ of integers where d_i is the "degree" of the ith node encountered in a depth-first, or preorder, traversal of a tree. The "degree" of a node here means the number of children of that node. For example, the depth-first degree sequence $\langle 320003000 \rangle$ corresponds to the tree,

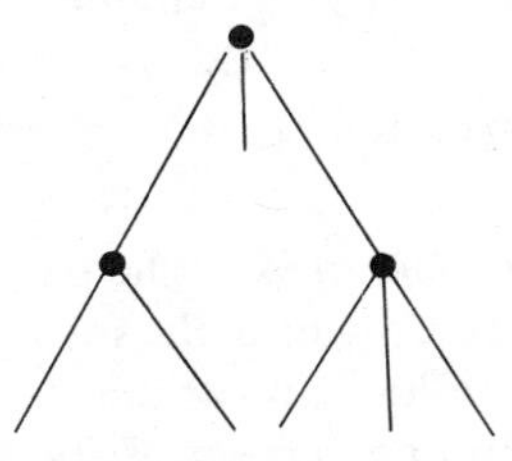

a. For any DFDS(n), $\Sigma_{j=1}^{n}[d_j] = ?$ Why?

b. Can the partial sums $\Sigma_{j=1}^{m}[d_j] = m - 1$, $m < n$? How big can d_1, d_2, d_i be?

c. Use the properties in a and b to write a recursive definition to generate the set of all DFDS(n)'s for trees with n nodes.

10. *A Depth-First Degree Sequence—Obtain a Nested Definition.* With $D =$ a DFDS$(n) = \langle d_1, \ldots, d_n \rangle$, $h(i)$ is the index in D that represents the last node in the subtree with root i represented by the depth-first degree sequence $\langle d_i, d_{i+1}, \ldots, d_{h(i)} \rangle$. Write a *nested* recursive definition for $h(i)$.

11. *Nodes at Level L in a Tree—Obtain a Definition.* Give a recursive definition for a function g which computes the number of nodes at depth k (the root is at depth 0) in a tree represented by a given depth-first degree sequence D. (Your definition may use h of the previous question or a modest extension of h as a primitive, but this is not necessary.)

12. *Getting a Backtrack Set—Exercises.*

a. Derive an equivalent backtrack set definition for the following set:

$$\text{comb}(n, m) = \{\langle v_1, \ldots, v_m \rangle | v_j \in N^n \text{ and } v_j > v_{j-a}, 0 < a < j \text{ and } j, a \in N^m\}$$

b. Derive an equivalent backtrack set definition for the set of all combinations of partitions of n into p parts in which no two components of a partition are the same, that is,
part$(p, n) = \{ v_1 + \cdots + v_p \,|\, n = v_1 + \cdots + v_p, v_i > v_{i-1}, : i \in N_2^p$ and $v_1 \geq 1\}$.

13. *Getting Backtrack Set and Corresponding Recursive Definition—Exercises.* Consider the following definition in which:

a. v_i is in N^n and

b. This is the condition called end k in the text. k is the largest index such that there is a v_k which satisfies condition C. (Condition C will be described in the definition.)

$$f(n, d) = \{\langle v_1, \ldots, v_k \rangle | C = (\text{for } i, j > 1: v_j > v_{j-1}; |v_j - v_i| \leq d)\}.$$

Give an equivalent definition in backtrack form and a corresponding recursive definition for this set.

14. *Backtrack-Set Definition to Recursive Definition—Exercises.* Give a recursive definition for the functions defined below, where G is a weighted graph; a and b are vertices of G.

a. $f(G, a, b) = \{\langle v_1, \ldots, v_k \rangle | C = (v_1 = a, \text{ for } j > 1: (v_{j-1}, v_j) = \text{an edge of } G; \text{ if } i \neq j, \text{ then } v_i \neq v_j; v_k = b)\}$

b. $f(G)$ is the vector of edges with smallest total cost in the following set:

$$\text{set}(G) = \{\langle e_1, \ldots, e_k \rangle | C = (e_j \text{ is an edge in } G, (e_1, \ldots, e_k) \text{ is a spanning tree in } G)\}$$

A spanning tree of connected graph G is a connected subgraph of G which contains all vertices of G and is acyclic.

15. *Update Functions—Exercises.* In each of the following cases, if there is a useful update function, give it; otherwise indicate why there is none.

a. v is vector
$o_i(\ldots,v,\ldots) = \ldots,v\|a_i,\ldots$
$\text{op}(\ldots,v,\ldots) = \text{maximum(components of } v)$

b. v is vector
$o_i(\ldots,v,\ldots) = \ldots,v - v_i,\ldots$
$\text{op}(\ldots,v,\ldots) = \text{maximum(components of } v)$

c. v is a vector
$o_i(\ldots,v,\ldots) = \ldots,v\|a_i,\ldots$
$\text{op}(\ldots,v,\ldots) = \text{all components of } v \text{ with property } P.$

16. *Update Functions Improvement in Recursive Definition.* Give a recursive definition equivalent to Definition 3.4.14 which has been simplified by including an update function.

17. *State Change to State Sequence.* Given the binary number state sequence definition,

$$\left[\begin{array}{ll} f(n) = \langle\,\rangle & :n = 0 \\ f(n) = \langle 0\rangle|*|f(n-1)\|\langle 1\rangle|*|f(n-1)) & :n \in N \\ \text{Initially } n \in N. & \end{array}\right.$$

Give the corresponding state change sequence.

CHAPTER FOUR

IMPLEMENTATION BY SUBSTITUTIONAL EVALUATION—SPACE EFFICIENCY

Implementations as Simulations of the $O-w-t$ NETWORK

In this chapter we study algorithms which simulate O–w–t networks—the parallel computation model of the substitutional evaluation process. The algorithms differ from each other in the order in which they simulate network blocks. All such algorithms are constrained to simulate a (child) block, say b, only after having simulated all (parent) blocks whose outputs are b's inputs. Broadly, the algorithms that simulate all the blocks in one stage of the O network before those of the next stage are called *breadth-first*, while those that always simulate the leftmost still unsimulated child block and only return to parent blocks when there are no unsimulated child blocks are called *depth-first*. Breadth-first simulation moves evenhandedly through the network, trying to produce all nonterminals before the terminals, and then all terminals one after the other, while depth-first moves leftward down the network to a terminal rapidly, then, intermixing only necessary nonterminals, moves to reach another terminal. These orderings are considered in [Nilsson 80] for passing through a state space, and they have gained currency as orders of passing through a finite graph largely through the work of [Tarjan 72]. Other orderings are possible, but these two and close variants are the ones studied here. The order of simulation has little effect on the speed of the algorithm, but can have considerable effect on its temporary storage needs.

All the networks simulated in this chapter are acyclic. Only algorithms that simulate each block at most once are considered; if the result of a block simulation is needed more than once, it is stored rather than recomputed.

We consider the implementation of a number of classes of recursive definition. Each class is given as a schema, that is, a general form of equations set whose primitives are constrained only by certain general properties. For each such definition schema we give a corresponding algorithm schema, that is, a program written in terms of the primitives of the corresponding definition, and elementary programming operations. Similar implementations of classes of recursive definitions are given in [Auslander 78] and [Strong 71].

Each algorithm in Section 4.1 is listed together with the definition it implements—algorithm on the left, definition on the right. Algorithms are given for standard form definitions with [eq. (2.2.3)] and without a $p(X)$ term. Applicable constraints on the primitives of the definition, such as, "w is associative," are also listed. The algorithm notation is that given in section 1.2. A review of the more exotic of these constructs may be helpful.

OUT $\leftarrow X$	Printout X.
0_w	A member of the domain of $w \ni \forall X$ in that domain, $w(X, 0_w) = w(0_w, X) = X$, that is, 0_w is the identity for w.

If Z is a list and X a variable containing a list entry, then

$Z[i]$	ith entry in Z, with the top of Z being the first ($Z[1]$).
$Z \leftarrow$ psh- X	Push contents of X into Z.
$X \leftarrow$ pop- Z	Remove top entry from Z and place in X.
$Z \leftarrow$ queue- X	Put contents of X at the bottom of queue Z.

If Z is a vector with n components, $\langle X_1, \ldots, X_n \rangle \leftarrow Z$ is equivalent to $X_1 \leftarrow Z_1; \ldots; X_n \leftarrow Z_n$.

4.1 BASIC BREADTH-FIRST ALGORITHMS

Breadth-first simulation of an O–w–t network procedes through the O network stage by stage. A queue holds all the argument structures computed at each succeeding stage. In general, during the run of the algorithm, terminal as well as nonterminal argument structures appear in the queue at different times. The terminals must remain there until the entire O network is simulated. Since the w network is the mirror image of the O network, its proper simulation requires that the structure of the O network also be preserved in the queue. This can be done by parentheses, which will delineate groups of argument structures with a common parent. If, as is frequently true, the w function is associative, then a record of parenthesization is unnecessary. If w is also commutative, it can be performed on terminal argument structures as generated; the order in which structures appear as arguments of w is of no concern. So, if w is associative and commutative, only nonterminal argument structures need be queued. Algorithm 4.1.1 implements definitions in which w satisfies these constraints.

Algorithm 4.1.1: Breadth-First—Queues Children—w Associative and Commutative

Implementation	*Definition, Constraints, and Notes*
$W \leftarrow 0_w$	$f(X) = t(X) \quad :T(X)$
$X \leftarrow d$	$f(X) = w(f(o_1(X)), \ldots, f(o_{m(X)}(X)), p(X)) \quad :\sim T(X)$
	Initially $X = d \in D$.
$Z \leftarrow \langle\, \rangle$	
FOR ever DO	
WHILE($\sim T(X)$) DO	w is *associative* and either *commutative*
$W \leftarrow w(W, p(X))$ *	or meets the conditions in note 2.
FOR $I = 1$ to $m(X)$ DO	DEF . f must terminate.
$Z \leftarrow$ queue- $o_I(X)$	
END	Z is a queue.
$X \leftarrow$ pop- Z	$w(0_w, X) = X$ for any X in domain of w.
END	
$W \leftarrow w(W, t(X))$	
IF $Z = \langle\, \rangle$ THEN DONE	
$X \leftarrow$ pop- Z	
END	

Notes

1. The starred line (*) should be excluded if $p(X)$ is not present in the definition.
2. If a definition does not involve $p(X)$, we can replace the commutativity constraint with the condition that whenever a terminal A is to the left of a terminal B, the path from the root of A in the O network is no longer than that to B.
3. If w is concatenation or the union of sets which are known to be disjoint, then for $r = t$ or p,

$$\text{OUT} \leftarrow r(X)$$

can replace

$$W \leftarrow w(W, r(X))$$

In Example 4.1.1 the operation of this algorithm is traced on a hypothesized O–w–t network with initial input a, with O block outputs b, c, d, e, f, g, and with w block outputs $1, 2, \ldots, 7$. The changing contents of variables X, W and the queue Z are given in the table that accompanies the network. Contrary to expectations perhaps, the w network is not the mirror image of the O network. Because w is associative, the form shown is equivalent to the mirror image, but more accurately reflects the action of the algorithm.

Example 4.1.1

Hypothesized network

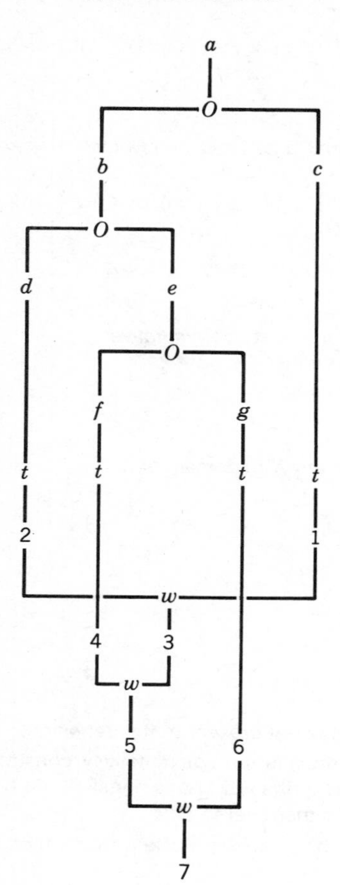

Trace

X	Z	W
a		0_w
a	bc	0_w
b	c	0_w
b	cde	0_w
c	de	$w(0_w, 1) = 1$
d	e	$w(1, 2) = 3$
e		3
e	fg	3
f	g	$w(3, 4) = 5$
g		$w(5, 6) = 7$

4.2 BASIC DEPTH-FIRST ALGORITHMS

Like their breadth-first relatives, depth-first algorithms simulate the blocks of the O–w–t network one at a time. The main difference between the two is in the ordering of this simulation. In depth-first order, a block is simulated before any of its right siblings, and after its left siblings and all their descendants have been simulated. Simulations of w and O blocks are interspersed in such a way that each w block is simulated immediately after all its inputs have been simulated. Properties of the w function determine how this is accomplished. Where breadth-first uses a queue, depth-first implementations use a stack to save O block outputs not yet used. These different storage modes produce the different order of simulation in the two types of implementation.

We will consider a number of depth-first algorithms. These are partitioned into two groups. Those in the first group all handle any O function, but differ in handling less and less constrained w functions. The algorithms in the second group are applicable only if the O function satisfies the *uniform inverse* constraint. A definition with a uniform inverse can be implemented with a compact depth-first algorithm, which is very efficient in storage usage. The inverse constraint is modest and often satisfied. Like those in the first group, algorithms in the second group differ from each other in the nature of the w functions they handle.

4.2.0.1 Basic Depth-First Algorithm with w Associative

For the first depth-first algorithm it is assumed that w is associative. This removes the need to save information about the structure of the w network and gives an implementation similar to the breadth-first algorithm, Algorithm 4.1.1. The breadth-first case required more than just associativity of w, a guarantee either that terminal values are produced in a left-to-right order or that the order does not matter (commutativity). In the depth-first case associativity alone is sufficient because depth-first sequencing produces terminal argument structures in left-to-right order.

In Algorithm 4.2.1, X starts with the initial argument structure. As long as X is nonterminal, all outputs of the block to which X is the input, that is, the children of X, are pushed into Z, rightmost first and continuing until the second child becomes the top entry of Z. The leftmost, that is, first child becomes the new value of X. If X is a terminal argument structure, its terminal value $t(X)$ is gathered up by the w function into the cumulative result held in W, that is, $W = w(t(X), W)$, stack Z is popped into X, and the process repeated. The algorithm ends when Z is empty.

Algorithm 4.2.1: Depth-First–Pushes Children—w Associative

Implementation	*Definition, Constraints, and Notes*
$W \leftarrow 0_w$	$f(X) = t(X)$ $\quad :T(X)$
$X \leftarrow d$	$f(X) = w(f(o_1(X)), \ldots, f(o_{m(X)}(X)))$ $\quad :\sim T(X)$
$Z \leftarrow \langle\,\rangle$	
	Initially $X = d \in D$.
FOR every DO	
WHILE($\sim T(X)$) DO	w is associative.
FOR $I = m(X)$ to 2 DO	DEF . f must terminate.
$Z \leftarrow$ psh- $o_I(X)$	
END	$w(0_w, X) = X$ for any X in domain of w.
$X \leftarrow o_1(X)$	Z is a stack or push-down list.
END	
$W \leftarrow w(W, t(X))$	
IF $Z = \langle\,\rangle$ THEN DONE	
$X \leftarrow$ pop- Z	
END	

Notes. If w is concatenation or the union of sets which are known to be disjoint, then

$$\text{OUT} \leftarrow t(X)$$

can replace

$$W \leftarrow w(W, t(X))$$

An example of the operation of this algorithm on a hypothesized network is given in Example 4.2.1. The network is shown on the left with $a, b, \ldots, g$ being hypothesized outputs of O blocks; 1, 2, 4, 6 of t blocks; and 3, 5, and 7 of w blocks. The trace of the contents of the variables X and W as well as the stack Z is given on the right.

Example 4.2.1

Hypothesized network

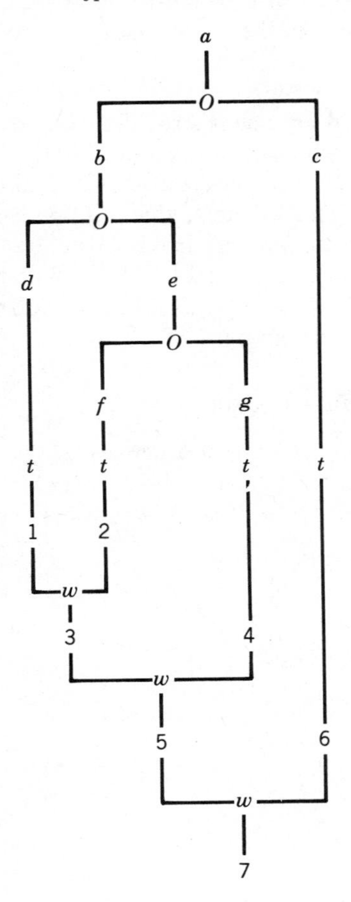

Trace

X	Z	W
a		0_w
b	c	0_w
d	ec	0_w
d	ec	$w(0_w, 1) = 1$
e	c	1
f	gc	1
f	gc	$w(1, 2) = 3$
g	c	3
g	c	$w(3, 4) = 5$
c		5
		$w(5, 6) = 7$

4.2.1 Depth-First Algorithms for Unconstrained w Function

The next three implementations will handle *any* w function, and will do so in the context of a more elaborate recursive line than that to which Algorithm 4.2.1 applies. This requires the inclusion of a new stack, V, to hold w network outputs, as a companion to the Z stack which holds O network outputs. The three implementations require progressively less and less storage in Z. The last requires no such stack at all. For each of the three algorithms a simplified version, with advantages over Algorithm 4.2.1, for handling associative w functions is easily developed. The simplified version for the last of the three implementations is usable whenever a uniform inverse is explicitly given.

4.2.1.1 Basic Depth-First Algorithm Pushing Parents and Children

To handle an arbitrary w function, a record of the structure of the O network must be available when the time comes to simulate its mirror-image w network so that correct argument can be presented to each simulated w block. In the next algorithm this is accomplished by stacking the parent together with the children in stack Z. The parent is distinguished from the children by a flag: * for children and) for parent. When a parent, known by its flag, is popped from Z, then w is to be applied to its evaluated children. The m function of this parent gives the number of evaluated children involved. These are found in the stack V from which this number of values are now popped to ARG, where w is applied to them and the result pushed back onto V. When first produced, terminal values are pushed onto V. Thus V always contains those t and w block outputs computed thus far which have not been used as inputs in computing other w block outputs.

Note. For all subsequent algorithms given in Section 4.2 it is assumed that the initial argument structure is not terminal. If this assumption is not warranted, a simple addition, the details of which are left to the reader, will permit such initial argument structures.

Algorithm 4.2.2: Depth-First—Pushing Parents and Their Children—w Unconstrained

Implementation	*Definition, Constraints, and Notes*
$X \leftarrow d$	$f(X) = t(X)$ $\quad :T(X)$
$Z \leftarrow \langle\ \rangle$	$f(X) = w(f(o_1(X)), \ldots, f(o_{m(X)}(X)), p(X)) \quad :\sim T(X)$
$V \leftarrow \langle\ \rangle$	Initially $X = d \in D$.
FOR ever DO	
WHILE($\sim T(X)$) DO	DEF . f must terminate.
$Z \leftarrow$ psh- [), X]	
FOR $I = m(X)$ to 2 DO	Z is a stack or push-down list.
$Z \leftarrow$ psh- [$*$, $o_I(X)$]	V is a stack or push-down list.
END	

```
      X ← o₁(X)
    END
    V ← psh- t(X)
    ⟨SYM, X⟩ ← pop- Z
    WHILE(SYM = )) DO
      ARG ← pop.m(X)- V          Means pop the top m(X) entries in V to ARG.
      V ← psh- w(ARG, p(X))
      IF Z = ⟨ ⟩ THEN DONE
      ⟨SYM, X⟩ ← pop − Z
    END
  END
```

Again we give a hypothesized network with a trace in Example 4.2.2. This time the contents of the variable X and the stacks Z and V are followed during simulation.

Example 4.2.2

Hypothesized network

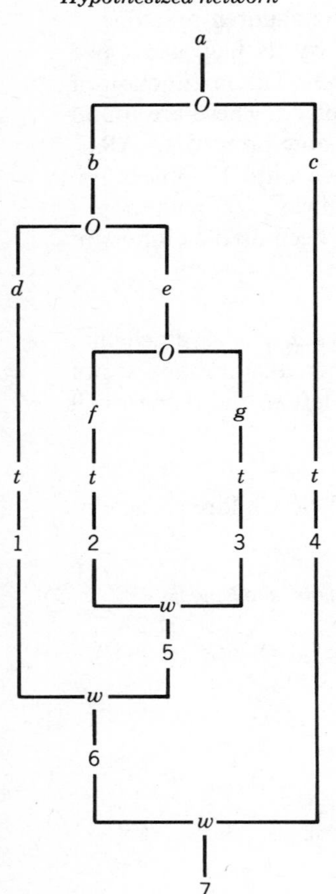

Trace

SYM	X	Z	V
	a		
	b	[*, c][), a]	
	d	[*, e][), b][*, c][), a]	
	d	[*, e][), b][*, c][), a]	1
*	e	[), b][*, c][), a]	1
	f	[*, g][), e][), b][*, c][), a]	1
	f	[*, g][), e][), b][*, c][), a]	21
*	g	[), e][), b][*, c][), a]	21
	g	[), e][), b][*, c][), a]	321
)	e	[), b][*, c][), a]	321
ARG = 2, 3; w(ARG) = w(2, 3) = 5			
	e	[), b][*, c][), a]	51
)	b	[*, c][), a]	51
ARG = 1, 5; w(ARG) = w(1, 5) = 6			
	b	[*, c][), a]	6
*	c	[), a]	6
	c	[), a]	46
)	a		46
ARG = 6, 4; w(ARG) = w(6, 4) = 7			

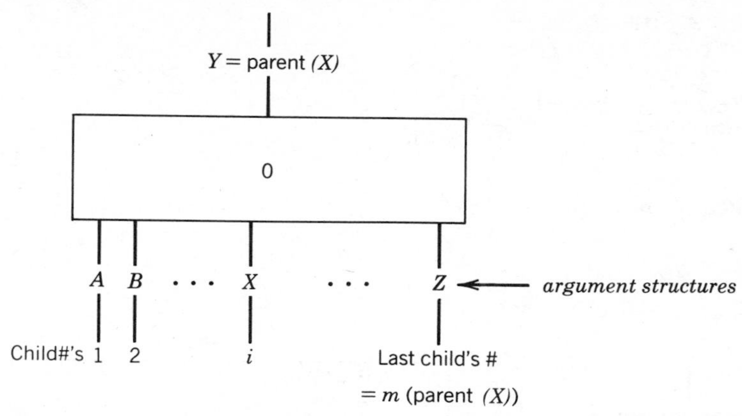

FIGURE 4-1 The uniform inverse. A uniform inverse exists if, given any O block output X, it is possible to compute a unique parent (input Y) and a unique child# of X (i), as shown.

4.2.1.2 Basic Depth-First Algorithm Pushing Parent and Child#

In the next algorithm each entry in stack Z holds information about one child and its parent. The parent is kept with the identity of its most recently generated child in Z until all its children have been processed. The child itself need never actually be pushed into Z since it would be immediately popped out into X. Z contains only parents, each flagged with a child number. If, when a parent is popped from Z, its accompanying flag indicates that it is a last child, then it is used, as in the previous algorithm, to determine the number of evaluated children on which to perform w, that evaluation is done, and the result is placed in stack V. But if the indication is that still other children are to be generated, the next child will be generated. In other respects the algorithm is similar to the previous one.

Algorithm 4.2.3: Depth-First—Pushing Parent and Child#—w Unconstrained

Implementation	*Definition, Constraints, and Notes*
$X \leftarrow d$	$f(X) = t(X)$:$T(X)$
$Z \leftarrow \langle\ \rangle$	$f(X) = w(f(o_1(X)), \ldots, f(o_{m(X)}(X)), p(X))$:~ $T(X)$
$V \leftarrow \langle\ \rangle$	Initially $X = d \in D$.
FOR ever DO	
WHILE(~ $T(X)$) DO	DEF. f must terminate.
$Z \leftarrow$ psh- $\langle 1, X \rangle$	
$X \leftarrow o_1(X)$	
END	Z is a stack or push-down list.
$V \leftarrow$ psh- $t(X)$	V is a stack or push-down list.
$\langle$CHILD#, PARENT$\rangle \leftarrow$ pop- Z	
WHILE(CHILD# = m(PARENT)) DO	
ARG $\leftarrow$ pop.m(PARENT)- V	
$V \leftarrow$ psh- w(ARG, p(PARENT))	

```
      IF Z = ⟨ ⟩ THEN DONE
      ⟨CHILD#, PARENT⟩ ← pop- Z
    END
    X ← o_CHILD#+1(PARENT)
    Z ← psh-⟨CHILD# + 1, PARENT⟩
END
```

A trace of this algorithm follows.

Example 4.2.3

Hypothesized network

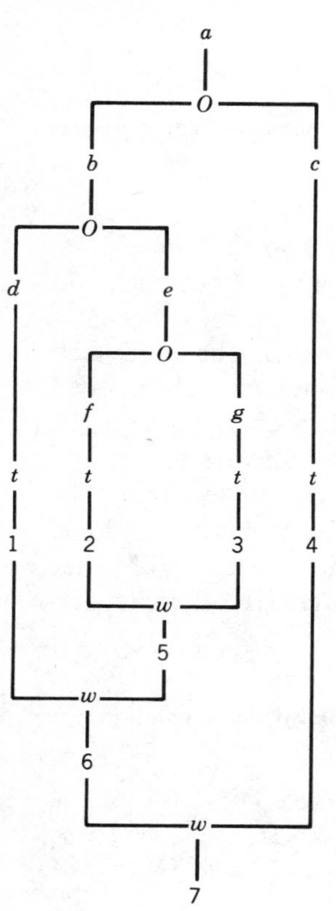

Trace

X	Child#, Parent	Z	V
a			
b		[1, a]	
d		[1, b][1, a]	
d	[1, b]	[1, a]	1
e	[1, b]	[2, b][1, a]	1
f	[1, b]	[1, e][2, b][1, a]	1
f	[1, e]	[2, b][1, a]	21
g	[1, e]	[2, e][2, b][1, a]	21
g	[2, e]	[2, b][1, a]	321
ARG = 2, 3; w(ARG) = w(2, 3) = 5			
g	[2, b]	[1, a]	51
ARG = 1, 5; w(ARG) = w(1, 5) = 6			
g	[1, a]		6
c	[1, a]	[2, a]	46
c	[2, a]		46
ARG = 6, 4; w(ARG) = w(6, 4) = 7			
c	3[2, a]		7

4.2.2 Algorithms Using the Uniform Inverse

The next version of the depth-first algorithm is applicable to definitions whose O function has a *uniform inverse*. A definition DEF.f has a uniform inverse if there are two functions, called *child#* and *parent*, both of which are defined over the domain (Δ_f − initial domain). For each argument structure X in its domain, child#(X) is an integer giving which child, X is of its parent, the 1st, or the 2nd, or..., or the last. Parent(X) is the argument structure in (Δ_f − terminal domain), say Y, for which $o_{\text{child}\#(X)}(Y) = X$. Figure 4-1 illustrates these definitions. Stack Z can be completely eliminated from a depth-first implementation of any definition with a uniform inverse. In the depth-first algorithm, Algorithm 4.2.3, X is the most recently generated argument structure. Z is used to provide the following: the parent [parent(X)], the index giving which child X is of that parent [child#(X)], and whether that index indicates that X is the last child of that parent [child#(X) = m(parent(X))?]. All this information can be computed without the Z stack if there is a uniform inverse. Additional properties of the uniform inverse will be studied later, but the algorithm that uses the child# and parent functions, Algorithm 4.2.4, which is otherwise similar to Algorithm 4.2.3, is presented now.

4.2.2.1 Depth-First Algorithm with Uniform Inverse with w Unconstrained

Algorithm 4.2.4: Depth-First—Uniform Inverse—*w* Unconstrained

Implementation	*Definition, Constraints, and Notes*
$S \leftarrow d$	$f(X) = t(X)$:$T(X)$
$X \leftarrow d$	$f(X) = w(f(o_1(X)), \ldots, f(o_{m(X)}(X)), p(X))$:~ $(T(X)$
$V \leftarrow \langle\ \rangle$	Initially $X = d \in D$.
FOR ever DO	
	DEF . f must terminate
first child { WHILE(~ $T(X)$) DO	
$X \leftarrow o_1(X)$	V is a stack or push-down list.
*******	[[V ← psh- p(X)]]
END	
terminal { $V \leftarrow$ psh- $t(X)$	
WHILE(child#(X) = m(parent(X))) DO	
backup { $X \leftarrow$ parent(X)	
ARG ← pop.$m(X)$- V	
$V \leftarrow$ psh- w(ARG, $p(X)$)	[[V[1] ← w(ARG, V[1])]]
IF $X = S$ THEN DONE	
END	
next sib { $X \leftarrow o_{\text{child\#}(X)+1}$(parent($X$))	
END	

Notes

1. If the $p(X)$ term is not in the recursive line, the $p(X)$ in w(ARG, $p(X)$) on the left should be removed.
2. The algorithm also works if the double bracketed lines on the right replace the respective lines on the left. This provides a way of computing $p(X)$ when the argument structure X is first encountered and saving it for use later when that argument structure is recreated during backup.

This algorithm not only produces $f(d)$, but also explicitly produces $f(X)$ for each X in $\Delta_f(d)$ along the way. The value of $f(X)$, for each X in $\Delta_f(d)$, is established in the backup section of the algorithm in the lower WHILE DO just prior to the statement $X \leftarrow$ parent(X). This is a significant feature.

The operation of this algorithm is now traced on an example O–w–t network, Example 4.2.4. The numbered double lines (=) cutting the hypothesized network indicate the successive contents of the V list as the algorithm simulates the blocks of the network. The trace in this case gives the value of some key functions of variables as well that of key variables.

Example 4.2.4

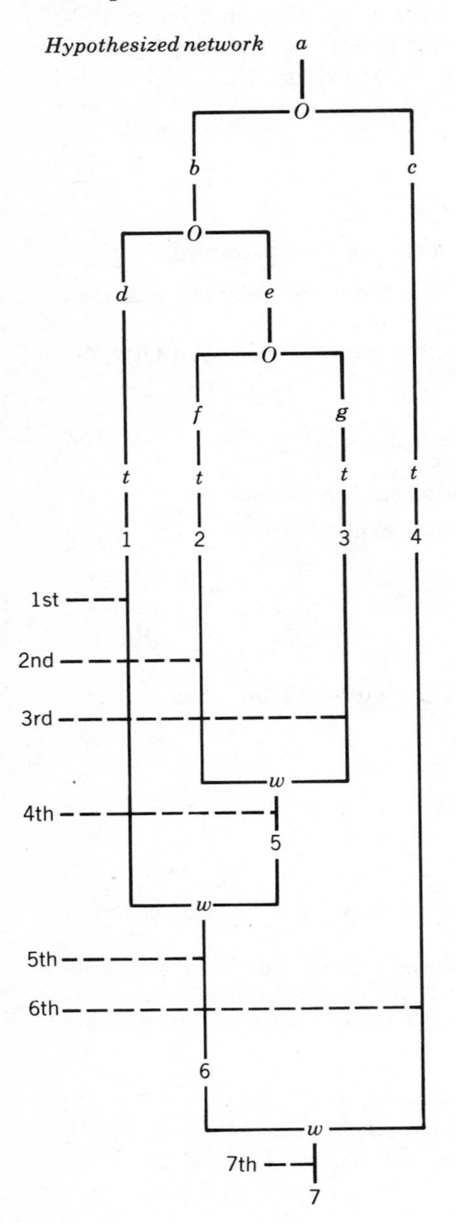

Trace

X	child#(X), parent(X)	V
a		
b		
d		
d	[1, b]	1
e		1
f		1
f	[1, e]	21
g		21
g	[2, e]	321
	ARG = 2, 3; w(ARG) = w(2, 3) = 5	
e	[2, b]	51
	ARG = 1, 5; w(ARG) = w(1, 5) = 6	
b	[1, a]	6
c		6
c	[2, a]	46
	ARG = 6, 4; w(ARG) = w(6, 4) = 7	
a		7

As has been noted, Algorithm 4.2.4, is an adaptation of Algorithm 4.2.3 to use the uniform-inverse functions, the *child# and parent*, is virtually a direct replacement for the *CHILD#* and *PARENT* variables in Algorithm 4.2.3. Understanding Algorithm 4.2.4 can be based on understanding of Algorithm 4.2.3. We also give an explanation of the workings of the algorithm, which provides a justification, independent of any relation to Algorithm 4.2.3. In doing so we will introduce definitions of general use.

In the algorithm only a single argument structure X is active at any moment. The algorithm starts with X being the initial argument structure d. It then moves into the *first-child section*. Here o_1, or the *first-child* function, is repeatedly applied to argument structure X to produce the next value of X. This continues until X becomes terminal, sending the algorithm into the *terminal value section*. There the terminal value, $t(x)$, is saved in V, and activity then passes through the *last-child test*, and, if X is not a last child, on to the *next-sib section*. There the *next-sib* function sets X equal to its next (rightward) sibling; *next-sib*(X) is alternatively defined as the next child after X of the parent of X. Control then passes back to the *first-child section*.

If X is determined to be a last child (by the *last-child test*, which asks whether $\text{child\#}(X) = \#$ of children of the parent of $X = m(\text{parent}(X))$), then there is no next sib. The algorithm then enters the *backup section*. Now the top entries of the V list are guaranteed to be the $f(Y)$'s for all of the children Y of the parent of X. These are popped from the V list to become arguments of the w function. The result of this application is $f(\text{parent}(X))$, and replaces the popped entries at the top of the V list. The current value of X is replaced by parent(X), and the *last-child test* again applied. The algorithm terminates when X again contains d, at which point the V list contains precisely $f(d)$.

Summarizing the notation developed above,

next-sib(Y) is equivalent to the function $o_{\text{child\#}(Y)+1}(\text{parent}(Y))$.
(Y *is a last-child*)? is equivalent to the predicate $\text{child\#}(Y) = m(\text{parent}(Y))$.
first-child(X) is equivalent to the function $o_1(X)$.

4.2.2.2 Inverse Depth-First Algorithm with w Associative

If in addition to having a uniform inverse, a definition has an associative w function, then along with retiring stack Z in favor of the uniform inverse functions, as in Algorithm 4.2.4, we can reduce stack V to a single-entry variable. Thus we get Algorithm 4.2.5 with no explicit stack at all.

Algorithm 4.2.5: Depth-First—Uniform Inverse—w Associative

	Implementation	*Definition, Constraints, and Notes*
1.	$S \leftarrow d$	$f(X) = t(X)$:$T(X)$
2.	$X \leftarrow d$	$f(X) = w(p(X), f(o_1(X)), \ldots, f(o_{m(X)}(X)))$:~ $T(X)$
3.	$V \leftarrow 0_w$	Initially $X = d \in D$.
4.	FOR ever DO	
5.	WHILE(~ $T(X)$) DO	
6.	$V \leftarrow w(V, p(X))$*	w is associative.

```
 7.        X ← first-child(X)        DEF . f must terminate.
 8.     END
 9.     V ← w(V, t(X))
10.     WHILE(X is last-child) DO
11.        X ← parent(X)
           [with X being a last-child]
12.        IF X = S THEN DONE
13.     END
14.     X ← next-sib(X)
        [with X being a non-last-child]
15.  END
```

Notes

1. If w is concatenation or the union of disjoint sets, then

$$\text{OUT} \leftarrow t(X)$$

can replace

$$V \leftarrow w(V, t(X))$$

2. If $p(X)$ is the last argument of w, line 6 can be moved between lines 10 and 11. If w is commutative, the algorithm also works as is if $p(X)$ appears as any argument of w.
3. If the $p(X)$ term is not in the recursive definition, line 6 may be removed.

4.2.2.3 Basic Depth-First Inverse Algorithm with w Locally Associative — Computes Intermediate Values

Like the previous one, the next algorithm takes advantage of constraints on w. It is designed for a locally associative w function (first introduced in Section 2.6.0.3) and therefore is applicable to the subcase when w is associative. Repeating the definition of locally associative: w is *locally associative* if it is equivalent to the composition of two functions n and v, where v is associative, n is any unary function, and either

$$w(x_1, \dots, x_p) = n\big(v(x_1, \dots, x_p)\big)$$

or

$$w(x_1, \dots, x_p) = v\big(n(x_1), \dots, n(x_p)\big)$$

The algorithm given is for the first of these alternatives. The changes necessary for the second are given in a note. The algorithm not only produces $f(d)$, but also explicitly produces $f(X)$ for each X in $\Delta_f(d)$. This also occurred in Algorithm 4.2.4, in which w is unconstrained, though not in Algorithm 4.2.5, in which, with the exception of $f(d)$, the value of $f(X)$ for each X in $\Delta_f(d)$ is lost. There is no such loss in the next algorithm, even though, like Algorithm 4.2.5, it takes advantage of w's (local) associativity; "local" makes the difference.

Algorithm 4.2.6: Depth First—Uniform Inverse–Local Associativity—Computes Intermediate Values

Implementation	*Definition, Constraints, and Notes*
$S \leftarrow d$	$f(X) = t(X) \qquad :T(X)$
$X \leftarrow d$	$f(X) = n(v(f(o_1(X)),\ldots,f(o_{m(X)}(X)),p(X))) :\sim T(X)$
$V \leftarrow \langle\ \rangle$	Initially $X = d \in D$.
FOR every DO	
WHILE($\sim T(X)$) DO	v is associative.
	n is unary.
	DEF . f must terminate.
	V is a stack or push-down list.
$V \leftarrow$ psh- 0_v	$V \leftarrow$ psh- $p(X)\ldots a$
$X \leftarrow$ first-child(X)	
END	
$V[1] \leftarrow v(V[1], t(X))$	
WHILE(X is last-child) DO	
ARG $\leftarrow$ pop- V	
$X \leftarrow$ parent(X)	
$FX \leftarrow n(v(\text{ARG}, p(X)))$	$FX \leftarrow n(\text{ARG})\ldots b$
IF $X = S$ THEN DONE	
$V[1] \leftarrow v(V[1], FX)$	
END	
$X \leftarrow$ next-sib(X)	
END	

Notes

1. An alternative form of the algorithm is given if lines a and b on the right replace the corresponding lines on the left. If there is no $p(X)$ term in the definition, the left side of line a and the right side of line b should be used.
2. The algorithm works for the alternate form of local associativity, that is, if the recursive line is

$$f(X) = v(n(f(o_1(X))),\ldots,n(f(o_{m(X)}(X))),n(p(X)))$$

with v associative and n a unary function, provided the existing occurrences of n are removed and those of 0_v, $t(X)$, and $p(X)$ on the right side of assignments are, respectively, replaced with $n(0_v)$, $n(t(X))$, and $n(p(X))$.

In Example 4.2.5 0 is an abbreviation for 0_v. The value of a function, say $v(x, y)$ will be represented by its Polish prefix representation vxy, so $v32$ is the value of $v(3,2)$ and $vv32pa$ that of $v(v(3,2), p(a))$. The hypothesized network is for a definition with a $p(X)$ term as in Eq. (2.2.3). The basic network for such a line is in Figures 2-9 and 2-10.

Example 4.2.5

Hypothesized network

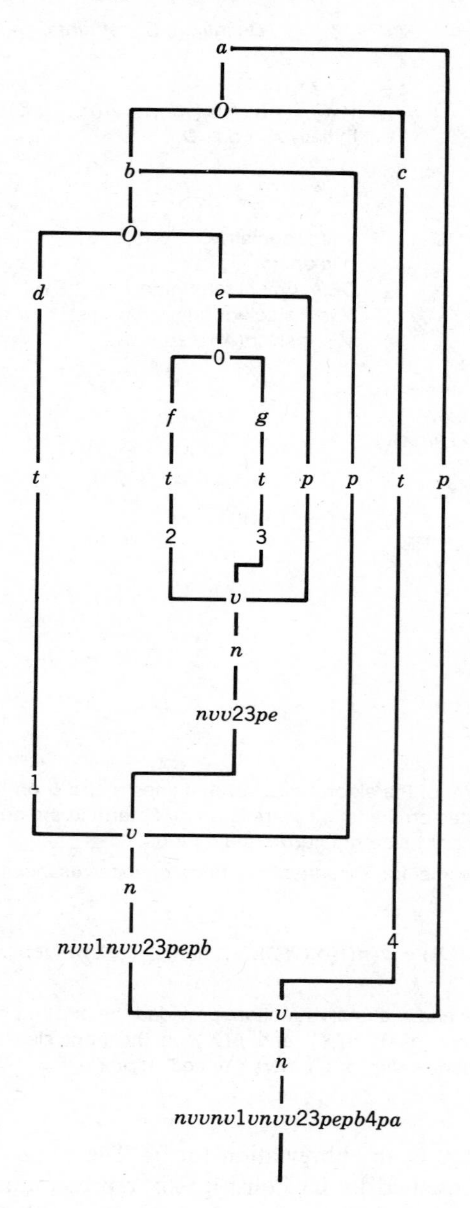

Trace

Step of algorithm	X	$\leftarrow V$ list—			
	a				
First-child	b				0
First-child	d			0	0
Terminal next-sib	e			$v01 = 1$	0
First-child	f		0	1	0
Terminal, next-sib	g		$v02 = 2$	1	0
Terminal	g		$v23$	1	0
Backup	e	$FX = nvv23pe$		$v1nvv23pe$	0
Backup	b	$FX = nvvnvv23pe1pb$		$v0nvv1nvv23pepb$ $= 0nvv1nvv23pepb$	
Next-sib	c			$nvv1nvv23pepb$	
Terminal	c			$vnvv1nvv23pepb4$	
Backup	a	$FX = nvvnvv1nvv23pepb4pa$			

A deeper understanding of this algorithm can be obtained by further study of the history of the V list. Any algorithm can be viewed as a mechanism for constructing and evaluating a function. During its execution, the construction–evaluation passes through a number of states. These may be characterized by the *partially developed function*, which has been constructed up to that point. For example, consider an algorithm to compute $\min(x_1, \ldots, x_n)$ by moving the x_i's into a list L, and each time two entries are accumulated in L, replacing them with their minimum. In execution the algorithm will pass through the state in which x_1 has just been moved into L, represented by the partially developed function $\min(x_1$. Then x_2 will be pushed into L, represented by $\min(x_1, x_2$. Then, letting $\min(x_1, x_2) = X$, a single entry of X at the top replaces the x_1 and x_2 in L, so our partial function is $\min(X$. Next x_3 is pushed into L, and the partial function is $\min(X, x_3$, and so on.

In Algorithm 4.2.6 the V list contents can be interpreted as a partially developed composition of the locally associative n and v functions. That partially developed function correlated to the V list is the *interpreted V list*. This interpretation is important because there are n and v functions which can be simplified when found in certain partially developed compositions. For example, if n is $-$ and v is min, as in the study of games (Chapter 5), then $-(\min(1, -\min(-1, x), y$ can be simplified to $-\min(1, y$ since $x \geq -1$. Returning to Algorithm 4.2.6, 0 (meaning 0_v) is entered in the V list each time the first-child part of the algorithm is executed. This creates a new nv term with a 0 argument nested as the next argument of the existing interpreted V list. So each time the algorithm passes through a *first child*, that is, goes down a level in the O network and pushes 0 into the V list, an $nv0$ [meaning $n(v(0]$ is nested at the end of the current partially developed function or interpreted V list. Except at the very end, the interpreted V list will always end with nvx, where x is called the end argument, after each pass through a section of the algorithm. When a *terminal* value, say c, is encountered, it is added as an argument to the end of the existing interpreted V list and thus as a next argument of a v function. Since v is associative, the function v of the end argument and the added c is immediately evaluated; the result replaces the existing end argument. When a backup is encountered, it is interpreted as placing the argument corresponding to $p(X)$ and closing the parenthesis at the end of the interpreted V list. This results in performing the v function on the end argument and the added $p(X)$ argument and the performing the n function on the result. In Example 4.2.6 we show the progress of the V list and its V list interpretation. In the example, "$\alpha \rightarrow \beta$" indicates that interpretation α simplifies to interpretation β.

Example 4.2.6

Section of algorithm	V list history			Interpretation	
first-child			0	$nv(0$	
first-child		0	0	$nv(0\ nv(0$	This is one step above
terminal, next-sib		1	0	$nv(0, nv(0, 1 \rightarrow nv(0, nv(1$	
first-child	0	1	0	$nv(0, nv(1, nv(0$	This is one step above
terminal	2	1	0	$nv(0, nv(1, nv(0, 2 \rightarrow nv(0, nv(1, nv(2$	
next-sib terminal,	$v23$	1	0	$nv(0, nv(1, nv(2, 3 \rightarrow nv(0, nv(1, nv(v23$	
backup	$v1nvv23pe$		0	$nv(0, nv(1, nv(v23, pe) \rightarrow nv(0, nv(1, nvv23pe \rightarrow nv(0, nv(vnvv23pe$	
backup	$v0nvv1nvv23pepb = nvv1nvv23pepb$			$nv(0, nv(v1nvv23pe, pb) \rightarrow nv(0, nvv1nvv23pepb \rightarrow nv(nvv1nvv23pepb$	
next-sib	$nvv1nvv23pepb$				
terminal	$nvv1nvv23pepb4$			$nv(nvv1nvv23pepb, 4 \rightarrow nv(vnvv1nvv23pepb4$	
backup	$nvvnvv1nvv23pepb4pa$			$nv(vnvv1nvv23pepb4, pa) \rightarrow nv(vnvv1nvv23pepb4pa$	

This example illustrates how in general when the V list contains

$$|V_{\text{top}}|V_{\text{top}-1}|V_{\text{top}-2}| \cdots |V_{\text{bottom}}|$$

it is appropriate to interpret the computation represented as being

$$w(v_{\text{bottom}} \cdots w(v_{\text{top}-2} w(v_{\text{top}-1} w(v_{\text{top}} \cdots$$

4.2.2.4 Proof That Depth-First Uniform Inverse Algorithm Implements Nonnested Recursive Definition

We have explained (that is, intuitively justified) each algorithm in this section. However, no actual proof has been given. One is given now. We prove that Algorithm 4.2.5 implements a simplified form of recursive non-nested definition, one having only two references to f on the right. The proof is of the kind first discussed in Chapter 2. We substitute the algorithm or implementing function to be shown equivalent to the recursive definition into that definition and verify the resulting equalities. Managing the substitution of something as cumbersome as an algorithm is of some interest. The same kind of proof is applicable to the other depth-first algorithms in this section, though it is not directly applicable to the breadth-first algorithms.

Let the recursive line be

$$f(X) = w(f(o_1(X)), f(o_2(X)))$$

where X, $o_1(X)$, $o_2(X)$ are nonterminal. Let the algorithm be Algorithm 4.2.5 and let ALG(X) be the value in V when the algorithm terminates at line 12: (DONE), having started initially with variable $X = X$.

For d any member of DEF.f's initial domain, we will show that ALG(d) $= f(d)$, assuming that ALG($o_1(d)$), ALG($o_2(d)$), $f(o_1(d))$, $f(o_2(d))$ all terminate.

The proof is given for the recursive line in Figure 4-2. In the left column we trace the algorithm run when the initial argument structure is d. In the center column we trace the algorithm run when the initial argument structure is $o_1(d)$. A third trace on the right is shown for the initial argument structure $o_2(d)$. The effect on V is shown at the bottom of each column. A trace is a listing of the program line numbers (see Algorithm 4.2.5) executed, and at critical points, the contents of variables S, X, and V. Of course an algorithm cannot be traced with a nonspecific input without making some assumptions. We assume that if the initial value of X is $o_1(d)$, the algorithm terminates and ALG($o_1(d)$) $= A$; this determines properties of the trace, shown in the middle column. With similar assumptions and with ALG($o_2(d)$) $= B$, the trace for X with initial value $o_2(d)$ is given in the rightmost column. The left-hand trace for initial value $X = d$ must be consistent with the assumptions incorporated in the middle and right columns, because the algorithm starting with $X = d$ must pass through states in which $X = o_1(d)$ and $o_2(d)$. In the middle trace

First we give the recursive line. Then we give results of substituting ALG(d) for $f(d)$ on both sides of that line. At that point the equality becomes tentative. Then we verify that equality by tracing the algorithm:

$f(d)$	$= w$ (	$f(o_1(d))$,	$f(o_2(d))$)
ALG(d)	$=?\, w$ (	ALG($o_1(d)$),	ALG($o_2(d)$))
(trace)		(trace)	(trace)
‖		‖	‖

Left trace:

$1, 2, 3, 4, 5, 6, 7,$
$5, [S = d,\ X = o_1(d),\ V = 0_w]$

—
—
Z1 ←−−− The Same Steps −−−−−
—
—
$12, [S = d,\ X = o_1(d),\ V = A]$

$13,$
$14(X \leftarrow o_2(d)),$
15
$4, [S = d,\ X = o_2(d),\ V = A]$

—
—
Z2 ←−−− The Same Steps −−−−−−−−−−−
—
—
$12, [S = d,\ X = o_2(d),$
$V = w(\ldots w(w(A, a), b)\ldots)]$
$= w(\ldots w(w(A, w(0, a)), b)\ldots)]$
$= w(A, B)$ (by assoc.)]

$13, 9, \leftarrow$
$11(X \leftarrow d),$
12 DONE

Middle trace:

$1, 2, 3, 4,$
$5, [S = o_1(d),\ X = o_1(d),\ V = 0_w]$

—
— RUNS
Z1 UNTIL
— DONE
—
$12, [S = o_1(d),\ X = o_1(d),\ V = A$(by hyp.)]

*What occurs in this stretch is independent of S except at line 12. X can never $= d$ before it is $o_1(d)$.

‖

Right trace:

$1, 2, 3,$
$4, [S = o_2(d),\ X = o_2(d),\ V = 0]$

—
—
Z2 RUNS
— UNTIL
— DONE
$12, [S = o_2(d),\ X = o_2(d),$
$V = w(\ldots w(w(0, a), b)\ldots)]$
$= B$ (by hyp.)]

The comment at * also applies here

‖		‖	‖
$w(A, B)$	$= w($	$A,$	B)

FIGURE 4-2 Proof of uniform inverse Algorithm 4.2.5.

after passing lines 1, 2, 3, 4, and arriving at line 5 with $[S = o_1(d),\ X = o_1(d),\ V = 0_w]$ the algorithm must run to completion at line 12, passing through a sequence of lines $Z1$, with the final result $V = A$. Line 5 with $[S = d,\ X = o_1(x),\ V = 0_w]$ will be reached in the left trace, after passing through steps 1, 2, 3, 4, 5, 6, 7, as shown in the figure. Since, except for S, the key variables are in the same states as reached in the middle trace after passing lines 1, 2, 3, 4, the left trace must now also pass through the $Z1$ sequence of lines, and arrives at 12 with $V = A$. (S affects only termination, and in the middle column in $Z1$, X can never be d when the algorithm is at line 12 because then it would quickly take the value $o_1(d)$ again and would never terminate.) In the middle trace there was termination at line 12 after $Z1$, so X must equal $S = o_1(d)$ at that point. Since, on the left, $S = d \neq o_1(d)$ at this point, there will be no termination on the left. Instead the left trace continues to line 12 as shown. The continuing trace on the left, which soon thereafter results in X being equal to $o_2(d)$, is correlated with the trace on the extreme right, which starts with $X = o_2(d)$. The trace on the left is finally shown to produce the same result in V as is obtained by applying the w function to the two results obtained in the middle and extreme right traces. From the algorithm, V starts with 0_w; on the next assignment to V, V becomes $w(0, y)$ for some y, and by termination time V must have the form $w(\ldots w(w(0, y), z)\ldots)$. We assume in particular that V's value is $B = w(\ldots w(w(0, a), b)\ldots)$ after the right-hand trace. From this assumption and the fact that $V = A$ after the middle trace, it follows that $V = w(\ldots w(w(A, w(0, a)), b)\ldots)$ after the left trace, and this in turn means $V = w(A, B)$ because w is associative.

4.2.2.5 Algorithms for Implementing Simple Nested Definitions

As shown in Section 2.6.0.1, some nested definitions have equivalent nonnested versions. Therefore the implementations already developed for the latter are applicable to the former. Both nested definitions, Definitions 2.6.2 and 2.6.4, are equivalent to the standard nonnested definition with the $p(X)$ term; the first is for w associative, the second for w locally associative. These are implemented by depth-first implementations, Algorithms 4.2.6 and 4.2.5, respectively. Thus *Algorithm 4.2.6 implements Definition 2.6.4.*

In the cases above there is no advantage to using a nested definition when an equivalent nonnested one is available. Nesting is useful when we wish to express efficiencies unexpressible without it, as illustrated in section 3.10.0.1.

Consider the generalization of Definition 2.6.4.

Definition 4.2.1

$$
\left[\begin{array}{ll}
g(V, X) = v(V, t(X)) & :T(X) \\
g(V, X) = r(V, X) & :Q(V, X) \text{ and } \sim T(X) \\
g(V, X) = v\big(V, n\big(v\big(g\big(\ldots g\big(g(0_v, o_1(X)), & \\
\qquad o_2(X)\big), \ldots, o_{m(X)}(X)\big), p(X)\big)\big)\big) & :\sim Q(V, X) \text{ and } \sim T(X) \\
\text{Initially } V = 0_v,\ X = d,\ d \in D. &
\end{array}\right.
$$

The second line is the only way in which this definition differs from Definition 2.6.4. The effect of this extra line is to partition the set $\{\sim T(X)\}$ into two sets $\{Q(V, X) \text{ and } \sim T(X)\}$ and $\{\sim Q(V, X) \text{ and } \sim T(X)\}$, depending on V. Without the addition, the definition requires a recursive call for each argument structure in both parts of the partition, whereas with it the recursive calls in one part are replaced by the generally more efficient nonrecursive call. This is the efficiency expressible in nested definitions, which cannot be expressed directly in a nonnested one. *It is not necessary to use a nonrecursive algorithm to take advantage of such time efficiency. The direct implementation of the recursive definition by a recursive procedure will do so.* Nevertheless a nonrecursive implementation may have other efficiencies. We can obtain one by simply augmenting Algorithm 4.2.6 to account for the partition of $\sim T(X)$, and the second line in Definition 4.2.1. This implementation has the time efficiency advantage of the second line, as well as the space advantage inherent in a uniform inverse implementation.

Algorithm 4.2.7: Depth-First Uniform Inverse for Nested Definition 4.2.1—Local Associativity—Intermediate Values

Implementation

```
S ← d
X ← d
V ← ⟨ ⟩

FOR ever DO
  WHILE( ~ T(X)) DO
    V ← psh − 0_v
    X ← o_1(X)
  END
  IF Q(V, X) THEN V[1] ← r(V, X)
              ELSE V[1] ← v(V[1], X)
  WHILE(X is last-child) DO
    ARG ← pop − V
    X ← parent(X)
    FX ← n(v(ARG, p(X)))
    IF X = S THEN DONE
    V[1] ← v(V[1], FX)
  END
  X ← next-sib(X)
END
```

Definition

$g(V, X) = v(V, t(X)) \quad : T(X)$

$g(V, X) = r(V, X) \quad : Q(V, X) \text{ and } \sim T(X)$

$g(V, X) = v(V, n(v(g(\ldots g(0, o_1(X)), \ldots, o_{m(X)}(X)), p(X))))$

$\quad :\sim Q(V, X) \text{ and } \sim T(X)$

Initially $V = 0_v$, $X = d$, $d \in D$.

This algorithm implements a definition similar to those in Theorem 2.6.2. It contains an added line to minimize recursive calls; many kinds of lines added to these definitions for the same purpose are as easily implemented.

4.3 THE UNIFORM INVERSE — A KEY PROPERTY FOR STORAGE EFFICIENCY

The uniform inverse is a generalization of the properties which allow one to eliminate stacking when removing recursion. In [Aho, Hopcroft, Ullman 85] there are some good examples of such stack elimination.

Though Algorithms 4.2.2 through 4.2.6 differ considerably, they can all be viewed as special cases of the uniform inverse algorithm. The uniform inverse, introduced in Section 4.2, Figure 4-1, thus embodies an important unifying concept. In this section it will be more carefully defined and its properties given in greater detail.

Consider the following standard nonnested definition.

Definition 4.3.1

$$\left[\begin{array}{ll} f(X) = t(X) & :T(X) \\ f(X) = w\big(f(o_1(X)),\ldots, f(o_{m(X)}(X)), p(X)\big) & :\sim T(X) \\ \text{Initially } X = d \in D. & \end{array}\right.$$

A definition in the form of Definition 4.3.1 is said to have a *weak uniform inverse* if for each $d \in D$ and for each $X \in (\Delta_f(d) - D)$ there are two functions called the *parent* and the *child#* with the following properties.

Definition 4.3.2: Weak Uniform Inverse

1. parent(X, d) has the range $(\Delta_f(d) -$ terminal domain) and is such that $o_i(\text{parent}(X, d)) = X$ for one and only one $i \in N^{m(\text{parent}(X, d))}$. Furthermore, parent$(X, d)$ is the only member Y of $(\Delta_f(d) -$ terminal domain) for which $o_i(Y) = X$ for any i in $N^{m(Y)}$. In place of parent we sometimes use o^{-1}.
2. child#(X, d) is such that $o_{\text{child\#}(X)}(\text{parent}(X, d)) = X$. This function gives the "one and only one i" mentioned in the parent definition. In place of child# we sometimes use i.

An alternative definition follows.

Definition 4.3.3: Weak Uniform Inverse Alternative
parchld(X, d) is a function with $d \in D$ and $X \in (\Delta_f(d) - D)$. parchld(X, d) is the unique pair $\langle Y, i\rangle$ with $Y \in (\Delta_f(d) -$ terminal domain) and $i \in N$ such that $o_i(Y) = X$. The first component of parchld(X, d) is designated *parent*(X, d), the second *child#*(X, d). If parchld(X, d) exists for DEF.f, then DEF.f has a *weak uniform inverse*.

Definition 4.3.4: Strong Uniform Inverse
If parent(X, d) and child#(X, d) are both independent of d, then DEF.f has a (*strong*) *uniform inverse*.

Important properties of the uniform inverse are given next.

As in Section 1.2, if $I = \langle i_1, \ldots, i_n\rangle$, then $o_I(X)$ will be short for $o_{i_n}(\ldots(o_{i_2}(o_{i_1}(X)))\ldots)$. Furthermore, if for each argument structure X in $\Delta_f(d)$ there is only one vector I such that $o_I(d) = X$, then we say that $\Delta_f(d)$ has *no duplicates*.

Lemma 4.3.1: DEF.f has a weak uniform inverse if and only if for each $d \in$ the initial domain of DEF.f, $\Delta_f(d)$ has no duplicates.
(In Definition 2.4.1 for Fibonacci numbers, for example, $\Delta_f(n)$, $n \geq 2$, has duplicates, so Definition 2.4.1 does not have a weak inverse.)

PROOF: The "only if" part of this fact is simply verified. The "if" part is verified by showing how, given that $\Delta_f(d)$ has no duplicates, the uniform inverse functions child# and parent can be implemented. Given an argument structure X and the initial argument structure d, one need only start with d and systematically, in a breadth-first sequence, generate the argument structures in $\Delta_f(d)$ until X is obtained. During this generation a list of O function indices as well as a list of the ancestor argument structures of the currently generated argument structure are saved. When X is found, its child# and parent need only be popped out of the appropriate list. (This is certainly an inefficient way to compute the inverse functions; we do so only to show their existence; those typically encountered in practice are much simpler). ∎

Lemma 4.3.2: If the starting domain of DEF.f is countable, then DEF.f has a strong uniform inverse if and only if Δ_f has no duplicates.

PROOF: Again the "only if" part is easily verified—it does not require the countability restriction. The "if" part amounts to showing how the child# and parent functions can be constructed, and is similar to that used in the justification of Lemma 4.3.1. In this case, however, because of "strong" in the theorem, only X, the argument structure to be found in a systematic generation is known—the starting argument structure d is not known. Therefore we need a systematic way of generating the starting argument structures d as well as members of the $\Delta_f(d)$ for each X. Because the starting domain is countable, this is simple to do, although it can also be done in many instances with a noncountable domain. Let the members of the initial domain be $\langle d_1, d_2, d_3, \ldots \rangle$. Let stage i of $\Delta_f(d_j)$ be all argument structures that can be

i \ j	1	2	3	4
1	1	3	6	10
2	2	5	9	
3	4	8		
4	7			

(a)

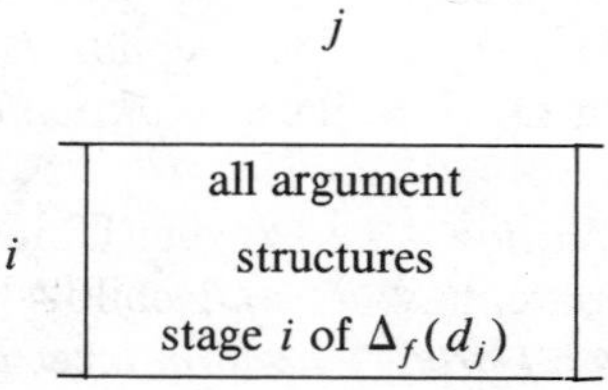

i \ j	
	all argument structures stage i of $\Delta_f(d_j)$

(b)

FIGURE 4-3 Order in which to explore argument-structures for child# and parent. (a) Each square represents a set of argument structures whose meaning is given in (b). The order in which they should be explored is given in the squares.

produced in i applications of the O functions applied to d_j. Now let argument structures be generated in the order given in Figure 4-3. While argument structures are generated in this diagonal ordering, a record is maintained with each argument structure of the index of the O function used to generate it, as well as of its parent. Each newly generated argument structure is checked to determine if it equals X. When this occurs, generation can be terminated. The child# and parent of X, having been recorded, are available. ■

If DEF.f has a weak uniform inverse, a definition with a strong uniform inverse can be easily created by adding d, the initial argument structure, to the argument of f in DEF.f.

Lemma 4.3.3: Adding a stack of child#'s to the argument of f will produce a definition with a weak uniform inverse, that is, if the original definition is Definition 4.3.1, then one with the form below has a guaranteed weak uniform inverse.

$$\left[\begin{array}{ll} f(X, I) = t(X) & :T(X) \\ f(X, I) = w(f(o_1(X), \langle 1\rangle \| I), \ldots, & \\ \qquad f(o_{m(X)}(X), \langle m(X)\rangle \| I), p(X)) & :\sim T(X) \\ \text{Initially } X = d \in D,\ I = \langle\ \rangle. & \end{array}\right.$$

PROOF: Given an argument structure (X, I) and the initial argument structure d, the fact that $I = \langle i_1, \ldots, i_n\rangle$ is the sequence of child#'s such that $O_I(d) = X$, implies that

$$\text{child\#}(X, I) = i_1 \text{ and parent}(X, I) = \left(o_{I-i_1}(d), I - i_1\right). \quad ■$$

Computing the inverse in the way described in the lemma requires considerable time and space. In practice we can usually find argument structure modifications which yield new, but equivalent definitions with inverse functions that are much easier to compute. These lemmas as well as other properties of the uniform inverse are illustrated in subsequent sections.

4.4 FINDING THE UNIFORM INVERSE, PLUGGING IN, OPTIMIZING

In our idealized design framework step 1 is to design a recursive definition, say D. In step 2 one of the applicable available algorithms is chosen, say A. (Among the depth-first algorithms we need consider only those using a uniform inverse, since all others are expressible as one of these.) Step 3 is to determine the uniform inverse functions *child#* and *parent* for D. For step 4 these functions together with the primitive functions m, o_i, p, w, t, and predicate T, as well as argument structures d, and X as given in D are *plugged in* to A. This section is devoted to study of the uniform inverse as it affects the plug-in process. Though straightforward, this process has some unusual aspects. Our algorithms are written consistently, so that a variable, X, always holds the active argument-structure. However, X typically consists of a number of components, that is, $X = \langle X_1, X_2, \ldots, X_n \rangle$. The plugging-in process must account for these separate components. So an assignment *of X to another variable*, as $Z \leftarrow X$, in the algorithm should be $Z_1 \leftarrow X_1$; $Z_2 \leftarrow X_2; \ldots; Z_n \leftarrow X_n$. Thus the partition of X propagates to a partition of Z. An assignment *to* X implies that the variable whose value is assigned can also be partitioned into n components. A function of X, say $p(X)$, becomes $p(X_1, \ldots, X_n)$. If p is independent of some of these components, they can be removed from p's argument list.

As a result of the substitutions that constitute plugging in, some simple optimizations become evident. Optimization is step 5 and includes, for example,

Removing assignments like $X \leftarrow X$.

Replacing every occurrence of a variable whose value never changes by that value.

Replacing "WHILE ($X = X$) DO" with "FOR ever DO."

Replacing "IF $X = X$ THEN *statement*" with "*statement*."

Replacing "IF $X + 1 > 2$ THEN" with "IF $X > 1$ THEN."

Replacing the sequence of statements:

```
X ← 0;
  WHILE DO (X < n);
    X ← X + a;
  END
```

with the single statement

```
X ← n
```

provided the condition $\text{rem}(n/a) = 0$ holds.

4.4.1 Finding the Uniform Inverse—Binary *n*-Tuple Example

A first example of the last three steps of the procedure *inverse determination, plugging in, and optimization* is provided by the implementation of yet another definition for enumerating all binary n-tuples. This definition differs from previous ones in its representation of the binary n-tuples. Instead of a 0–1 string, growing by concatenation, as in Definition 2.2.1, it represents the n-tuples by a fixed (maximum) length ($= n$) vector v, initialized to all 0's with a pointer j, indicating the current length. The reasoning leading to this definition is, as before, that all n-tuples can be grown by repeated stages in which 0 and 1 are each concatenated to each of the tuples at one state to produce those at the next stage. The notation used in assigning a value to a vector component is defined in Section 1.2.

Definition 4.4.1: Binary n-Tuples—Fixed-Length Vector

$$\begin{array}{lll} 1. & \left[\begin{array}{l} f(v, j) = \{v\} \\ f(v, j) = f(v, j+1) \cup f(v_{j+1} \leftarrow 1, j+1) \\ \text{Initially } v = 0^n,\ j = 0. \end{array}\right. & \begin{array}{l} : j = n \\ : j < n \\ \ \end{array} \\ 2. & & \end{array}$$

In this version of binary n-tuple generation $o_1(v, j)$ is particularly simple. Because of the *all* 0 initialization $o_1(v, j)$ adds 1 to j but leaves v unchanged. More important to our current concern, this definition has a uniform inverse. To obtain it, we first record the dependence of the components of the child argument structure (*indicated with underlines*) on those of the parent (*not underlined*). These equations, being just another way of writing the O function, are called *O function equations*. The O function equations are then solved to give the dependence of the parent argument structure on its child. This is essentially the inverse function. Let i = child# $(\underline{v}, \underline{j})$. Then the O function equations

$$\begin{array}{ll} 1. & \left[\begin{array}{ll} \underline{j} = j + 1 & \\ \underline{v} = v & \text{if } i = 1 \\ \quad = v_{j+1} \leftarrow 1 & \text{if } i = 2 \end{array}\right. \\ 2. & \end{array}$$

There are two equations for $\underline{j}$ and $\underline{v}$ in three unknowns j, v, and i. This would not be enough for a unique solution for j, v, and i if these were linear equations. However, these equations are not linear ($\leftarrow$ is not found in linear equations), and they are solvable.

In this case each equation can be solved independently. Solving Eq. 1, we get $j = \underline{j} - 1$. If $i = 2$ in Eq. 2, since v_{j+1} is set to 1 in going from parent to child, it ought to be reset to 0 in returning to the parent. This is also valid if $i = 1$, because v_{j+1} is never changed, and resetting it to 0 is harmless. (We are assuming that components of v beyond the jth are 0.) The initialization conditions guarantee that the algorithm starts with this assumption being true. It continues to be true because whenever the parent state of $\langle v, j \rangle$ is computed in a depth-first implementation and j is decremented, v_j is reset to 0.

In summary we have as our uniform inverse solution the following solutions for parent and child#.

$$\left[\begin{array}{l} j = \underline{j} - 1 \\ v = \underline{v}_{\underline{j}} \leftarrow 0 \\ i = \underline{v}_{\underline{j}} + 1 \end{array}\right.$$

Solving O function equations is a good start toward an algorithm for determining inverses. Because of the diverse mixture of primitive functions (for example, concatenation, addition, conditionals), in these equations, there is no general approach that guarantees a solution. Furthermore even if the O function equations have no unique solutions the absence of an inverse is not ensured. (An example is given in Problem 2, Section 4.7.) Definition 4.7.2 there defines $f(v, j)$ with v a vector and j an integer. The relation (last component of v) = (second argument j) holds initially, and also holds in going from parent to child in the recursive line. This relation is important in determining the uniform inverse, even if it is not derivable solely from the equations relating child to parent. Although solution methods for solving a set of equations can be augmented to account for such dependencies in seeking an inverse, there is a way which is almost always easier. Namely study some small example evaluations represented by their O network and generalize the principle by which the parent and child# of an argument structure X are obtained from X alone.

Definition 4.4.1 for binary n-tuples, with v initially $\langle 000\rangle$ and j initially 0, gives the O network, represented horizontally in Example 4.4.1, with the value of j given by the position of the underline, so, for example, 011, is the second (child# $= 1 + 1$) child of $(\underline{0}00)$, which in turn is the first (child# $= 0 + 1$) child of $\underline{\cdot}(000)$.

Example 4.4.1: Sample L Network—Binary n-tuples—Definition 4.4.1

j:0 123

$\underline{\cdot}\langle 000\rangle$ — O
- $\langle \underline{0}00\rangle$ — O
  - $\langle 0\underline{0}0\rangle$ — O
    - $\langle 00\underline{0}\rangle$ 1st child
    - $\langle 00\underline{1}\rangle$ 2nd child
  - $\langle 0\underline{1}0\rangle$ — O
    - $\langle 01\underline{0}\rangle$ 1st child
    - $\langle 01\underline{1}\rangle$ 2nd child
- $\langle \underline{1}00\rangle$ parent — O
  - $\langle 1\underline{0}0\rangle$ 1st child of parent
  - $\langle 1\underline{1}0\rangle$ 2nd child of parent

The uniform inverse can be induced from the above example. Observe the change to the argument structure when returning from children $\langle 1\underline{0}0\rangle$ and $\langle 1\underline{1}0\rangle$ to their parent $\langle \underline{1}00\rangle$. The position of the underline moves left 1 place ($j = j - 1$), and the underlined position $v_{\underline{j}}$ is set to 0 before that leftward move ($\underline{v}_{\underline{j}} \leftarrow 0$). The same changes take place moving from any child to its parent in the example network, so we seem to have found the inverse. If formal justification is required, this can be verified by other means. This seems a simpler approach than solving the O function equations, although perhaps not as easy to automate.

With parent and child# functions determined we can procede with plugging in and optimizing. Since w is associative, we plug in to Algorithm 4.2.5. Since argument structure $X = (v, j)$ in Definition 4.4.1, V and J are plugged in for all references to X according to the prescription given at the beginning of the chapter. (Our convention is to use the same name for variables in the algorithm as used in the corresponding definition, in lowercase in the former and uppercase in the latter.) As called for by Definition 4.4.1, in Algorithm 4.2.5, $o_1(X)$ is replaced by assigning $J + 1$ to J, while leaving V unchanged; $m(X)$ references are replaced by 2, and so on, for all the primitives in the definition and algorithm. Algorithm 4.2.5 contains the variable S which is not in Definition 4.4.1. S saves the initial values of $X = (v, j)$ to detect termination when these initial values recur, though we need only save j, since j's return to its initial value is sufficient to establish termination. The (X is a last child?) predicate and next-sib(v, j) plug-ins are obtained as follows:

$$
\begin{aligned}
(X \text{ is a last child?}) &\equiv (\text{the number of children of parent}(X) = \text{child}\#(X)?) \\
&\equiv \big(m\big(\text{parent}(v, j\big) = \underline{v}_{\underline{j}} + 1?\big) \\
&\equiv \big(2 = \underline{v}_{\underline{j}} + 1?\big) \\
&\equiv \big(1 = \underline{v}_{\underline{j}}?\big)
\end{aligned}
$$

$$
\text{next-sib}(v, j) = o_{\text{child}\#(v, j)+1}(\text{parent}(v, j))
$$

[child#(v, j) can only be 1 or 2, so we only need next-sib(v, j) if child#$(v, j) = 1$ or equivalently if child#(v,j) + 1 = 2.]

$$
\begin{aligned}
&= o_2\,(\text{parent}(v, j)) \\
&= o_2\big((\underline{v} \text{ with } \underline{v}_{\underline{j}} = 0,\ \underline{j} - 1)\big) \\
&= \big((\underline{v}_{\underline{j}} \leftarrow 0;\ v_{(j-1)+1} \leftarrow 1, (\underline{j} - 1) + 1\big)\big) \\
&= \big((\underline{v}_{\underline{j}} \leftarrow 1,\ \underline{j})\big)
\end{aligned}
$$

The result of plugging in are shown on the left in Example 4.4.2. To produce the fixed-length vector V initialized to 0's, we have made our first use of a declaration. After plugging in, modest optimization at the level discussed at the beginning of this section is applied, with the result shown on the right in the example.

Example 4.4.2: Inverse Implementation of Definition 4.4.1—Binary n-Tuples

Plugging In

```
N ← n; S ← 0
J ← 0;
DCL(V[N] vector) INIT(0)
FOR ever DO
  WHILE(J < N) DO  )
    J ← J + 1      }
  END              )
  OUT ← V
  WHILE(V[J] + 1 = 2) DO
    V[J] ← 0
    J ← J − 1
    IF J = S THEN DONE
  END
  V[J] ← 1
END
```

Optimizing

```
N ← n
J ← 0;
DCL(V[N] vector) INIT(0)
FOR ever DO
 (
 { J ← N
 (
  OUT ← V
  WHILE(V[J] = 1) DO
    V[J] ← 0
    J ← J − 1
    IF J = 0 THEN DONE
  END
  V[J] ← 1
END
```

Optimization Notes

1. Optimizing removes a WHILE loop by setting the variable J, incremented in the loop, to its final value N.
2. Strictly speaking, assigning only the J part of the initial argument structure 0 to S is optimization, since S never changes and can be replaced by 0.

The algorithm on the right starts with an n-tuple of 0's say binum_0. binum_1 is produced by "adding 1" to binum_0, and in general binum_i is produced by "adding 1" to binum_{i-1}. In this way the sequence of 2^n binary n-tuples is produced. That "adding 1" is the appropriate description of what the algorithm in Example 4.4.2 does, is justified by tracing its actions. The algorithm proceeds by scanning the current binary n-tuple in V from the extreme right, by setting J to N, until it encounters the first 0, changing each 1 to a 0 as it goes. This is done in the WHILE ($V[J] = 1$) loop. When the first 0 is encountered, it is changed to a 1 on exit from the loop. These are the essentials of adding 1 to a binary number. It is remarkable that starting from the recursive definition, Definition 4.4.1, in whose origin there was no notion of addition, an algorithm based on adding 1 was produced by the mechanical and generally applicable process of plugging the primitives of that definition into a fixed form algorithm and doing some minor simplifications. When algorithms for other enumeration definitions are similarly developed, analogously striking implementations result.

4.4.2 Finding the Uniform Inverse — Unary Nonnested Definitions

The process, obtain inverse/plug in/optimize, can be used to obtain a general implementation for a class of recursive definitions.

Consider the standard nonnested definition, Definition 4.3.1, simplified to have only one f call appearance on the right of its recursive line

Definition 4.4.2: Unary Nonnested

$$\left[\begin{array}{ll} f(X) = t(X) & :T(X) \\ f(X) = w(f(o_1(X)), p(X)) & :\sim T(X) \\ \text{Initially } X = d \in D. & \end{array}\right.$$

If this definition terminates, then it has a weak uniform inverse. Clearly child#$(X, d) = 1$, and termination implies that if X and Y are both in $\Delta_f(d)$, then $X \neq Y$. This, in conjunction with Lemma 4.3.1, implies that parent(X, d) exists. The algorithm will be designed under the assumption that, in fact, there is a strong uniform inverse. (If not, the addition of d to the argument would provide a definition with one.) The uniform inverse implementation obtained by straight plugging in to Algorithm 4.2.4, followed by simple optimization, is given in Example 4.4.3.

Example 4.4.3: Inverse Implementation of Definition 3.4.3—Unary Definitions

Plugging In

```
X ← d; S ← d
FOR ever DO
  WHILE(~T(X)) DO
    X ← o1(X)
  END
  V ← t(X)
  WHILE(X is a last sib) DO
    X ← parent (X)
    ARG ← V
    V ← w(ARG, p(X))
    IF X = S THEN DONE
  END
  X ← next-sib (X)
END
```

Optimizing

```
X ← d; S ← d

WHILE(~T(X)) DO  \  This often reduces
  X ← o1(X)       }  to a simple
END               |  assignment
V ← t(X)         /   statement
FOR ever DO
  X ← parent (X)

  V ← w(V, p(X))
  IF X = S THEN DONE
END
```

Optimization Notes

1. In plugging the unary definition into Algorithm 4.2.4, the pops and pushes are unnecessary because the V list can never have more than one entry.
2. Child# $(X) = m(\text{parent}(X)) = 1$, so X is always a last-sib and the WHILE loop controlled by this condition becomes FOR ever DO.
3. Since V never has more than one entry, the variable ARG, which accumulates arguments for the w function, is unnecessary.

Here is a general optimization especially significant for unary definitions. Let $c(X) = o_1(\ldots o_1(o_1(X))\ldots) = p$ applications of o_1, where p is the smallest integer so that $T(c(X))$ holds. If the definition terminates, there is such a function. The upper WHILE loop ends when X satisfies the predicate $T(X)$. Thus, that loop and the assignment following it can be replaced with $X \leftarrow t(c(X))$. If $c(X)$ is a constant independent of X, $t(c(X))$ requires no computation. This is also true if $t(c(X))$ is constant. For example, given the unary definition for the factorial

Definition 4.4.3: Factorial

$$\begin{bmatrix} f(n) = 1 & :n = 1 \\ f(n) = f(n-1) * n & :n > 1 \\ \text{Initially } n \in N & \end{bmatrix}$$

$c(X) = 1$ and $t(1) = 1$, so the upper WHILE and the following assignment can be replaced with $X \leftarrow 1$. Also for this example parent$(n) = n + 1$, and there is no difficulty in plugging the other primitives of this definition into the unary definition algorithm above, getting an optimal implementation for the factorial definition.

4.4.3 Finding the Uniform Inverse — Combination Examples

Consider implementation of a recursive definition for all combinations of m of the first n integers. Each combination will be represented by a vector with m different components drawn from N. From a vector whose first j positions are assigned, the generating principle produces child vectors in each of which the first $j + 1$ positions are assigned. The first j of these are the same as those of the parent. The principle requires that the $(j + i)$th component take a value equal to (the jth component + i), with i ranging up to a value chosen to ensure that m vector components are assigned. In addition to the growing vector v, the definition has arguments j, the index of the last assigned component of v, and e. j and e aid in deciding the largest legitimate value of the $(j + 1)$th component, as illustrated in Example 4.4.4, the O network for all combinations of two of the first four integers.

Example 4.4.4: Sample O Network—Combinations—m out of n—Definition 4.4.4

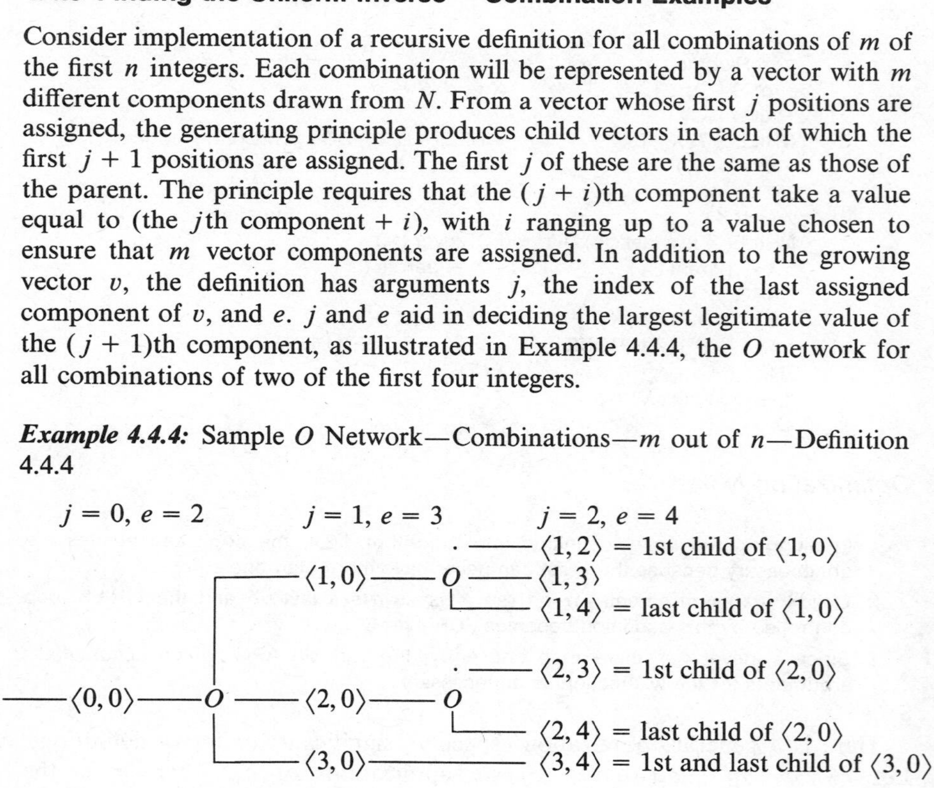

```
j = 0, e = 2          j = 1, e = 3          j = 2, e = 4
                                   . ———— ⟨1,2⟩ = 1st child of ⟨1,0⟩
               ┌——— ⟨1,0⟩ ——— O ——— ⟨1,3⟩
               │                 └———— ⟨1,4⟩ = last child of ⟨1,0⟩
               │
               │                   . ———— ⟨2,3⟩ = 1st child of ⟨2,0⟩
——⟨0,0⟩——— O ——— ⟨2,0⟩ ——— O
               │                 └———— ⟨2,4⟩ = last child of ⟨2,0⟩
               └——— ⟨3,0⟩ ——————————— ⟨3,4⟩ = 1st and last child of ⟨3,0⟩
```

Definition 4.4.4 embodies the generating principle of the O network of Example 4.4.4. It uses the assignment notation first introduced in Definition 4.4.1.

Definition 4.4.4: Combination m of First n Integers

$$\left[\begin{array}{ll} c(v, j, e) = \{v\} & : j = m \\ c(v, j, e) = \bigcup_{i=1}^{i=e+1-v_j} \left[c\left(v_{j+1} \leftarrow v_j + i,\ j+1, e+1\right)\right] & : j < m \\ \text{Initially } v = \langle v_1, v_2, \ldots, v_m\rangle,\ v_j \in N,\ j = 0,\ 0 \le e = n - m,\ n, m \in & \\ N \text{ and invariant.} & \end{array}\right.$$

We now solve for the inverse of this definition. The O function equations are

$$\left[\begin{array}{l} \underline{j} = j + 1 \\ \underline{e} = e + 1 \\ \underline{v} = \left[v_{j+1} \leftarrow v_j + i\right] \end{array}\right.$$

Getting the inverse for j and e is easy but not for v. v_1 through v_j are the same in parent and child, but v_{j+1} is not. v_{j+1} was set to $v_j + i$. Thus strictly speaking, to get the parent, v_{j+1} must be reset to the value it had before this change. However, v_{j+1} need not have the same value it did when the child was first created, because the components of v beyond the jth are not truly part of the argument structure. This makes sense because, unlike binary n-tuples, Definition 4.4.1, the correctness of the definition does not depend on the values of the vector beyond the current pointer j. This is evident from the initialization and the fact that the O function always assigns a new value to the $(j+1)$th component. So we have solutions for parent and child#.

$$\left[\begin{array}{l} j = \underline{j} - 1 \\ e = \underline{e} - 1 \\ v = \underline{v} \ \left(\text{with } v_{j+1} \text{ reset to its value at the parent}\right) \\ i = v_{j+1} - v_j = \underline{v}_j - \underline{v}_{j-1} \end{array}\right.$$

By the argument above this can be simplified by removal of the parenthesized qualification on the v equation.

In this example we were able to replace the uniform inverse with a simpler function because the components of v beyond the jth are of no concern, and resetting them is unnecessary. In general, in getting an inverse of an argument structure $Z = o_i(X)$, the function parent(Z) may be made Y, provided Y

behaves as X would behave in the algorithm using that inverse. Y is then *ui-equivalent* to X for that algorithm. If Y is ui-equivalent to X for one of our uniform inverse implementations of a standard nonnested recursive definition, we say Y ui $\equiv X$. Definition 4.4.5 gives the conditions under which Y ui $\equiv X$.

Definition 4.4.5: ui-Equivalence

$$Y \text{ ui} \equiv X \text{ if and only if}$$

1. $T(Y) = T(X)$
2. If $T(X)$, then
   $t(Y) = t(X)$
3. If $\sim T(X)$, then
   a. $m(Y) = m(X)$ and
   b. for $i = 1$ to $m(X)$: $o_i(Y)$ ui $\equiv o_i(X)$
4. If X is not an initial argument structure, then
   a. child#(Y) = child#(X) and
   b. parent(Y) ui $\equiv$ parent(X)

In summary

Proposition 4.4.1: Set Encoding Principle
In any uniform inverse implementation given here the computation of parent(X) or $o_i(X)$ produce an argument structure ui-equivalent to parent(X) or $o_i(X)$, respectively, without affecting the validity of the algorithm.

Returning to the combinations example. The inverse solution determines the functions and predicates necessary for plugging in, to Algorithm 4.2.5 in this case. These include the first-child, next-sib, and backup functions, as well as the $\sim T(X)$ and (X is a last child) predicate. To compute the (X is a last child) predicate, for example, we need $m(\text{parent}(v, j, e))$. We need to determine if

$$X \text{ is a last child} \equiv \text{the number of children of parent}(X) = \text{child\#}(X)$$

$$\equiv m(\text{parent}(v, j, e)) = e + 1 - v_j$$

$$\equiv m(\text{parent}(v, j, e)) = \underline{e} - 1 + 1 - \underline{v}_{j-1}$$

$$\equiv m(\text{parent}(v, j, e)) = \underline{e} - \underline{v}_{j-1}$$

An alternative way to find the uniform inverse is to generalize from a small sample O network such as that for an evaluation of Definition 4.4.4 given in Example 4.4.3.

The outcome of plugging the primitive and inverse functions for Definition 4.4.4 into Algorithm 4.2.4 (since w of Algorithm 4.2.4 = union of Definition 4.4.4 is associative) and optimizing follows.

Example 4.4.5: Inverse Implementation of Definition 3.4.4—Combinations—m out of n

Plugging In

```
N ← n; M ← m
DCL(V[N] vector); V[0] ← 0
J ← 0; S ← 0
E ← N − M
FOR ever DO
  WHILE(J −M) DO
    V[J + 1] ← V[J] + 1
    J ← J + 1
    E ← E + 1

  END

  OUT ← V
  WHILE(V[J] − V[j − 1] = E − V[j − 1]) DO
    J ← J − 1
    E ← E − 1
    IF J = S THEN DONE
  END
  V[J] ← V[J] + 1
END
```

Optimizing

```
N ← n; M ← m
DCL(V[N] vector); V[0] ← 0
J ← 0
E ← N − M
FOR ever DO
  WHILE(J < M) DO
    V[J + 1] ← V[J] + 1
    J ← J + 1

  END

  E ← N
  OUT ← V
  WHILE(V[J] = E) DO
    J ← J − 1
    E ← E − 1
    IF J = 0 THEN DONE
  END
  V[J] ← V[J] + 1
END
```

Optimization Notes. In addition to straightforward algebraic simplification, the optimization depends on the fact that the terminal values of J and E increase and decrease together. $J = m$ and $E = n$ at the end of the first WHILE loop. Since E is not used in that loop, it need not be changed as long as it is given correctly as N on exit.

4.5 MODIFICATIONS TO CREATE A UNIFORM INVERSE — THE UNIFORM INVERSE COVERS ALL THE OTHER DEPTH-FIRST ALGORITHMS

Given a nonnested recursive definition has no uniform inverse, it is always possible to design an equivalent one which does. Therefore the uniform inverse implementations, Algorithms 4.2.4 through 4.2.6, can be applied to any nonnested recursive definition. Algorithm 4.2.3, which is depth-first, but not in the uniform inverse group, is also applicable to any such definition, but it requires a stack Z. The uniform inverse algorithms do not require an *explicit* stack and will be shown never to require more memory than required by Algorithm 4.2.3.

Consider any standard nonnested definition:

Definition 4.5.1: Standard Nonnested Again

$$\left[\begin{array}{ll} f(X) = t(X) & :T(X) \\ f(X) = w\big(f(o_1(X)), \ldots, f\big(o_{m(x)}(X)\big), p(X)\big) & :\sim T(X) \\ \text{Initially } X = d \in D. & \end{array}\right.$$

We can construct an equivalent definition with an inverse by adding an argument Z, a vector in which child#(X) and X itself are saved. Each of Z's components are a pair [child#(X), X]. $[Z_1]_1$ refers to the first, or child#, part, and $[Z_1]_2$ to the second part of Z_1. The result is

Definition 4.5.2: Standard Nonnested with Guaranteed Uniform Inverse

$$\left[\begin{array}{ll} f(X, Z) = t(X) & :T(X) \\ f(X, Z) = w\big(f(o_1(X), \langle[1, X]\rangle \| Z), \ldots, & \\ \qquad\qquad f\big(o_{m(x)}(X), \langle[m(X), X]\rangle \| Z\big), p(X)\big) & :\sim T(X) \\ \text{Initially } X = d \in D,\ Z = \langle\ \rangle. & \end{array}\right.$$

It is easy to see that Definitions 4.5.2 and 4.5.1 are equivalent. They define the same function because with Z removed, they are identical, and Z has nothing to do with the value of $f(X, Z)$ in Definition 4.5.2. However, Definition 4.5.2 is guaranteed to have a uniform inverse. The O function equations,

$$\left[\begin{array}{l} \underline{X} = o_i(X) \\ \underline{Z} = \langle[i, X]\rangle \| Z \end{array}\right.$$

The solutions for parent and child# are

$$\left[\begin{array}{ll} X = [\underline{Z}_1]_2 & (Z_1 \text{ is the top entry in } Z) \\ Z = \underline{Z} - \underline{Z}_1 & \\ \text{child\#}(X, Z) = [\underline{Z}_1]_1 & \end{array}\right.$$

Now all this can be plugged into the uniform inverse Algorithm 4.2.4. The result is shown in Example 4.5.1 on the left. Then with simple rearrangements and renaming of variables, including, most importantly, using CHILD# for $[Z_1]_1$ and PARENT for $[Z_1]_2$, the equivalent algorithm shown on the right is obtained. The algorithm on the right is in fact identical to Algorithm 4.2.3.

Example 4.5.1: Inverse Implementation of Definition 4.5.3—Standard Nonnested Definition with Vector Added for Inverse

Plugging In

```
X ← d; Z ← ⟨ ⟩
S ← ⟨X, Z⟩
FOR ever DO
  WHILE (~T(X)) DO
    Z ← psh- ⟨1, X⟩
    X ← o_1(X)
  END
  V ← psh- t(X)

  WHILE([Z_1]_1 = m([Z_1]_1) DO
    ARG ← pop.m([Z_1]_2) − V
    V ← psh- w(ARG, p([Z_1]_2)
    X ← [Z_1]_2
    ← pop- Z
    IF ⟨X, Z⟩ = S THEN DONE
  END
  X ← o_{[Z_1]_1+1}([Z_1]_2)
END
```

Optimizing and Renaming

```
X ← d; Z ← ⟨ ⟩

FOR ever DO
  WHILE (~T(X)) DO
    Z ← psh- ⟨1, X⟩
    X ← o_1(X)
  END
  V ← psh- t(X)
  ⟨CHILD#, PARENT⟩ ← pop- Z
  WHILE(CHILD# = m(PARENT)) DO
    ARG ← pop.m(PARENT)- V
    V ← psh- w(ARG, p(PARENT))
    X ← PARENT
    IF Z = ⟨ ⟩ THEN DONE
    ⟨CHILD#, PARENT⟩ ← pop-- Z
  END
  X ← o_{CHILD#+1}(PARENT)
END
```

Thus, applying the uniform inverse algorithm, Algorithm 4.2.4, to any standard nonnested definition, altered to make certain it has a uniform inverse, will at worst result in Algorithm 4.2.3. However, it is seldom necessary to add a vector, which holds the entire ancestry of argument structures and indices as in Definition 4.5.2 to obtain an inverse. Modest augmentation is often enough, as illustrated by the following examples.

4.5.1 Modifications to Create a Uniform Inverse—Tower of Hanoi Example

Consider the definition for the Tower of Hanoi puzzle, first given in Definition 3.1.1, and repeated here.

$$\left[\begin{array}{ll} h(n, i \to j) = \langle i \to j\rangle & :n = 1 \\ h(n, i \to j) = h(n-1, i \to k) \| h(1, i \to j) \| h(n-1, k \to j) & :n \notin N \\ \text{Initially } n \in N;\ i, j, k \text{ are each different members of } \{1,2,3\} & \end{array}\right.$$

This definition almost has an inverse. If the child# of an argument structure is known, the parent can be computed. But the child# cannot be computed. The following example of a partially evaluated O network demonstrated this.

Example 4.5.2

```
  n = 4            n = 3              n = 2              n = 1
                                                   ┌─────1 → 3
                              ┌───────1 → 2— O ────┼─────1 → 2
                              │      ‾‾‾‾‾‾        └─────3 → 2
           ┌───────1 → 3— O ──┼───────1 → 3
           │                  │                    ┌──────
           │                  └───────2 → 3— O ────┼──────
           │                                       └──────
 1 → 2— O ─┼───────1 → 2
           │                  ┌───────3 → 1
           │                  │
           └───────3 → 2— O ──┼───────3 → 2
                              │
                              └───────1 → 2
                                     ‾‾‾‾‾‾
```

The underlined argument structures are identical, yet they are, respectively, a 1st and a 3rd child. Therefore by Lemma 4.3.1, the child# cannot be determined from the argument structure alone. But, generalizing from this example gives the function pm(child#, $i \to j$) for the move component of the parent in terms of the child#,

$$\mathrm{pm}(1, i \to j) = i \to k$$

$$\mathrm{pm}(2, i \to j) = i \to j$$

$$\mathrm{pm}(3, i \to j) = k \to j$$

So we propose adding just enough information to the argument structure to determine the child#. We try stacking child#'s. This is not quite enough, although if an argument structure is a first or third child, then the first component, n, of its parent is one less than the n of the child. But this fails for a second child, because in such an argument structure n is 1, independent of its value in its parent. However, with the addition of the list of child#'s, a second child is easily identified, so $n = 1$ in the second child can be changed to $n - 1$. Then n in the parent is always 1 more than it is in the child, independent of whether it is a first, second, or third child.

Definition 4.5.3: Tower of Hanoi—Definition with Inverse

$$\left[\begin{array}{ll} h(n, i \to j, Z) = \langle i \to j \rangle & :n = 1 \text{ or } Z_1 = 2 \\ h(n, i \to j, Z) = h(n-1, i \to k, \langle 1 \rangle \| Z) & \\ \qquad \| h(n-1, i \to j, \langle 2 \rangle \| Z) & \\ \qquad \| h(n-1, k \to j, \langle 3 \rangle \| Z) & :n > 1 \text{ and } Z_1 = 1 \text{ or } 3 \\ \text{Initially } n \in N;\ i,\ j \text{ in } \{1, 2, 3\}, \text{ and } i \neq j. & \end{array}\right.$$

The uniform inverse is

$$\text{child\#}(n, i \to j, Z) = Z_1$$

$$\text{parent}(n, i \to j, Z) = \left(n + 1, \text{pm}(\text{child\#}, i \to j), Z - Z_1\right)$$

where pm(child#, $i \to j$) is as defined above.

4.5.2 Modifications to Create a Uniform Inverse — Merge – Sort Example

Definition 4.5.4 of $s(v, i, j)$ for merge–sort is based on the discussion in Section 3.2. It does not have an inverse.

$s(v, i, j)$ is interpreted as being a vector in which the components $v_i, v_{i+1}, \ldots, v_j$ of v are in nondecreasing order.

Definition 4.5.4: Merge–Sort with Explicit Indexing

$$\left[\begin{array}{ll} s(v, i, j) = v_i & :i = j \\ s(v, i, j) = \text{merge}\big(s(v, i, \lceil (i+j)/2 \rceil - 1), & \\ \qquad\qquad s(v, \lceil (i+j)/2 \rceil, j)\big) & :i < j \\ \text{Initially } v \text{ is vector of keys of length } n \text{ (invariant)}, \; i = 1, \; j = n. & \end{array}\right.$$

The O function in this definition *splits* the vector v to be sorted into two subvectors which are as close to equal as possible. This way of partitioning ensures the lowest worst-case cost in the number of comparisons. This statement is justified in the next section, after which the inverse properties of the definition will be explored.

4.5.2.1 Merge – Sort Complexity, Optimality

The merge function referred to here is a *linear merge*. It merges two sorted lists, $A = \langle a_1, \ldots, a_n \rangle$ and $B = \langle b_1, \ldots, b_m \rangle$, repeatedly moving the minimum of the smallest members of A and B to the output. It requires at least $m + n - 1$ comparisons worst case to empty A and B into a completely sorted output because when

$$a_i < b_i < a_{i+1} < b_{i+1} \text{ for } i = 1 \text{ to } n - 1$$

$m + n - 1$ comparisons are executed by a linear merge. On the other hand, it requires no more than $m + n - 1$ comparisons, as established by the following argument. Merging A and B is tantamount to establishing for all $m * n$ pairs, $a_i \in A$, $b_j \in B$ whether or not $a_i > b_j$. After $k = x + y$ comparisons, where x and y are, respectively, the number of entries that have been moved from lists A and B to the output, we will still have to merge a sorted list A' of size $(n - x)$ with B' of size $(m - y)$ and thus still need to establish $(n - x)*(m - y)$ comparisons. When $k = n + m - 1 = x + y$, either $x \geq n$ or

$y \geq m$, so $(n - x)*(m - y) \leq 0$. Therefore either A or B has been emptied and no more comparisons are needed. Thus

Lemma 4.5.1: $m + n - 1$ comparisons are sometimes necessary and always sufficient for a linear merge of two sorted lists of sizes n and m.

To determine the complexity of merge–sort, we construct a definition for $c(n)$ = (worst-case cost in comparisons) to merge–sort n keys. For this purpose the transformation described in Section 3.9 can be applied to Definition 4.5.4. v is dropped as an argument since it doesn't affect the cost; merge is replaced by + because the costs of sorting the two vectors to be merged must be added. The split operation partitions a vector of length n into two parts of size j and $n - j$. Instead of fixing these so $j = \lfloor n/2 \rfloor$, we choose j to minimize $c(n)$. We get the worst case for $c(n)$ by assuming that each merge requires the maximum possible number of comparisons $n - 1$ to do a linear merge on two sorted partitions whose total length is n. This cost is reflected in the factor $n - 1$ in the recursive line.

Definition 4.5.5: Complexity of Merge–Sort

$$\left[\begin{array}{ll} c(n) = 0 & :n = 0 \text{ or } 1 \\ c(n) = \text{MIN}_{j=1}^{n-1}[c(j) + c(n-j)] + n - 1 & :n > 1 \end{array}\right.$$

The exact solution for $c(n)$ is

$$c(n) = n * k - 2^k + 1$$

where $k = \lceil \log_2(n) \rceil$. We will not prove this but instead will prove an upper bound on $c(n)$, namely,

$$c(n) \leq n \log_2 n \qquad \text{when } n > 0$$

This illustrates a useful technique for proving inequalities in recursively defined functions.

The inequality we have chosen in this case is plausible, because if j is $n/2$, the recursive line is roughly

$$c(n) = 2 * c(n/2) + n = 2 * (2 * c(n/4) + n/2) + n$$

$$c(n) = 4 * c(n/4) + 2 * n = 2^k * c(n/2^k) + k * n$$

and when $n/2^k = 1$, $k = \log_2 n$, and

$$c(n) = \log_2 n * n$$

Assume for $j < n$, that $c(j) \leq j \log_2 j$.

$$c(n) \leq \text{MIN}_{j=1}^{n-1}[j \log_2 j + (n - j)\log_2(n - j)] + n - 1$$

It is easy to show that a minimum occurs when $j = \lceil n/2 \rceil$,

$$c(n) \leq \lceil n/2 \rceil \log_2(\lceil n/2 \rceil) + \lfloor n/2 \rfloor \log_2(\lfloor n/2 \rfloor) + n - 1$$

If n is even,

$$c(n) \leq 2((n/2)(\log_2(n) - 1)) + n - 1$$

$$c(n) \leq n \log_2(n) - 1$$

and therefore

$$c(n) \leq n \log_2(n) - 1 \leq n \log_2(n)$$

If n is odd,

$$\begin{aligned} c(n) &\leq ((n+1)/2)\log_2((n+1)/2) \\ &\quad + ((n-1)/2)\log_2((n-1)/2) + n - 1 \\ &\leq ((n+1)/2)(\log_2(n+1) - 1) \\ &\quad + ((n-1)/2)(\log_2(n-1) - 1) + n - 1 \\ &\leq ((n+1)/2)(\log_2(n+1)) + ((n-1)/2)(\log_2(n-1)) - 1 \\ &\leq (n/2)(\log_2(n+1) + \log_2(n-1)) \\ &\quad + (1/2)(\log_2(n+1) - \log_2(n-1)) - 1 \end{aligned}$$

Because $\log_2(n)$ increases less than linearly,

$$|\log_2(n+1) - \log_2(n)| \leq |\log_2(n-1) - \log_2(n)|$$

$$\begin{aligned} (n/2)(\log_2(n+1) + \log_2(n-1)) &\leq (n/2)(\log_2(n) + \log_2(n)) \\ &\leq n \log_2(n) \end{aligned}$$

and

$$(1/2)(\log_2(n+1) - \log_2(n-1)) \leq 1 \qquad \text{if } n > 1$$

(Try it at $n = 2$, it get smaller thereafter.) Therefore,

$$\begin{aligned} c(n) &\leq ((n+1)/2)(\log_2(n+1)) + ((n-1)/2)(\log_2(n-1)) - 1 \\ &\leq n \log_2(n) \end{aligned}$$

4.5.2.2 Creating the Inverse

Definition 4.5.4 does not have an inverse. In Definition 4.5.6 the addition of vectors x and y, containing the stacked ancestors of arguments i and j, respectively, provides sufficient information for a reasonably efficient uniform inverse.

Definition 4.5.6: Merge–Sort with Inverse

$$\left[\begin{array}{lr} s(v, i, j, x, y) = v_i & :i = j \\ s(v, i, j, x, y) = \mathrm{mg}(s(v, i, \lceil (i+j)/2 \rceil - 1, \langle i \rangle \| x, \langle j \rangle \| y), & \\ \qquad s(v, \lceil (i+j)/2 \rceil, j, \langle i \rangle \| x, \langle j \rangle \| y)) & :i < j \\ \text{Initially } v \text{ is vector of keys of length } n,\ i = 1,\ j = n,\ x = \langle\ \rangle,\ y = \langle\ \rangle. & \end{array}\right.$$

The resultant inverse is

$$\begin{aligned} \text{child\#}(v, i, j, x, y) &= 1 \qquad \text{if } i = x_1 \\ &= 2 \qquad \text{otherwise} \\ \text{parent}(v, i, j, x, y) &= (v, x_1, y_1, x - x_1, y - y_1) \end{aligned}$$

From these the other functions involving the inverse which are required for the implementation can be derived. For example,

$$\text{next-sib}(v, i, j, x, y) = (v, j + 1, y_1, x, y)$$

To achieve the given complexity, merges must be done in the order determined by the O network, so we cannot take advantage of the associativity of merge. Thus we use Algorithm 4.2.4 modified as per note 2, for nonassociative w functions with a $p(X)$ term. We can implement Definition 4.5.6 by simply plugging in to Algorithm 4.2.4. Doing so requires the subvectors formed by simulating the merge operations in the w network to be kept in the stack V, thus involving considerable storage and unnecessary movement of keys. In terms of the O–w–t network for this example, we see that

1. The inputs to the w network each come from a different place in the input structure v.
2. The output of each of the first-level w blocks can be accommodated in the space occupied by the inputs, and if they are made to occupy that space, the same statement will be true on the next level of w blocks, etc.

These conditions imply that we can use a global data structure (GV *in this case*), *and replace the original function* (*merge*) *with one that acts on this structure.* Furthermore since the change is not in the values of the keys, but in their ordering, we use linking to represent that order. Therefore, the action on the global structure involves link changes. Instead of stacking the subvectors of keys, we stack pointers to these subvectors. GV consists of an array of pointer locations pointing to cells holding keys that can be linked together. Initially there is a pointer to each key arranged in a vector v with v_i pointing to the ith key. Now when two keys, pointed to by v_a and v_b, are merged at the first stage of the w network, the result is two keys in size order. The order is maintained by linking the two keys in a list. The new merge action function, called mg, in the w network has as arguments pointers to the lists it will merge. It outputs a pointer to the beginning of the linked list resulting from the merge. By convention the pointer location of the result is v_a, where a is the first pointer argument of mg, and $a < b$. Either input pointer could be reused for the result, because merged lists are never referred to anywhere else in the w network. The possibility of reusing pointer locations this way, not recopying subvectors but just relinking them, is a result of properties 1 and 2 discussed above.

With this arrangement, when a last child is detected, the V stack will contain indices of pointer locations. The top members of V will be popped, the vectors pointed to from these merged, and the result pointed to from one of those two pointer locations. That location is pushed back into the stack. So the V stack contains only pointer locations, that is, indices in the vector v. The other intermediate results of merges of subvectors are kept in the links that connect these subvectors in the data structure GV. The primitive operation, which merges the two lists and returns the pointer location, is called mg, defined as follows.

$$\text{mg}(\text{GV}, \text{indp1}, \text{indp2})$$

1. *Performs the action*: merges the list pointed to from $v(\text{indp1})$ with that pointed to by $v(\text{indp2})$ in the data structure GV.
2. *Has the value i.*

The merge–sort function sa(GV, i, j, x, y), will be recursively defined using mg, and is interpreted as follows:

$$\text{sa}(\text{GV}, i, j, x, y)$$

1. *Performs the action*: the series of actions (mg merges) which will result in forming a list in GV of all keys originally pointed to by $v[i]$ through $v[j]$. L is now pointed to from $v[i]$ and its keys are linked together in the order of nondecreasing key values.
2. *Has the value i.*

The resultant Definition 4.5.7 is almost identical to Definition 4.5.6. The difference is that the argument GV replaces v.

Definition 4.5.7: Merge–Sort with Inverse 1

$$\left[\begin{array}{ll} \text{sa}(\text{GV}, i, j, x, y) = i & :i = j \\ \text{sa}(\text{GV}, i, j, x, y) = \text{mg}(\text{GV}, \text{sa}(\text{GV}, i, \lceil (i+j)/2 \rceil - 1, \langle i \rangle \| x, \langle j \rangle \| y), & \\ \qquad \text{sa}(\text{GV}, \lceil (i+j)/2 \rceil, j, \langle i \rangle \| x, \langle j \rangle \| y)) & :i < j \\ \text{Initially GV} = \text{global data structure}, \ i = 1, \ j = n, \ x = \langle\ \rangle, \ y = \langle\ \rangle. & \end{array}\right.$$

Algorithm 4.5.1 is an implementation of Definition 4.5.7 using the generic depth-first inverse Algorithm 4.2.4 [for a definition with no $p(X)$ term].

Algorithm 4.5.1: Merge—Sort Algorithm Implementing Definition 4.5.7

```
[I, J, X, Y] ← [1, n,⟨ ⟩,⟨ ⟩]
V ← ⟨ ⟩
FOR ever DO
  WHILE(J > 1) DO
    [I, J, X, Y] ← [I, ⌈(I + J) / 2⌉ − 1, ⟨I⟩||X, ⟨J⟩||Y]
  END
  V ← psh- I
  WHILE(J = Y[1])
    [I, J, X, Y] ← [X[1], Y[1], X − X[1], Y − Y[1]]
    [ARG1, ARG2] ← pop.2- V
    V ← psh- mg(GV, ARG1, ARG2)
    IF X = ⟨ ⟩ THEN DONE
  END
  [I, J] ← [J + 1, Y[1]]
END
```

A trace of Algorithm 4.5.1 is shown in Example 4.5.3, sorting the keys $\langle 10, 17, 13, 8 \rangle$ as represented in Figure 4-4.

Example 4.5.3

Trace of Algorithm 4.5.1, using V list, and Algorithm 4.5.2, without any V list, sorting the $\langle 10, 17, 13, 8 \rangle$ as represented in Figure 4-4.

Action	I	J	X	Y	V	mg mg′	Fig. 4-4
Initially	1	4	$\langle\ \rangle$	$\langle\ \rangle$	$\langle\ \rangle$		A
First-child	1	2	$\langle 1 \rangle$	$\langle 4 \rangle$			
First-child	1	1	$\langle 11 \rangle$	$\langle 24 \rangle$	$\langle 1 \rangle$		
Next-sib	2	2	$\langle 11 \rangle$	$\langle 24 \rangle$	$\langle 21 \rangle$		
Backup	1	2	$\langle 1 \rangle$	$\langle 4 \rangle$		$\text{mg}(v, 1, 2)$	B
					$\langle 1 \rangle$	mg′(GV,,,1, 2)	
Next-sib	3	4	$\langle 1 \rangle$	$\langle 4 \rangle$			
First-child	3	3	$\langle 31 \rangle$	$\langle 44 \rangle$	$\langle 41 \rangle$		
Next-sib	4	4	$\langle 31 \rangle$	$\langle 44 \rangle$	$\langle 341 \rangle$		
Backup	3	4	$\langle 1 \rangle$	$\langle 4 \rangle$		$\text{mg}(v, 3, 4)$	C
					$\langle 31 \rangle$	mg′(GV,,,3, 4)	
Backup	1	4	$\langle\ \rangle$	$\langle\ \rangle$		$\text{mg}(v, 1, 3)$	D
					$\langle 1 \rangle$	mg′(GV,,,3, 4)	

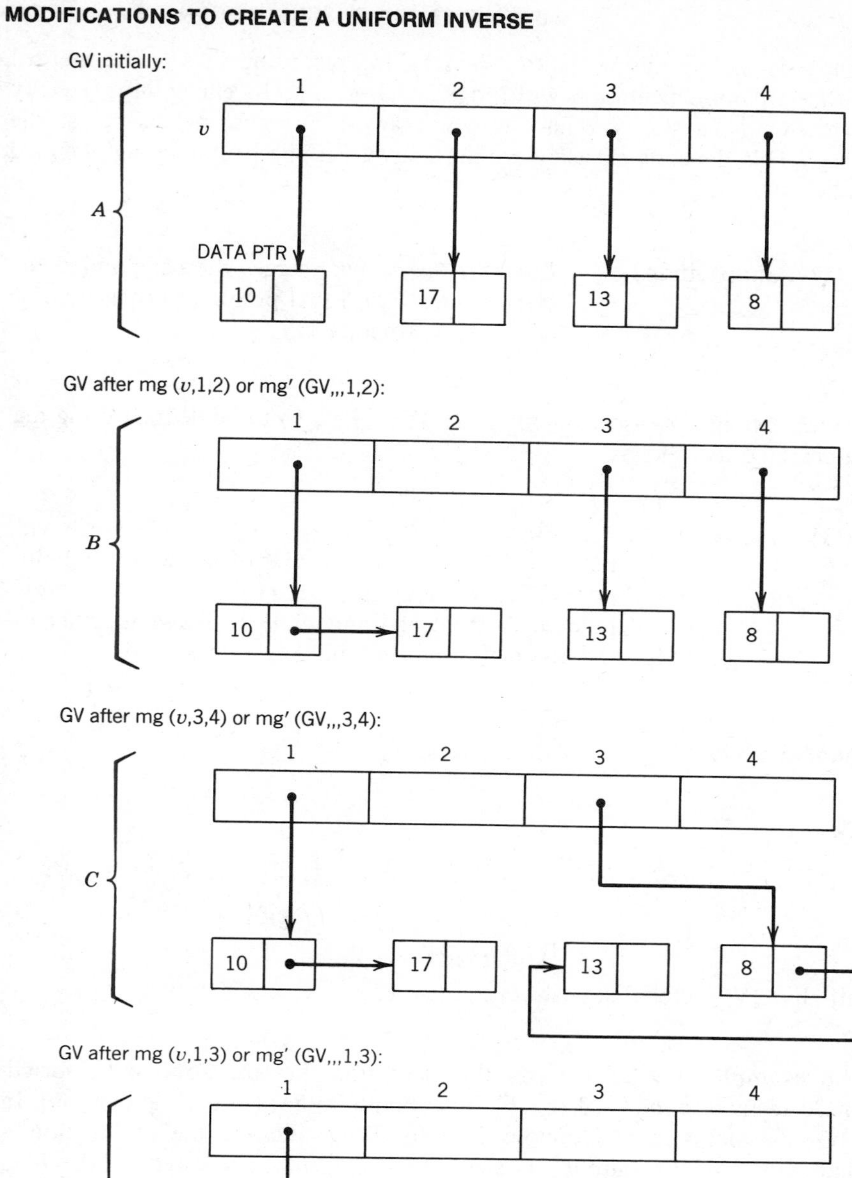

FIGURE 4-4 Illustrative example of data structure GV.

There is an alternative to the V stack for recording lists to be merged allowing an implementation with no V stack at all. (In effect the necessary information is in stacks x and y, which are in any case necessary for the inverse.) This depends on using an alternative merging primitive mg′, defined as follows:

mg′(GV,,,indp1, indp2) — *Performs the action*: merges the list pointed to from v(indp1) with that pointed to by v(indp2) in the data structure GV.

The recursive merge—sort function sa′(GV, i, j, x, y) to be defined using mg′ is interpreted as follows:

sa′(GV, i, j, x, y) — *Performs the series of actions*: (mg′ merges) which will result in forming a list, in GV of all keys originally pointed to by $v[i]$ through $v[j]$. That list is now pointed to from $v[i]$ and its keys linked together in the order of nondecreasing key values.

Definition 4.5.8: Merge–Sort with Inverse

$$\left[\begin{array}{ll} \text{sa}'(\text{GV}, i, j, x, y) = [\text{noop}] & :i = j \\ \text{sa}'(\text{GV}, i, j, x, y) = \text{mg}'(\text{GV}, \text{sa}'(\text{GV}, i, \lceil (i+j)/2 \rceil - 1, \langle i \rangle \| x, \langle j \rangle \| y), & \\ \qquad \text{sa}'(\text{GV}, \lceil (i+j)/2 \rceil, j, \langle i \rangle \| x, \langle j \rangle \| y), i, & \\ \qquad \lceil (i+j)/2 \rceil & :i < j \\ \text{Initially GV = global data structure, } i = 1,\ j = n,\ x = \langle\ \rangle,\ y = \langle\ \rangle. & \end{array}\right.$$

An example may help clarify this definition. Let the keys to be sorted, pointed to from v, be $\langle 5, 3, 4, 2, 6 \rangle$ [$v(1)$ points to 5, $v(2)$ to 3, and so on]. In the O–w–t network of Example 4.5.4 for the evaluation the O function is called *split*. At the output of each mg′ the resulting change to the data structure GV is shown. Each mg′ works on data structure GV in the state in which it finds it, using the single input line coming from a p-block. The dashed lines indicate the time order in which the mg′ are to be executed. An mg′ should not be executed until all those mg′ from which its dotted inputs come are executed.

Example 4.5.4: Example of O–w–t network for Definition 4.5.8 (read left to right)

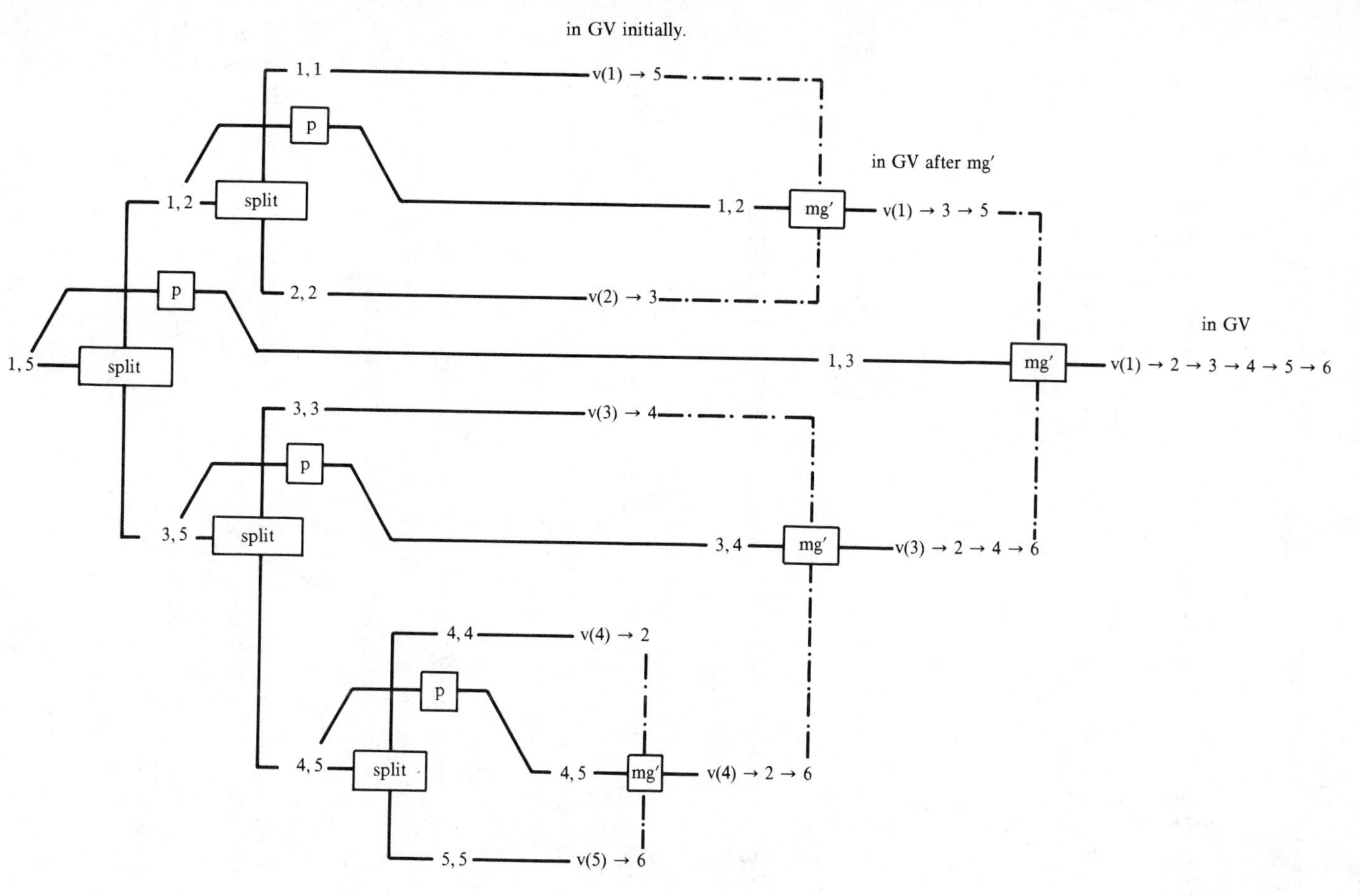
in GV initially.
1, 1
v(1) → 5
p
1, 2
split
1, 2
mg′
in GV after mg′
v(1) → 3 → 5
2, 2
v(2) → 3
p
1, 5
split
1, 3
mg′
in GV
v(1) → 2 → 3 → 4 → 5 → 6
3, 3
v(3) → 4
p
3, 5
split
3, 4
mg′
v(3) → 2 → 4 → 6
4, 4
v(4) → 2
p
4, 5
split
4, 5
mg′
v(4) → 2 → 6
5, 5
v(5) → 6

Algorithm 4.5.2 is an implementation of Definition 4.5.8 using the generic depth-first inverse implementation, Algorithm 4.2.4 [for a definition *with* $p(X)$ term].

Algorithm 4.5.2: Merge-Sort Algorithm Implementing Definition 4.5.8

```
[I, J, X, Y] ← [1, n, ⟨ ⟩, ⟨ ⟩]
FOR ever DO
   WHILE (J –I) DO
      [I, J, X, Y] ← [I, [(I + J) / 2] – 1, ⟨I⟩||X, ⟨J⟩||Y]
   END
   WHILE (J = Y[1])
      mg'(GV,,,X[1], I)
      [I, J, X, Y] ← [X[1], Y[1], X – X[1], Y – Y[1]]
      IF X = ⟨ ⟩ THEN DONE
   END
   [I, J] ← [J + 1, Y[1]]
END
```

4.6 NOTES ON THE EFFICIENCY OF IMPLEMENTATIONS OF ENUMERATIONS

All SE implementations for a given definition simulate the same O–w–t networks, so their running time cannot differ greatly. Nevertheless there are small, sometimes significant, time differences due to different ways the implementations manage the simulation. To simulate the O–w–t network, each of its blocks must be simulated. The time complexity is

number of O block outputs $*$ complexity per O block output +
number of w block outputs $*$ complexity per w block +
number of t blocks $*$ complexity of each t block

In an actual implementation there is some handling time for moving arguments in and out of stacks and queues. In most implementations this time can be included in the time per block.

Assume that the complexity of the O and w blocks is $O(1)$. Then the time required would be

O(number of O block outputs) +
O(number of t blocks $*$ complexity of each t block)

Usually, the number of O block outputs as well as the, usually related, number of t blocks can be counted. For example, if an O network has n levels and every block at every level has two outputs, summing the O block outputs at each level from n to 1, we get

$$2^n + 2^{n-1} + \cdots + 2^1 = 2 * 2^n - 2 = O(2^n)$$

We develop a somewhat more general, though still very simple, approach to counting O block outputs. Partition the O block outputs into n sets, called *olevels*. The number of outputs at the jth olevel is $q(j)$. These are usually, though not necessarily, related to the levels of the O network. For example, assume an O network has n levels, and every block in the first $n - 1$ levels has two outputs; at the nth level half the O blocks have two, and half have zero, outputs. We can let the jth olevel be the outputs of the jth level with $q(j) = 2^j$, or alternatively we can make olevel 1 empty, and for $1 < j < n$ let the jth olevel equal the outputs of the $(j - 1)$th O network level, so $q(j) = 2^{j-1}$ and $q(n) = 2^n$. The alternative ways of defining olevels for a given O network can affect the ease of counting.

We obtain a bounds on the number of outputs in terms of the numbers of olevels, assuming that the number of outputs in the last, or nth, olevel $q(n)$ is related to the number in the $(n - i)$th olevel $q(n - i)$:

Lemma 4.6.1: If $c * q(n - i) \le q(n)/\alpha^i$ for $n \ge i \ge 1$, with $\alpha > 1, c > 0$, then TOT is the total number of outputs at all olevels $\le q(n) * \alpha/(c * (\alpha - 1)) = O(q(n))$.

PROOF

$$c * q(n - i) \le q(n)/\alpha^i \qquad \text{for } n \ge i \ge 1$$

Then

$$c * \text{TOT} = c * \sum_{i=0}^{n-1} [q(n - i)] \le \sum_{i=0}^{n-1} [q(n)/\alpha^i]$$

$$\le q(n) * \sum_{i=0}^{n-1} [1/\alpha^i]$$

but

$$\sum_{i=0}^{n-1} [1/\alpha^i] \le \alpha/(\alpha - 1) = \sum_{i=0}^{\infty} [1/\alpha^i]$$

$$\text{TOT} \le q(n) * \alpha/(c * (\alpha - 1))$$

$$= O(q(n))$$

Lemma 4.6.1 may now be directly applied to an O network with n levels, and in which every block has $m > 1$ outputs. We let olevel j equal all the

	olevel j	1	2	3			n
a	$q(j) =$ # of outputs	1	$2*1$	$3*2*1$	$\cdots$	$n-1* \ldots *2*1$	$n*n-1* \ldots *2$
b	$q(n) =$ # of outputs	n	$n(n-1)$	$n(n-1)(n-2)$	$\cdots$	$n(n-1)(n-2)\ldots 2$	$n(n-1)(n-2)\ldots$

FIGURE 4-5 Number of outputs in O network per level for permutations.

outputs of the O network at level j. Then

$$q(n-i) = m^{n-i}$$

$$m^{n-i} \leq m^n/m^i$$

So
$$q(n-i) \leq q(n)/m^i \qquad \text{for } n \geq i \geq 1$$

Therefore by Lemma 4.6.1, with $\alpha = m$,

$$\begin{aligned} \text{TOT} &\leq q(n)*\alpha/(\alpha-1) \\ &\leq m^{n+1}/(m-1) \\ &= O(m^n) \end{aligned}$$

(It is sufficient to have a lower bound m on the number of block outputs.)

For a permutation generation algorithm in which there are n levels of O block outputs, with the jth having $j!$ outputs, if we let the jth olevel equal all outputs at level j, (see Figure 4-5a). Then

$$\begin{aligned} q(n-i) = (n-i)! &= n!/(n*(n-1)* \cdots *(n-i+1)) \\ &= q(n)/(n*(n-1)* \cdots *(n-i+1) \\ &\leq q(n)/2*2* \cdots *2 \qquad \text{for } n-i \geq 1 \\ &\leq q(n)/2^i \qquad \text{for } n \geq n-i \geq 1 \end{aligned}$$

So by Lemma 4.6.1,

$$\begin{aligned} \text{TOT} &\leq q(n)*\alpha/(\alpha-1) \\ &\leq q(n)*2/(2-1) \\ &\leq n! \end{aligned}$$

Consider a permutation generation algorithm in which there are n levels of O block outputs, with the jth having $n(n-1)(n-2)\ldots(n-j)$ outputs (see Figure 4-5b).

Let the jth olevel equal the outputs at the jth level,

$$\begin{aligned} q(n-i) &= n!/i! \\ &= q(n)/i*(i-1)* \cdots *2*1 \\ &\le q(n)/2^{i-1} \end{aligned}$$

Thus $\quad cq(n-i) \le q(n)/\alpha^i \quad$ for $n \ge n-i \ge 1, \alpha = 2, c = \frac{1}{2}$

so

$$\text{TOT} \le q(n)*\alpha/c(\alpha-1)$$

$$\text{TOT} \le 4n! = O(n!)$$

We have often dealt with n level O networks in which the total number of O and t block outputs is $O(q(n))$, and each t block prints vectors, grown one component per level. The total cost will be $O(n(q(n))$ because

$$O(\text{number of } O \text{ block outputs}) = O(q(n)) +$$

$$O(\text{number of } t \text{ blocks} * \text{complexity of each } t \text{ block}) = O(nq(n))$$

In some applications the terminal argument structure vectors need not be printed out; rather, a pointer to the vector is passed to another routine. Even if printing is not required, however, as long as the argument structures have to be moved in and out of queues or stacks at each O block simulation (as in the nonuniform inverse standard depth-first and breadth-first algorithms), the complexity will still be $O(nq(n))$. However, if printout is not required and there is a uniform inverse implementation in which each t block and inverse function is $O(1)$, a complexity of $O(q(n))$ can be achieved.

Another case in which $O(q(n))$ is sufficient for simulating an n level O network in which the number of O and t block outputs are $O(q(n))$ is if the argument structure does not contain any growing structures. If the outputs from the t blocks *and all parts of the argument structure* are of fixed size, independent of any initial argument, which incidentally implies that there is no uniform inverse, then the movement in and out of stacks associated with each O block output will not increase the order of the complexity, which will then be $O(q(n))$, or, more generally, O(number of O block outputs). Implementation of the Tower of Hanoi solution according to Definition 3.1.1 provides an example.

Many of the definitions we have considered, including those for binary numbers, combinations, and partitions, require only $O(1)$ time per O block output. So the preceding conclusions are applicable. However, this is not true for the permutation definitions developed thus far. A permutation definition for which it is true is now developed.

For permutations of the first n integers we have the definition

$$\left[\begin{array}{ll} f(v,b) = v & :b = \{\ \} \\ f(v,b) = \bigcup_{i=1 \text{ to } |b|} (v \| b_i, b - b_i) & :b \neq \{\ \} \\ \text{Initially } v = \langle\ \rangle,\ b = \langle 1, 2, \ldots, n\rangle,\ n \in N. & \end{array}\right.$$

The inverse is

parent(v, b) = remove l, the last component of v and insert l into b between two components b_i and b_{i+1}, where $b_i < l < b_{i+1}$.

child#(v, b) = number of components to the left of the position into which l was inserted in b.

Although both computing the O function and doing the (child = last child?) test have complexity $O(1)$, determining the parent as well as the next-sib requires a scan of the string and so is O(length of v).

Instead of continually removing members of b, unused members of b can be linked. h is the pointer to the first location in b connected by this series of links. To produce each child, a linked member of b is unlinked, the value found there concatenated to v, and the sequence of links repaired.

The array b together with the sequence of links of its unused members is designated D. To be able to back up, it is necessary to store the indices in b of the entries immediately to the left and right of each component in v along the linked list when that component was added to v. By convention the index to

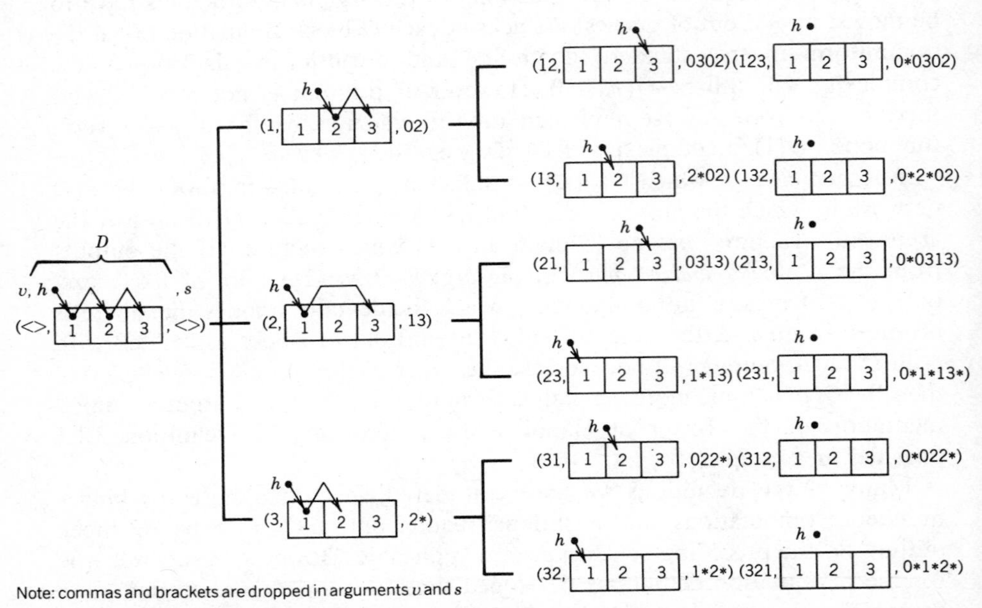

FIGURE 4-6 O network for permutation example $O(1)$. Commas and brackets ($\langle$, $\rangle$) are omitted in listing nonempty v and s arguments.

the left of the first entry on the linked list is 0, to the right of the last entry on the list it is $*$. These left-right pairs, associated with members of v, are stacked in the third argument, s.

The example in Figure 4-6 indicates that with these additions to the argument structure, the inverse can be computed in $O(1)$ time, and Definition 4.6.1 is the definition incorporating them.

Definition 4.6.1: Recursive Definition for Permutation with $O(1)$ Child Generation Cost

D is a data structure consisting of an array, each component or cell of which has a data and a pointer field, and a cell h containing a pointer to the beginning of a list, linked using the pointers, of unused members of D.

data $(D(i))$ is the data in the ith cell of D
ptr$(D(i))$ is the pointer in the ith cell of D
$|D|$ is the number of cells in the unused list.

Initially $D = D_{\text{initial}}$ and in D_{initial}

$$\text{data}(D(i)) = i, \qquad \text{ptr}(D(i)) = i + 1 \text{ for } 0 \le i < n,$$

$$\text{ptr}(D(n)) = *, \qquad h = 1 = \text{ptr}(D(0))$$

$$\text{link}^{-1}(h) = h, \quad \text{link}^{0}(h) = \text{cell to which } h \text{ points.}$$

$$\text{link}^{i}(h) = \text{cell to which pointer in cell link}^{i-1}(h) \text{ points.}$$

$$\begin{cases} f(v, D, s) = \{v\} & : h = \text{nil} \\ f(v, D, s) = \bigcup_{i=1 \text{ to } |D|} f\big(v \| \langle \text{data}(\text{link}^{i-1}(h))\rangle, \\ \qquad \text{ptr}(\text{link}^{i-2}(h)) \leftarrow \text{link}^{i}(h), \langle \text{link}^{i-2}(h), \\ \qquad \text{link}^{i}(h)\rangle \| s\big) & h \ne \text{nil} \\ \text{Initially } v = \langle\,\rangle,\ D = D_{\text{initial}},\ s = \langle\,\rangle. \end{cases}$$

Inverse: Let $s_1 = \langle x, y \rangle$ and

last-child: $y = *$

parent:
- v: $v - v_{\text{last}}$
- D: $\text{ptr}(D(x)) \leftarrow v_{\text{last}}$
  $\text{ptr}(D(v_{\text{last}})) \leftarrow y$
- s: $s \leftarrow s - s_1$

next-sib:
$\text{ptr}(D(x)) \leftarrow v_{\text{last}}$
$\text{ptr}(D(v_{\text{last}})) \leftarrow \text{ptr}(D(y))$
$x \leftarrow v_{\text{last}}$
$v_{\text{last}} \leftarrow y$
$y \leftarrow \text{ptr}(D(y))$

Algorithm 4.6.1: Permutations with $O(1)$ Uniform Inverse

Implementation	*Notes*
$v \leftarrow \langle\rangle$	
$D \leftarrow D_{\text{initial}}$	
$h \leftarrow 1$	
$s \leftarrow \langle\rangle$	
FOR ever DO	
WHILE($h \neq$ nil) DO	
[first-child]:	
$v \leftarrow v \Vert \text{data}(h)$	
$y \leftarrow \text{ptr}(h)$	
$s \leftarrow \langle 0, y\rangle \Vert s$	$s_1 = (x, y)$
$h \leftarrow y$	
END	
OUT $\leftarrow v$	
WHILE(*last-child*: $y = *$) DO	
[*parent*]:	
$v \leftarrow v - v_{\text{last}}$	
$\text{ptr}(D(x)) \leftarrow v_{\text{last}}$	$D(o) = h, \text{ptr}(h) = h$
$\text{ptr}(D(v_{\text{last}})) \leftarrow y$	
$s \leftarrow s - s_1$	
IF $v = \langle\rangle$ THEN DONE	
END	
[*next-sib*]:	
$\text{ptr}(D(x)) \leftarrow v_{\text{last}}$	
$\text{ptr}(D(v_{\text{last}})) \leftarrow \text{ptr}(D(y))$	
$x \leftarrow v_{\text{last}}$	
$v_{\text{last}} \leftarrow y$	
$y \leftarrow \text{ptr}(D(y))$	
END	

In concluding this section we note some properties of the inverse and its complexity, which are illustrated by this development.

Interestingly, although all the inverse-related functions used in the uniform inverse algorithm, including parent, next-sib, and last child, can be computed in $O(1)$ time, computing the child# requires more than $O(1)$ time. Stacking child#'s would give an $O(1)$ time if this were needed.

This illustrates the following general relations between the complexities of the different functions that enter into the uniform inverse algorithm.

1. If o_i, parent, child#, and $m(X)$ can be computed in $O(1)$ time, then so can last-child, parent, next-sib, and o_1.
2. If last-child, next-sib, parent, and o_1 can be computed in $O(1)$ time, then it does not follow that child# can be computed in $O(1)$ time. It does follow that it can be computed in $O(\text{child\#})$ time.

Though unproven, it also appears to be true that:

1. If the complexity of $o_i(X)$ is $O(1)$, then, if an inverse exists, the inverse child#(X) and parent(X) can be constructed, in $O(1)$ time.

2. If $o_1(X)$ can be computed in $O(1)$ time and $o_i(X)$ can be computed in $O(i)$ time, and the $O(i)$ time is used to find the data in $o_i(X)$ after which $O(1)$ time is enough for processing, then an inverse can be computed in $O(1)$ time if it exists.

4.7 PROBLEMS

1. *Storage Complexity Queues and Stacks.* Given an O network which forms a uniform binary tree of depth n whose leaves are O blocks with terminal value outputs, how large does the queue in Algorithm 4.1.1 get when applied to a function that generates such an O network? How large does the stack in Algorithm 4.2.1 get?
2. *Find Uniform Inverse—Exercise.* Consider the following definitions.

Definition 4.7.1: Permutations of the Components of vector b = First n Integers

$$\left[\begin{array}{ll} p(v, j, b) = \{v\} & : j = n \\ p(v, j, b) = \bigcup_{i=1 \text{ to } n \,\&\, b_1 \neq 0} \left[p(v_{j+1} \leftarrow b_i, j+1, b_i \leftarrow 0) \right] & : j < n \\ \text{Initially } v = 0^n,\ j = 0,\ b = \langle 1, \ldots, n \rangle,\ n \in N. & \end{array}\right.$$

In general in the recursive line of Definition 4.7.2, $v = \langle v_1, \ldots, v_k \rangle$ and $j = v_k$, while if $k = 0$, $j = 1$.

Definition 4.7.2: Partitions

$$\left[\begin{array}{ll} f(v, j, r) = \{v \| \langle r \rangle\} & : \lfloor r/2 \rfloor - j < 0 \\ f(v, j, r) = \bigcup_{i=0}^{[r/2]-j} [f(v \| (j+i), j+i, r-(j+i)] & \\ \qquad \cup \{v \| \langle r \rangle\} & : \lfloor r/2 \rfloor - j \geq 0 \\ \text{Initially } v = \langle\,\rangle,\ j = 1,\ r \text{ in } N. & \end{array}\right.$$

Find the uniform inverse for Definitions 4.7.1, 4.7.2, and 3.4.2.

3. *Write Uniform Inverse Algorithm—Exercise.* Write a uniform inverse algorithm to implement Definition 4.7.1 by plugging into the depth-first algorithm, Algorithm 4.2.4 or Algorithm 4.2.5.
4. *Trace Uniform Inverse Algorithm—Exercise.* Put the algorithm in Problem 3 in a programming language of your choice and run it on a small value of n (< 7).
5. *Nonexistence of Uniform Inverse.* Justify the assertion that the Chinese ring puzzle definition, Definition 3.8.7, does not have a uniform inverse.

Alter that definition minimally so that the result still gives the solution to the puzzle and does have a uniform inverse.

6. *Uniform Inverse for a Class of Definitions.* Any recursive definition for backtrack selection of the form of Definition 3.4.2 has an inverse. Give the inverse in terms of the primitives in Definition 3.4.2.
7. *Global Variables.* Given a recursive definition which does not assume any global variables, what property(s) in that definition indicate that a global variable can be used?
8. *Complexity of Uniform Inverse Computation.* Justify the two assertions at the end of this chapter on the complexity of the uniform inverse in simple cases. Also try to find counterexamples for the two conjectures given there.

CHAPTER FIVE

IMPLEMENTATION BY SUBSTITUTIONAL EVALUATION — TIME EFFICIENCY

Implementations we have considered take advantage of properties such as uniform inverse and associativity to minimize the use of stacks and other temporary memory. Nevertheless, they all still simulate *every* block in the O–w–t network, and therefore cannot differ appreciably in running time. Time savings will be achieved only if some of the simulation is avoided, that is, some of the blocks in the O–w–t network are *pruned*. The possibility for pruning must arise from properties of the recursive definition. We consider two general classes of such properties.

5.1 PRUNING AND COMPARABILITY

5.1.1.1 Special Properties of the w Function

Pruning is possible if the w function of a recursive definition is *truncatable*. For example, the "or" function has a value of 1 if any one of its arguments is 1; the other arguments, examined or not, are irrelevant. Again, if all arguments of a "maximum" function are known to be $\leq B$, then if B appears as an argument, further arguments are irrelevant. Such functions, which can, under appropriate circumstances, be evaluated without any knowledge of all their arguments, are called truncatable. The w function used in game analysis, considered later in this chapter, provides an example.

5.1.1.2 Comparability

Another possibility for pruning depends on the existence of a relation, called *comparability*, between argument structures in the substitutional domain.

Consider a substitutional evaluation of a definition DEF.f. Assume that at some point in the evaluation $f(X)$ and $f(Y)$ appear in the developing sentential form E in some specified positional relation to each other. We say that argument structures X and Y are comparable in DEF.f if the substitution of the expression $q(f(X))$ for $f(Y)$ in E and continuation will not change the final output. q is assumed to be a simple primitive function, that is, the work necessary to evaluate $q(f(X))$, given the value of $f(X)$, is less than that necessary for a substitutional evaluation of $f(Y)$. This definition is illustrated in Figure 5-1. Equivalently, argument structures X and Y are said to be comparable in DEF.f if the output of an O–w–t network containing X and Y in a specified positional relation to each other is unchanged when, as illustrated in Figure 5-1, $q(f(X))$ replaces all occurrences of the w block input $f(Y)$ in the network.

Comparability comes in many forms. If argument structures X and Y in an O–w–t network are *equal*, then obviously they are comparable, independent of the relative positions of X and Y, and the function q from the comparability definition is the identity. In fact, only the information and definition control components of X and Y need be equal; program control components can differ, and the above conclusions still hold. If in addition the w function is a selection function whose value is not affected by repetition of arguments, such as minimum, whose value is equal to one of its arguments, then no q is needed and the subnetworks which compute $f(Y)$ can be removed.

Although similarities weaker than equality can produce comparability in a definition, these can often be made equalities by a transformation to an equivalent definition. Such transformations will be described in Chapter 6. When similarities cannot be thus transformed, they must be confronted directly. In the minimum-select monotonic class of recursive definitions, Section 5.4, there are many examples of this kind.

All algorithms in this chapter which take advantage of comparability, whether originating in equality or not, are SE based. When comparability is due to equality, however, there is a powerful alternative approach based on ESE. This ESE approach will be developed in Chapter 6.

The concept of comparability is given and used to describe and classify simplifications in [Cohen 79].

To make use of comparability, the basic breadth- and depth-first algorithms are typically augmented with a *table*, which records argument structures generated during simulation. Newly generated argument structures are compared with those in the table for possible pruning. With each argument structure X in the table, it may be necessary to store $f(X)$ when it becomes available. Consider how the table could be used in augmenting the basic implementations if comparability is based on equality. In Algorithms 4.2.1 through 4.2.6, code would be added to store any argument structure not already in the table every time new argument structures are generated. Also code to store $f(X)$ should be added every time $f(X)$ (a w block output) is computed, typically in the backup section. (If w is a selection function, such

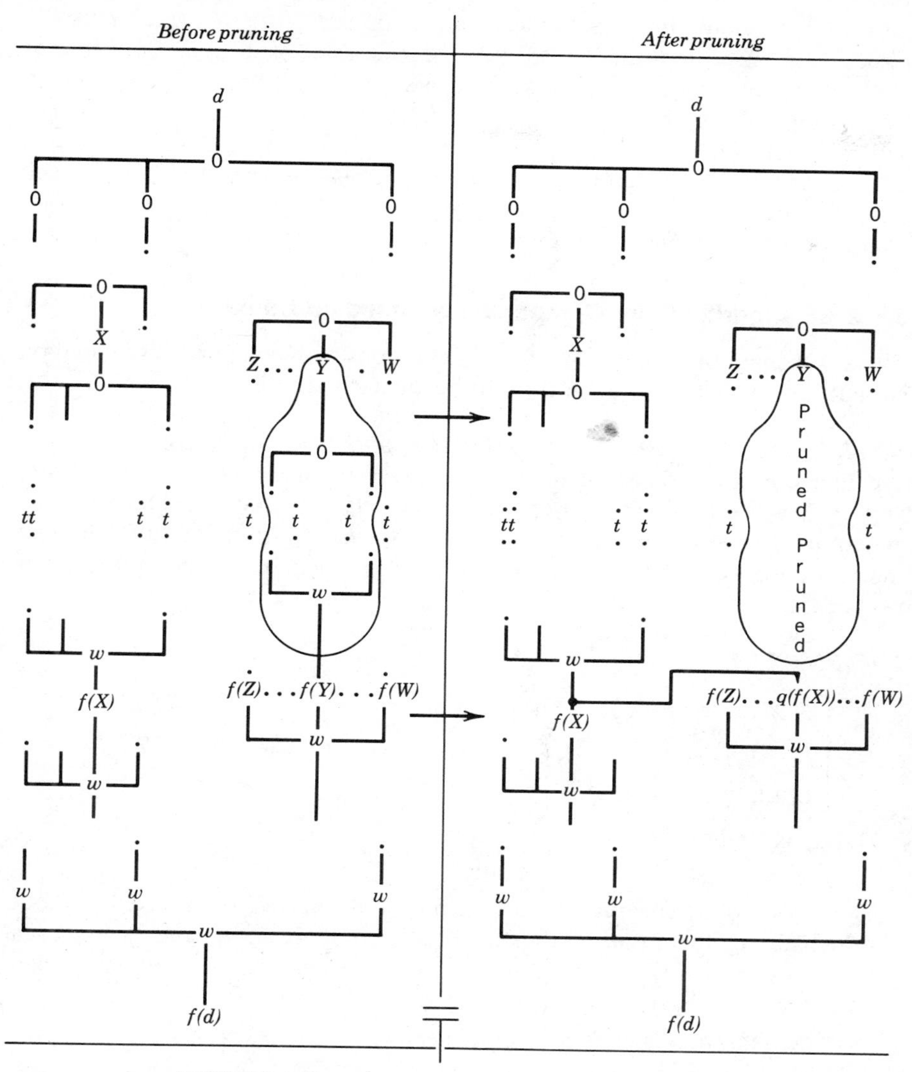

FIGURE 5-1 Comparable argument structures and pruning.

as minimum or union, $f(X)$'s need not be stored.) When an argument structure X, equal to one previously stored, is detected, the value $f(X)$ stored with X may, if w is associative, be "w'ed" in with the cumulative result of "w-ing" all w block outputs, or, if w is not associative, saved, usually in a stack of still unused w block outputs. In either case the sequencing of the algorithm must be continued as though a terminal value had been encountered. When the nature of the comparability requires saving $f(X)$, or some information about $f(X)$, along with argument structure X, $f(X)$ must be available in order to prune in response to a recurrence of X. So a recurrence of X is not really comparable unless it appears after $f(X)$ has been determined. Thus comparability may depend on the relative position as well as the value of argument structures.

Pruning may be counterproductive. Saving and reprocessing $f(X)$ each time X occurs may be as costly as regenerating $f(X)$. The Tower of Hanoi definition provides such an example.

5.1.2 SE Algorithms for Comparability Based on Equality

There is a small but important class of definitions with comparability based on equality for which the SE algorithm is particularly simple.

5.1.2.1 Completely Prunable Definitions and Their Algorithms

We first consider a class of definitions, *completely prunable definitions*, with very strong comparability properties. This class includes definitions that generate all vertices reachable from a given vertex in a graph, and for finding the minimum of numbers stored at each vertex reachable from a given set of vertices.

Definition 5.1.1

$$\left[\begin{array}{ll} f(X) = w(p(X)) & :T(X) \\ f(X) = w\big(f(o_1(X)), \ldots, f\big(o_{m(X)}(X)\big), p(X)\big) & :\sim T(X) \\ \text{Initially } X = d \in D. & \end{array}\right.$$

1. Δ_f is finite.
2. w is associative and commutative.
3. Any duplicated argument of w may be removed or inserted without changing the value of the function, that is, if α, β, and ϕ are each arbitrary strings of arguments and x an individual argument, then $w(\alpha, x, \beta, x, \phi) = w(\alpha, x, \beta, \phi)$.
4. The definition is equation complete.

Such functions as union, intersection, and minimum satisfy these constraints.

It will be convenient to assign an integer from 1 to $n = |\Delta_f|$ to each argument X in Δ_f. Let i be the integer assigned to X, and x_i a variable assigned to i. Then let

1. $f(X) \equiv x_i$,
2. $f(o_j(X)) \equiv x_{i_j}$, where i_j is the integer assigned to $o_j(X)$.
3. $p(X) = c_i$.

Then the definition can be rewritten.

Definition 5.1.2

$$\{x_i = w(x_{i_1}, \ldots, x_{i_{n(i)}}, c_i) | i \in N^n\}$$

The implementation of any definition in this highly restricted class is characterized very simply in terms of its dependency graph. For each variable x_i there is a vertex v_i in the dependency graph, G and a vector $rc(i)$. $rc(i)$ has a component for each vertex reachable from vertex v_i by a path of length ≥ 0. If v_{i_j} is the jth vertex reachable from v_i then the jth component of $rc(i)$, $rc(i)_j$, is the constant c_{i_j} which is the $p(x)$ on the right side of the equation for x_{i_j}.

Theorem 5.1.1: Definition 5.1.2 is satisfied by $x_i = w(\mathrm{rc}(i)_1, \mathrm{rc}(i)_2, \ldots, \mathrm{rc}(i)_{\mathrm{last}})$.

PROOF: Substitute the dependency path interpretation $x_i = w(\mathrm{rc}(i)_1, \mathrm{rc}(i)_2, \ldots, \mathrm{rc}(i)_{\mathrm{last}})$ for x_i throughout Definition 5.1.2 and show that the resultant equalities are valid. The result of the substitution is

$$\begin{aligned} &w(\mathrm{rc}(i)_1, \mathrm{rc}(i)_2, \ldots, \mathrm{rc}(i)_{\mathrm{last}}) \\ &\quad =?\, w\Big(w(\mathrm{rc}(i_1)_1, \mathrm{rc}(i_1)_2, \ldots, \mathrm{rc}(i_1)_{\mathrm{last}}), \\ &\qquad w(\mathrm{rc}(i_2)_1, \mathrm{rc}(i_2)_2, \ldots, \mathrm{rc}(i_2)_{\mathrm{last}}), \ldots, \\ &\qquad w(\mathrm{rc}(i_{n(i)})_1, \mathrm{rc}(i_{n(i)})_2, \ldots, \mathrm{rc}(i_{n(i)})_{\mathrm{last}}), c_i\Big), \end{aligned}$$

On the right of this equation $w(w$ variables $1), \ldots, w($variables $n(i))$ can be replaced by $w($variables $1, \ldots,$ variables $n(i))$ because of the associativity and commutativity of w.

$$\begin{aligned} &w(\mathrm{rc}(i)_1, \mathrm{rc}(i)_2, \ldots, \mathrm{rc}(i)_{\mathrm{last}}) \\ &\quad =?\, w(\mathrm{rc}(i_1)_1, \mathrm{rc}(i_1)_2, \ldots, \mathrm{rc}(i_1)_{\mathrm{last}}, \\ &\qquad \mathrm{rc}(i_2)_1, \mathrm{rc}(i_2)_2, \ldots, \mathrm{rc}(i_2)_{\mathrm{last}}, \ldots, \\ &\qquad \mathrm{rc}(i_{n(i)})_1, \mathrm{rc}(i_{n(i)})_2, \ldots, \mathrm{rc}(i_{n(i)})_{\mathrm{last}}, c_i) \end{aligned}$$

In consequence of the "remove duplicates" property 3 of w, this tentative equality can be given the following interpretation:

> w of all constants associated with vertices reachable from x_i by a path of length zero or more $=?$ w of all constants associated with vertices reachable by a path of length zero or more from some neighboring variable x_{i_j} of x_i in the dependency graph, together with c_i.

And the tentative equality is now clearly true. ■

For completely prunable definitions then it is only necessary to simulate each argument structure associated with a vertex of the dependency graph once, that is, any recurrence of an argument structure can be pruned. Virtually any legitimate order of simulation of the O network, depth-first, breadth-first, or other, can be adopted as the basis for these definitions. This will be demonstrated in the following paragraphs.

5.1.2.2 SE Algorithms to Implement Completely Prunable Definitions

To characterize the algorithms implementing completely prunable definitions we need a relation between vertex ordering in the dependency graph and in the O network of these definitions. The legitimate vertex orderings in these structures can be characterized by the order in which a fluid entering at a starting vertex would arrive at the other vertices.

COMPLETE CUTOFF FLOW

Think of a directed graph as representing an interconnected set of pipes (edges) with flow restricted by one-way valves to following the direction of the edges. Let increasing amounts of a fluid be introduced at vertex d and flow out toward the remaining vertices but

1. The flow cannot simultaneously reach a node from more than one inward edge.
2. As soon as the fluid reaches any vertex v along one inward edge, valves are shut so as to prevent fluid reaching v from any other of its inward edges.

This is called *complete cutoff flow* from d.

Theorem 5.1.2: Under complete cutoff flow from d the fluid will reach vertex v in digraph G if and only if there is a path from d to v in G.

PROOF: Assume that on the contrary there is a path, p, in G from d to v and that the flow does not reach v. There must be a first vertex on p from d to v which the flow never reaches. Let it be v', which is preceded by u', the last vertex on p which the flow reaches. There is an edge from u' to v' so flow must be prevented in that edge. But flow is only prevented on incoming edges of vertices which have been reached by the flow. Therefore v' must have been reached by the flow, which is a contradiction. ■

This theorem is significant because the ordering implied by complete cutoff flow in the dependency graph of a recursive definition is equivalent to the order of simulating blocks in the O network for that definition when every repeat of an argument structure is pruned. The flow in the O network corresponding to the complete cutoff flow moves along *frontier paths*. The concept of frontier paths will be significant henceforward. Before giving the formal definition we give an example of frontiers.

Example 5.1.1: Consider the Following Portion of an O-Network, with *Frontiers* F_1 through F_5. (A frontier is a cut across the network.)

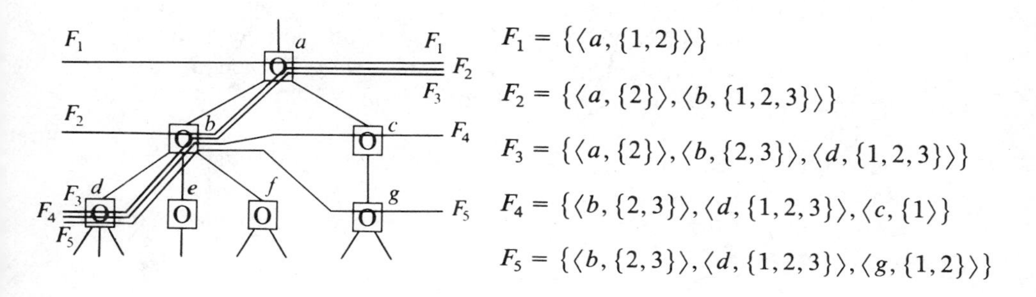

$$F_1 = \{\langle a, \{1,2\}\rangle\}$$

$$F_2 = \{\langle a, \{2\}\rangle, \langle b, \{1,2,3\}\rangle\}$$

$$F_3 = \{\langle a, \{2\}\rangle, \langle b, \{2,3\}\rangle, \langle d, \{1,2,3\}\rangle\}$$

$$F_4 = \{\langle b, \{2,3\}\rangle, \langle d, \{1,2,3\}\rangle, \langle c, \{1\}\rangle\}$$

$$F_5 = \{\langle b, \{2,3\}\rangle, \langle d, \{1,2,3\}\rangle, \langle g, \{1,2\}\rangle\}$$

In each case, F_{i+1} *follows directly* from F_i:

F_2 from F_1 by $\langle a, \{1,2\}\rangle \to \langle a, \{2\}\rangle$ and $\langle b, \{1,2,3\}\rangle$, since $b = o_1(a)$,
F_3 from F_2 by $\langle b, \{1,2,3\}\rangle \to \langle b, \{2,3\}\rangle$ and $\langle d, \{1,2,3\}\rangle$,
F_4 from F_3 by $\langle a, \{2\}\rangle \to \langle c, \{1\}\rangle$, since $|\{2\}| = 1$, and
F_5 from F_4 by $\langle c, \{1\}\rangle \to \langle g, \{1,2\}\rangle$.

Thus F_1, F_2, F_3, F_4, F_5 constitute a *frontier path*. (Note also that expansions need not have been of the first child; $\langle a, \{1,2\}\rangle$ could have been expanded into $\langle a, \{1\}\rangle$ and $\langle c, \{1\}\rangle$ instead.)

FRONTIER PATH

A *frontier* of an O tree is a multiset of (argument structures, number) pairs (X, Z), where X is an argument structure which appears as input and/or output of O blocks in that tree, and Z is the set of indices of children of X that neither are on the frontier, nor ancestors of argument structures on the frontier.

Not all such multisets of pairs are frontiers of a given O tree. The following recursive definition must also be satisfied by the multiset in order for it to be a frontier.

1. $\{(X, Z)\}$ is a *frontier* if X is the initial argument structure and $Z = \{1, \ldots, m(X)\}$.
2. Let (X, Z), X not terminal, be a member of frontier F'. Let $j \in Z$.
   a. If $|Z| > 1$, replace (X, Z) in F' with two entries, $(o_j(X), \{1, \ldots, m(o_j(X))\})$ and $(X, Z - j)$.

b. If $|Z| = 1$, replace (X, Z) in F' with one entry, $(o_j(X), \{1, \ldots, m(o_j(X))\})$.

The result F is a *frontier*. F is said to *follow directly* from F'.

A frontier F partitions all argument structures in the O tree into three sets —the ancestors of argument structures of F, members of F, and the descendants of members of F. (The ancestors of (X, Z) are all the ancestors of X, its descendants are the members of $\{o_k(X) | k \in Z\}$ and all their descendants.) A sequence of frontiers $F_1, F_2, \ldots, F_n$ is a *frontier path* if each F_i follows directly from F_{i-1}. The concepts of frontier and "follows directly" are illustrated in example 5.1.1.

An algorithm which simulates an O tree by sequencing through a frontier path and immediately prunes any (X, Z) with $Z = \{1, \ldots, m(X)\}$ if an identical (X, Z) pair has previously appeared on the frontier, is called a *simple pruning algorithm*.

CONNECTION BETWEEN DEPENDENCY GRAPH AND O TREE

We now show that the sequence of frontier paths in simulation of an O tree is essentially a sequence of snapshots of a complete cutoff flow in the corresponding dependency graph.

Given a nonnested recursive definition DEF.f, with d an initial argument structure and v any other argument structure in $\Delta_f(d)$, we have

A *dependency graph path p from d to v* is a vertex path through vertices identified by their argument structure in the dependency graph for DEF.f from initial argument structure d to v.

An *O-tree path from d to v* is a vertex path in the O network for DEF.f, from initial root argument structure d to v, interpreting O block argument structure inputs (outputs) as vertices. Two successive vertices in this path will be an input and output argument structure of the same O block.

The next three lemmas establish the connection between a simple pruning algorithm and a complete cutoff flow. The proofs follow from the definitions above.

Lemma 5.1.1: For DEF.f, with initial argument d, p is a dependency graph path if and only if it is an O tree path.

Lemma 5.1.2: Simulating the O network by a sequence of frontier paths simulates the flow through the dependency graph.

Lemma 5.1.3: For a nonnested definition DEF.f, a simple pruning algorithm on an O tree with initial argument structure d simulates a complete cutoff flow in a dependency graph for DEF.f from vertex d. Further, both terminate if the dependency graph for DEF.f is finite.

PROOF: Consider the simple pruning algorithm as it sequences through an O tree on a frontier path. Each step on the frontier path simulates the advance of the flow along an edge. Whenever an argument structure X reoccurs in a member of the frontier in the form $(X, \{1, \ldots, m(X)\})$, it is pruned and removed from the frontier. This simulates the shutting of valves into the argument structure (vertex) after flow first reaches it in complete cutoff flow. That means that (X, Z) with Z a *proper* subset of $\{1, \ldots, m(X)\}$ will never be removed. Once flow has been allowed to progress from a vertex, it will progress along each output edge which does not go to an already activated vertex.

If there are only a finite number of *different* argument structures X (the dependency graph is finite), and each must appear as a member $(X, \{1, \ldots, m(X)\})$ in the O tree before it appears as a member of any pair (X, Z) in which Z is a proper subset of $\{1, \ldots, m(X)\}$, then eventually there must be no members on the frontier. At this point clearly no previously unseen argument structures can be generated. ■

Because complete cutoff flow reaches every argument structure once (Theorem 5.0.1) and the simple pruning algorithm simulates that flow (Lemma 5.0.3), every argument structure reachable in the O tree will be reached by a simple pruning algorithm. All the breadth- and depth-first algorithms covered in Chapter 4 sequence on frontier paths, modified to do simple pruning, are simple pruning algorithms and will handle any completely prunable definition —even definitions which do not terminate. Despite nontermination, pruning guarantees that each different argument structure will be considered nonterminal at most once, and there are only a finite number of argument structures. Of course with nontermination the definition may be satisfied in more than one way. The solution obtained with these algorithms is usually the desired one. In summary

Theorem 5.1.3: If a definition is completely prunable, then it can be implemented with a depth-first inverse or breadth-first implementation with pruning of every recurrence of an argument structure, or by any frontier path sequence with simple pruning simulation of the O tree.

If a definition is completely prunable, its w function is associative and commutative, and reoccurrences of arguments can be ignored. The w functions for which this is true include minimum, maximum, and union, all of whose complexities are normally $O(1)$ per pair of arguments. Therefore the complexity of these algorithms will usually be determined by

the number of O block outputs, each simulated by the simple-pruning algorithm
**** the cost per output***
or, in the most usual case, when that cost is O(1), the number of O block outputs.

In Section 5.2 we will look in some detail at the definition for generating all vertices reachable from a given vertex in a digraph. This will give a prime example of a completely prunable definition, and simple prunable algorithms. Before doing this, however, we consider an alternative, compact way to express implementations involving any kind of pruning, including that in simple prunable algorithms.

5.1.3 Expressing Implementation Refinements by Nested Recursion

Theorems 2.6.2 and 2.6.3 show that there is a nested definition equivalent to any nonnested one. In Section 3.10.0.1 it was further shown how nesting can be used to express implementation details not otherwise expressible. With nesting, not only can functions otherwise undefinable be defined, but implementation efficiencies made possible by comparability can be expressed at definition level.

It is because nesting imposes an order on the evaluation of a definition that it can specify implementation details. Depth-first order is implicit in any nested definition, so it is possible to specify, in the definition itself, that repeated occurrences of comparable argument structures are not to be processed. This nested definition may then be directly implemented as a recursive procedure. In contrast to a nonnested definition, in a nested one the following properties can be specified:

1. The order in which f calls (on the right of the recursive line) are to be executed.
2. How an f call will depend on the result of f calls specified as preceding it.

Given a nonnested definition DEF.f, we now have two general ways for taking advantage of comparability.

1. Augmenting one of the general algorithms for DEF.f by addition of a table to record processed argument structures and incorporation of tests to avoid reprocessing of these argument structures.
2. Transforming DEF.f into a nested form, DEF.g, in which the very structure of the definition imposes an order (depth-first) on the generation of argument structures. We will use definitions of the form developed in Section 3.10.0.1 for this purpose.

In this chapter we will see a number of examples of both of these ways of expressing efficiencies.

5.2 APPLICATION OF COMPARABILITY TO GRAPH CONNECTIVITY

Many graph algorithms are considered as examples in this chapter. Most of these are covered in the general algorithm references already given. Additional valuable references are [Even 79; Tarjan 83; Mehlhorn 84].

Consider the connectivity problem. Find all vertices reachable from a given vertex x by a path of length ≥ 0 in digraph G. Here is a definition for $f(X)$ = the set of all vertices reachable from vertex x in G. (Compare to Definition 2.3.1.)

Definition 5.2.1: Reachable Vertices Graph (No Inverse)

$$\left[\begin{array}{ll} f(x) = \bigcup_{i=1}^{i=N(x)} [f(n_i(x))] \cup \{x\} & :N(x) \text{ in } N_0 \\ \text{Initially } x \text{ is a vertex} = \text{“begin” in } G,\ G \text{ is invariant.} & \end{array}\right.$$

Because of the following convention, a single line of the definition serves as both a terminal line, when $N(x) = 0$, and a recursive line, when $N(x) > 0$, $\bigcup_{i=1}^{i=0}$ [expression in i] = { }.

Definition 5.2.1 terminates if and only if G is acyclic. However, even if it doesn't terminate, the definition may, as indicated in Section 5.1.2, still form the basis for a useful algorithm.

This definition is completely prunable, Theorem 5.1.3 is applicable, and thus there are breadth- and depth-first algorithms augmented for pruning available to implement it. Comparable argument structures are certainly expected in evaluating instances of this definition. For every path from "begin" to vertex x there is a distinct and comparable occurrence of argument structure x in the substitutional evaluation. There is no need to continue evaluating the definition from more than one of them. On the other hand, since there are such duplicates, the definition lacks a uniform inverse. So in particular all algorithms (other than those using the uniform inverse, Algorithms 4.1.1 through 4.2.3), once modified for pruning, will implement this definition.

The breadth-first algorithm, modified by the inclusion of table (TBL) in which processed vertices are stored so that their subsequent occurrences can be detected and pruned, is given next.

5.2.1 Breadth-First Algorithm for Reachability

Following is the basic breadth-first algorithm, Algorithm 4.1.1, modified to take advantage of equal argument structures.

Algorithm 5.2.1: Augmented Breadth-First—Reachable Vertices

Augmented Algorithm

```
W ← { }
X ← 'begin'
Z ← ⟨ ⟩
TBL('begin') ← 1                        *1

FOR ever DO
  WHILE(N(X) ≠ 0) DO
    W ← W ∪ {X}
    FOR I = 1 to N(X) DO
      IF TBL(n_I(X)) = 0 THEN           *2
        DO
        TBL(n_I(X)) ← 1                 *3
        Z ← queue − n_I(X)
      END
    END
    IF Z = ⟨ ⟩ THEN DONE                *4
    X ← pop − Z
  END
  W ← W ∪ {X}
  IF Z = ⟨ ⟩ THEN DONE
  X ← pop − Z
END
```

Optimized

```
Z ← psh − d
TBL('begin') ← 1

WHILE(Z ≠ ⟨ ⟩) DO
  X ← pop − Z
  FOR I + 1 to N(X) DO
    IF TBL(n_i(X)) = 0 THEN
      DO
      TBL(n_I(X)) ← 1
      Z ← queue − n_I(X)
    END
  END
END
```

Notes

1. On the left the * lines are added to the standard breadth-first algorithm to provide a table (*1), and to update (*3) and check (*2) that table, and provide an additional check for termination (*4).
2. On the right the algorithm is optimized, assuming that a record of which vertices were reached in the form of positions marked 1 in TBL is a sufficient output. The while condition has been changed, allowing removal of a number of other lines.

5.2.1.1 Complexity of Breadth-First Reachable Algorithm

There are at most n O blocks, one per vertex, in the pruned O network simulated by this algorithm. There is one output per edge from a vertex. The total number of outputs is no more than the total number of edges ($= e$) in the graph. Using the appropriate graph representation, the cost per O block output is $O(1)$. Thus the total cost, or the algorithm complexity, is $O(e)$. This argument holds equally well for the depth-first implementation we will give in Section 5.2.2.

The complexity can also be argued directly from the right side of Algorithm 5.2.1. There is one execution of the inner FOR for each edge of the given graph G. The outer WHILE allows entry to the inner FOR at most once for each vertex in G. This follows since the inner FOR is executed only for vertices in the queue WHILE($Z = \langle\ \rangle$), and only vertices which have not previously been in the queue can get into it [if TBL($n_I(X)$) = 0]. Thus the inner FOR will be executed at most once from each edge in G. The cost to get each edge ($n_i(X)$) depends on the representation, but can be made $O(1)$. (A similar but fuller complexity argument is given for the uniform inverse case later.)

5.2.2 Uniform Inverse Algorithms for Reachability

Although Definition 5.2.1 lacks a uniform inverse, a simple modification changes that and adds the uniform inverse algorithms to the others which apply to reachability. Add the reverse path vector p from "begin" to x, and a stack c of the indices of children encountered on p to the argument structure. The reverse path vector p is a vector of all vertices in the path with p_1 being the most recently encountered.

Definition 5.2.2: Reachable Vertices—Acyclic Digraph (with Inverse)

$$\left[\begin{array}{l} f(p, c, x) = \bigcup_{i=1}^{i=N(x)} [f(\langle x\rangle \| p, \langle i\rangle \| c, n_i(x))] \cup \{x\} \\ \text{Initially } p = \langle\ \rangle,\ c = \langle\ \rangle,\ x \text{ is a vertex in } G. \end{array}\right.$$

The reverse path vector p and index stack c are *program control components*. As might be expected of such components, their inclusion does not destroy the completely prunable property—Theorem 5.1.3 still applies. This follows from the equivalence of Definition 5.1.1 in the theorem to one which is the same except for added program control components (see Problem 1, Section 5.8). So the uniform-inverse algorithms, Algorithms 4.1.1 through 4.2.3, are also applicable to reachability equivalently defined by Definition 5.2.2. Although this definition does not terminate if the graph G has cycles, the algorithm will still work.

For implementing Definition 5.2.2 we give Algorithm 5.2.2, an augmentation of Algorithm 4.2.5 for definition which, like this one, has a w function ($\cup$) that is both associative and commutative. A table is added to keep track of the vertex part of argument structures already simulated in the O network. The parent part of the uniform inverse is easy to compute, namely, parent(p, c, x) = ($p - p_1, c - c_1, p_1$). The ((p, c, x) is last child?) and next-sib(p, c, x) functions depend on the child# part of the inverse. There is a uniform inverse, so these can be computed. The predicate ((p, c, x) is last child?) determines whether x is the last neighbor of p_1; in terms of the primitive graph functions it is ($c_1 = N(p_1)$?). Next-sib(p, c, x) is the next neighbor of p_1 after x and is given by the primitive graph function $n_{c_1+1}(p_1)$. We also need first-child(p, c, x) in graph terms; it is given by ($\langle x\rangle \| p, \langle 1\rangle \| c, n_1(x)$). Although all these functions exist and can be defined in terms of primitive graph functions, the details of their implementation depend on, and therefore should guide, the graph representation. A brief review of some available representations and their properties is given in Section 5.2.2.2.

The adaptation of Algorithm 4.2.5 incorporating TBL to record visited vertices, for implementing Definition 5.2.2 is as follows.

Algorithm 5.2.2: Augmented Depth-First—Reachable Vertices

```
  Plug into Algorithm 4.2.5               Augmentation for Comparability
P ← ⟨ ⟩
C ← ⟨ ⟩
X, S ← 'begin'
FOR ever DO
  WHILE(N(X) > 0) DO                 ← WHILE(N(X) > 0 & TBL(X) = 0) DO
    OUT ← X                                                                  *
    * * * * * * * * * * * * *        ←     TBL(X) ← 1
    P ← ⟨X⟩||P
    C ← ⟨1⟩||C
    X ← n_1(X)

  END
    * * * * * * * * * * * * *        ← IF TBL(X) = 0 THEN                    *
    * * * * * * * * * * * * *        ←   DO                                  *
    OUT ← X                          ←     OUT ← X                           *
    * * * * * * * * * * * * *        ←     TBL(X) ← 1
    * * * * * * * * * * * * *        ←   END                                 *
  WHILE(C(1) = N(P[1])) DO

    X ← pop − P
      ← pop − C
    IF X = S THEN DONE

  END
  C(1) ← C(1) + 1
  X ← n_C(1)(P[1])

END
```

Notes

1. Since this algorithm is based on Algorithm 4.2.5, which requires special provisions to work when the initial argument structure is terminal, this algorithm requires special provisions to work for a graph with one vertex.
2. If and only if a vertex X is reachable and outputted, TBL(X) will be marked with a 1. Thus TBL contains a record of all outputted vertices, and actually outputting them is redundant. If the form of record given in TBL is adequate, all * lines in the algorithm may be removed.

There is also an augmented implementation based on Algorithm 4.2.6, in which not only are all vertices reachable from the "begin" vertex computed, but so too, along the way, are all vertices reachable from each vertex reachable from "begin."

5.2.2.1 Complexity of the Uniform Inverse Algorithm for Reachability

Consider Algorithm 5.2.2 operating on graph G before augmentation. The O network which simulates that algorithm has the following notable property. Let the vertex, or third argument of the argument structure, which is the input to that block, be x, and the vertex similarly associated with an output of that block be y; then $\langle x, y\rangle$ is an edge in G. Thus an edge is associated with each O block output. The same edge $\langle x, y\rangle$ is associated with all O blocks having x as the vertex argument of its input. Any vertex which can be reached by two or more different paths from "begin" will be input to two or more O blocks. However, if Algorithm 5.2.2 is augmented as indicated on the right of its description, the resultant will not simulate any O block whose input involves a vertex that was the vertex argument of any previously simulated O block input. Thus the augmented algorithm, though it may put the same vertex, say x, into X more than once, will not simulate outputs of an O block with input x more than once. This is enforced by the test in the upper WHILE. This test is executed after each O block output is simulated, whether by the first-child or next-sib function. With e the number of edges in G, at most e O block outputs are simulated. Thus if the graph representation allows (see below) computation of each O block output in $O(1)$ time, that is, allows each next neighbor and each WHILE to be executed in $O(1)$ time, then the time complexity is $O(e)$. The lower WHILE is not used to compute O block outputs, and needs to be accounted for. Each passage through that loop pops a vertex from p; thus the number of invocations of the WHILE is $\leq$ the number of vertices on p, which in turn is $\leq$ the total number of vertices, and so is $O(n)$. Thus the complexity of the algorithm is $O(e + n)$.

5.2.2.2 Graph Representations, the Complexity of the Uniform Inverse and Other Graph Primitives

The complexity of the primitive functions in the standard algorithms for nonnested recursive definitions involving a graph G, depends on how G is represented. Representations (data structures) in which these primitive functions require $O(1)$ time are available. Assume that graph G has n vertices and e edges. For every pair of vertices x, y with y an outward neighbor of x, we store the identity of the next neighbor of x beyond y. To be able to access this information easily, we could put it in an *adjacency array*, a two-dimensional matrix M, with $M(y, x) =$ next-neigh(y, x). [$M(y, 0) =$ the first neighbor of vertex y.] This requires $O(n^2)$ memory. Alternatively we could, with a memory cost of $O(e)$, hash each vertex neighbor pair and store the next neighbor at the resultant address, giving $O(1)$ average retrieval time.

More commonly the graph is represented by an *adjacency list*, as illustrated in Figure 5-2. This representation contains a base vector NEIGH, with one component corresponding to each vertex of the graph. NEIGH(x) contains a pointer to a linked list having one node for each neighbor of the vertex x. Now finding $n_1(x)$ requires following the pointer at NEIGH(x). This representation can be used efficiently to obtain inverses provided that lists of

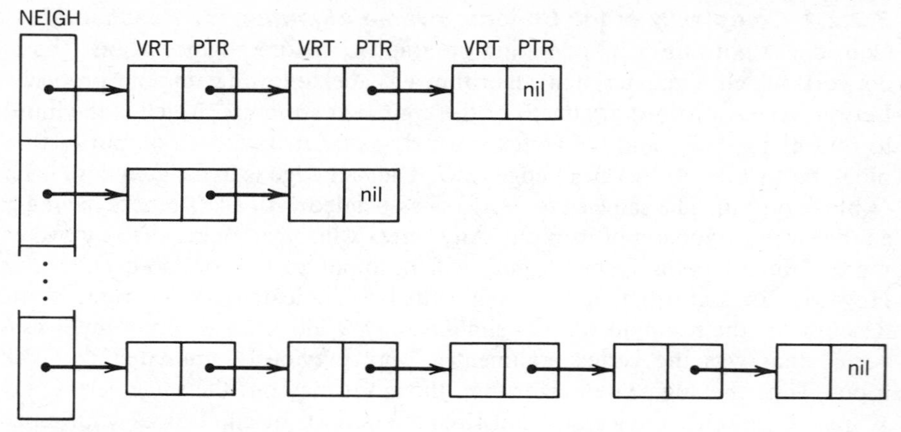

FIGURE 5-2 Adjacency list. The functions for referring to adjacency list data structure are

$$\text{PTR}^i[C[x])] \text{ is PTR(PTR(}\ldots\text{PTR(}C[x])\ldots)) \text{ with } i \text{ occurrences of PTR}$$

$$\text{VRT}^i(C[x]) \text{ is VRT(PTR(PTR(}\ldots\text{PTR(}C[x])\ldots)) \text{ with } i \text{ occurrences of PTR}$$

parents and child#'s is part of the argument structure, as are p and c, respectively, in Definition 5.2.2. Each entry in c can be replaced by the address of the node in the linked list pointed to by NEIGH($p[1]$), where the identity of the next child of p_1 to be processed is found. With this information available in the new c list, finding the next and the last neighbor are $O(1)$. A better alternative is to replace list c with array C, with one entry per vertex. $C[p_1]$ contains the address of the linked list node containing next child of p_1 to be processed. This alternative is incorporated into recursive Definition 5.2.3, which uses these functions for explicitly referring to the adjacency list in Figure 5-2.

Definition 5.2.3: Reachable Vertices—Acyclic Digraph (with Inverse)

$$f(p, C, x) = \bigcup_{i=1}^{i=N(x)} [f(\langle x\rangle \| p, C[x] \leftarrow \text{PTR}^i(C[x]), \text{VRT}^i(C[x])) \cup \{x\}]$$

Initially $p = \langle\ \rangle$, PTR[$C(v)$] = NEIGH(v) for all vertices v in G, x is vertex in G = "begin," G is invariant.

The inverse, in terms of adjacency list primitives is quite simple. The parent of x is p_1, p's parent is $p - p_1$; array C requires no alteration on backing up to its parent because it contains the correct child numbers for every vertex in p, and is needed for that alone. For other vertices it may have the wrong information the next time they are reached, but it will be reset at that time. $C[x]$ always contains the pointer to next-sib of x. It is nil if x is a last sib.

In future considerations of graph algorithms we seldom give descriptions in such detail, though complexity estimates require the ability to do so. Instead the existence of the primitive predicates in Section 1.2 are assumed, and although list c and table C are not explicitly included, we will assume existence of the following predicates, each implementable in constant time:

(x = last-neigh(y)?), true if x is the last neighbor of y,

and

nxt-neigh(y, x) ≡ the next vertex adjacent to y after x.

5.2.2.3 Nested Recursive Definition Formulation for Reachability

When there is comparability, a standard depth-first algorithm augmented with a table can prune redundant computations and improve time efficiency. Alternatively the same efficiency can be directly expressed in the formulation of a nested recursive definition.

The following nested definition is modeled on Definition 3.10.5(s) with additions for efficiency. Starting with Definition 5.2.1, we obtain an equivalent nested definition, and add a relation corresponding to $R(V, X)$ and a line corresponding to $g(v, X) = V$ to ensure that vertices encountered in evaluating one f call are not pursued in the evaluation of successive f calls. This is accomplished by recording the vertices which have served as arguments of f calls in a table V, and passing V as an argument to subsequent f calls. V serves the same purpose here as TBL served in Algorithm 5.2.2. V, like TBL, V is a one-dimensional array with one entry for each vertex in the given graph. In the definition, read $V[x] \leftarrow 1$ as: "assign 1 to the table V entry for vertex x." To see how the nested reachability definition is an adaptation of Definition 3.10.5(b), its recursive line is repeated.

$$g(V, X) = g(\ldots g(g(w(V, p(X)), o_1(X)), o_2(X)), \ldots, o_{m(X)}(X))$$
$$:\sim R(V, X) \text{ and } \sim T(X)$$

and some of the correspondence of its terms with the those or the nested reachability definition, Definition 5.2.4, are given,

$X \equiv$ vertex x

$o_i(X) \equiv n_i(x)$

$w \equiv \cup$

$p(X) \equiv \{x\}$

$T(X) \equiv N(x) = 0$

$R(V, X) \equiv (V[x] = 1)$

$w(V, p(X)) \equiv$ (set represented by V) $\cup \{x\}$

$\equiv V[x] \leftarrow 1$

Definition 5.2.4: Nested Definition—Reachable Vertices

$$g(V,x) = V[x] \leftarrow 1 \qquad N(x) = 0$$

$$g(V,x) = V \qquad :V[x] = 1 \text{ and } N(x) > 0$$

$$g(V,x) = g(\ldots g(g(V[x] \leftarrow 1, n_1(x)), n_2(x)) \ldots n_{N(x)}(x)) \qquad :V[x] = 0 \text{ and } N(x) > 0$$

Initially x is a vertex in V is a table with one entry for each vertex in G, all entries are 0.

The actual recursive procedure for Definition 5.2.4, which in its direct translation to programming language embodies the major efficiencies of Algorithm 5.2.2, is as follows:

Algorithm 5.2.3: Nested Recursive Procedure—Reachable Vertices

```
PROC g(X, V)
  IF N(X) = 0 THEN DO
    V[X] ← 1
    g ← V & RETURN
  END
  IF (V[X] = 1) THEN g ← V & RETURN
  V[X] ← 1
  FOR I = 1 TO N(X) − 1 DO
    V ← g(V, n_I(X))
  END
  g ← g(V, n_N(X)(X))
  RETURN
ENDg
```

After the first call there is a call of g for each edge out of a reachable vertex. So the number of calls is the number of edges out of the vertices reachable from the vertex "begin" + 1. There is a constant amount of computation required for each call, so the complexity is $O(e)$.

5.3 COMPARABILITY WITHOUT EQUALITY IN THE REACHABILITY PROBLEM

When applied to graphs with cycles, neither Definition 5.2.1 nor Definition 5.2.2 terminate, but they do yield simple equality comparability. These definitions are easily modified to get termination, but the resultant definitions have a more complex form of comparability. The algorithms we developed work even without termination, but this instance of a complex form of comparability provides a good example for study. Definition 5.3.1 includes such a modification. It enumerates all *complete paths*, paths which continue as far as possible without forming a cycle. A vertex is reachable from the "begin" if and only if it is on a complete path from "begin." Each complete path is finite, so the definition will terminate. (Again the reverse path vector is used to represent paths.)

Definition 5.3.1: Reachable Vertices—Digraph Terminating Definition

$$\left[\begin{array}{ll} f(p,c,x) = \{\ \} & : x \in \{p\} \\ f(p,c,x) = \bigcup_{i=1}^{i=N(x)} [f(\langle x\rangle \| p, \langle i\rangle \| c, n_i(x))] \cup \{x\} & : x \notin \{p\} \\ \text{Initially } p = \langle\ \rangle,\ x \text{ is a vertex in } G. & \end{array}\right.$$

Despite the similarity to Definition 5.2.2, argument p is not a program control component here, since it appears in the terminal condition.

Since there is a uniform inverse, the implementation is based on Algorithm 4.2.5. We also have comparability. Two argument-structures (p_1, c_1, x) and (p_2, c_2, x) are comparable because all vertices found by continuing the reverse path $P_a = \langle x\rangle \| p_1$, not already in P_a will also be found in continuing $P_b = \langle x\rangle \| p_2$, and similarly with P_a and P_b interchanged. To make use of this kind of comparability we need the following definitions for argument-structures in an O-network.

An argument structure X *surrounds A but not B* if

1. X is not an ancestor of B and
2a. X is a descendant of A or
2b. X is an ancestor of A or
2c. X is a descendant of an ancestor of A.

Theorem 5.3.1: If (p_1, c_1, x) and (p_2, c_2, x) are argument structures in the O network for an evaluation of Definition 5.3.1, then any vertex appearing in a descendant of (p_2, c_2, x) will also appear in an argument structure which surrounds (p_1, c_1, x) but not (p_2, c_2, x) in the O network.

Assume vertex y is in both a descendant, D_2, of (p_2, c_2, x), and a descendant of an ancestor, A_1, of (p_1, c_1, x). If z is a vertex in A_1 then z is also in an argument structure on the O-network path from (p_2, c_2, x) to D_2.

PROOF: Figure 5-3 illustrates the references in the proof. If the theorem is untrue, then there must be some "first" argument structure, say (a, u, q), which is a descendant of (p_2, c_2, x) such that q *does not* appear in any argument structure that surrounds (p_1, c_1, x) but not (p_2, c_2, x). (a, u, q) is the "first" such descendant of (p_2, c_2, x) in that all vertices in argument structures on the path from (p_2, c_2, x) to (a, u, q) *do* appear in argument structures which surrounds (p_1, c_1, x) but not (p_2, c_2, x). Thus in particular if

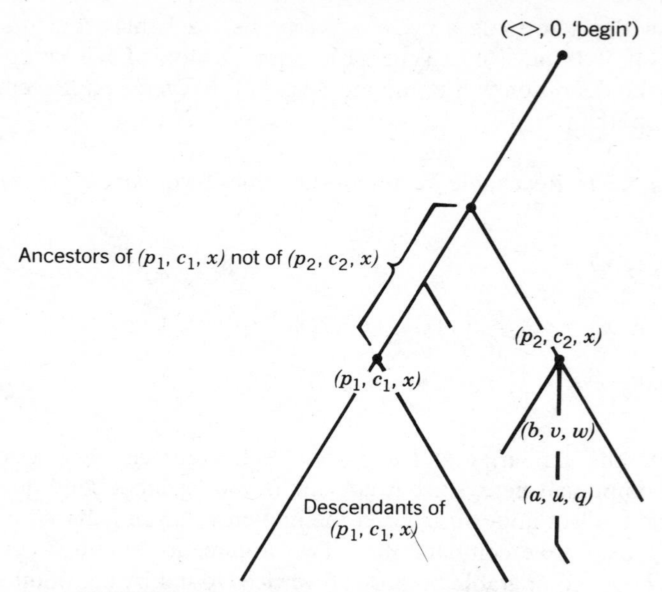

FIGURE 5-3 Comparability in reachability definition.

(a, u, q)'s parent is (b, v, w), then w does appear in an argument structure which surrounds (p_1, c_1, x) but not (p_2, c_2, x).

Since w is in the parent of (a, u, q), there must be an edge from w to q in the graph for which the O network is constructed. Also since w is in an argument structure, say D, which surrounds (p_1, c_1, x) but not (p_2, c_2, x), q is in a child or an ancestor argument structure of D (but not of (p_2, c_2, x)) and thus also in an argument structure which surrounds (p_1, c_1, x) but not (p_2, c_2, x). This contradicts the assumption that q was not in any such argument structure.

The last sentence of the theorem follows from the fact that a vertex, say r, in a first descendant of (p_2, c_2, x) on any path, which is not also in a descendant of (p_1, c_1, x), must be in one of (p_1, c_1, x)'s ancestors. Because r is in a first descendant of (p_2, c_2, x), there is an edge from x to r in the graph. If that does not lead to an argument structure containing r as a descendant of (p_1, c_1, x), it must be because r is already in p_1, and therefore in an ancestor of (p_1, c_1, x). ■

Note that the proof of the theorem does not depend on the relative position in the O network of (p_1, c_1, x) and (p_2, c_2, x). It is still valid if (p_1, c_1, x) is to the right of (p_2, c_2, x) instead of to its left, as in the illustration above.

Comparability in this definition is more complex than simple equality, though its effect is virtually the same. The information components of comparable argument structures are, in fact, equal, while the child#, c, and the path p parts differ. c is a program–control component and thus does not affect the function value. p is an information–control component which affects the terminal condition, but only so as to effectively remove argument structures from the substitutional domain which in any case would be removed by the comparability condition. It removes consideration of any second occurrence of the same vertex on a path from "begin." Theorem 5.1.3 can be generalized to cover this kind of comparability. The proof is explored in Problem 1 (Section 5.8).

5.3.1.1 Algorithm for Reachability in Cyclic Graphs

A depth-first inverse implementation of Definition 5.3.1 is based on inverse Algorithm, 4.2.5, as was the implementation, Algorithm 5.2.2, of Definition 5.2.2. The addition in Definition 5.3.1 assuring termination must be reflected in the algorithm. Here also there is pruning; after an argument structure with third component x is encountered, all subsequent argument structures with x as a third component need not be pursued. The upper WHILE is where the possibility of a cycle as represented by the reappearance of a vertex in a path must be accounted for. The implementation of the nonterminating case in Algorithm 5.3.1, is then the same as Algorithm 5.2.2, except for the upper WHILE as shown below.

Algorithm 5.3.1: Modifications to Algorithm 5.2.2 for Reachability in Cyclic Graphs

$$\text{WHILE}(N(X) > 0 \text{ and } X \notin P \;\&\; \text{TBL}(X) = 0) \text{ DO}$$

Notice however that $\text{TBL}(X) = 0$ only if $X \notin P$. So ($N(X) > 0$ and $\text{TBL}(X) = 0$) if and only if $N(X) > 0$ and $X \notin P$ and $\text{TBL}(X) = 0$).
Therefore this upper WHILE can be simplified to

$$\text{WHILE}(N(X) > 0 \;\&\; \text{TBL}(X) = 0) \text{ DO}$$

so in fact Algorithm 5.3.1 implementing Definition 5.3.1 is identical to Algorithm 5.2.2 implementing Definition 5.2.2.

It follows also that the nested recursive procedure, Algorithm 5.2.3, and the breadth-first algorithm, Algorithm 5.2.1, implement Definition 5.3.1.

Our development for reachable vertices has been for digraphs. Virtually identical approaches are applicable to (undirected) graphs.

5.4 MINIMUM-SELECT MONOTONIC FUNCTIONS

Definitions of the standard nonnested form in which w is a min-sel function, and both the t and O functions are monotonically nondecreasing, have an interesting and useful kind of comparability. This class of *minimum-select monotonic* Definitions is specified in Definition 5.4.1.

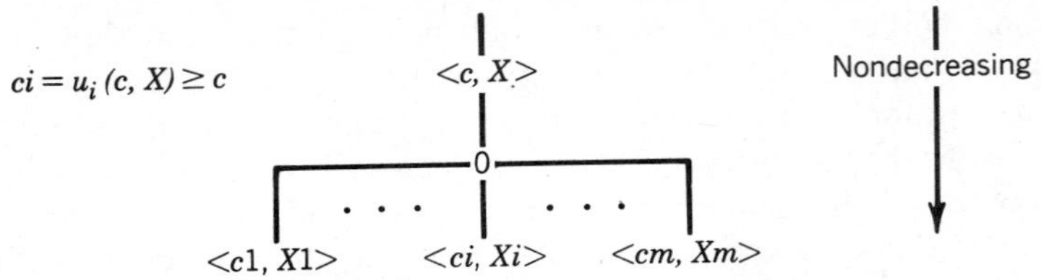

FIGURE 5-4 Minimum-select monotonic condition.

Definition 5.4.1: Minimum-Select Monotonic Definitions

$$\left[\begin{array}{ll} f(c, X) = \langle c, t(c, X)\rangle & :T(c, x) \\ f(c, X) = \text{MIN-SEL}_{i=1}^{i=m(c,X)}[f(u_i(c, X), v_i(c, X))] & :\sim T(c, X) \\ \text{Initially } c = 0,\ X = d \in D. & \end{array}\right.$$

In which u has the property $u_i(c, X) \geq c$.

c is referred to as the cost and X the argument structure with that cost. In terms of the O network these constraints imply that the cost component of a "child" argument structure [corresponding to an argument on the right of the recursive line $= u_i(c, X)$] is always larger than or equal to the cost component of its parent [corresponding to the argument on the left of the recursive line $= (c, X)$]. Figure 5-4 illustrates this relation. A similar *monotonicity* condition holds in the terminal line.

If, in Definition 5.4.1, the $\geq$ property of u is changed to $>$ by at least b, b being a number greater than 0, incorporated as an invariant in the argument structure, then the definition is minimum-select *proper* monotonic.

One form of comparability in minimum-select monotonic definitions occurs between identical argument structures, because of the minimum-select w function. Only one of any such pair need be pursued in the implementation. Another form of comparability results from the fact that MIN-SEL chooses the smallest cost component. Assume that (c_1, X_1) is terminal, (c_2, X_2) may or may not be, and $c_1 \leq c_2$. The cost component of f is $\leq c_1 \leq c_2$. Therefore $c_1 \leq c_k$ for all terminal argument structures (c_k, X_k), which are descendants of (c_2, X_2). So (c_2, X_2) and all its descendants can be pruned in favor of (c_1, X_1).

Branch and bound (B & B) and *best-first* (B-F) are the names given to implementations which take advantage of this second kind of comparability as well as the min-sel's associativity (often it may also be treated as though it is commutative) in minimum-select monotonic definitions. The first kind of comparability will reemerge later in considering specialized versions of these definitions.

5.4.1 Branch and Bound Algorithm (B & B)

B & B is an adaptation of the uniform inverse algorithm, Algorithm 4.2.5, to minimum-select monotonic definitions. It achieves time saving by augmenting that algorithm using the fact that the cost of any terminal argument structure is an upper bound on the cost of the final answer. This bound, W, is maintained in the B & B, and updated as each terminal argument structure is generated. W can be used to eliminate any argument structure, Y, whose cost is equal to or greater than W. Furthermore all descendants of Y have costs at least as great as Y's, because of monotonicity, so that the entire subtree rooted at Y can be pruned.

Algorithm 5.4.1: Branch and Bound Algorithm for Definition 5.4.1

```
[c, X] ← [0, d]
S      ← [c, X]
W      ← [∞, ⟨ ⟩]                              W = [W1, W2]

FOR ever DO
  WHILE( ~ T([c, X]) & c < W1) DO
    [c, X] ← [u1(c, X), v1(c, X)]
  END
  IF c < W1  THEN W < ⟨c, t(c, X)⟩
  WHILE ([c, X] is a last child) DO
    [c, X] ← parent(c, X)
    IF [c, X] = S THEN DONE
  END
  [c, X] ← next-sib(c, X)
END
```

Notes

1. The algorithm stops, as do all depth-first based algorithms considered thus far, by backing up to the initial argument structure.
2. The algorithm picks out the leftmost pair which satisfies the min-sel condition.

This algorithm will terminate for a minimum-select monotonic definition (Definition 5.4.1) *even if that definition does not terminate*, provided only that

1. The definition is proper monotonic, and
2. Either
   a. There is an integer $n(d)$ for each d in the initial domain D, such that if I is a sequence of $n(d)$ 1's, then $o_I(d)$ terminates, or
   b. (With slight modification of the algorithm), for each d in D there is an upper bound $r(d)$ on the cost of all solutions, and W_1 is initialized to $r(d)$.

Termination of the algorithm is then guaranteed by the fact that

1. for any positive integer C, any sequence of applications of O functions, the first of which applies to the initial argument structure, must produce a cost component $\geq C$ after a finite number of steps, and
2. there is such a finite upper bound on the cost component.

So Algorithm 5.4.1 will terminate with an answer that satisfies Definition 5.4.1 even if the latter doesn't terminate. However, because the definition may be satisfied by more than one answer, the value returned may not be the desired answer.

5.4.1.1 Complexity of B & B

In very general terms, the worst-case complexity of B & B is

of O block outputs simulated * (cost to generate each O block output, that is, compute u_i and v_i)

+ # of t block outputs simulated * cost to generate each t block output

In the worst-case this is no different than the complexity of the unaugmented algorithm on which B & B is based. However, the number of O blocks simulated will be reduced on average with the use of the running bound in B & B.

5.4.1.2 Nested Recursive Version of B & B

The algorithm above implements Definition 5.4.1, but it also implements the following nested recursive definition even more directly. This definition results from applying the transformation given by Definition 3.10.5 to Definition 5.4.1. Whereas Definition 5.4.1 contains no explicit reference to the bound (W in Algorithm 5.4.1), the following definition does.

Definition 5.4.2: Nested Definition for Branch and Bound
Let $W = \langle W_1, W_2 \rangle$ and $uv_i(c, X) \equiv u_i(c, X), v_i(c, X)$.

$$\left[\begin{array}{ll} f(W, c, X) = \langle c, t(c, X) \rangle & :T(c, X) \\ f(W, c, X) = W & :W_1 \leq c \text{ and } \sim T(c, X) \\ f(W, c, X) = f(\ldots f(f(W, uv_1(c, X)), & \\ \qquad uv_2(c, X)) \ldots uv_{m(c, X)}(c, X)) & :\sim T(c, X) \text{ and } W_1 > c \\ \text{Initially } W = \langle \infty, \langle\ \rangle \rangle,\ c = \text{ initial cost (usually 0)},\ X = d \in D. & \end{array}\right.$$

5.4.2 Optimal Search or Best-First Algorithm (B-F)

Algorithm 5.4.2 is the B-F algorithm. It is based on the breadth-first algorithm, Algorithm 4.1.1. The queue Z becomes a linked list in B-F. In both algorithms the first entries in Z are the children of the initial argument structure. At each stage of both algorithm, an entry of Z, say X, is replaced by all its children argument structures. In breadth-first X is the first, while in B-F it is the minimum cost, or minimum-selected, entry of Z. This is why a list rather than a queue is used in B-F. In B-F, if this minimum-selected argument structure is nonterminal, it is replaced by its children in Z, if it is terminal, it is the answer and the algorithm ends. This is unlike the breadth-first algorithm, which stops when queue Z is empty. Thus in B-F only descendants of minimum-selected argument structures need be generated, all others are pruned. The abrupt ending of B-F is legitimate because of the comparability found in minimum-select monotonic definitions.

Algorithm 5.4.2: Best-First Algorithm for Definition 5.4.1

```
Z ← ⟨ ⟩
[c, X] ← [0, d]

WHILE( ~ T(c, X))
  FOR I = 1 to m(c, X) DO
    [c', X'] ← [u_I(c, X), v_I(c, X)]
    Z ← link – [c', X']
    [links the pair [c, X] into Z which
    is a linked list of such pairs]
  END
  [c, X] ← delete-(Min-Sel(Z), Z)
  [this line finds the min-sel pair in
  Z, removes it from Z, and puts it in [c, X]]
END
ANSWER ← [c, t(c, X)]
```

```
Z ← ⟨ ⟩
[c, X] ← [0, d]
U ← ∞
WHILE(c ≠ U))
  FOR I = 1 TO m(c, X) DO
    [c', X'] ← [u_I(c, X), v_I(c, X)]
    IF c' < U THEN
      DO
        IF T(c', X') THEN (UD ← X'
        U ← c')
      END
      ELSE Z ← link – [c', X']
  END
  [c, X] ← delete-(Min-Sel(U, Z), Z)
END
ANSWER ← [U, UD]
```

Notes

1. A somewhat more effective algorithm, which limits the size of list Z, is given on the right. It takes advantage of the upper bound on the cost of the answer given by the cost of the smallest cost terminal argument structure, kept in U. The terminal argument with cost U is kept in UD. An argument with cost greater than the bound is never placed in Z.
2. In the algorithm on the right, "← delete-" min-selects over both U and Z. Additional savings in space can be attained at some cost in time by scanning the list Z and deleting from Z any argument structure with cost greater than or equal to U, each time a minimum is selected.

3. The algorithm picks a pair that satisfies the min-sel condition but not necessarily the leftmost one.

If one adjusted the B-F algorithm on the left to continue until Z was empty, then the smallest, next smallest, etc., cost argument structures would all be produced in order.

Like B & B, and for analogous reasons, the B-F algorithm terminates for minimum-select monotonic definition, Definition 5.4.2, even if that definition does not terminate provided only that

1. The definition is proper monotonic.
2. For each d in the initial domain D there is at least one sequence of integers I, such that $O_I(d)$ terminates. The caution concerning solutions to nonterminating definitions noted for B & B is also applicable here.

So B-F terminates with the correct answer as long as the monotonic minimum-select definition terminates or the above conditions hold. If, however, equal argument structures occur, considerable time may be expended in reevaluating identical argument structures. At the cost of possibly large amounts of memory, an addition could easily be made to B-F to avoid this.

5.4.2.1 Complexity of B-F

Since minimum members of a changing set in Z are repeatedly computed in the B-F algorithm, a HEAP is a reasonable alternative data structure for Z. (See Problem 7, Section 5.8.) In very general terms the worst-case complexity of B-F without and with a HEAP is

1. For Z a linked list,

   # of O block outputs simulated * cost to generate each O block output

   + # of O blocks simulated * average number of entries in Z

2. For Z a HEAP,

   # of O block outputs simulated * (cost to generate each O block output

   + cost to make an entry in the HEAP)

   + # of O blocks simulated * cost to remove minimum from,

   and reestablish HEAP

5.4.2.2 Breadth-First Implementation with Pure Information Components and the Solution Structure of B-F

The minimum-select Definition 5.4.3 is similar to the selection Definition 2.3.5 considered in Section 2.3.3.1 where efficiencies for such definitions are discussed. The expanded version of the minimum-select definition given below is even closer to that initial definition.

Definition 5.4.3: Expanded Minimum-Select Monotonic Definition

$$\left[\begin{array}{ll} f(c, X, Y) = \langle c, t(c, X), t(Y)\rangle & :T(c, X) \\ f(c, X, Y) = \text{MIN-SEL}_{i=1}^{i=m(c, X)}\ [f(u_i(c, X), & \\ \qquad v_i(c, X), w_i(c, X, Y))] & :\sim T(c, X) \\ \text{Initially } c = 0,\ X = d1 \in D1,\ Y = d2 \in D2. & \end{array}\right.$$

In which u has the property $u_i(c, X) \geq c$.

Like the earlier definition, this expanded form of the minimum-select monotonic definition produces a number of components, selected independent of their last, pure information component. (Although there are three, rather than two, components as in the earlier case, the argument still holds. The first two here correspond to the first there.) The match with the earlier case is sufficient to justify the conclusion that the last component Y in Definition 5.4.3 can be replaced with a simple queue to record only the solution structure, instead of the complete Y component. After obtaining the solution for this revised definition, the value of Y is easily constructed using this solution structure as described in Section 2.3.3.1, Definitions 2.3.5 and 2.3.6.

In the depth-first B & B implementation of a minimum-select definition the queue would be implemented directly. The B-F implementation is breadth-first, and it is possible to save these queues using common storage in a treelike structure. A brief discussion of memory savings in breadth-first implementations will provide a general setting.

Depth-first generally uses less memory than breadth-first implementations in all their variations, including B-F and B & B. However, memory savings are often possible in the breadth-first implementation. In the standard breadth-first implementation a set of argument structures is kept in queue Z. For some argument structures it is not necessary to store each argument structure in Z separately; combined storage is an option.

If every child of an argument structure has an identical argument, we can arrange to keep one copy of that argument rather than one for each entry in Z, provided that we are dealing with a straightforward breadth-first implementation. A more important use is associated with the occurrence of pure information components.

Assume the argument-structure is of the form (X, Z) where Z is a queue. Assume further that the ith child has the form $(o_i(X), \langle Z \| q_i(X)\rangle)$. In place of the usual breadth-first queue (or list) for saving argument-structures, we use a queue (or list), Q, to hold the X component and a tree T for the Z components of the argument-structure. In T there is a pointer from each node to its parent. Each queue stored in T is found as the nodes on a path from a leaf to the root of T. The typical operation in a breadth-first algorithm is the replacement of an argument structure by its children. In this arrangement that replacement involves replacing the X component in Q with a set of entries, namely its children $o_i(X)$; "replacing" the Z component in T involves replacing the current leaf in T on the path corresponding to Z by an internal node and a set of leaves, corresponding to the child components $q_i(X)$, each pointing back to the internal node now at the end of Z in T.

Now we specialize this discussion to B-F. Remember that even when a definition does not have a pure information component, it still has a solution structure. A solution structure like that in Definition 2.3.8 or 2.3.9 of Section 2.3.3.1 added to an argument structure can trace the history of any evaluation. It is this history that constitutes the solution structure. Maintaining this structure allows omission of any components dependent on it; these may be reconstructed from the structure as needed.

B-F is a basically breadth-first implementation. Assume we want the solution structure, whether or not a pure information component is explicitly included in the definition. Then a good way to do this efficiently is to maintain a tree, analogous to T above, called PATHTREE, of solution structure queue components (explicitly present or not) as discussed above. B-F finally chooses an optimal or minimum selected terminal structure. The path in PATHTREE from this finally selected argument structure back to the root of PATHTREE gives the solution *path* for the B-F implementation.

For possible addition to the B-F algorithm we consider the design of PATHTREE in somewhat greater detail.

The solution path consists of all of the minimum-selected argument structures found during the execution of B-F. PATHTREE contains a node for each minimum-selected argument structure and its children, with pointers from each child to its minimum-selected parent. We assume PATHTREE can be uniquely addressed by the argument structures it contains. This will be difficult to do if identical argument structures can occur, so assume that there are no such duplicates, because either they are not generated by the definition or they are removed in an augmented B-F with a table recording all argument structures. (This table could be combined with PATHTREE.) The root of PATHTREE is initialized to contain and to be addressed by the initial argument structure; in its "parent" component it contains the nil pointer. If X is minimum-selected on the solution path, then it must be followed on that path by one of its children, so a node is created in PATHTREE at addresses of X's children. Each child points to the X node, that is, to PATHTREE(X). The X entry itself is available either because it was once the child of a minimum-selected argument structure, or because it is the initial argument structure. By the time the solution is obtained, this construction creates a tree in PATHTREE with nodes corresponding to the minimum-selected argument structures and their children each, excluding the initial argument structure, pointing to one other node. The solution path is the sequence of argument structures in the path of PATHTREE nodes found by following the pointers from the final solution node back to the root.

Remembering our assumption that there are no identical argument structures. The alterations in Algorithm 5.4.2 to construct the table is

1. After line "$[c, X] \leftarrow [0, d]$" add

$$\text{PATHTREE}(0, d) \leftarrow \text{nil}$$

2. After line "$[c', X'] \leftarrow [u_I(c, X), v_I(c, X)]$" add

$$\text{PATHTREE}(u_I(c, X), v_I(c, X)) \leftarrow X'$$

When B-F has completed its run, the information in PATHTREE is adequate to construct the solution path. This will be explicitly illustrated in Section 5.4.2.5 when we consider Dijkstra's algorithm.

5.4.2.3 B-F Is an Optimal Algorithm

The argument structures maintained at any time in the B-F algorithm are those on a frontier. In B-F an argument structure is always replaced with all its children, with no intervening changes in the frontier. Therefore the second member of every frontier element in (X, Z) (frontier definition, Section 5.1.2.2), is $\{1, \ldots, m(X)\}$. Thus X is enough to specify an element so we consider a frontier to consist of a multiset of argument structures, containing the X's only.

The B & B algorithm is based on the fact that the cost component of any terminal argument structure is an upper bound on the cost component of the function value. This same upper bound, together with an evolving lower bound, is used in the B-F algorithm. The lower bound is the minimum of the cost component of the argument structures on the frontier. This follows from the fact that all terminal argument structures have cost $\geq$ that of some argument structure in the frontier. The "all" part of this assertion follows from the definition of frontier, and the "$\geq$" part follows from monotonicity.

The B-F algorithm passes through stages. At each stage of the algorithm, list Z contains the value of each argument structure on a frontier, and thus a lower bound on the result. Also the smallest cost component of the terminal argument structure on the frontier gives an upper bound. When the two bounds are equal, the algorithm stops.

Over a wide class of algorithms B-F will be shown to be best by a significant efficiency measure. In doing so, the following property of the B-F algorithm is used.

Lemma 5.4.1: Assume that the final frontier ff computed in the B-F algorithm has a minimum cost component $= c$. Then the cost component of the parent of any member of ff is $\leq c$ ($<$ if the definition that B-F implements is proper-monotonic).

PROOF: Let C be the terminal in ff with cost c. The cost of C's parent is clearly $\leq c$ ($<$ if proper monotonicity holds). If Y is any member of ff and became a member after C, then Y's parent must have cost $< c$, or else Y would not have been produced. If Y became a member before C, then Y's parent must have cost $\leq$ that of C's parent, that is, cost $\leq c$ ($<$ if proper monotonicity holds). ■

Let A be the class of algorithms having the following properties.

1. Evaluate minimum-select monotonic definitions by examining argument structures in $\Delta_f(d)$.
2. Generate an argument structure X only after having generated all ancestors of X.
3. Can only determine the cost components of argument structure X by generating it.
4. In deciding from which argument structure to generate children next, as well as in deciding whether a terminal argument structure is the answer, use only cost components.

The B-F algorithm is in A. It is related to other members of A as follows.

Theorem 5.4.1: Given a proper monotonic minimum-select definition with initial argument structure d, B-F will examine no more members of $\Delta_f(d)$ than any other algorithm in A.

PROOF: Suppose the theorem is not true, that is, an algorithm in A, say $A1$, examined fewer members of $\Delta_f(d)$ than B-F and produced, the correct answer $\langle c, Z \rangle$, where c is the cost of a terminal argument structure. Since $A1$ does not examine all the argument structures which B-F does, it must fail to examine one of the argument structures on the final frontier examined by B-F. If it examined all of them, it would also have had to examine all their ancestors, which is all the argument structures B-F must examine. Say Y is the frontier argument structure which $A1$ fails to examine. Now $A1$ must have looked at some ancestor of Y (d is such an ancestor). Let anc(Y) be the most closely related ancestor of Y which it did examine. Since the cost of the parent of $Y < c$, by Lemma 5.4.1, the cost of anc(Y) is clearly $< c$ and any of the argument structures that descend from anc(Y) up to and including Y could be $< c$ and terminal and still be consistent with everything examined by $A1$. If this had been so, $A1$ would have given the same result and have been wrong. Therefore $A1$ is not an algorithm which always correctly computes the minimum-select monotonic function. ■

Note that whether or not the upper bound, U is used in Algorithm 5.4.2 to avoid placing argument structures with higher cost in argument structure list Z will not affect the number or identity of the argument structures minimum-selected from Z, nor the number examined as children of minimum-selected argument structures. It will only contribute to reducing the average size of Z and thus the complexity of minimum-selecting a member of Z. So in developing the theorem, one can make the simplifying assumption that the upper bound is not used.

This theorem can be extended to apply even when comparabilities other than those incorporated in B-F exist, if pruning for them is added to the B-F algorithm. If comparability is based directly on complete equality of argument structures, the extension is direct. Consider a more complex form of comparability. Suppose two argument structures (c_1, X_1) and (c_2, X_2) are such that $X_1 = X_2$ and $c_1 \leq c_2$, and further for each descendant of $(c_1, X_1), (c, Y)$, there is a similarly positioned descendant of $(c_2, X_2), (c', Y')$, with $c \leq c'$ and $Y = Y'$. In this case we can prune (c_2, X_2) and all its descendants. Here too B-F, augmented to do this pruning, can be shown to be optimal. Dijkstra's algorithm, considered in Section 5.4.2.5, uses such an augmented B-F.

Since B-F is "optimal," why should we ever use B & B? While B-F generates no more O block outputs than any other member of A, including B & B, it requires additional time to find the minimum cost components each time it tries to generate a new frontier by replacing a parent with its children. Other members of A, B & B for example, only need time to generate O block outputs. (Updating the upper bound and computing parents is usually subsumed in the generation time.) So despite its "optimality," it is not obvious that B-F is always faster than other members of A. Furthermore B-F is a big user of temporary memory.

The bulk of the temporary storage requirement in B-F holds argument structures on a frontier. In a simple implementation, storage is needed for all of them. In a more sophisticated implementation using the upper bound provided by the least costly terminal argument structure found in a frontier, any argument structure with a cost greater than or equal to that bound can be removed from the frontier. In either case the least upper bound on the storage required for the frontier equals the number of terminal argument structures in the O network. Thus the storage requirement, as for other breadth-first based algorithms, is on the order of the greatest width of the O network. The comparable storage for B & B, like other depth-first algorithms, is on the order of the depth of the O network. Thus complexity is proportional to depth for B-F, while B & B's is at least of the same order and can be as large as exponential in that depth. So generally B-F is at least as good timewise, but requires much more storage than B & B.

The expected good time behavior of B-F and good space behavior of B & B suggest a mixed form of implementation, BFBB, in which, while size is kept limited, an entire frontier of argument structures is kept in temporary storage so that a lower bound can be computed each time the frontier changes. Also an upper bound based on the least costly terminal argument structure found thus far is kept. Thus whenever the lower bound in the frontier is equal to the upper bound, BFBB can, like B-F, terminate early with answer equal to the argument structure that generated the upper bound. The temporary storage requirement for the frontier is limited by simulating the B-F algorithm until the limit is approached. At that point BFBB simulates a depth-first approach like B & B, using an explicit stack rather than a computed inverse to back up. This damps the growth of the frontier.

5.4.2.4 Comparability and Minimum-Select Monotonic Definitions

As we have noted, B-F remains optimal when in addition to comparability based on the cost component of the minimum-select argument-structure, there is comparability based on the equality of entire argument structures, or perhaps on the equality of the noncost component only. We assume this latter form of additional comparability exists and consider alterations in the B & B and B-F algorithms to take advantage of it. More particularly assume that when there are two argument structures (c_1, X) and (c_2, X) with $c_1 < c_2$, then(c_2, X) and all its descendants can be pruned in favor of (c_1, X).

To take advantage of this comparability in the B & B algorithm, a table TB is added, with TB(X) containing c, the minimum cost found thus far, in association with X in an argument structure (c, X). TB(X) gives an upper bound on the cost associated with data argument X in any argument structure. When an argument structure (d, X) is encountered, it can be pruned if $d \geq \text{TB}(X)$; otherwise d would replace TB(X) and (d, X) would be pursued.

In B-F, no more than one representative of the argument structures on the frontier with a given data argument X is kept in argument list Z. This is (c, X), where c is the minimum cost of all argument structures on the frontier which contains X. When (c, X) is minimum-selected and deleted from the new Z, c is the lowest cost that can ever be associated with data X. This leads to the following alterations in the updating of Z. When a child (b, Y) of a minimum-selected argument structure is found then

If there is no (x, Y) in Z, then add (b, Y) to Z.
If there is an (x, Y) in Z and $x > b$, replace (x, Y) with (b, Y) in Z.
If there is an (x, Y) in Z and $x \leq b$, then do nothing with (b, Y).
If some (x, Y) has been minimum-selected, then do nothing with (b, Y).

Thus after minimum-selecting of (x, Y), no data structure (b, Y) will ever again appear in Z. If then there are a finite number of possible data parts Y in the substitutional domain, the algorithm must terminate. *It will terminate even if the corresponding definition does not terminate and even if conditions for termination of pure B-F for nonterminating definitions* (*section* 5.4.2) (i.e., *proper monotonicity*) *do not hold*.

For each occurrence of a data argument Y it is necessary to determine if Y is in Z. For efficiency it is well to be able to address the Y entry in Z associatively with Y. If the number of possible different data arguments is small, say m, then this is reasonably done by indexing into a table of size m. There must also be a list or HEAP which includes all the entries in Z so that new minima may be found with reasonable efficiency. In summary we need a data structure which can act as an addressable list or HEAP, in which minimum-selected members can be found and effectively removed for finding subsequent minimum-selected members. Furthermore a record must be kept of data arguments of minimum-selected argument structures for comparison with subsequent occurrences of the same data. This can be accomplished a number of ways, one of which will be given in the minimum path algorithm.

5.4.2.5 Dijkstra's Shortest Path Algorithm, Comparability and B-F

An example of a definition of a minimum-select monotonic definition which produces comparability based on equality of its data arguments is provided by the following definition for the cost of the shortest path, designed for a graph G in which all edge costs $c_i(v)$ are greater than 0, and there is at least one path from vertex "begin" to vertex "goal." It uses the same primitives as Definition 2.12.4, but is developed by a growth, rather than a recursive, approach. It is based on the generalization of an O network constructed by growing each possible path together with its cost.

Definition 5.4.4: Cost of Shortest Path

$$\left[\begin{array}{ll} f(c, v) = c & : v = \text{"goal"} \\ f(c, v) = f(\infty, \text{"goal"}) & : v \neq \text{"goal" and } N(v) = 0 \\ f(c, v) = \text{MIN-SEL}_{i=1}^{N(v)} f(c + c_i(v), n_i(v)) & : v \neq \text{"goal" and } N(v) \neq 0 \\ \text{Initially } c = 0,\ v = \text{a vertex} = \text{"begin" in } G,\ G \text{ is invariant; } c_i(v) > 0 & \\ \text{for all } i \text{ and } v. & \end{array}\right.$$

This definition does not terminate. If G has a cycle, then argument structures with vertices on that cycle will be generated endlessly on a path in the O network. Nevertheless, the algorithm is still applicable because the definition is minimum-select proper monotonic, and there is at least one path from "begin" to "goal" in G. Furthermore it can be shown that the definition is uniquely satisfied. In addition to comparability based on cost alone, there is comparability based on equality of vertices. If two argument structures have the same second argument (vertex reached), only the one with the smaller (or equal) first component (cumulative cost) need be developed further. We will take advantage of both kinds of comparability.

Algorithm 5.4.3 is an adaptation of the B-F, Algorithm 5.4.2, with additions to take advantage of the fact that argument structures with the same vertex argument are comparable, as discussed in Section 5.4.2.4 is. In effect it is a combination of the B-F algorithm, with some use of the lower bound, and the breadth-first algorithm, Algorithm 5.2.1, for reachable vertices in a digraph. The argument structure list Z, of B-F becomes an array with links connecting active frontier entries. The array structure provides fast addressing by vertex identity. Its use is now possible because the number of possible argument structures is the number of vertices. The entire frontier is not kept, but the minimum cost thus far to reach *each* vertex v is kept at $Z(v)$. In a second array TBL, also indexed by the vertices, the vertices which have been minimum-selected, are marked.

The vertex data on the path (= the path in the graph) through the O network from the root ("begin") to the minimum-selected answer (= "goal") give the actual minimum path, which need not be unique. As discussed in Section 5.4.2.2, it can be easily reconstructed by keeping a small amount of additional information in the form of a tree. The PATHTREE array is that tree. It can be added if the actual minimum path is to be computed this way.

Algorithm 5.4.3: Cost of Shortest Path
Z and TBL, and optionally PATHTREE, are each one-dimensional arrays equal in size to the number of vertices in G. Z is also completely linked, with each entry containing ∞ initially. TBL initially contains all 0's. Vertices are represented as integers 1 through n, so they can be used as indices in an array.

```
[C, V] ← [0, 'begin']
Z('begin') ← 0
← delete- ('begin', Z) [Remove 'begin' from the linked entries in Z]
TBL (V) ← 1
WHILE(V ≠ 'goal') DO
  FOR I = 1 TO N(V) DO
    IF TBL (n_i(V)) = 0 THEN
      IF C + c_i(V) < Z(n_i(V))) THEN
        DO
          Z(n_i(V)) ← C + c_i(V)
          ************            ← PATHTREE (n_i(V)) ← V
        END
  END
  [C, V] ← delete- (Min-Sel(Z), Z)
  TBL(V) ← 1
END
  [Answer is in Z]
```

Note. If the path is to be computed, the * line should be replaced as shown.

On completion this algorithm has not only computed the shortest path cost from "begin" to "goal" but also from "begin" to each vertex which is reachable with a cost less than or equal to that required to reach the "goal." Each position marked 1 in TBL corresponds to such a vertex. The shortest path cost to reach vertex v is recorded in $Z(v)$. If the WHILE condition allowed continuation of the algorithm until all vertices reachable from "begin" were marked 1 in TBL, the algorithm gives the minimum cost from "begin" to every vertex in G.

Since a tree of possible paths with pointers toward its root vertex, "begin," has been kept in PATHTREE, the minimum path can be printed (as a reverse path vector) by the following algorithm.

Algorithm 5.4.4

```
V ← 'goal'
WHILE (PATHTREE(V) ≠ 0) DO
  OUT ← V
  V ← PATHTREE (V)
END
```

Algorithms 5.4.3 and 5.4.4 together are essentially Dijkstra's shortest path algorithm

5.4.2.6 Complexity of Minimum Path Algorithm

Recognizing that this algorithm is a special case of the B-F algorithm, we can argue complexity from the general complexity equations for B-F for an indexed linked list as used above, for a standard binary HEAP (Section 5.4.2.1), or for the more recently invented F-HEAP or comparable structures [Tarjan 83] [Fredman Tarjan 84] [Fredman et al., 86].

\# of O block outputs simulated $\leq$ number of edges $= O(e)$
Cost to generate each O block output $= O(1)$
\# of O blocks simulated $=$ # of minimum-selects $= O(n)$

1. For Z a linked list, (or a HEAP) $|Z| \leq$ number of vertices $= O(n)$,

$$\text{complexity} = O(e) * O(1) + O(n) * O(n)$$
$$= O(e + n^2) = O(n^2)$$

2. For Z a HEAP, cost to make or change an entry in the HEAP $= O(\log n)$, and cost to remove minimum from, and reestablish HEAP $= O(\log n)$,

$$\text{complexity} = O(e) * (O(1) + O(\log n)) + O(n) * O(\log n))$$
$$= O(e \log n + n \log n) = O(e \log n)$$

3. For Z an F-HEAP, cost to make or change an entry in the HEAP $= O(1)$, and cost to remove minimum from, and reestablish HEAP $= O(\log n)$,

$$\text{complexity} = O(e) * (O(1) + O(1)) + O(n) * O(\log n))$$
$$= O(e + n \log n)$$

Alternatively we can argue the complexity more directly from the given shortest path algorithm. Again n is the number of vertices and e the number of edges. Once a vertex is unlinked or deleted from the linked list in Z and assigned a 1 in TBL, and its outward neighbors are examined in the indexed FOR loop, it can never again be selected from the list in Z. [This incidentally implements the recursive line for $N(v) = 0$.] Thus, at worst, each vertex gets deleted from the list in Z once. Associated with each vertex selection from Z the following action must occur to make the selection.

1. A minimum must be taken over the members of the list in Z, costing $O(n)$ comparisons. This is done at most n times with total cost of $O(n^2)$.
2. Each neighbor of a selected vertex (at most n for each vertex) is looked at in the FOR loop. The total number of all the neighbors of all vertices, and thus the total cost, is $O(e)$

Thus the worst-case time complexity is $O(n^2)$. Similar arguments can be given when a HEAP is used.

5.5 MORE COMPLEX EXAMPLES OF COMPARABLE ARGUMENT STRUCTURES — ARTICULATION (CUT) POINTS AND DEPTH-FIRST GRAPH SEARCH

An efficient algorithm for finding articulation points, biconnected components, strongly connected components, and other graph properties based on tracing through a graph in a depth-first order, was developed in [Tarjan 72], it is also described in [Aho, Hopcroft, Ullman 75] as well as other algorithm books. It seems that it should fit well within our framework. After we establish the appropriate recursive definition it does.

Although our development of an articulation point algorithm is somewhat more involved than for previous algorithms, it follows the same basic pattern. First a recursive definition of the function(s) to be computed is developed. In this instance, that definition depends on Tarjan's way of identifying articulation points in terms of a trace of paths in a graph. The validity of this approach is justified, and a definition in standard nonnested form with an associative w function obtained. It is shown to produce comparable argument structures. Our goal is not to determine simply whether a single vertex is an articulation point, but rather to determine the set of articulation points of the graph, that is, to determine for every vertex in the graph, whether or not it is an articulation point. For this purpose the definition we construct must be evaluated at each member of its substitutional domain. The definition will be implemented by an algorithm based on Algorithm 4.2.6 designed to satisfy these requirements. The resultant algorithm is then augmented to prune comparable argument structures.

The final algorithm we develop is well known, and could be given immediately. However, tracing the development in stages within our framework will illuminate significant problem properties, and alternatives for their use in improving algorithms, which make the final algorithm elegant and efficient.

Consider an undirected connected graph G. The vertex y is an articulation point of G if there are two other vertices $u \neq x$ such that all paths between u and x in G contain y. In such a case we say y is an *articulation point with respect to u and x*.

To develop a definition which identifies articulation points we need to be a little indirect. It might be expected that the recursive line would be based on a substitutional domain in which the function value at a vertex y would depend on its values at y's neighbors. But it is difficult to see how the decision as to whether y is an articulation point can be determined by knowledge of whether y's neighbors are articulation points. Note however that

Lemma 5.5.1: If there is no pair of neighbors of y such that y is an articulation point with respect to them, then y is not an articulation point.

If y is an articulation point and u is any neighbor of y, there is another neighbor, x, of y such that every path from x to u is through y, that is, y is an articulation point with respect to u and x.

PROOF: If y is an articulation point, then it is so with respect to some pair of vertices, say a and b. There is at least one path, say $p1$, from a to y and one, say $p2$, from b to y. Clearly if a' is on $p1$ and b' is on $p2$, then y is an articulation point with respect to a' and b'. In particular choosing a' to be the neighbor of y on $p1$, and b' to be the neighbor of y on $p2$ satisfies the lemma. So if there is no pair of neighbors with respect to which y is an articulation point, it is not an articulation point.

The second part is proven by contradiction. Assume that y is an articulation point, u is a neighbor, and there is no other neighbor x such that y is an articulation point with respect to u and x. Since y is an articulation point, there must still be some pair of neighbors of y, say $x1$ and $x2$, neither equaling u, such that y is an articulation point with respect to them. Since y is not an articulation point with respect to $x1$ and u, there is a path $p1$ from $x1$ to u, and for analogous reasons path $p2$ from u to $x2$, neither of which contains y. Therefore $p1 \| p2$ is a path from $x1$ to $x2$, which does not go through y. So y cannot be an articulation point with respect to $x1$ and $x2$. ∎

Lemma 5.5.1 motivates the following sequence of definitions solving the articulation point problem.

5.5.1 Description of Articulation Points in Terms of Return Vertices of Trails

Whenever we refer to a path, subsequently we are referring to a *vertex path*; a vector $\langle x_1, \ldots, x_n \rangle$ in which $\{x_{j-1}, x_j\}$ is an edge of G for $j \in \{2, 3, \ldots, n\}$,

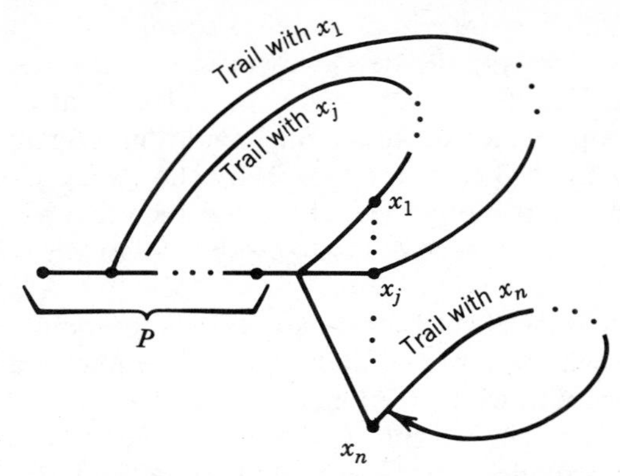

FIGURE 5-5 Trails. The trails shown thru x_1 and x_j have retrun vertices in p, while the trail thru x_n does not.

and $x_j \neq x_i$ if $j \neq i$. A path is a sequence of vertices with no repeats, that is, it cannot contain a cycle. A *trail* is similar to a path, except that its last vertex may be a repeat.

More precisely, *a trail* in a graph or digraph G is a sequence of vertices $\langle v_1, \ldots, v_n \rangle$, in which (v_i, v_{i+1}) is an edge of G and either v_n has out-degree 0 or there exists j, $1 \leq j < n \ni v_j = v_n$; and for all other i and j, $v_i \neq v_j$ if $i \neq j$.

In this latter case, v_n is the *return vertex* of the trail. In an undirected graph no vertex has degree 0, so v_n is always the same as some previous vertex in a trail.

Let $p = \langle p_1, \ldots, p_{\text{last}} \rangle$, with last ≥ 1, be a *path* in G and y an outward neighbor of p_{last} such that $p \| \langle y \rangle$ is a path, that is, no vertex is repeated. Consider all trails in G with prefix $p \| \langle y \rangle \| \langle x \rangle$, where x is an outward neighbor of y [$p \| \langle y \rangle \| \langle x \rangle$ may itself be a trail]. If any such trail has a return vertex in p, y cannot be an articulation point with respect to p_{last} and x. On the other hand if no such trail has return vertex in p, then y is an articulation point with respect to p_{last} and x. In further consequence of Lemma 5.5.1, *if for each neighbor of y, other than p_{last}, at least one trail can be found with a return vertex in p, then y is not an articulation point* (with respect to any vertices whatsoever). Such trails are illustrated in Figure 5-5, while their connection to articulation points is restated formally in Theorem 5.5.1.

Define Trail(q), where q is a path or trail, to be the set of all trails with prefix q. So Trail($p \| \langle y \rangle \| \langle x \rangle$) is all trails with prefix $p \| \langle y \rangle \| \langle x \rangle$. RtVrt($p$, y, x) is the set of all return vertices of trails in Trail($p \| \langle y \rangle \| \langle x \rangle$). Let $\{q\}$ be the set of vertices on path (trail) q. The conditions under which y, with neighbors x_j, is an articulation point is given in terms of trails in Theorem 5.5.1.

Theorem 5.5.1: y is an articulation point if for an arbitrarily chosen path $p \| \langle y \rangle$ there exists an outward neighbor x of $y \ni [\text{RtVrt}(p, y, x) \cap \{p\}] =$

$\{\ \}$. y is not an articulation point if for an arbitrarily chosen path $p\|\langle y\rangle$ there is no outward neighbor x of y for which $[\text{RtVrt}(p, y, x) \cap \{p\}] = \{\ \}$.

PROOF: If y is not an articulation point then there is a path, $\langle p_{\text{last}}\rangle\|\alpha\|\langle x\rangle$, where α is a path without y. Then there is a trail $\langle p_{\text{last}}, y, x\rangle\|\alpha\|\langle p_{\text{last}}\rangle$ so $\text{RtVrt}(p, y, x)$ cannot be empty—a contradiction.

The only if part is also straightforward. For each outward neighbor x of y there is a trail with return vertex in p. This trail contains a cycle which includes p_{last}, y, and x. Therefore two disjoint paths run from x to p_{last}. So y is not an articulation point with respect to p_{last} and any x which is an outward neighbor of y, and therefore, by Lemma 5.5.1, y is not an articulation point. ■

Theorem 5.5.1 establishes the significance of the return vertices and motivates the development of a recursive definition for $\text{RtVrt}(p, y, x)$.

If x is in p, then certainly

$$\text{RtVrt}(p, y, x) = \{x\} \text{ because Trail}(p\|\langle y\rangle\|\langle x\rangle) \text{ can only contain the single trail } p\|\langle y\rangle\|\langle x\rangle.$$

If on the other hand x is not in p, then

$$\begin{aligned}\text{RtVrt}(p, y, x) &= \text{all return vertices of the trails in Trail}(p\|\langle y\rangle\|\langle x\rangle\|\langle n_i(x)\rangle) \\ &= \text{RtVrt}(p\|\langle y\rangle, x, n_i(x)) \text{ for all neighbors } n_i(x).\end{aligned}$$

Thus we get Definition 5.5.1 for $\text{RtVrt}(p, y, x)$.

Definition 5.5.1: Return Vertices

$$\left[\begin{array}{ll}\text{RtVrt}(p, y, x) = \{x\} & :x \in \{p\} \\ \text{RtVrt}(p, y, x) = \bigcup_{i=1}^{N(x)} [\text{RtVrt}(p\|\langle y\rangle, x, n_i(x))] & :x \notin \{p\} \\ \text{Initially } p = \langle\ \rangle,\ y = \text{zilch},\ x = \text{a vertex in } G. & \end{array}\right.$$

Note. Zilch is such that $\langle\text{zilch}\rangle = \langle\ \rangle$, $\{\text{zilch}\} = \{\ \}$.

The existence of articulation points can be gracefully inferred when $\text{RtVrt}(p, y, x)$ is computed for each (p, y, x) in Δ_f. The following rule for determining whether y is an articulation point with respect to p_{last} and x is justified by Theorem 5.5.1. We have

Proposition 5.5.1: Interpretation Rule for Definition 5.5.1
The vertex y is an articulation point with respect to p_{last} and x *if* $\text{RtVrt}(p, y, x)$ does not include any vertex in p. Further y is an articulation point *only if* there is a neighbor x of y such that $\text{RtVrt}(p, y, x)$ does not include any vertex in p.

Notice that although Definition 5.5.1 is used to solve for $f(\langle\ \rangle, \text{zilch}, x)$, this solution has no meaning under the interpretation rule, Proposition 5.5.1. If this were all we were looking for, the definition would be pointless. It is the value of the definition at every other argument structure in Δ_f, that is, at every vertex that can be reached from the initial vertex in G that we want. Algorithm 4.2.6 is designed to evaluate definitions at each of these argument structures and so is an appropriate one to use. Note further that $f(\langle\ \rangle, x, n_i(x))$ also has no useful interpretation, essentially because the path $p = \langle\ \rangle$. Thus it cannot tell us whether x is an articulation point. Fortunately p is $\langle\ \rangle$ only if x is the initial vertex, "begin." For this "root" case, as we will see, a special though simple condition is available for detecting articulation points. Alternatively one could add an extra dummy vertex d and an edge to "begin" in the given graph and implement $f(\langle\ \rangle, \text{zilch}, d)$.

5.5.2 Comparability in the Articulation Point Definition — Back and Tree Edges

Definition 5.5.1 will be the basis for our implementation. To determine if x is an articulation point find any path $p \| \langle y \rangle \| \langle x \rangle$ and then evaluate $f(p, y, x)$, which involves, in turn, evaluation of $f(p \| \langle y \rangle, x, n_i(x))$ for each neighbor $n_i(x)$ of x. Since a vertex x can usually be reached by many paths, a straightforward depth-first implementation of Definition 5.5.1 would involve evaluation of $f(p, y, x)$ for many different paths $p \| \langle y \rangle \| \langle x \rangle$ for the same x. This is unnecessary since the result with different paths will be equivalent in determining whether x is an articulation point. So two argument structures of the form (p, y, x), are comparable if they have equal x (third, or vertex) arguments. It turns out that in a depth-first implementation of Definition 5.5.1, the second and subsequent occurrences of an argument structure with a given vertex argument need not be evaluated at all. Interestingly, the relative positions of two comparable argument structures determine which of the two can be pruned. The next theorem establishes the condition on the position relation between comparable argument structures that justifies removal of one.

In the proof of this theorem it is necessary to consider the O network produced by Definition 5.5.1 for a graph G. For this purpose, and to correlate the notation used here with that in the literature, we develop some definitions. The reader may wish to refer to Figure 5-6 for illustrations of these definitions.

DEPTH-FIRST SEARCH DEFINITIONS

An *abbreviated O network* as the name implies, contains the O network for $f(p, y, x)$ in an abbreviated form. It has the structure of the O network, but has the vertex (3rd) argument of the O block inputs in place of the O block itself. The part of the abbreviated O network in which vertices are drawn from nonterminal argument structures is the *Forward Net*. Nodes in the Forward Net with no children are *end nodes*, and their vertices, *end vertices*. A vertex path from the root "begin" to an end vertex in the Forward Net is a *complete*

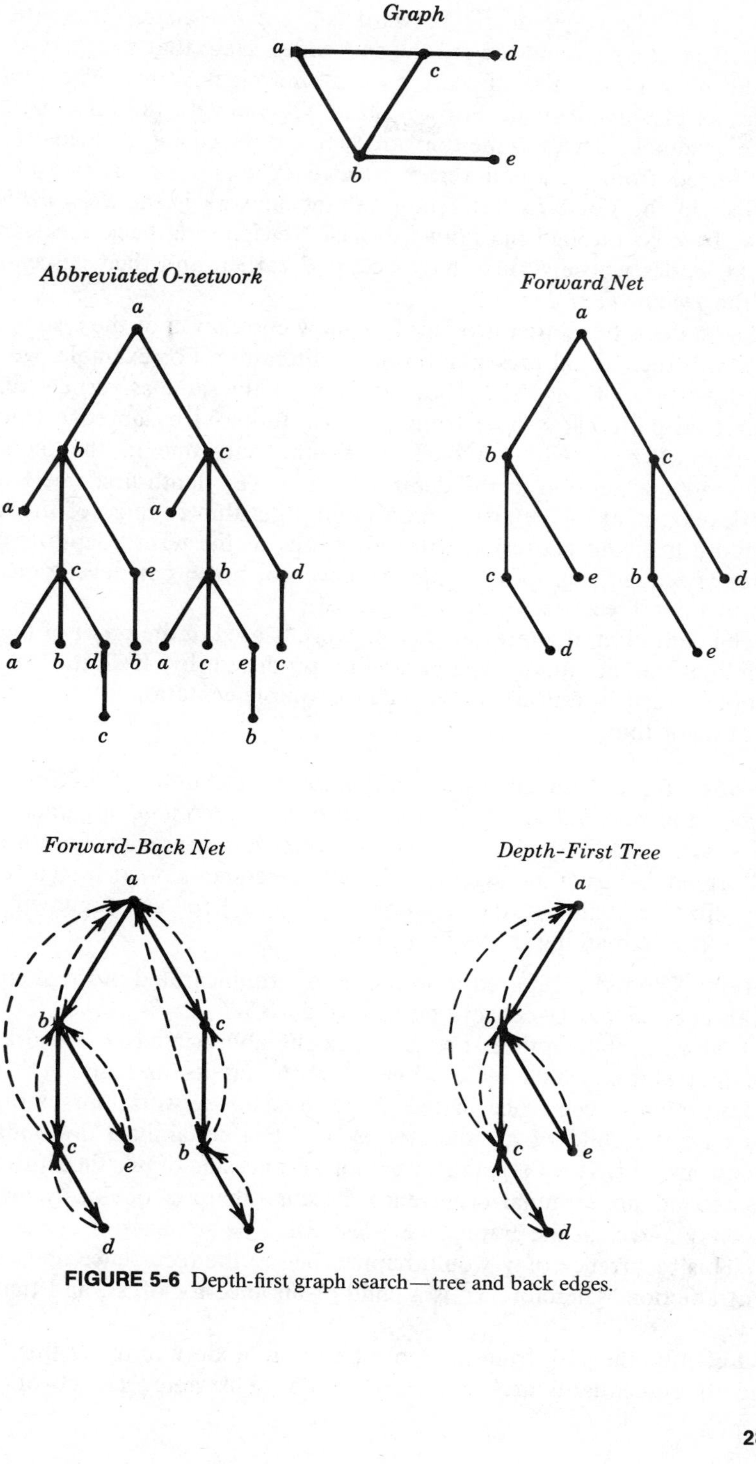

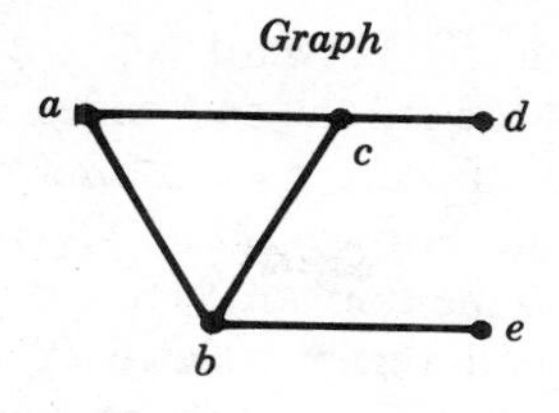

Abbreviated O-network

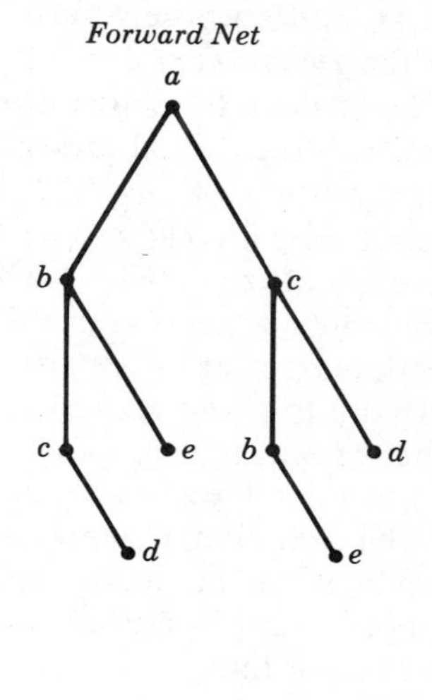

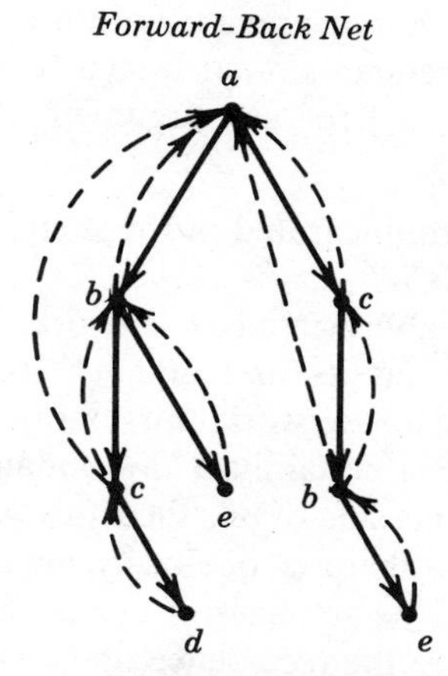

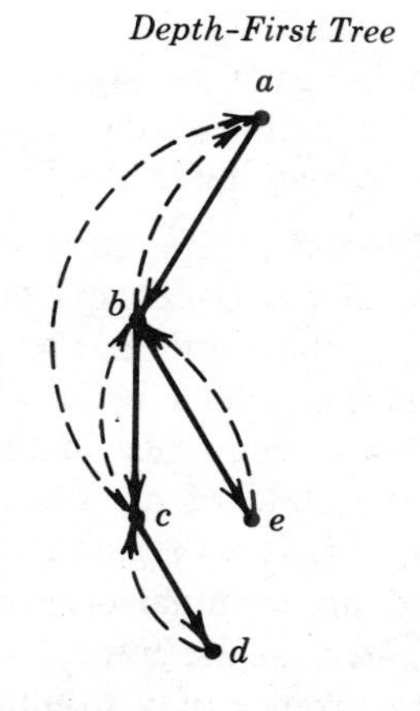

FIGURE 5-6 Depth-first graph search—tree and back edges.

path of G. The edges in the Forward Net are *tree edges*. They are directed away from the root and toward the end nodes. Note that each terminal vertex in the O network also appears as a nonterminal vertex. The nonterminal vertex is preserved in the Forward Net. To complete the information about these terminal vertices to the Forward Net, a directed edge, called a *back edge*, is directed from each end vertex occurrence to the occurrence of terminal vertices in the Forward Net. The resultant network is the *Forward-back Net*. Now if we go through the Forward-Back Net ignoring back edges and prune out all nodes whose vertices have occurred earlier, and their descendants, we get the *Depth–First Tree*.

These definitions are introduced to allow correlation of the results as stated here with their usual presentation in the literature. For example, we spoke of return vertices of $p\|\langle y\rangle\|x$. These vertices are the same as vertices reached by a directed path $p\|\langle y\rangle\|\langle x\rangle$ from the root, followed by an edge from x to a member of p, which is a back edge—the only one in the path—in the forward-back net and in the depth-first tree. The depth-first tree is the basis for development of the articulation point algorithm in most of the literature. Partially to avoid confusion with other uses of the term "depth-first," as an order of sequencing, for example, we have not based our development on the Depth-First Tree structuring of a graph.

The definition of abbreviated O network is used in the proof of the theorem detailing the nature of comparability produced by Definition 5.5.2. The comparability is crucial to the efficient implementation of the articulation point algorithm.

Theorem 5.5.2: In a depth-first evaluation of Definition 5.5.2 for *undirected graphs* the removal of all occurrences of a *nonterminal* argument structure (p, y, x) containing a vertex x which has occurred previously in a nonterminal argument structure, together with all descendants of such recurrences, will not affect the value of the function $f(p', y', x')$ for any argument structure (p', y', x') remaining in the O network.

PROOF: The terms referred to in the proof are illustrated in the sketch of part of an abbreviated O network given in Figure 5-7.

Following the depth-first sequence in the abbreviated O network, let x be the first vertex which recurs. Let x^1 represent its first and x^2 its second (nonterminal) occurrence in the abbreviated O network. First note that x^2 must be the child of an ancestor of x^1. It is certainly a descendant of an ancestor of x^1. If x^2's parent y is not an ancestor of x^1, then this would be y's second nonterminal occurrence, because there is obviously an edge between y and x in the graph, so y must either be an ancestor or descendant of x^1. This recurrence of y would happen before the recurrence of x, which is a contradiction. Therefore x^2 is a child of an ancestor of x^1, and that ancestor is y.

Let p be the path from the root "begin" to node y in the O tree. Now if x^2 and its descendants are removed from the abbreviated O network, the only

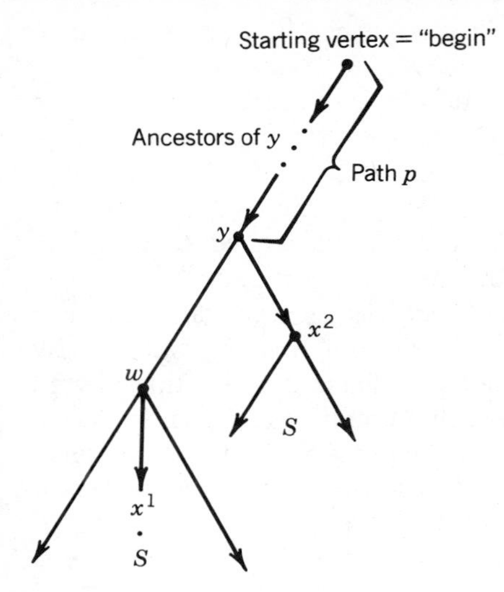

FIGURE 5-7 Comparability in the articulation point definition.

vertex in the preserved part of the network which could be affected with respect to the function $f(p, y, x)$ is y (and thus its ancestors). Only these could be return vertices of the trail $p\|\langle y\rangle\|\langle x\rangle$, given by back edges from descendants of x^2 in the O tree, which might be lost by the removal of x^2.

Let the child of y on the network path from y to x^1 be vertex w. *As can easily be shown, because the graph is undirected, the set of vertices consisting of* x *and all descendants of* x^2 *is the same as the set consisting of* w *and all its descendants.* Therefore removal of x^2 and its descendants will in no way affect which members of the path from the starting vertex to y can be a return vertex of a trail that goes through a child of y. The same vertices are in the subtree headed by w, as in the removed subtree rooted at x^2. Obviously the result for vertices in the abbreviated O network goes over directly to the corresponding argument structure in the O network. ■

In the proof of Theorem 5.5.2 we demonstrated the italicized statement, repeated here as a corollary.

Corollary 5.5.2.1: Each occurrence of vertex x after its first is either a terminal occurrence or a child of an ancestor of the first occurrence.

It follows from this corollary that since there is an occurrence of x (x^2 in figure 5-7) as a child of y, one of the children of the first occurrence of x (x^1) is also y—a terminal occurrence. Thus in a depth-first implementation the edge (x, y) would have been examined from x to y before the same edge was examined from y to x, revealing a second occurrence of x (x^1) in the argument structure. Therefore

Corollary 5.5.2.2: If in a depth-first evaluation of Definition 5.5.2 any edge having been examined in one direction is made nonexistent in the reverse direction, then there will be no repeated occurrences of any nonterminal vertex.

From this corollary it follows that a variation on depth-first implementation in which each edge is removed from the graph representation immediately after it is processed will accomplish the pruning allowed by Theorem 5.5.2. (Theorem 5.5.2 refers to the first recurrence. However, once the network is so pruned, the next occurrence becomes the first in the altered network. So pruning of subnetworks continues based on the same principle.)

Notice that in Theorem 5.5.2 the set of vertices in argument structures surrounding the one containing x^1 but not x^2 not only include, but actually equal, those surrounding x^2 and its descendants. This is to be contrasted with the weaker result in Theorem 5.3.1, where this notation is first introduced. Unlike Theorem 5.3.1, in Theorem 5.5.2 x is required to be the *first* occurrence of an argument structure. The more complex requirement on each argument structure in Theorem 5.5.2 leads to more restrictive conditions for pruning. The restriction on the "first recurrence" in depth-first sequence is important.

5.5.2.1 Improving the Definition — Indices Instead of Sets of Return Vertices

Definition 5.5.1 can be implemented, using an augmented version of Algorithm 4.2.6, to accommodate comparability, but we can do better. $\mathrm{RtVrt}(p, y, x)$ is used to decide whether y is an articulation point with respect to p_{last} and x, and this depends on whether $\mathrm{RtVrt}(p, y, x)$ contains a member of p. However, it is not necessary to know the identity of every vertex in $\mathrm{RtVrt}(p, y, x)$ to know whether one is in p. The vertices on each trail in $\mathrm{Trail}(p\|\langle y\rangle\|\langle x\rangle)$ are assigned an index. Since each such trail starts with $p\|\langle y\rangle\|\langle x\rangle$ or $\langle p_1, \ldots, p_k, y, x\rangle$, the indices of these vertices would run from 1 to k in p, then to $k+1$ for y, and finally to $k+2$ for x. Assume that instead of $\mathrm{RtVrt}(p, y, x)$ we compute

$$f(p, y, x) = \text{lowest index of a vertex in } \mathrm{RtVrt}(p, y, x)$$

There is still sufficient information to determine whether $\mathrm{RtVrt}(p, y, x)$ contains a member of p and thus whether y is an articulation point with respect to p_{last} and x. Clearly, $\mathrm{RtVrt}(p, y, x)$ contains a member of p if and only if

$$f(p, x, y) \leq \text{index of } p_{\mathrm{last}} = \mathrm{ind}(p_{\mathrm{last}}, p)$$

The recursive definition for $f(p, y, x)$ is similar to that for $\mathrm{RtVrt}(p, y, x)$.

1. If p contains x, then $f(p, y, x)$ is the index of x in $p = \mathrm{ind}(x, p)$.
2. If p does not contain x, then $f(p, y, x)$ depends on the neighbors of x in a way completely analogous to that given in the recursive line of Definition 5.5.1. It is the minimum index of return vertices on the trails, with prefixes extended to the neighbors of x, $\mathrm{Trail}(p\|\langle y\rangle, x, n_i(x))$;

$i = 1$ to $N(x)$, that is

$$\mathrm{Min}_{i=1}^{N(x)}[f(p\|\langle y\rangle, x, n_i(x))]$$

So we have

Definition 5.5.2: Lowest Index of a Vertex in RtVrt(p, y, x)—Articulation Points

$$\left[\begin{array}{ll} f(p, y, x) = \mathrm{ind}(x, p) & :x \in \{p\} \\ f(p, y, x) = \mathrm{MIN}_{i=1}^{N(x)}[f(p\|\langle y\rangle, x, n_i(x))] & :x \notin \{p\} \\ \text{Initially } p = \langle\ \rangle,\ y = \text{zilch},\ x = \text{vertex 'begin' in } G. & \end{array}\right.$$

This change does not affect the comparability property demonstrated for Definition 5.5.1.

5.5.2.2 Improvements in Argument Representation — Updates

Definition 5.5.2 can be improved further without destroying its comparability property. Arguments p and y are combined into one, pth $= p\|\langle y\rangle$. To implement Definition 5.5.2, the ind(x, p) = ind(x, $p\|\langle y\rangle$) = ind(x, pth) function requires a search through pth for x, which takes time proportional to the length of pth. To speed up that computation arguments PY and j are added together with an update function. j is the index of the last vertex to be added to pth. PY is an array with one entry for every vertex in the graph. The number of nonzero entries in PY is $j - 1$. If v is the ith vertex in path pth, then PY[v] = i. So PY[v] gives ind(x, pth) directly. It is updated when a new vertex x is concatenated to pth by storing $j + 1$ at PY[x]. With the introduction of PY and j, the use of pth is diminished. It is, however, still useful in obtaining the inverse necessary for the implementation. The revised definition incorporating the array PY, j and pth is Definition 5.5.3.

Definition 5.5.3: Articulation Points

$$\left[\begin{array}{l} f(\mathrm{PY}, j, \mathrm{pth}, x) = \mathrm{PY}[x] \qquad :\mathrm{PY}[x] > 0 \\ f(\mathrm{PY}, j, \mathrm{pth}, x) = \mathrm{MIN}_{i=1}^{N(x)}[f(\mathrm{PY}[x] \leftarrow j + 1, j + 1, \mathrm{pth}\|\langle x\rangle, n_i(x))] \\ \qquad :\mathrm{PY}[x] = 0 \\ \text{Initially PY} = 0^n,\ j = 0,\ \mathrm{pth} = \langle\ \rangle,\ x = \text{vertex 'begin' in } G. \end{array}\right.$$

Proposition 5.5.2: Interpretation Rule for Definition 5.5.2

Vertex $\mathrm{pth}_{\mathrm{last}}$ is an articulation point with respect to $\mathrm{pth}_{\mathrm{last\text{-}1}}$ and x *if* $f(\mathrm{PY}, j, \mathrm{pth}, x) > \mathrm{PY}(\mathrm{pth}_{\mathrm{last}})$. And $\mathrm{pth}_{\mathrm{last}}$ is an articulation point *only if* there is a neighbor x of $\mathrm{pth}_{\mathrm{last}}$ such that $f(\mathrm{PY}, j, \mathrm{pth}, x) > \mathrm{PY}(\mathrm{pth}_{\mathrm{last}})$.

An inverse exists and will be used in the implementation. pth contains the vertices encountered in reaching x. Its parent is obtained by removing $\mathrm{pth}_{\mathrm{last}}$, or viewing pth as a stack with $\mathrm{pth}_{\mathrm{last}}$ on top, popping pth. The vertex popped from pth is x's parent. $j-1$ is the parent of j. The parent of PY is PY with PY[x] set to 0. Thus there is a parent function. Instead of the child# function we compute the last-sib and next-sib functions directly using a graph representation, which allows efficient computation of a last-neigh predicate and next-neigh function (see Section 5.2.2.2). Notice that all the children of a given parent have the same PY, j, and pth arguments, so these are unchanged in obtaining the next-sib.

To implement Definition 5.5.3 and determine whether each vertex of G is an articulation point, $f(\text{PY}, j, \text{pth}, x)$ must be computed for each vertex. w ($=$ min) is associative and commutative, so we might choose Algorithm 4.2.5, but *we want the answer for every vertex*, and thus for every nonpruned argument in the substitutional domain, so we choose Algorithm 4.2.6 for our implementation. The result is shown on the left, while the modifications necessary to take advantage of the comparability of Definition 5.5.3 are given on the right of Algorithm 5.5.1.

Algorithm 5.5.1: Depth-First Articulation Point Algorithm 1

```
        Plug in to Algorithm 4.2.6                  Modifications for Comparability
V ← ⟨ ⟩
DCL(PY[n], vector) INIT(0)
J ← 0
PTH ← ⟨ ⟩
X, S ← 'begin'
FOR ever DO
  WHILE(PY[X] = 0) DO                          ← WHILE(PY[X] = 0 & T[X] = 0) DO
    * * * * * * * * * * * * * *                ← T[X] ← 1
    V ← psh- ∞
    PY[X] ← J + 1
    J ← J + 1
    PTH ← psh- X
    X ← n₁(X)
  END
  * * * * * * * * * * * * * *                  ← IF PY[X] ≠ 0 THEN
  V[1] ← min(V[1], PY[X])
  WHILE(X = last-neigh(PTH[last]) DO
    FX ← pop- V
    X ← pop- PTH
    PY[X] ← 0
    J ← J − 1
    IF X = S THEN DONE
    V[1] ← min(V[1], FX)
  END
  X ← next-neigh(PTH[last], X)
END
```

Notes

1. In this algorithm popping PTH means removing PTH[last] and pushing means adding an entry at the end of PTH to create a new PTH[last].
2. A test for articulation points can be inserted directly after the "$J \leftarrow J - 1$" line. It reads: IF FX $\geq$ PY[PTH[last]] then [PTH[last] is an articulation point with respect to PTH[last-1] and X]

5.5.2.3 Algorithm Improvements Not Expressible in the Definition

A slightly different use of PY and J than prescribed by Definition 5.5.3 ensures that they not only give an efficient way of determining the index of a vertex, but also record every vertex encountered. Definition 5.5.3 works if indices are assigned in any nondecreasing order, not necessarily consecutively, to vertices added to PTH and recorded in PY. Thus J may be used to globally number all vertices, not just along each path PTH, in the order in which they are encountered, no longer decrementing J, nor resetting PY entries to 0 when backing up and removing a vertex from PTH. Thus PY[v] = 0 only if v was never encountered. But if PY entries are not reset to zero, the terminal and nonterminal tests as given by Definition 5.5.3 will not work. However, once the algorithm is altered to eliminate vertex recurrences, the terminal condition is again testable. It is again possible to determine whether a vertex X, just acquired, is in the path recorded in PTH. This is done as follows.

1. If PY[X] = 0, then vertex X has not been previously visited and so is not in the path PTH.
2. The next two conditions are justified by Corollary 5.5.2.1, which says: Each occurrence of the vertex x after its first is either a terminal occurrence or a child of an ancestor of the first occurrence. Therefore,

   a. If X is terminal, then X was seen earlier than PTH[last] and must be in path PTH. So PY[X] < PY[PTH[last]].

   b. If X is not terminal, then X must be a child of an ancestor of its first occurrence. That ancestor is in fact PTH[last] and has a lower index than X. So PY[X] > PY[PTH[last]].

In summary the proposed changes are

1. Remove the table T from all its occurrences, including in the upper WHILE.
2. Remove the backup of J and PTH. The upper WHILE can now fail because PY[X] $\neq$ 0 or because either

   a. PY[X] < PY[PTH[last]], implying X was seen before PTH[last] and must be in the path PTH, in which case we wish to take the action indicated at exit from the WHILE, or

   b. PY[X] > PY[PTH[last]], implying X was seen before on a path from PTH[last], not in path PTH, but on some path represented in the O network to the left of PTH. In this case we do not need to take the action indicated at the WHILE, but it will not change anything if we do.

The result is given in Algorithm 5.5.2. Though an improvement, its worst-case complexity is no better than that of Algorithm 5.5.1.

[The changes we have are too extensive to fit the set encoding principle (Proposition 4.4.1) but suggest an extension of that principle.]

Algorithm 5.5.2 is no longer a straightforward implementation of Definition 5.5.3, which does not allow J to grow continuously nor PY to contain vertices not on the path represented in PTH. In fact it is no longer the implementation of any reasonable nonnested definition, but rather implements a nested definition related to Definition 5.5.3. This nested definition will be displayed later as Definition 5.5.5.

Algorithm 5.5.2: Depth-First Articulation Point Algorithm 2

```
Modifications to Algorithm 5.5.1
V ← ⟨ ⟩
DCL(PY[n], vector) INIT(0)
J ← 0
PTH ← ⟨ ⟩
X, S ← 'begin'
FOR ever DO
   WHILE(PY[X] = 0) DO
      V ← psh- ∞
      PY[X] ← J + 1
      J ← J + 1
      PTH ← psh- X
      X ← n1(X)
   END
   * * * * * * * * * * * * * * *                IF PY[X] < PY[PTH[last]] THEN
   V[1] ← min(V[1], PY[X])
   WHILE(X = last-neigh (PTH[last]) DO
      FX ← pop- V
      X ← pop- PTH
      IF X = S THEN DONE
      V[1] ← min(V[1], FX)
   END
   X ← next-neigh (PTH[last], X)
END
```

Notes

1. The test on the right is not necessary since $\min(V[1], \mathrm{PY}[X]) = V[1]$ if that test is not satisfied. However, if one is modifying the definition to keep track of biconnected components, the test will allow us to eliminate repetition of edges in such components.
2. As justified by Corollary 3.9.2, if after a graph edge is traversed in one direction it is effectively removed and never seen in the other direction, Algorithm 5.5.2 will work. The condition on the right is always satisfied, and is not needed. A graph representation which allows this to be done efficiently is the adjacency multilist [Horowitz Sahni, p. 291].

In Example 5.5.1 the algorithm is traced in operation on a small graph. "begin" of the algorithm $= d$ of the example graph. The effect on PTH, V, and PY on encountering vertex X is given on each line. To correlate V and PTH, both are shown growing from right to left, stackwise, with $V[1]$ and PTH[last] both on the left. There is a caret ($\wedge$) under $V[1]$. (t) means X is in a terminal argument structure; (b) means backup. On the right, computations for updating the V list on termination or backup are shown, for backup the value of FX is given and marked ** when an articulation point is detected.

Example 5.5.1

Graph

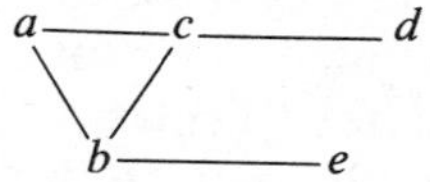

a — c — d
a — b
c — b
b — e

Vertex	Neighbors
a	b, c
b	a, c, e
c	a, b, d
d	c
e	b

Trace

	X	← PTH ← V				PY a	b	c	d	e	
	a					0	0	0	0	0	
1st-neigh of a =	b				a ∞	1	0	0	0	0	
1st-neigh of b =	a (t)			b ∞	a ∞	1	2	0	0	0	
next-neigh of b =	c			b 1	a ∞	1	2	0	0	0	
1st-neigh of c =	a (t)		c ∞	b 1	a ∞	1	2	3	0	0	
next-neigh of c =	b (t)		c 1	b 1	a ∞	1	2	3	0	0	$V[1] = \min(\infty, 1)$
next-neigh of c =	d		c 1	b 1	a ∞	1	2	3	0	0	$V[1] = \min(2, 1)$
1st-neigh of d =	c (t)	d ∞	c 1	b 1	a ∞	1	2	3	4	0	
last-neigh of d =	c (b)	d 3	c 1	b 1	a ∞	1	2	3	4	0	$V[1] = \min(3, \infty)$
last-neigh of c =	d (b)		c 1	b 1	a ∞	1	2	3	4	0	$FX = 3$ * * $V[1] = \min(1, 4)$
~ last neighbor	c			b 1	a ∞	1	2	3	4	0	$FX = 1$ $V[1] = \min(1, 1)$
next-neigh of b =	e			b 1	a ∞	1	2	3	4	0	
next-neigh of e =	b (t)		e ∞	b 1	a ∞	1	2	3	4	5	
last-neigh of e =	b (b)		e 2	b 1	a ∞	1	2	3	4	5	$V[1] = \min(2, \infty)$
last-neigh of b =	e			b 1	a ∞	1	2	3	4	5	$FX = 2$ * * $V[1] = \min(1, 2)$
~ last neighbor	b				a 1	1	2	3	4	5	$FX = 1$ $V[1] = \min(1, \infty)$
next-neigh of a =	c (b)				a 1	1	2	3	4	5	
last-neigh	a					1	2	3	4	5	$FX = 1$ $V[1] = \min(1, 3)$

The lower WHILE in the algorithm, just after FX is computed and the backup is completed, is a convenient place at which to detect whether PTH[last] (corresponding to argument y in Definition 5.5.1) is an articulation point. FX $= f(\text{pth}, \text{PY}, j, x)$ at that point. If FX < PY[PTH[last]], then the vertex PTH[last] *is not* an articulation point with respect to vertex X or any of its descendants and PTH[last-1], though it may be with respect to some other vertices. If on the other hand, FX ≥ PY[PTH[last]], then PTH[last] *is* an articulation point separating the vertices in PTH from descendants in the O-tree for Definition 5.5.2. This holds when any vertex other than the initial one is in PTH[last], that is, when |PTH| > 1.

Whether "begin" at the root of the O-network is an articulation point depends on conditions different from those for other vertices.

> ***The "begin" vertex is an articulation point if PY[z] = 0 for any sibling vertex, say z, of its first child vertex, say y, after the algorithm has first backed up to "begin".***

If $P[z] = 0$, then z is nowhere on any path with prefix ⟨'begin', y⟩. Therefore there is no way to reach z from the root by first going through the first child y. Therefore every path between y and z must pass through "begin."

5.5.2.4 Complexity of the Articulation Point Algorithm

This algorithm simulates one O block for each vertex reachable from "begin," in the graph G. The time needed to simulate an O block with vertex input v is proportional to the number of its outputs = the number of neighbors of v. The total number of O blocks in the O network equals the total number of neighbors of vertices in $G = 2 *$ the number of edges $= 2e = O(e)$. The internal complexity required to compute each output of an O block is

1. The cost of finding each first child is $O(1)$, plus
2. The cost of getting next-sib, which, given the appropriate graph representation, is also $O(1)$, plus
3. The cost of backing up at last-sib, computing the parent, and so on, which is $O(1)$. In addition during backup the test executed for an articulation point is $O(1)$.

Therefore $O(e)$ is the worst-case complexity of the algorithm.

5.5.2.5 The Comparability and Breadth-First Search

We have shown how a depth-first algorithm can take advantage of the articulation point definition comparability. This comparability *cannot* be used similarly in a breadth-first algorithm. The result of an attempt to do so is shown in Figure 5-8. An O–w–t network for Definition 5.5.1 operating on a small graph, with pruning of argument structures whose vertex arguments have appeared earlier, is shown for a depth- and breadth-first implementation. Vertex c is an articulation point, correctly identified in both cases, but the

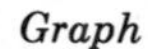

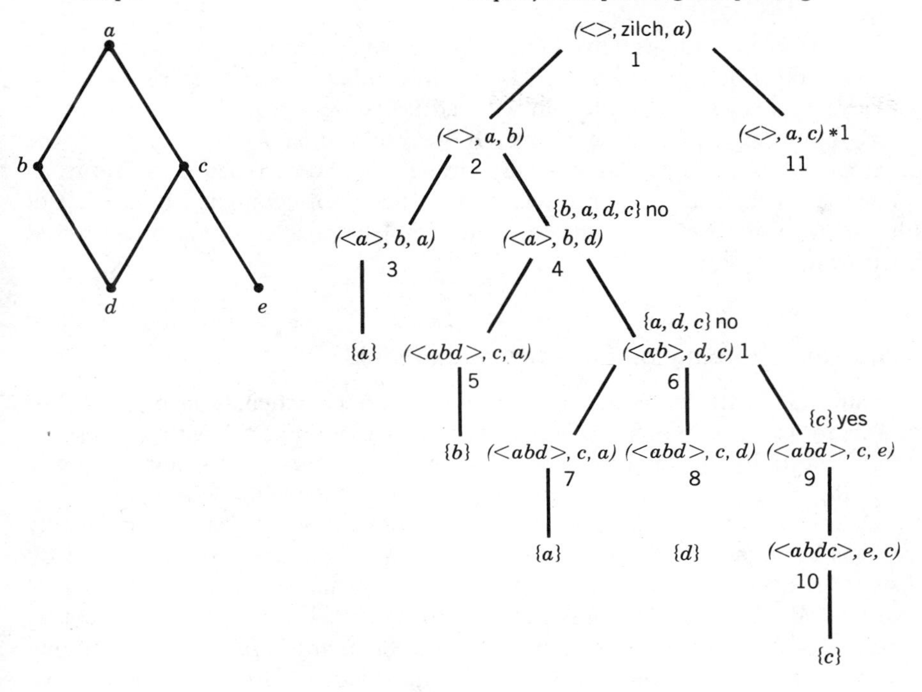

Breadth-first sequencing with pruning

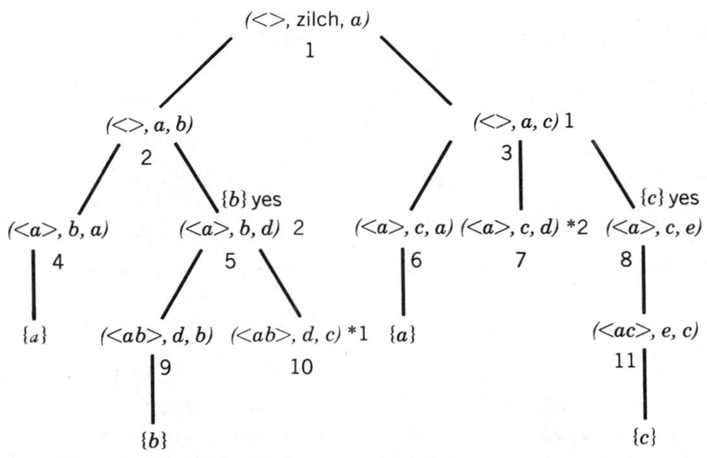

FIGURE 5-8 Depth-first works, but breadth-first does not. The number under the argument-structures indicates the sequence in which they are examined. $*i$ after argument structure X indicates that X has a vertex argument that duplicates one in an argument structure which occurred earlier in the sequence and has i written after it. $\{x, y, \ldots\}$ appearing above an argument structure X gives the value of $f(X)$. If followed by "yes"/"no," it means that an articulation point is/is not detected.

breadth-first case also concludes, incorrectly, that vertex b is an articulation point.

Even though comparability requires complete simulation of a subtree of the O network with vertex x at its root before other occurrences of x can be pruned, this does not mean that a breadth-first search algorithm cannot be used in finding articulation points. If we are willing to give up using the comparability at all, or if we only prune on a recurrence of a vertex x precisely when a subtree with x at its root has been completed to the left of the recurrence, we can still base our implementation on a breadth-first algorithm.

5.5.3 Nested Definition for Articulation Points

An alternative to depth-first implementation of the articulation point definition is to use a nested recursive equivalent of Definition 5.5.3 with a comparability test added to implement it directly with a recursive procedure. (Strictly speaking, Algorithm 5.5.2 implements a nested recursive definition.)

Definition 5.5.3 will be transformed to represent the way PY cumulatively records vertices and the continuous increase in j. The ground is prepared by first transforming to nested Definition 5.5.4 which allows access to the value of every f call. Definition 5.5.4 is synthesized by direct application of Theorem 2.6.3. [Local associativity is assumed in the theorem; here we have a purely associative function, min(...), which is equivalent to the locally associative function n(min(...)) where n is the identity.]

Definition 5.5.4: Articulation Points—Nested Definition I

$$
\left[
\begin{array}{ll}
g(W, \mathrm{PY}, j, \mathrm{pth}, x) = \min(W, \mathrm{PY}[x]) & :\mathrm{PY}[x] > 0 \\
g(W, \mathrm{PY}, j, \mathrm{pth}, x) = \min\big(W, g(\ldots g(g(\infty, \mathrm{PY}[x] \leftarrow j+1, & \\
j+1, \mathrm{pth}\|\langle x\rangle, n_1(x)), \mathrm{PY}[x] \leftarrow j+1, j+1, \mathrm{pth}\|\langle x\rangle, n_2(x)), \ldots, & \\
\qquad \mathrm{PY}[x] \leftarrow j+1, j+1, \mathrm{pth}\|\langle x\rangle, n_{N(x)}(x))\big) & :\mathrm{PY}[x] = 0 \\
\text{Initially } W = \infty, \mathrm{PY} = O^n, j = 0, \mathrm{pth} = \langle\ \rangle, x = \text{a vertex in } G, & \\
\qquad G \text{ has } n \text{ vertices.} &
\end{array}
\right.
$$

Terms $\mathrm{PY}[x] \leftarrow j+1$ and $j+1$ are repeated in the same positions in every recursive call in Definition 5.5.4. This ensures that at each call PY and j start with the same value. Since, however, the cumulative value is adequate, we can make the cumulative value of PY and j two new components of g. g is then a three-component vector. This is shown in Definition 5.5.5, where three arguments are grouped in vector $\langle W, \mathrm{PY}, j\rangle$, indicating that they are the components of g. This definition uses [min], an extension of the min function. [min] is a function on two arguments, the second an ordered triple. Its value is a triple given as follows:

$$[\min](x, \langle a, b, c\rangle) = \langle \min(x, a), b, c\rangle$$

Definition 5.5.5: Articulation Points—Nested Definition II

$$\left[\begin{array}{ll} g(\langle W, \mathrm{PY}, j\rangle, \mathrm{pth}, x) = \langle \min(W, \mathrm{PY}[x]), \mathrm{PY}, j\rangle & :\mathrm{PY}[x] > 0 \\ g(\langle W, \mathrm{PY}, j\rangle, \mathrm{pth}, x) = [\min](W, g(\ldots g(g(\langle \infty, PY[x] \leftarrow j+1, & \\ \qquad j+1\rangle, \mathrm{pth}\|\langle x\rangle, n_1(x)), & \\ \qquad \mathrm{pth}\|\langle x\rangle, n_2(x)), \ldots, \mathrm{pth}\|\langle x\rangle, n_{N(x)}(x))) & :\mathrm{PY}[x] = 0 \\ \text{Initially } W = \infty, \mathrm{PY} = 0^n, j = 0, \mathrm{pth} = \langle\ \rangle, x = \text{a vertex in } G, & \\ \quad G \text{ has } n \text{ vertices.} & \end{array}\right.$$

The fact that this definition now incorporates cumulative use of PY and j means that it already has pruning incorporated because now terminal condition $\mathrm{PY}[x] > 0$ applies in more cases than in Definition 5.5.4, and thus generally there are fewer recursive calls. Assume for the moment that this was not known. Then we might try to incorporate pruning by modifying the definition according to the prescription in Section 3.10.1. The terminal condition can be partitioned into two parts. One is $\mathrm{PY}[\mathrm{pth}_{\mathrm{last}}] < \mathrm{PY}[x]$, which implies x is a repeat but not in pth, so no change in W, PY, or j is required, that is, $g(\langle W, \mathrm{PY}, j\rangle, \mathrm{pth}, x) = \langle W, \mathrm{PY}, j\rangle$. On the other hand, if $\mathrm{PY}[\mathrm{pth}_{\mathrm{last}}] \geq \mathrm{PY}[x]$, then $g(\langle W, \mathrm{PY}, j\rangle, \mathrm{pth}, x) = \min(W, \mathrm{PY}[x]), \mathrm{PY}, j\rangle$. However, if $\mathrm{PY}[\mathrm{pth}_{\mathrm{last}}] > \mathrm{PY}[x]$, then $W = \min(W, \mathrm{PY}[x])$, so the partition is not necessary.

In Example 5.5.2, the definition is applied to the graph from Example 5.5.1. The evaluation is carried out by ESE approach (Process 2.7.1). The equations are given in groups, each of whose left sides are identical. The first equation in a group is the initial form of that equation. After a new equation is formed, its right side is examined. When a right side appearance of an f call for which there is no existing equation is found, a new equation is written for that f call. If the right side has no f call, it is back substituted, and the equation into which it is substituted is rewritten within its group. Then that rewritten equation's right side is scanned for an f call for which there is no existing equation.

Example 5.5.2: In this example pth grows by concatenation to the left, so $\mathrm{pth}_{\mathrm{last}}$ is the leftmost component of pth.

$$\begin{aligned} g(\langle \infty, \langle 00000\rangle, 0\rangle, \langle\ \rangle, a) &= [\min](\infty, g(g(\langle \infty, \langle 10000\rangle, 1\rangle, \langle a\rangle, b), \langle a\rangle, c)) \\ &= [\min](\infty, g(\langle 1, \langle 12345\rangle, 5\rangle, \langle a\rangle, c)) \\ &= [\min](\infty, \langle 1, \langle 12345\rangle, 5\rangle) \\ &= \langle 1, \langle 12345\rangle, 5\rangle \end{aligned}$$

$$
\begin{aligned}
g(\langle \infty, \langle 10000\rangle, 1\rangle, \langle a\rangle, b) &= [\min](\infty, g(g(g(\langle \infty, \langle 12000\rangle, 2, \langle ba\rangle, a), \langle ba\rangle, c), \\
&\qquad \langle ba\rangle, e)) \\
&= [\min](\infty, g(g(1, \langle 12000\rangle, 2, \langle ba\rangle, c), \langle ba\rangle, e)) \\
&= [\min](\infty, g(1, \langle 12340\rangle, 5, \langle ba\rangle, e)) \\
&= [\min](\infty, \langle 1, \langle 12345\rangle, 5\rangle) \\
&= \langle 1, \langle 12345\rangle, 5\rangle \\
g(\infty, \langle 12000\rangle, 2, \langle ba\rangle, a) &= \langle \min(\infty, 1), \langle 12000\rangle, 2\rangle \\
&= \langle 1, \langle 12000\rangle, 2\rangle \\
g(1, \langle 12000\rangle, 2, \langle ba\rangle, \underline{c}) &= [\min](1, g(g(g(\infty, \langle 12300\rangle, 3, \langle cba\rangle, a), \langle cba\rangle, b), \\
&\qquad \langle cba\rangle, d)) \\
&= [\min](1, g(g(1, \langle 12300\rangle, 3, \langle cba\rangle, b), \langle cba\rangle, d)) \\
&= [\min](1, g(1, \langle 12300\rangle, 3, \langle cba\rangle, d)) \\
&= [\min](1, \langle \underline{1}, \langle 12340\rangle, 4\rangle) \qquad ** \ \mathrm{PY}[b] = 1 \\
&= \langle 1, \langle 12340\rangle, 5\rangle \\
g(\langle \infty, \langle 12300\rangle, 3\rangle, \langle cba\rangle, a) &= \langle \min(\infty, 1), \langle 12300\rangle, 3\rangle \\
&= \langle 1, \langle 12300\rangle, 3\rangle \\
g(\langle 1, \langle 12300\rangle, 3, \langle cba\rangle, b) &= \langle \min(1, 2), \langle 12300\rangle, 3\rangle \\
&= \langle 1, \langle 12300\rangle, 3\rangle \\
g(1, \langle 12300\rangle, 3, \langle cba\rangle, \underline{d}) &= [\min](1, g(\langle \infty, \langle 12340\rangle, 4, \langle dcba\rangle, c)) \\
&= [\min](1, \langle \underline{3}, \langle 12340\rangle, 4\rangle) \qquad ** \ \mathrm{PY}[c] = 3 \\
&= \langle 1, \langle 12340\rangle, 4\rangle) \\
g(\langle \infty, \langle 12340\rangle, 4, \langle dcba\rangle, c) &= \langle \min(\infty, 3), \langle 12340\rangle, 4\rangle \\
&= \langle 3, \langle 12340\rangle, 4\rangle \\
g(1, \langle 12340\rangle, 4, \langle ba\rangle, \underline{e})) &= [\min](1, g(\infty, \langle 12345\rangle, 5, \langle eba\rangle, b)) \\
&= [\min](1, \langle \underline{2}, \langle 12345\rangle, 5\rangle) \qquad ** \ \mathrm{PY}[b] = 2 \\
&= \langle 1, \langle 12345\rangle, 5\rangle \\
g(\langle \infty, \langle 12345\rangle, 5\rangle, \langle eba\rangle, b) &= \langle \min(\infty, 2), \langle 12345\rangle, 5\rangle \\
&= \langle 2, \langle 12345\rangle, 5\rangle \\
g(\langle 1, \langle 12345\rangle, 5\rangle, \langle a\rangle, c)) &= \langle \min(1, 3), \langle 12345\rangle, 5\rangle \ \text{[c's second occurrence]} \\
&= \langle 1, \langle 12345\rangle, 5\rangle
\end{aligned}
$$

The underlined number, say i, in the $**$ equations gives the least index, of a return vertex of trails in Trail(pth, y, x), where $y = \mathrm{pth}_{\mathrm{last}}$, and x is the underlined vertex on the left of that equation. x is a child of $\mathrm{pth}_{\mathrm{last}}$ (the leftmost element in pth) in the argument on the left of that line. i is compared with PY[y] to determine if y is an articulation point. If $i < \mathrm{PY}[y]$, then y is *not* an articulation point with respect to $\mathrm{pth}_{\text{last-1}}$ and x, otherwise it is. $**$ lines indicate the presence of an articulation point. If $\mathrm{pth}_{\mathrm{last}} = \langle\ \rangle$ of the root vertex, $\langle a\rangle$ in the example, this test is meaningless and therefore is omitted. The information for determining whether the root is an articulation point is contained in the evaluation of the innermost g call on the right of the first equation. If PY there has an assignment to every vertex, then every vertex has been visited as a descendant of the first child of the root and the root is not an articulation point, otherwise it is.

5.5.4 Biconnected Components — Their Enumeration Using Articulation Points

Detection of articulation points is useful in finding *biconnected components*.

Assume G is a connected graph with more than one vertex. Two edges x and y are mutually *biaccessible* in G if $x = y$ or x and y are in a common cycle. The relation of biaccessibility is an equivalence relation, and thus partitions the set of edges of G into equivalence classes. Each subgraph consisting of all the edges in an equivalence class together with all vertices incident on those edges is a biconnected component of G. It follows that every edge of G is in exactly one of its biconnected components, whereas a vertex of G may be in more than one of its biconnected components.

An alternative equivalent definition is S is a biconnected component of G if and only if S is a subgraph of G and

1. For every pair of edges in S there is a cycle in S on which both appear, and
2. There is no edge e in G but not in S such that $S \cup \{e\}$ has property 1.

The justification of the equivalence of these two definitions is left to the reader.

The following lemma relating biconnected components and articulation points is basic to the biconnected component algorithm development. We define

$$\text{Comp}(p, y, x) = \text{the set of vertices in Trail}(p\|\langle y\rangle\|x\rangle) \text{ but not in } p\|\langle y\rangle$$

Lemma 5.5.2: If y is an articulation point with respect to p_{last} and x and none of the vertices in Comp(p, y, x) is an articulation point, then all the edges in trails of Trail$(p\|\langle y\rangle\|\langle x\rangle)$, other than those in $p\|\langle y\rangle$, are in the same biconnected component.

PROOF: First: (a) if v is a vertex in Comp(p, y, x), then there is a trail in Trail$(p\|\langle y\rangle\|\langle x\rangle)$ with return vertex y that includes v.

If v was not on any such trail, then, because y is an articulation point, there is no cycle which includes (y, x) and the vertex v. Therefore the only way to get between y and v is through x. So x is an articulation point—a contradiction.

From this it follows directly that every vertex in Comp(p, y, x) is on a cycle, including the edge (y, x), and then that every edge in Trail$(p\|\langle y\rangle\|\langle x\rangle)$, except those in $p\|\langle y\rangle$, is on such a cycle. If an edge (v, w) is in a trail, then v is in a cycle including (x, y), and w is also in such a cycle. Since there is an edge (v, w), there must be a cycle including (x, y) and (v, w). So all edges in Trail$(p\|\langle y\rangle\|\langle x\rangle)$ other than those in $p\|\langle y\rangle$ are in the same biconnected component. ■

If we remove from the hypotheses of Lemma 5.5.2 the requirement that Comp (p, y, x) not contain articulation points, we have the following lemma, whose proof is similar.

Lemma 5.5.3: If y is an articulation point with respect to p_{last} and x, then all the edges in trails of Trail($p\|\langle y\rangle\|\langle x\rangle$) other than those in $p\|\langle y\rangle$ are in different biconnected components than any other edges in G.

5.5.4.1 Development of the Recursive Definition for Biconnected Components

The biconnected component definition is a modestly modified version, justified by Lemmas 5.5.2 and 5.5.3, of articulation point definition, Definition 5.5.2. We need only add a facility to collect edges to Definition 5.5.2. When an articulation point y is found, all edges on trails in Trail($p\|\langle y\rangle\|\langle x\rangle$), but not in $\{p\|\langle y\rangle\}$, that are not already members of a biconnected component, form a new one.

The part of the definition used for collecting edges is a combination of the recursive definition for the function h, which records the edges of the biconnected components as they develop, and a second nonrecursive one for *OUT*, which uses h and gives the actual biconnected components. More precisely:

$h(p, y, x)$ = the set of edges, E, on the trails Trail($p\|\langle y\rangle\|\langle x\rangle$), not in $\{p\|\langle y\rangle\}$ and not already included in a biconnected component if y is not an articulation point.

= { } otherwise

OUT(p, y, x) = E, the biconnected component including edge (y, x), if y is an articulation point.

= { } otherwise (this means no output at all, and so will not be explicitly included in the following definitions).

Notice that when $h(p, y, x)$ is not empty, OUT(p, y, x) is, and vice versa.

Consider the edges in Trail($p\|\langle y\rangle\|\langle x\rangle$). First remove edges in $p\|\langle y\rangle$, next remove any subset which forms a biconnected component. What remains is designated $h(p, y, x)$. By Lemma 5.5.3 if y is an articulation point with respect to p_{last} and x, then

$$h(p, y, x) = \{\ \}$$

$$\mathrm{OUT}(p, y, x) = \bigcup_{i=1}^{N(x)} [h(p\|\langle y\rangle, x, n_i(x))] \cup \{(y, x)\}$$

If on the other hand y is not an articulation point, then

$$h(p, y, x) = \bigcup_{i=1}^{N(x)} [h(p\|\langle y\rangle, x, n_i(x))] \cup \{(y, x)\}$$

$$\mathrm{OUT}(p, y, x) = \{\ \}$$

The "$\cup$ $\{(x, y)\}$" can be dropped from these equations if we define $h(p, y, x) = \{(y, x)\}$ when $x \in \{p\}$, (i.e. when (p, y, x) is terminal), because every edge included in $h(p, y, x)$ by "$\cup \{(x, y)\}$ will already be there. However this is not done here, because it would effect the order of edge storage and the applicability of some efficiencies noted later.

In the construction of our first definition for biconnected components the existence of a test for articulation points is assumed, and is expressed as follows:

$$\begin{aligned} \text{art}(p, y, x) &= 1 \text{ if } y \text{ is an articulation point with respect to } p_{\text{last}} \text{ and } x \\ &= 0 \text{ otherwise} \end{aligned}$$

Definition 5.5.6: Biconnected Components—with art(p, y, x) Assumed Available

$$\left[\begin{array}{ll} h(p, y, x) = \{(y, x)\} & :x \in \{p\} \\ h(p, y, x) = \bigcup_{i=1}^{N(x)} [h(p\|\langle y\rangle, x, n_i(x))] \cup \{(y, x)\} & \\ & :\sim \text{art}(p, y, x) \text{ and } x \notin \{p\} \\ \left\{\begin{array}{l} h(p, y, x) = \{\ \} \\ \text{OUT}(p, y, x) = \bigcup_{i=1}^{N(x)} [h(p\|\langle y\rangle, x, n_i(x))] \cup \{(y, x)\} \end{array}\right\} & \\ & :\text{art}(p, y, x) \\ \text{Initially } p = \langle\ \rangle,\ y = \text{zilch},\ x = \text{a vertex} = \text{"begin" in } G. & \end{array}\right.$$

Upon improving the argument structure as in the articulation point definition, we get

Definition 5.5.7: Biconnected Components—with art(PY, j, pth, x) Assumed Available and an Improved Argument Structure

$$\left[\begin{array}{ll} h(\text{PY}, j, \text{pth}, x) = \{(\text{pth}_{\text{last}}, x)\} \qquad :\text{PY}[x] > 0 & \\ h(\text{PY}, j, \text{pth}, x) = \bigcup_{i=1}^{N(x)} [h(\text{PY}[x] \leftarrow j+1, j+1, \text{pth}\|\langle x\rangle, n_i(x))] & \\ \qquad\qquad \cup \{(\text{pth}_{\text{last}}, x)\} & \\ \qquad\qquad\qquad :\sim \text{art}(\text{PY}, j, \text{pth}, x) \text{ and } \text{PY}[x] = 0 & \\ \left\{\begin{array}{l} h(\text{PY}, j, \text{pth}, x) = \{\ \} \\ \text{OUT}(\text{PY}, j, \text{pth}, x) = \bigcup_{i=1}^{N(x)} [h(\text{PY}[x] \leftarrow j+1, j+1, \text{pth}\|\langle x\rangle, n_i(x))] \\ \qquad\qquad \cup \{(\text{pth}_{\text{last}}, x)\} \end{array}\right\} & \\ \qquad\qquad\qquad :\text{art}(\text{PY}, j, \text{pth}, x) & \\ \text{Initially PY} = 0^n,\ j = 0,\ \text{pth} = \langle\ \rangle,\ x = \text{a vertex in } G,\ G \text{ has } n \text{ vertices.} & \end{array}\right.$$

Eventually we compute the predicate *art* using Definition 5.5.3. Definition 5.5.7 and 5.5.3 both have the same argument structure (PY, j, pth, x), which changes in the same way in their recursive lines. They both depend on similar terminal and nonterminal conditions. They differ in that Definition 5.5.7 has one more terminal line. Their structures are sufficiently similar to allow *jamming* (Section 3.10). Jamming makes computation of art(PY, j, pth, x) possible using the recursive function f. The resultant definition has a two-component output, $\langle f(\text{PY}, j, \text{pth}, x), h(\text{PY}, j, \text{pth}, x)\rangle$. It will be represented by the interspersed definitions for the two components. The lines are grouped with the condition (terminal or nonterminal) of their applicability given on the first line of the group.

Definition 5.5.8: Biconnected Components—with Information for art(PY, j, pth, x) Included

$$\left\{\begin{array}{l} f(\text{PY}, j, \text{pth}, x) = \text{PY}[x] \\ h(\text{PY}, j, \text{pth}, x) = \{(\text{pth}_{\text{last}}, x)\} \end{array}\right\} :\text{PY}[x] > 0$$

$$\left\{\begin{array}{l} f(\text{PY}, j, \text{pth}, x) = \text{MIN}_{i=1}^{N(x)}[f(\text{PY}[x] \leftarrow j+1, j+1, \\ \qquad\qquad \text{pth}\|\langle x\rangle, n_i(x))] \\ h(\text{PY}, j, \text{pth}, x) = \bigcup_{i=1}^{N(x)} [h(\text{PY}[x] \leftarrow j+1, j+1, \\ \qquad\qquad \text{pth}\|\langle x\rangle, n_i(x))] \cup \{(\text{pth}_{\text{last}}, x)\} \end{array}\right\}$$

$$:\sim \text{art}(\text{PY}, j, \text{pth}, x) \text{ and } \text{PY}[x] = 0$$

$$\left\{\begin{array}{l} f(\text{PY}, j, \text{pth}, x) = \text{MIN}_{i=1}^{N(x)}[f(\text{PY}[x] \leftarrow j+1, j+1, \\ \qquad\qquad \text{pth}\|\langle x\rangle, n_i(x))] \\ h(\text{PY}, j, \text{pth}, x) = \{\ \} \\ \text{OUT}(\text{PY}, j, \text{pth}, x) = \bigcup_{i=1}^{N(x)} [h(\text{PY}[x] \leftarrow j+1, j+1, \\ \qquad\qquad \text{pth}\|\langle x\rangle, n_i(x))] \cup \{(\text{pth}_{\text{last}}, x)\} \end{array}\right\}$$

$$:\text{art}(\text{PY}, j, \text{pth}, x) \text{ and } \text{PY}[x] = 0$$

Initially PY $= 0^n$, $j = 0$, pth $= \langle\ \rangle$, $x =$ vertex begin in G, G has n vertices.

If $|\text{pth}| > 1$ then:

$$\text{art}(\text{PY}, j, \text{pth}, x) = \begin{cases} 1 & \text{if } f(\text{PY}, j, \text{pth}, x) \geq \text{PY}[\text{pth}_{\text{last}}] \\ 0 & \text{otherwise} \end{cases}$$

Although the lines of Definition 5.5.8 are in f, h groups, it should be remembered that each group is equivalent to a single line for a two component vector valued function whose argument structure is common to both f and h. Therefore the control flow of Algorithm 5.5.2 for articulation point is adequate for joint implementation of f and h. Additions produce the information associated with OUT. Unlike the f part, the h part of the definition has $\{(\text{pth}_{\text{last}}, x)\}$ as an argument of $\cup$ on the right. This corresponds to $p(X)$ in the generic unnested definition. At each point Algorithm 5.5.3 is designed to accommodate the component function f, h, or OUT having the most stringent requirements. Thus the algorithm pushes edge $(\text{pth}_{\text{last}}, X)$ into list Z in the upper (first-child) WHILE DO because h's $p(X)$ term requires this, even though neither f nor OUT contains a $p(X)$ term.

Note that, as with the test for articulation point, the algorithm does not work properly at the root. Also, some edges will be encountered in both directions and so will be stored twice. These difficulties are easily alleviated, as discussed in Sections 5.5.1 and 5.5.2.

Algorithm 5.5.3: Depth-First Biconnected Component Algorithm Modifications to Algorithm 5.5.2

```
V ← ⟨ ⟩
Z ← ⟨ ⟩
DCL(PY[n], vector)INIT(0)
J ← 0
PTH ← ⟨ ⟩
X, S ← 'begin'
FOR ever DO
  WHILE(PY[X] = 0) DO
    V ← psh- ∞
    Z ← psh- {(PTH[last], X)}
    PY[X] ← J + 1
    J ← J + 1
    PTH ← psh- X
    X ← n1(X)
  END
  IF PY[X] < PY[PTH[last]] THEN DO
    V[1] ← min(V[1], PY[X])
    Z[1] ← u(Z[1], {(PTH[last], X)})          Z ← psh- {(PTH[last], X)}
  END
  WHILE(X = last-neigh(PTH[last]))DO
    FX ← pop- V
                                              AEDGE ← (PTH[last], X)
    HX ← pop- Z
    X ← pop- PTH
    IF X = S THEN DONE
    V[1] ← min(V[1], FX)
    IF FX < PY[PTH[last]] THEN
```

```
                                     ←   OUTX ← { }
                                     ←   EDGE ← pop- Z
   Z[1] ← u(Z[1],HX)                 ←   WHILE EDGE ≠ (AEDGE) DO
   ELSE OUTX ← u(Z[1],HX)            ←     OUTX ← OUTX u EDGE
                                     ←     EDGE ← pop- Z
                                     ←   END
                                     ←   OUTX ← OUTX u AEDGE
  END
  X ← next-neigh(PTH[last], X)
END
```

Notes

1. On the left we have a straightforward plug-in into Algorithm 4.2.6 of Definition 5.5.8. This requires the implementation of the storage and union of sets in list Z. To implement this efficiently, an appropriate data structure is needed. A linked list for each set in Z with pointers to the start and end of each list would make the union operation efficient (constant time). However, the order of edge storage allows the simplification of 2 below.
2. On the right, a simpler technique for storing and retrieving edges is given. As on the left, edges are pushed into Z when first encountered. When backup occurs without an articulation point having been found, no operation corresponding to the set union operation on the left is necessary. When an articulation point is detected during backup, a little loop puts all edges in the stack Z down to (PTH[last], X) (PTH[last] is the articulation point) into biconnected component OUTX. Thus a result equivalent to that on the left is achieved without using an elaborate data structure.

A nested version of the biconnected definition, Definition 5.5.8, is obtained by applying the transformations given in Theorem 2.6.3 and noting the cumulative changes in PY and j and the extra line to check the second occurrence of a vertex. All of these things have been done previously for the articulation point f part of the definition; g is the nested version. (An extra terminal line is added because the definition of h has an additional line, and because we are interested in making distinctions unnecessary when only articulation points are sought. We use the version originally discussed following Definition 5.5.5.) f and h are transformed to get the g and e in the nested definition.

Definition 5.5.9: Nested Definition for Biconnected Components—Interspersed Form

$$\left[\begin{array}{l} \left\{\begin{array}{l} g(W,\mathrm{PY},j,\mathrm{pth},x) = \langle \min(W,\mathrm{PY}[x]),\mathrm{PY},j\rangle \\ e(T,\mathrm{PY},j,\mathrm{pth},x) = \langle T \cup \{(\mathrm{pth}_{\mathrm{last}},x)\},\mathrm{PY},j\rangle \end{array}\right\} \\ \qquad\qquad\qquad\qquad :\mathrm{PY}[\mathrm{pth}_{\mathrm{last}}] < \mathrm{PY}[x] \text{ and } \mathrm{PY}[x] > 0 \end{array}\right.$$

$$\left\{\begin{array}{l} g(W, \mathrm{PY}, j, \mathrm{pth}, x) = \langle W, \mathrm{PY}, j \rangle \\ e(T, \mathrm{PY}, j, \mathrm{pth}, x) = \langle T, \mathrm{PY}, j \rangle \end{array}\right\} : \mathrm{PY}[\mathrm{pth}_{\mathrm{last}}] > \mathrm{PY}[x] > 0$$

$$\left\{\begin{array}{ll} g(W, \mathrm{PY}, j, \mathrm{pth}, x) = & [\min](W, g(g(\ldots g(g(\infty, \mathrm{PY}[x] \leftarrow j+1, j+1, \\ & \mathrm{pth}\|\langle x\rangle, n_1(x)), \mathrm{pth}\|\langle x\rangle, n_2(x)), \ldots \\ & \mathrm{pth}\|\langle x\rangle, n_{N(x)}(x))) \\ e(T, \mathrm{PY}, j, \mathrm{pth}, x) = & [\cup](T, e(e(\ldots e(e(\{\ \}, \mathrm{PY}[x] \leftarrow j+1, j+1, \\ & \mathrm{pth}\|\langle x\rangle, n_1, (x)), \mathrm{pth}\|x, n_2(x)), \ldots \\ & \mathrm{pth}\|\langle x\rangle, n_{N(x)}(x)), \{(\mathrm{pth}_{\mathrm{last}}, x)\})) \end{array}\right\}$$

$$: \mathrm{PY}[x] = 0 \text{ and } \sim \mathrm{art}(\mathrm{PY}, j, \mathrm{pth}, x)$$

$$\left\{\begin{array}{ll} g(W, \mathrm{PY}, j, \mathrm{pth}, x) = & [\min](W, g(g(\ldots g(g(\infty, \mathrm{PY}[x] \leftarrow j+1, j+1, \\ & \mathrm{pth}\|\langle x\rangle, n_1(x)), \mathrm{pth}\|x, n_2(x)), \ldots \\ & \mathrm{pth}\|\langle x\rangle, n_{N(x)}(x))) \\ e(T, \mathrm{PY}, j, \mathrm{pth}, x) = & \langle\{\ \}, \mathrm{PY}, j\rangle \\ \mathrm{OUT}(\mathrm{PY}, j, \mathrm{pth}, x) = & [\cup](T, e(e(\ldots e(e(\{\ \}, \mathrm{PY}[x] \leftarrow j+1, j+1, \\ & \mathrm{pth}\|\langle x\rangle, n_1(x)), \mathrm{pth}\|\langle x\rangle, n_2(x)) \ldots \\ & \mathrm{pth}\|\langle x\rangle, n_{N(x)}(x)), \{(\mathrm{pth}_{\mathrm{last}}, x)\})) \end{array}\right\}$$

$$: \mathrm{PY}[x] = 0 \text{ and } \mathrm{art}(\mathrm{PY}, j, \mathrm{pth}, x)$$

Initially $W = \infty, T = \{\ \}, \mathrm{PY} = 0^n, j = 0, \mathrm{pth} = \langle\ \rangle$,

x = vertex 'begin' in G, G has n vertices.

If $|\mathrm{pth}| > 1$:

$$\mathrm{art}(\mathrm{PY}, j, \mathrm{pth}, x) = 1 \text{ if } g(\mathrm{PY}, j, \mathrm{pth}, x) \geq \mathrm{PY}[\mathrm{pth}_{\mathrm{last}}]$$
$$= 0 \text{ otherwise.}$$

To express the jammed definition B as a single, uninterspersed definition, the [min] operation of Definition 5.5.5, which has W as an argument, must be combined with a union operation $[\cup]$ with T as an argument. The jammed primitive function $[m - \cup]$, given below, combines these two. It takes two arguments, the first an ordered pair and the second an ordered quadruple, and returns an ordered quadruple as follows:

$$[m - \cup](\langle x_1, y_1\rangle, \langle x_2, y_2, z_2, w_2\rangle) = \langle \min(x_1, x_2), (y_1 \cup y_2), z_2, w_2\rangle$$

Definition 5.5.10: Nested Definition for Biconnected Components

$$B(\langle W, T, \mathrm{PY}, j\rangle, \mathrm{pth}, x) = \langle \min(w, \mathrm{PY}[x]), T \cup \{(\mathrm{pth}_{\mathrm{last}}, x)\}, \mathrm{PY}, j\rangle \quad :\mathrm{PY}[\mathrm{pth}_{\mathrm{last}}] < \mathrm{PY}[x] \text{ and } \mathrm{PY}[x] > 0$$

$$B(\langle W, T, \mathrm{PY}, j\rangle, \mathrm{pth}, x) = \langle W, T, \mathrm{PY}, j\rangle \quad :\mathrm{PY}[\mathrm{pth}_{\mathrm{last}}] < \mathrm{PY}[x] < 0$$

$$B(\langle W, T, \mathrm{PY}, j\rangle, \mathrm{pth}, x) = [m - \cup](\langle W, T\rangle, B(B(\ldots B(B(\infty, \{\ \}, \mathrm{PY} \leftarrow j+1, j+1, \mathrm{pth}\|\langle x\rangle, n_1(x)), \mathrm{pth}\|\langle x\rangle, n_2(x)), \ldots, \mathrm{pth}\|\langle x\rangle, n_{N(x)}(x)), \langle\infty, \{(\mathrm{pth}_{\mathrm{last}}, x)\}\rangle) \quad :\mathrm{PY}[x] = 0 \text{ and } \sim \mathrm{art}(\mathrm{PY}, j, \mathrm{pth}, x)$$

$$\left\{\begin{array}{ll} B(\langle W, T, \mathrm{PY}, j\rangle, \mathrm{pth}, x) = \langle A, \{\ \}\rangle & \\ (*) & \text{where } A \text{ is the first component of} \\ & [m - \cup](\langle W, \{\ \}\rangle, B(B(\ldots B(B(\infty, \{\ \}, \\ & \mathrm{PY} \leftarrow j+1, j+1, \mathrm{pth}\|\langle x\rangle, n_1(x)), \\ & \mathrm{pth}\|\langle x\rangle, n_2(x)), \ldots, \mathrm{pth}\|\langle x\rangle, n_{N(x)}(x)), \\ & \langle\infty, \{(\mathrm{pth}_{\mathrm{last}}, x)\}\rangle) \\ \mathrm{OUT}(\mathrm{PY}, j, \mathrm{pth}, x) & = \text{the second component of } (*) \end{array}\right\} \quad :\mathrm{PY}[x] = 0 \text{ and } \mathrm{art}(\mathrm{PY}, j, \mathrm{pth}, x)$$

Initially $W = \infty, T = \{\ \}, \mathrm{PY} = 0^n, j = 0, \mathrm{pth} = \langle\ \rangle,$
x = vertex 'begin' in G, G has n vertices,

If $|\mathrm{pth}| > 1$

$\mathrm{art}(\mathrm{PY}, j, pth, x) =$ 1 if the first component of $B(\langle W, T, \mathrm{PY}, j\rangle, \mathrm{pth}, x)$ is greater than or equal to $\mathrm{PY}[\mathrm{pth}_{\mathrm{last}}]$

$=$ 0 otherwise.

5.6 STRONGLY CONNECTED COMPONENTS IN A DIGRAPH

[Tarjan 83; Aho, Hopcroft, Ullman 75]

Undirected graphs have connected components, digraphs have *strongly connected components*. In this section, unless explicitly denied, graphs, paths, and cycles are directed.

Two vertices, x and y, in digraph G are *mutually accessible* if $x = y$ or there is a path from x to y *and* a path from y to x. The relation of mutual accessibility is an equivalence relation, and thus it partitions the set of vertices of G into equivalence classes. Each subgraph consisting of all the vertices in an equivalence class together with all edges from G connecting those vertices is a *strongly connected component* of G. Consequently every vertex is in exactly one, whereas an edge may not be in any strongly connected component.

An alternative equivalent definition is: S is a strongly connected component of G if and only if

1. There is a path between every pair of vertices in S, and
2. There is no vertex x which is in G but not in S such that $S \cup \{x\}$ has property 1.

From these definitions it follows that the strongly connected component which includes vertex x is the set of all vertices reachable from x, Fx, intersected with the set of all vertices that can reach x, Rx. Since both of these sets can be determined by the reachability algorithm, this may be an approach to a strongly connected components algorithm. Such an approach has the drawback that although Fx and Rx may both be large, requiring considerable time to accumulate, their intersection may be small, so that it may be necessary to repeat the process many times. A development analogous to that for biconnected components leads to a more efficient algorithm.

In the following development we assume that all vertices in the digraph, G, can be reached from the "begin" vertex. This is not necessarily true. To make the development generally applicable it is necessary either

1. To add an artificial "begin" vertex that reaches every vertex in G, or
2. To reenter the algorithm at an unreached vertex as necessary.

5.6.1 Development of Strong Component Definition in Terms of First Representative

Let p be a path of length zero or more, starting from an arbitrarily chosen vertex in G. x is the *first representative of a strongly connected component with respect to a path* p if and only if no return vertex of a trail in Trail($p \| \langle x \rangle$) is in p.

Lemma 5.6.1: x is a first representative of strongly connected component S with respect to p if and only if x is the first vertex in $p \| \langle x \rangle$ in the strongly connected component S.

PROOF: Assume that x is the first vertex in $p \| \langle x \rangle$ in the strongly connected component S. Then no member of p is in S. Thus none is the return vertex of a trail in Trail($p \| \langle x \rangle$), and so x is a first representative. This follows since if y in p were such a return vertex, there would be a directed path from x to y, and since, running through p, there also is a directed path from y to x, x and y would be in the same strongly connected component.

If on the other hand x is a first representative and no member of p is a return vertex of a trail in Trail($p \| \langle x \rangle$), no member of p can be in the same strongly connected component as x, so x is the first member of S to be found among the set of vertices in $p\|\langle x\rangle$. ■

This characterization of the first representative of a strongly connected component with respect to a path p is similar to that of an articulation point, and there is an analogous recursive definition for identifying first representatives similar to Definition 5.5.2. Differences are related to the three vertex definitions of articulation points as contrasted to the vertex-path definition for first representative.

The key lies in detecting when no member of p is a return vertex of a trail in Trail($p\|\langle x\rangle$). $f(p, x, j)$ is defined for that purpose. As in Definition 5.5.2, the vertices in p are indexed, and j is made equal to ind(x, $p\|\langle x\rangle$) or $|p| + 1$ in the recursive line by use of an update function,

$$f(p, x, j) = \text{the minimum of}$$

1. Indices of return vertices of trails in Trail($p\|\langle x\rangle$), and
2. j (= ind(x, $p\|\langle x\rangle$)).

j is incremented by 1 when p is extended by concatenation on the right of the recursive line in $f(p, x, j)$. j gives a bound on the smallest index of a vertex in p reached from $p\|\langle x\rangle$.

Note that, when used in conjunction with directed graphs, the primitive function $n_i(x)$ is the ith *outward* neighbor of x.

Definition 5.6.1: Strongly Connected Components—First Representative 1

$$\left[\begin{array}{ll} f(p, x, j) = \text{ind}(x, p) & :x \in \{p\} \\ f(p, x, j) = j & :x \notin \{p\} \text{ and } N(x) = 0 \\ f(p, x, j) = \text{MIN}_{i=1}^{N(x)}[f(p\|\langle x\rangle, n_i(x), j+1)] & :x \notin \{p\} \text{ and } N(x) \neq 0 \\ \text{Initially } p = \langle\,\rangle,\ x = \text{vertex in } G \text{ (directed)},\ j = 1. & \end{array}\right.$$

Proposition 5.6.1: Interpretation Rule for Definition 5.6.1
The vertex x is a first representative of a strongly connected component with respect to p if and only if the function $f(p, x, j) \geq j$.

An algorithm similar to Algorithm 5.5.1 could be given to implement Definition 5.6.3 and identify first representatives. More useful, however, is one we now develop, which produces the strongly connected components signaled by these representatives.

We augment Definition 5.6.1, which detects the first representatives of strongly connected components, with one to identify the sets of vertices which constitute the strongly connected components. This is done analogously to the way that the biconnected component definition was developed from that for articulation points.

First we recognize that

Lemma 5.6.2: Let x be the first representative of strongly connected component S with respect to path p, and Q the set of vertices (including x) reachable along trails starting with $p\|\langle x\rangle$ other than those in p. If no member of Q other than x is a first representative, then $S = Q$.

PROOF: Consider an O network path through descendants of (the argument structure containing) x up to, but not including, the first argument structure containing a vertex which starts a *new strongly connected component*. In fact S consists of all vertices found in descendants of x along such network paths. If y is a descendant of x reached by the network path $p\|\langle x\rangle\|q\|\langle y\rangle$, then y is certainly reached by a path from x in the graph. Since y is not the first member of another strongly connected component, a return vertex of a trail in $\text{Trail}(p\|\langle x\rangle\|q\|\langle y\rangle)$, say y^1, must be in $\langle x\rangle\|q$. y^1 is either x, in which case y is clearly in S, or it is descendant of x in q. But, like y, y^1 is not the first member of a strongly connected component. Let the path from x to y^1, a prefix of q, be q^1, that is, $p\|\langle x\rangle\|q^1\|\langle y^1\rangle$ is the path to y^1, $|q^1| < |q|$. A return vertex of a trail in $\text{Trail}(p\|\langle x\rangle q^1\|\langle y^1\rangle)$, say y^2, must be in $\langle x\rangle\|q^1$. Again this is either $x = y^2$, in which case y can reach x via y^1 and the argument is complete, or y^2 is a descendant of x. Note that y^1 is closer to x along p than y. The argument can be applied analogously to y^2, and the path q^2 from x to y^2, then, if necessary, to $y^3, \ldots,$ then to y^4, and so on. Eventually, for some finite, p, y^p is x, thus concluding the argument. ■

Based on this lemma we define $g(p, x, j)$ whose value is the set of vertices on trails of $\text{Trail}(p\|\langle x\rangle)$ and not in p, which also are not yet in a strongly connected component. $g(p, x, j)$ is jammed with $f(p, x, j)$, Definition 5.6.1, for the first representative to give Definition 5.6.2 for strongly connected components. We need the following function for that definition.

$$\text{rep}(p, x, j) = \begin{cases} 1 & \text{if } x \text{ is the first representative of a strongly connected component, that is, if } f(p, x, j) \geq j \\ 0 & \text{otherwise} \end{cases}$$

As in the previous section, the output function OUT is shown only when output is not empty.

Definition 5.6.2: Strongly Connected Components With Information for rep(p, x, j) Included

$$\begin{cases} \begin{Bmatrix} f(p,x,j) = \text{ind}(x, p\|\langle x\rangle) \\ g(p,x,j) = \{\ \} \end{Bmatrix} & : x \in \{p\} \\ \begin{Bmatrix} f(p,x,j) = j \\ g(p,x,j) = \{\ \} \\ \text{OUT}(p,x,j) = \{x\} \end{Bmatrix} & : x \notin \{p\} \text{ and } N(x) = 0 \\ \begin{Bmatrix} f(p,x,j) = \text{MIN}_{i=1}^{N(x)}[f(p\|\langle x\rangle, n_i(x), j+1)] \\ g(p,x,j) = \bigcup_{i=1}^{N(x)} [g(p\|x, n_i(x), j+1)] \cup \{x\} \end{Bmatrix} & : x \notin \{p\} \text{ and } N(x) > 0 \text{ and } \sim \text{rep}(p,x,j) \\ \begin{Bmatrix} g(p,x,j) = \{\ \} \\ \text{OUT}(p,x,j) = \bigcup_{i=1}^{N(x)} [g(p\|x, n_i(x), j+1)] \cup \{x\} \end{Bmatrix} & : x \notin \{p\} \text{ and } N(x) > 0 \text{ and rep}(p,x,j) \end{cases}$$

Initially $p = \langle\ \rangle$, x = vertex in G, $j = 1$.

$$\text{rep}(p,x,j) = \begin{cases} 1 & \text{if } f(p,x,j) \geq j \\ 0 & \text{otherwise} \end{cases}$$

5.6.2 Comparability in Strongly Connected Component Definition

In evaluating Definition 5.6.2 comparable argument structures are generated. Unlike the articulation point case, comparability here is for a *jammed* definition, which forms strongly connected components in conjunction with detecting first representatives. Both parts play a role in this comparability. As in the articulation point definition, comparability is position dependent, giving depth-first implementation its significance. In this case comparable argument structures are those with the same second (or vertex) component, though their other components may differ. Let (p_1, x, j_1) be the first, and (p_2, x, j_2) the second occurrence of two comparable argument structures, ordered by their occurrence in a depth-first implementation. When (p_2, x, j_2) is encountered, $f(p_1, x, j_1)$ has already been evaluated. Unlike the evaluation in the articula-

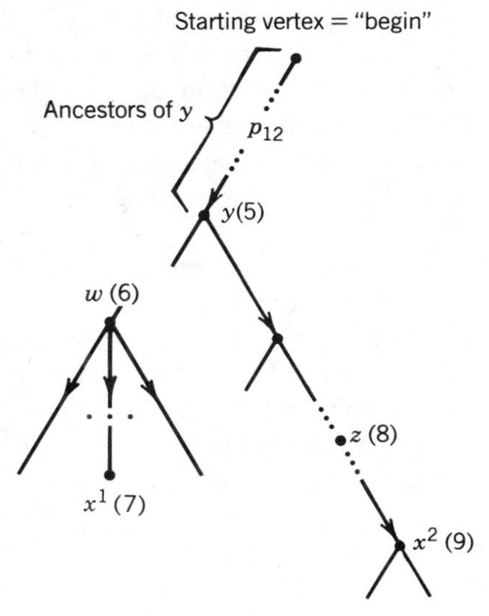

FIGURE 5-9 Comparability in strongly connected component definition.

tion point case, $f(p_2, x, j_2)$ must be evaluated (at least partially) even though f has already been evaluated for an argument structure with the identical vertex component. Fortunately the earlier evaluation of $f(p_1, x, j_1)$ and $\mathrm{OUT}(p_1, x, j_1)$ is useful in doing so.

Theorem 5.6.1 describes in detail how this comparability can be used for pruning in a depth-first implementation of Definition 5.6.2. Figure 5-9 is used extensively in the proof of the theorem.

Theorem 5.6.1: Consider a depth-first evaluation of Definition 5.6.2 (see Figure 5-9) in which x is the first vertex to reoccur in a nonterminal argument-structure. (p_1, x^1, j_1) represents the first and (p_2, x^2, j_2) the second occurrences of x in argument structures of the O-network. y is the youngest common ancestor of x^1 and x^2, and z is *any* vertex on the abbreviated O-network from y to x^2. Pruning of (p_2, x^2, j_2) together with all its descendants will not affect the value of $f(p', y', x')$ for any argument structure (p', y', x') remaining in the O network, provided that

1. If x has previously been included in a strongly connected component, then the algorithm behaves as though there were no argument structure (p_2, x^2, j_2), and
2. If x has not previously been included in a strongly connected component, then z is included in a strongly connected component with x and y, and otherwise the algorithm proceeds as if (p_2, x^2, j_2) were not there.

PROOF: As in the proof of Theorem 5.5.2, we refer here to the abbreviated O network which was first defined for Definition 5.5.1. The same definition holds for Definition 5.6.1. Referring to Figure 5-9, let w be the child of y on the network path from y to x^1. The removal of x^2 and its descendants could affect f computed at all ancestors of x^2. However, y and its ancestors are not affected by such a removal because of the presence of x^1, guaranteeing that all descendants of y remain its descendants. However, removal of x^2 still leaves the value of f for each vertex z on the path from y to x^2. z cannot be a return vertex of any trail from x^2, that is, a trail in Trail($p_2 \| \langle x^2 \rangle$), because z would then also be a return vertex on a trail from x^1 and so z, not x, would be the first vertex to recur—a contradiction. Therefore z cannot be reached from any of the vertices in the network proposed for pruning. Still y or one of y's ancestors (all of which are ancestors of z) may be a return vertex on a trail from x^1 and thus on a trail from x^2. *If* x^1 *is in a strongly connected component*, that is, if it or any of its ancestors was already found to be a first representative, then there is clearly no trail from x^2 on which z or any of its ancestors could be a return vertex, so trails from x^2 need not be pursued and argument structure (p_2, x^2, j_2) can be ignored. (Since this would mean that z and x were in the same strongly connected component, but since this is the first occurrence of z, x could not already have been placed in a strongly connected component). On the other hand, *if* x^1 *is not yet included in a strongly connected component*, that is, neither it nor any of its ancestors was already found to be a first representative, then no z can be a first representative since it must be on a cycle through x^2 through an ancestor of x^1, thence through y or an ancestor of y (otherwise x would already be in a strongly connected component). ■

Notice that if x is the first recurring vertex, then there is no path in the graph from x to vertices on the O network path from y to x^2, but it does not follow that x^2 is a child of y as in Theorem 5.5.2. This is because we are dealing here with *directed* graphs. On the other hand we are implicitly using the fact that the set of vertices surrounding the argument structure containing x^1 includes the set of all descendants of that containing x^2. This was shown in Theorem 5.5.1, which does refer to directed graphs. These comparisons are useful in understanding the conditions under which comparability is order dependent, as well as how it is affected by whether the graphs are directed or undirected. This determines which implementations are applicable. Although, unlike the articulation point problem, the strongly connected component problem is on a directed graph, comparability is between vertices that "first recur" in both. Thus a breadth-first algorithm is not, and the depth-first algorithm, Algorithm 4.2.6, is applicable to the strongly connected component problem, provided we make use of comparability in a straightforward way.

It is not clear at this point how the theorem can be used to guide pruning. On encountering x^2 it is easy to determine whether x is already in a strongly connected component. But how will we indicate that vertices like z on the path from y to x^2 are not to be first representatives when x is not already in a strongly connected component, without pursuing (p, x^2, j_2)? We want to do this by assigning an appropriate value to $f(p_2, x^2, j_2)$. The value of j_2 may be $>$, $=$, or $<$ than $f(p_1, x^1, j_1)$ if x is not yet in a strongly connected component, so straightforward use of $f(p_1, x^1, j)$ is not possible. However it can be shown that the minimum value of $f(p, r, j)$ for all vertices r, encountered since the first occurrence of x, can be used in place of $f(p_2, x^2, j_2)$. An alternative approach, used in the classical strongly connected component algorithm, replaces the indexing by a cumulative indexing as in the articulation point development. However, cumulative indexing cannot be expressed in a nonnested recursive definition. Before considering its use we introduce some simple improvements in the first representative definition.

5.6.3 Improvement in the Strongly Connected Components Definition

As in developing the articulation point algorithm, we can eliminate computation of the ind function by keeping array PY with PY[x] the index of x in $p\|\langle x\rangle$. The index $j = \text{ind}(p_{\text{last}}, p\|\langle x\rangle) + 1$ is similar, though one larger, than that used in the articulation point definition. To emphasize the relation between the definitions we have also changed the name of the path p to pth.

Definition 5.6.3: Strongly Connected Components, First Representative II

$$
\begin{aligned}
f(\text{PY}, j, \text{pth}, x) &= \text{PY}[x] && :\text{PY}[x] > 0 \\
f(\text{PY}, j, \text{pth}, x) &= j && :\text{PY}[x] = 0 \text{ and } N(x) = 0 \\
f(\text{PY}, j, \text{pth}, x) &= \text{MIN}_{i=1}^{N(x)}[f(\text{PY}[x] \leftarrow j, j+1, \text{pth}\|\langle x\rangle, n_i(x)] \\
& && :\text{PY}[x] = 0 \text{ and } N(x) \neq 0
\end{aligned}
$$

Initially $\text{PY} = 0^n$, $\text{pth} = \langle\ \rangle$, x = vertex begin in G, G has n vertices.

A similar improvement in the arguments of the other functions appearing in the jammed Definition 5.6.2 gives the following:

$$
\text{rep}(\text{PY}, j, \text{pth}, x) = \begin{cases} 1 & \text{if } x \text{ is the first representative of a strongly connected component } [f(\text{PY}, j, \text{pth}, x) \geq j] \\ 0 & \text{otherwise} \end{cases}
$$

Definition 5.6.4: Strongly Connected Components with Information for rep(PY, j, pth, x) Included

$$\begin{cases} \begin{Bmatrix} f(\text{PY}, j, \text{pth}, x) = \text{PY}[x] \\ g(\text{PY}, j, \text{pth}, x) = \{\ \} \end{Bmatrix} & :\text{PY}[x] > 0 \\ \begin{Bmatrix} f(\text{PY}, j, \text{pth}, x) = j \\ g(\text{PY}, j, \text{pth}, x) = \{\ \} \\ \text{OUT}(\text{PY}, j, \text{pth}, x) = \{x\} \end{Bmatrix} & :\text{PY}[x] = 0 \text{ and } N(x) = 0 \\ \begin{Bmatrix} f(\text{PY}, j, \text{pth}, x) = \text{MIN}_{i=1}^{N(x)}[f(\text{PY}[x] \leftarrow j, j+1, \langle x\rangle \| p\text{th}, n_i(x))] \\ g(\text{PY}, j, \text{pth}, x) = \bigcup_{i=1}^{N(x)} [g(\text{PY}[x] \leftarrow j, j+1, \langle x\rangle \| p\text{th}, n_i(x))] \\ \qquad \cup \{x\} \end{Bmatrix} & \\ \qquad :\text{PY}[x] = 0 \text{ and } N(x) > 0 \text{ and } \sim \text{rep}(\text{PY}, j, \text{pth}, x) & \\ \begin{Bmatrix} f(\text{PY}, j, \text{pth}, x) = \text{MIN}_{i=1}^{N(x)}[f(\text{PY}[x] \leftarrow j, j+1, \langle x\rangle \| p\text{th}, n_i(x))] \\ g(\text{PY}, j, \text{pth}, x) = \{\ \} \\ \text{OUT}(\text{PY}, j, \text{pth}, x) = \bigcup_{i=1}^{N(x)} [g(\text{PY}[x] \leftarrow j, j+1, \langle x\rangle \| p\text{th}, n_i(x))] \\ \qquad \cup \{x\} \end{Bmatrix} & \\ \qquad :\text{PY}[x] = 0 \text{ and } N(x) > 0 \text{ and rep}(\text{PY}, j, \text{pth}, x) & \end{cases}$$

Initially pth = $\langle\ \rangle$, PY = 0^n, $j = 1$, x = vertex "begin" in G, G has n vertices.

$$\text{rep}(\text{PY}, j, \text{pth}, x) = \begin{cases} 1 & \text{if } f(\text{PY}, j, \text{pth}, x) = j \\ 0 & \text{otherwise.} \end{cases}$$

5.6.3.1 The Strongly Connected Component Algorithm

The implementation of jammed Definition 5.6.4 is based on depth-first Algorithm 4.2.6. Modifications are needed.

1. To account for the comparability proven in Theorem 5.6.1.
2. To number the index j of x in pth$\|\langle x\rangle$ recorded in PY *cumulatively*, as was done for the articulation point algorithm, and as is required to take advantage of comparability. This also records vertices that have been processed, and results in a terminal condition adjustment as described in Section 5.5.2.3. This modification is necessary because cumulative indexing cannot be expressed in the nonnested recursive definition. Furthermore since j is not backed up, it will not necessarily have the index of vertex x(X in the alogithm). This index is in PY[X].

3. With indexing made cumulative, it becomes legitimate (referring to Figure 5-9) to assign j_1 to $f(p_2, x^2, j_2)$ and otherwise ignore argument structure (p_2, x^2, j_2) when vertex x recurs and x is not already assigned to a strongly connected component. By doing this we guarantee that the return vertex found for every vertex z on path p_2 from y to x^2 of that figure will be less than the index assigned to that vertex, so none will be a first representative. Furthermore for y and its ancestors this index assignment will have no effect on return vertices, which is legitimate since at this point the effect of trails from x^1 is already in force. In Figure 5-9 we show possible values of j (in parentheses) for a cumulative indexing. Note that $j_1 = 7$, $j_2 = 9$. Suppose $f(p_1, x, j_1) = 5$, then x is not included in a strongly connected set. When $f(p_2, x, j_2)$ is assigned the value $j_1 = 7$, then neither x^2 nor z can become a first representative.

Algorithm 5.6.1 works only for a graph G with no neighborless vertex, that is, no vertex without outward neighbors. The algorithm then implements Definition 5.6.4, where lines applicable when $N(x) = 0$ are never invoked. If a graph has neighborless vertices, then adding an edge from each such vertex to itself (or from each such vertex to an artificial "sink" vertex with a loop to itself) will give a graph with the same strong components on which the algorithm works. Furthermore if a vertex in the given graph cannot be reached from the "begin" vertex, then it cannot be placed in any strongly connected component. This problem can be remedied by adding an additional "source" vertex to the given graph with edges to all other vetices and making it the "begin" vertex for the álgorithm.

Algorithm 5.6.1: Depth-First Strongly Connected Component Algorithm
PTH is a path vector in reverse order—new vertices are added on its left. The last vertex in the represented path is PTH[last]. WL and TL are lists. An entry in WL is a number requiring a fixed amount of storage per entry. An entry in TL is a set and so may require an amount of storage per entry, which varies widely. This can be handled with a list of pointers from which members of the corresponding set are chained.

Plug into Algorithm 4.2.6	*Modifications for Comparability for Accumulating Strongly Connected Components*

```
WL, TL ← ⟨ ⟩
DCL(PY[n]) INIT(0)
J ← 1
PTH ← ⟨ ⟩
X ← 'begin'
FOR ever DO
  WHILE (PY[X] = 0) DO
    WL ← psh- ∞
    TL ← psh- { }
    PTH ← psh- X
```

```
    PY[X] ← J
    X ← n1(X)
    J ← J + 1
  END
  ***************                          ← IF CC[X] = 0 THEN DO
  WL[1] ← min(WL[1], PY[X])
  ***************                          ← END
  WHILE(X = last-neigh(PTH[1])) DO
    ARG ← pop- WL
    CMP ← pop- TL
    X ← pop- PTH
    FXWL ← min(PY[X], ARG)
    FXTL ← CMP ∪ {X}
    IF X = 'begin' THEN DONE
    WL[1] ← min(WL[1], FXWL)
    IF FXWL ≥ J
    THEN DO
    ******                                 ← FOR EACH X in FXTL DO
    ******                                 ←    CC[X] ← 1
    ******                                 ←    END
    OUT ← FXTL
    END
    ELSE TL[1] ← ∪(TL[1], FXTL)
   END
   X ← next-neigh(PTH[1], X)
 END
```

As with the articulation point algorithm, and for the same reason, the number of O network block outputs simulated is $O(e)$. The internal complexity associated with each block output is similarly a constant. In addition, output of the strongly connected components involves outputting each vertex once, which is $O(n)$.

Furthermore, because pruning and cumulative indexing can be expressed in a nested recusive definition, the strongly connected component algorithm can be implemented directly with a nested recursive definition. The development is analogous to that for biconnected components and is not given here.

5.7 TWO-PERSON ZERO-SUM LOOPLESS GAMES [Nilsson 80, sec. 3.4; Knuth 75]

We have seen how the existence of comparable argument structures in the O network can contribute to the design of an efficient algorithm. An even simpler form of comparability, with similar uses, can occur between outputs in the w network. For example, if w is the AND function, its value is determined to be 0 as soon as one of its arguments is known to be 0—no more arguments need be examined. The OR function has a similar property with the roles of 1 and 0 interchanged. A more complex example is provided by the function $-(\min(a, b, \ldots))$. The values of certain compositions of this function are independent of some of its arguments, which is significant since it is the w function in the recursive definition used here in evaluating zero-sum two-person games. We will study the properties of compositions of the $-$min function in the context of a study of efficient evaluation of such games.

Consider a game with two alternately moving players. The player who moves in state s is *mover*(s), the other is *nonmover*(s). Mover(s) may choose any of a finite number $m(s)$ of moves. A move changes the state of the game. If the game state is s, then mover(s) can change the game to any of the states $n_i(s)$, $i = 1$ to $m(s)$. The *starting state* is the set of conditions under which the game is to start. No sequence of states through which a game can pass as a result of legitimate moves can contain two occurrences of the same state. For each starting state, there is a finite number of states the game can possibly reach. Thus every sequence of moves must bring the game to a final state in which no further move is allowed. Each final state s has a numerical value for mover(s) designated val(s); its value for the other player is $-$val(s). Given the value of the game to the mover in every possible final state, the value of the game to the mover in every other state can be computed. The value of the game in state s is $g(s)$. This equals the value of the game to mover(s) in the final state at which the game eventually arrives if each player plays so as to maximize the value to himself or herself of the final state. Since the sum of the values of the game to both players in any state is zero, the value of the game to nonmover(s) is $-g(s)$.

For the class of games having the properties given above we now formulate a recursive definition for $g(s)$.

5.7.1 Definition for the Game Analysis

The best move for mover(s) is the one that puts the game in a state $n_a(s)$ whose value to the opponent $= g(n_a(s))$ is the smallest over all states $n_j(s)$; $j = 1$ to $m(s)$. If that is done, then the resultant value of the game, $g(s)$, to mover(s) is $-g(n_a(s)) = -(\min(g(n_j(s))))$; $j = 1$ to $m(s)$. The w function for the recursive line is the locally associative function $-(\min(x_1, \ldots, x_n))$, abbreviated $\text{nmin}(x_1, \ldots, x_n)$. (It is interesting to note that $-(\min(x_1, \ldots, x_n)) = \max(-x_1, \ldots, -x_n)$, sometimes used as an alternative. Both of these formulations of the w function are based on viewing the value of the game alternately at each move from the point of view of a different player, namely, the mover. An alternative is to always view the game from the same player, say A's, point of view. In this alternative the w function alternates between max and min, respectively, as A or A's opponent is the mover.)

Enough has been said to justify the following definition.

Definition 5.7.1: Two-Person Zero-Sum Game

$$\left[\begin{array}{ll} g(s) = \text{val}(s) & :s \text{ is final} \\ g(s) = \text{nmin}\big(g(n_1(s)), \ldots, g(n_{m(s)}(s))\big) & :s \text{ is not final} \\ \text{Initially } s \text{ is the starting state of the game.} & \end{array}\right.$$

We assume the definition has a uniform inverse (it can always be altered to have one). Thus the implementation can be based on a uniform inverse depth-first algorithm. The function nmin is locally associative (see Section 4.2), and so Algorithm 4.2.6 is applicable. But nmin has an additional useful property called *truncatability*, which allows an augmentation of Algorithm 4.2.6 that will often result in pruning.

5.7.2 Property Allowing Simplification of Game Definition — Truncatability of nmin — Alpha – Beta Property

Consider a game whose O–w–t network, corresponding to Definition 5.7.1, is that given in Figure 5-10. As in any such network, the O network part builds all game states that can be reached by legitimate sequences of player moves from the initial state to a terminal state, while the w network computes the value of each state X as a function (nmin) of the game values of its children if

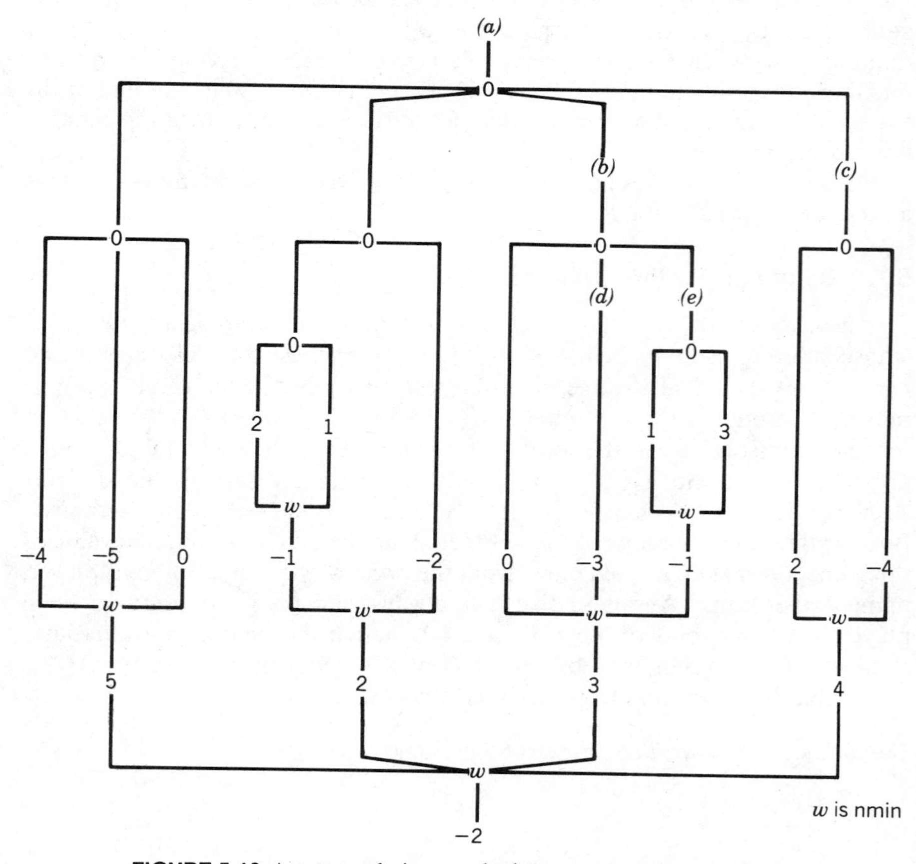

FIGURE 5-10 A game analysis—standard O–w–t network representation.

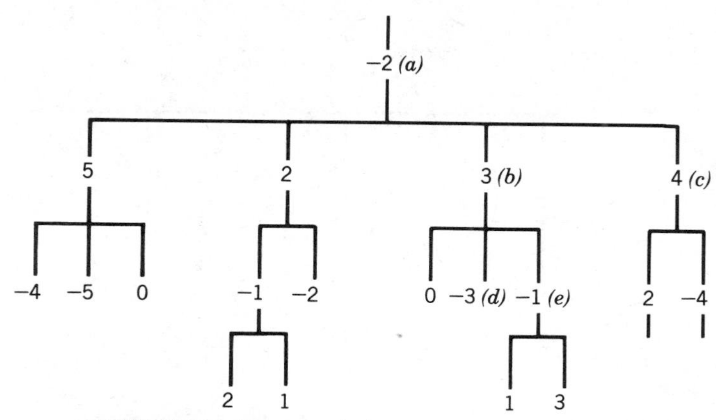

FIGURE 5-11 A game analysis—game tree representation.

X is nonterminal, or as a given final state value if X is terminal. (The t functions are omitted.)

The traditional way to represent the O–w–t for a game is the *game tree*. The game tree is the result of amalgamating the O and w networks, the w network is inverted and superimposed on the O network. Each node in the game tree has a two-part label.

1. A game state (the O network part). The root node is labeled with the initial state of the game, each other node n at an ith child is labeled with the state resulting from the ith move from the state represented at the parent of n.
2. The value to the mover in that state of the game (the inverted w network part) derived by propagation via the nmin function of values of terminal game states upward from the leaves of the game tree.

Figure 5-11 gives the game tree which is equivalent to the O–w–t network of Figure 5-10.

Let us examine the evaluation of the node labeled a.

1. The value at b, 3, has no effect on a, because a lower value, 2, has already been calculated among a's children, and the nmin function will "choose" that lower value. In other words, 3 is *greater than the current partial minimum of a's children*.
2. That last statement is true exactly because a value whose negative is greater than or equal to the current partial minimum of a's children exists among b's children. If b's children are evaluated left to right, then the value at e need not be known. And correspondingly, the entire subtree rooted at e in the O network need not be expanded.

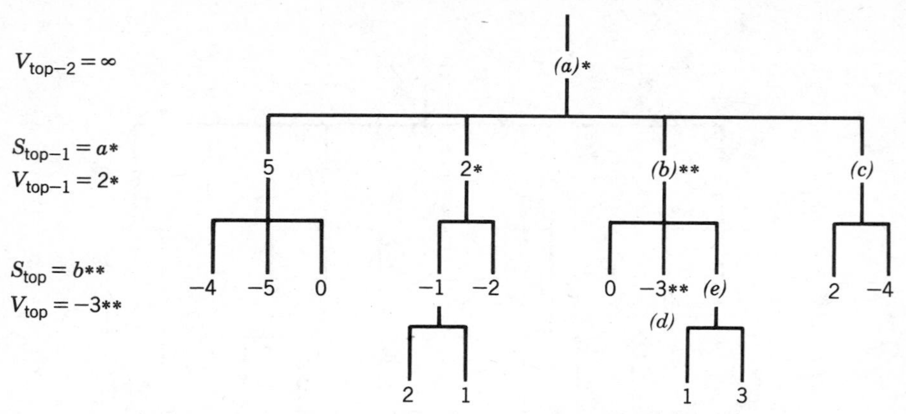

FIGURE 5-12 V list just before evaluation of node e; [condition (Cut) is met].

This analysis does not depend on the particular game represented, and thus is a result of properties of the w function nmin. We now develop these properties algebraically.

In Section 4.2.2.3 we traced the contents of the V list and considered its interpretation as an expression of a partially developed composition of w functions, called an *interpreted V list*. (A review of that development will help in what follows.) When Algorithm 4.2.6 is run with the locally associative function nmin, the interpretation of the contents of an n-member V list on entering the upper WHILE is

$$\text{nmin}(V_{\text{top}-n+1}, \ldots, \text{nmin}(V_{\text{top}-1}, \text{nmin}(V_{\text{top}}$$

It may be helpful to notice that $V_{\text{top}-i}$ in the interpreted V list contains the so far computed minimum of the values of the game state at the ith level above the level of X in Algorithm 4.2.6. We digress briefly to give a more detailed interpretation of the interpreted V list in game terms.

The game tree and the corresponding V list entries in Figure 5-12 should make this correspondence clear. On the left it shows the V list entries $V_{\text{top}-i}$ together with the same state, $S_{\text{top}-i}$, whose value that entry bounds, when the depth-first game tree analysis has progressed to state e, as shown on the right. Notice that $S_{\text{top}-i}$ is the state of the game corresponding to the parent of the value of X (in Algorithm 4.2.6) when the V list is popped and $V_{\text{top}-i}$ becomes the top of the V list. (In Figure 5-12, S_{top} is the game state at b and $S_{\text{top}-1}$ is the state at a.) Then $V_{\text{top}-i}$ is the value of the game to the nonmover in state $S_{\text{top}-i}$ based upon complete evaluation of 0 or more children game states of $S_{\text{top}-i}$. Notice it is the *non*mover in the above proposition, since the nonassociative part n of the nmin function has yet to be applied on $V_{\text{top}-i}$. *The value of the game in state* $S_{\text{top}-i}$ *cannot be any greater than* $V_{\text{top}-i}$ *to the nonmover*$(S_{\text{top}-i})$ because the mover in that state can already move to force the nonmover's to be $V_{\text{top}-i}$ and would only choose another move if it could lower that value. Similarly *the value of the game cannot be any less than* $-V_{top-i}$ to the mover in state S_{top-i}.

Returning now to the interpreted V list,

$$\text{nmin}(V_{top-n+1}, \ldots, \text{nmin}(V_{top-1}, \text{nmin}(V_{top}$$

As terminal and backed up values are minimized into V_{top}, V_{top} must finally become $V_{top} - e$ for some $e \geq 0$. The evolution of the interpreted V-list is

1. $\text{nmin}(V_{top-n+1}, \ldots, \text{nmin}(V_{top-1}, \text{nmin}(V_{top}$ eventually becoming
2. $\text{nmin}(V_{top-n+1}, \ldots, \text{nmin}(V_{top-1}, -(V_{top} - e)$, which equals
3. $\text{nmin}(V_{top-n+1}, \ldots, \text{nmin}(V_{top-1}, -V_{top} + e$

Now if

$$(\text{CUT})\ V_{top-1} \leq -V_{top}$$

then $-V_{top} + e \geq V_{top-1}$ and line 3 equals

4. $\text{nmin}(V_{top-n+1}, \ldots, \text{nmin}(V_{top-1}$

Thus, as soon as condition (CUT) becomes true, the V list corresponding to line 1 may be backed up to that corresponding to line 4. This is the algebraic statement we have been looking for. [It is possible to argue the same result in terms of the game interpretation of the V list, as follows. (Consider as an example Figure 5-12, where $S_{top} = b$, $S_{top-1} = a$.)

$$\text{value to nonmover } (S_{top-i}) \leq V_{top-i};$$

in particular, true if $i = 1$.

But the state S_{top} is a possible next state from state S_{top-1}, so: *value* to nonmover(S_{top-1}) $\leq$ value to mover(S_{top}). Therefore *value* to nonmover(S_{top-1}) $\leq$ minmax(V_{top-1}, final value to mover(S_{top})). Also we know that

$$\begin{aligned} \text{final value to mover } (S_{top}) &\leq -V_{top} \\ &= -(V_{top} - e), \text{ so:} \end{aligned}$$

$$\text{value to nonmover } S_{top-1} \leq \min(V_{top-1}, -V_{top} + e)$$

$$\text{and if } V_{top-1} \leq -V_{top},$$

$$\text{then value to nonmover } (S_{top-1}) \geq V_{top-1}$$

and is independent of value to mover(S_{top}).]

Even if condition (CUT) is not met, it may still be possible to back up, due to what is called the *deep cutoff* conditions. An argument justifying deep cutoff is outlined in Figure 5-13, in which the following notation is used. V_{top-i} is the

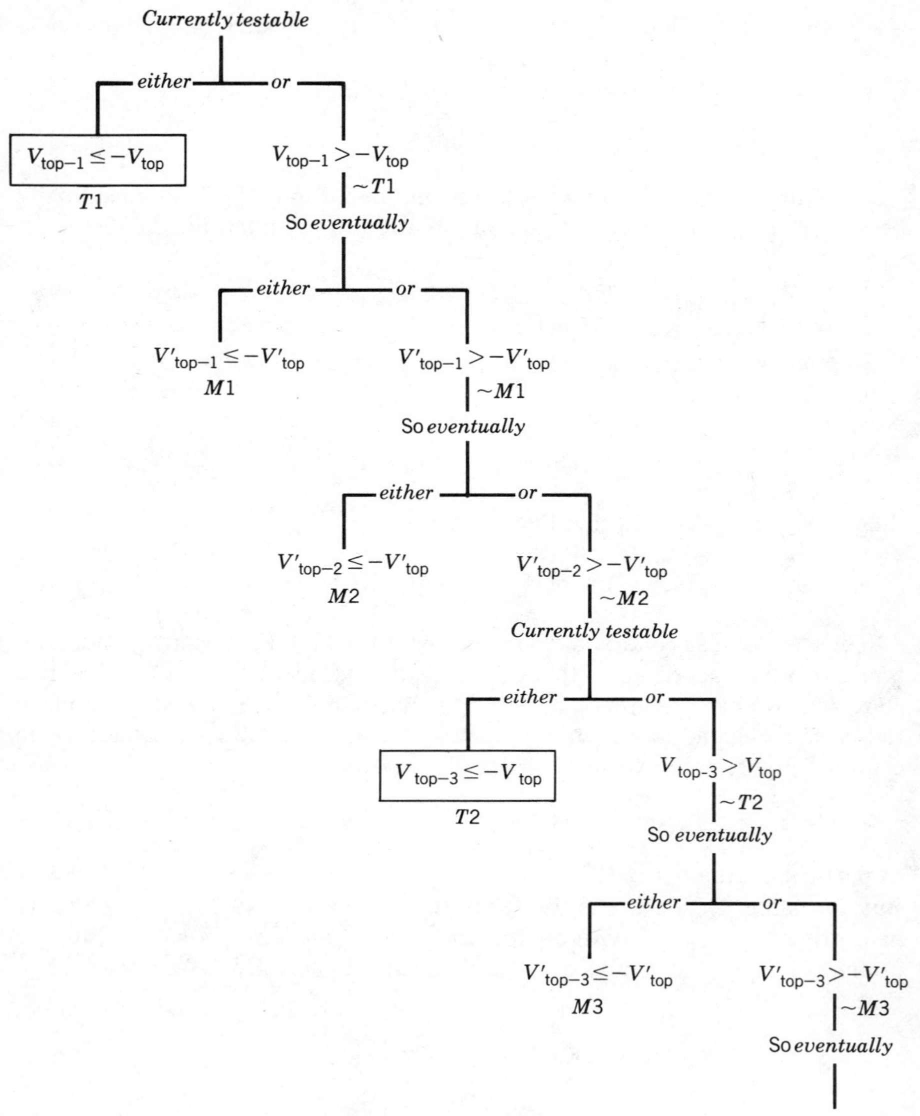

FIGURE 5-13 Argument for deep cutoff.

current value in stack V at position i down from the top. V'_{top} is the value at the top of stack V when all values that are to be, have been minimized into the top, that is, $V'_{top} = V_{top} - e$. It follows that $V'_{top} \leq V_{top}$. V'_{top-i} is the minimum of V_{top-i} together with all values that are to be minimized into that position, other than the one which depends on the current value of V_{top}. More precisely, V'_{top-1} is defined as follows. Let nmin$(V'_{top}) = U_0$. When the V list is backed up, U_0 must be minimized with V_{top-1}, as must several other arguments, say $U_1, \ldots, U_m$. That is, after U_0 is formed, the interpreted V list will have a top $-$ 1 entry, which is the result of the expressions $\text{nmin}(V_{\text{top}-1}, U_0, U_1, \ldots, U_m)$. We define $V'_{\text{top}-1}$ to be

$$V'_{\text{top}-1} = \min\big((V_{\text{top}-1}), U_1, \ldots, U_m\big), \text{ so}$$

$$V_{\text{top}-1} = \text{nmin}\big(V'_{\text{top}-1}, U_0\big)$$

$V'_{\text{top}-i}$ is the generalization of the above definition. The expression that determines the value at top $- i$ when that value has become the top of the interpreted V list and is ready to be backed up is given by the expression

$$\text{nmin}\big(V_{\text{top}-i}, \underline{\text{nmin}\big(, \ldots, \text{nmin}(V'_{\text{top}}), \ldots, \big)} U_1, \ldots, U_m\big)$$

Let U_0 be the underlined portion of the expression so we have

$$\text{nmin}\big(V_{\text{top}-i}, U_0, U_1, \ldots, U_m\big)$$

By definition, $V'_{\text{top}-i} = \min\big((V_{\text{top}-1}), U_1, \ldots, U_m\big)$

So the top $- i$ entry will become $\text{nmin}(V'_{\text{top}-i}, U_0)$ Clearly $V'_{\text{top}-i} \leq V_{\text{top}-i}$.

Refer to Figure 5-13 starting at the top. Given the V list we can make tests whose outcome may result in popping the top entry without having to examine many of the argument structures. First we can compare $V_{\text{top}-1}$ and $-V_{\text{top}}$: *either* the first is less than or equal to the second ($T1$) *or* it is not ($\sim T1$). The first of these outcomes ($T1$) means we can pop the top entry. It is testable, and if it is true, then popping is legitimate. For that reason it is displayed boxed. The next place in the figure testability occurs is where $V_{\text{top}-3}$ is compared with $-V_{\text{top}}$. None of the intervening conditions, although they exhaust the possibilities, can actually be tested. (It is because all possibilities are exhausted, and all would result in popping, that we can say that popping is legitimate if $V_{\text{top}-3} \leq -V_{\text{top}}$.) Returning to the first comparison, if $V_{\text{top}-1} > -V_{\text{top}}$, then after the top value is completed and becomes V'_{top} it will *eventually* be compared with $V'_{\text{top}-1}$, containing all contributions to the eventual value at position top $-$ 1 except $-V'_{\text{top}}$. The two possible outcomes of this comparison are labeled $M1$ and $\sim M1$. The figure is further developed in this way. The boxed conditions are the only ones shown that are actually testable. If any such test succeeds, then the interpreted V list can be *backed up* to give

$$\ldots \text{nmin}\big(V_{\text{top}-3}, \text{nmin}\big(V_{\text{top}-2}, \text{nmin}\big(V_{\text{top}-1}$$

We cannot back up any further because we do not know precisely which of the intervening nontestable conditions were met. At each *either–or* branch one or the other of the conditions must be met. Though we do not know which, we are guaranteed that when a *testable* condition $T1$ or $T2$ is met, the test process can end with a backup of the top entry of the V list. All possibilities are included in the paths which terminate at this testable condition or above it.

Although whether any of the untestable conditions on these paths are met is unknown, a backup of the top entry is consistent with all possibilities. The reasoning is based on Figure 5-13. We can see from Figure 5-13 that $T1$ or $\sim T1$ is *true*, and $T1$ is *Testable*.

If $\sim T1$ is *not true*, then $T1$ is *true* and we can back up.

If $\sim T1$ is *true*, then $M1$ or $\sim M1$ must be true, so

$M1$ or $\sim M1$	is *true*
$M1$ or ($M2$ or $\sim M2$)	is *true*
$M1$ or ($M2$ or ($T2$ or $\sim T2$))	is *true*
$M1$ or $M2$ or $T2$ or $\sim T2$	is *true*, and $T2$ is *testable*.

If $\sim T2$ is *not true*, then ($M1$ or $M2$ or $T2$) is *true*, and since $M1$, $M2$, and $T2$ each allow backup, we can back up.

This form of reasoning would continue as follows.

If $\sim T2$ is *true*, then since $M3$ or $\sim M3$ must be *true*,

$M1$ or $M2$ or $M3$ or $\sim M3$	is *true*
$M1$ or $M2$ or $M3$ or $M4$ or $T3$ or $\sim T3$	is *true*.

If $\sim T3$ is *not true*, then ($M1$ or $M2$ or $M3$ or $M4$ or $T3$) is *true*, and since $M1$, $M2$, $M3$, $M4$, and $T3$ each allow backup we can back up.

This argument can be extended to every entry in the V lists following the given pattern, using Figure 5-13, with the conclusion that V list entry positions 1, 3, 5, ... from the top are testable. Thus,

$$(\text{CUT}')\ V_{\text{top}-i} \leq -V_{\text{top}} \qquad i \text{ is odd and } i < n$$

The top entry on the interpreted V list can be backed up if this condition is met for any i.

It follows that

$$(\text{CUT}'')\ \text{Min}_{i \text{ odd},\, i<n}\left[V_{\text{top}-i}\right] \leq -V_{\text{top}}$$

The left side of (CUT″) can be computed in an update fashion in coordination with the update of the V list. For that purpose define another list called *BETA* as follows.

***Definition** 5.7.2*

$$\text{BETA}_{\text{top}} = \text{MIN}_{i \text{ odd},\, i<n}\left[V_{\text{top}-i}\right]$$

It follows that

$$\text{BETA}_{\text{top}} = \min\left(V_{\text{top}-1}, \text{MIN}_{i \text{ odd},\, i<n}\left[V_{\text{top}-i-2}\right]\right)$$

So we have

$$\text{BETA}_{\text{top}} = \text{Min}\left(V_{\text{top}-1}, \text{BETA}_{\text{top}-2}\right)$$

Therefore, in the V-BETA Algorithm 5.7.1, when an ∞ is pushed onto the V list to form a new V_{top}, and a new BETA_{top} is correspondingly formed, it is accomplished as follows:

$$\text{BETA} \leftarrow \text{psh-Min}(V_{top}, \text{BETA}_{top-1})$$

$$V \leftarrow \text{psh-}\infty$$

In terms of BETA the condition (CUT″) under which the top can be backed up is

Proposition 5.7.1: Cutoff or Pruning Condition for Game Definition 5.7.1
The V list can be backed up if

$$\text{BETA}_{top} \leq -V_{top}$$

It is instructive to interpret BETA and Proposition 5.7.1 in terms of the game tree. If $\text{BETA} \leq -V_{top}$ then the V-list can be backed up, which means S_{top} will not be entered. But $\text{BETA} = V_{top-i} \geq$ value to nonmovers(S_{top-i}) for some odd i. Therefore if BETA holds, an upper bound for nonmover(S_{top-i}) for some odd i is less than or equal to $-V_{top}$, the lower bound for mover(S_{top}), and S_{top} will not be entered.

Let X be mover(S_{top-i}) for i odd; the same player makes all these moves. X wishes to keep the value to the other player Y as small as possible and knows how to keep it $\leq$ BETA. But if X allows Y into state S_{top}, Y can get a value for the game which is $\geq -V_{top}$. Since $-V_{top} \geq$ BETA, X will not let the game into state S_{top}.

5.7.3 Augmenting Game Analysis Algorithm to Incorporate Speedups Made Possible by Alpha – Beta Property

Algorithm 5.7.1 is an adaptation of Algorithm 4.2.6, implementing the game definition and augmented to take advantage of properties of nmin. This allows backing up the push-down lists early, which is equivalent to pruning the simulated O–w–t network. The BETA list, containing the minimum of odd entries in the V lists, is added and is pushed/popped whenever and wherever the V list is pushed/popped in Algorithm 4.2.6. We now consider in detail the opportunities for backup at critical points during the run of Algorithm 4.2.6, and the augmentations and changes needed at those points.

5.7.3.1 Notes on Augmenting Algorithm 4.2.6 to Include Cutoff

Focus attention on the activity at the top of the interpreted V and BETA lists. Assume the top few entries in each to be

$$\ldots \text{nmin}(V_{top-2}, \text{nmin}(V_{top-1}, \text{nmin}(V_{top},$$

$$\ldots\ \text{BETA}_{top-2},\ \text{BETA}_{top-1},\ \text{BETA}_{top},$$

1. Assume that $\text{BETA}_{\text{top}} > -V_{\text{top}}$, that is, cutoff condition, proposition 5.7.1, is not met. Suppose now that a next argument of the top nmin in the interpreted V list, a', is received. a' then becomes the value of the game to mover(X). This event could occur in Algorithm 4.2.6 either at the point at which a terminal value is processed, or just after backup in the lower DO WHILE. If $a' \geq V_{\text{top}}$, then all remains unchanged. If, however, $a' < V_{\text{top}}$, then a' replaces V_{top}. This would happen in Algorithm 4.2.6 even without alteration, since V_{top} would be replaced with $\min(V_{\text{top}}, a')$. However, now we wish to detect any replacement in V_{top} since it may result in BETA_{top} no longer being $> -V_{\text{top}}$ $(= -a')$ now. If $\text{BETA}_{\text{top}} >$ new value of $-V_{\text{top}}$ $(= -a')$, then we are done with input a'; if not, V is backed up and its top becomes $\ldots\text{nmin}(V_{\text{top}-1}$. We already know that $\text{BETA}_{\text{top}-1} > -V_{\text{top}-1}$, for otherwise it would not be in the list. So again we are done dealing with a'.

2. If a last sibling is detected, the V list must is backed up so that $\ldots\text{nmin}(V_{\text{top}-1}, -V_{\text{top}}$ becomes the top of the interpreted V list. But the fact that V_{top} survived in the V list implies that

$$\text{BETA}_{\text{top}} > -V_{\text{top}}$$

and the definition of BETA_{top}, Definition 5.7.2, implies that $\text{BETA}_{\text{top}} \leq V_{\text{top}-1}$. Therefore

$$-V_{\text{top}} < V_{\text{top}-1}$$

Therefore $-V_{\text{top}}$ is the minimum of $-V_{\text{top}}$ and $V_{\text{top}-1}$, and the negative of this value is placed in the V list. So the top of the interpreted V list becomes $\ldots\text{nmin}(-V_{\text{top}}$. This might better be expressed as $\ldots\text{nmin}(-\text{OLD}.V_{\text{top}}$, since the new $V_{\text{top}} = -$the former value of V_{top}.

3. Although $\text{BETA}_{\text{top}-1} > -V_{\text{top}-1}$ since $-V_{\text{top}-1}$ survived, the new value replacing $V_{\text{top}-1}$ $(= -\text{OLD}.V_{\text{top}})$ is less than it. So $\text{BETA}_{\text{top}-1}$ may not be > than − this new value $[= -(-\text{OLD}.V_{\text{top}})]$. It must be tested. If $\text{BETA}_{\text{top}-1} > -(-\text{OLD}.V_{\text{top}})$, then the handling of the last sibling is complete; if not, another backup is necessary, leaving $\text{nmin}(V_{\text{top}-2}$ (with the knowledge that $\text{BETA}_{\text{top}-2} > -V_{\text{top}-2}$) on top of the interpreted V list.

4. Finally, the V and BETA lists acquire additional members when the first child of a nonterminal argument structure X, $o_1(X)$, is generated, and only then. At this time ∞ is pushed onto the V list, as prescribed by Algorithm 4.2.6, and $\min(V_{\text{top}}, \text{BETA}_{\text{top}-1})$ is pushed onto the BETA list. (For the BETA list this amounts to making $\text{BETA}_{\text{top}} = \text{Min}(V_{\text{top}-1}, \text{BETA}_{\text{top}-2})$, as required by Definition 5.7.2. For the V list this update comes directly from Algorithm 4.2.6 and the fact that $\infty = 0_{\text{min}}$. Since the value of BETA_{top} is computed from previous values of V and BETA, both of these lists are initialized with dummy ∞ entries. This allows the first entries in BETA to be made by the same computation used for later entries.

Now these alterations are incorporated directly into the uniform inverse depth-first algorithm, Algorithm 4.2.6, to give Algorithm 5.7.1.

Algorithm 5.7.1: *V*-BETA Algorithm

Because the need to back up can occur any time the top of the V list is changed, a subroutine for this purpose is given first. It also tests for termination. It returns V_{top} in ARG, called OLD.V_{top} in the previous discussion. It is similar to the first four lines of the backup part of Algorithm 4.2.6, with the popping of BETA added wherever the V list is popped. FX ← -ARG has been moved so as to give the game's value in the start state only. Due to cutoffs, it is generally incorrect elsewhere.

```
PROC: backup( )
        ARG ← pop- V
            ← pop- BETA
        X   ← parent(X)
        IF X = S THEN DO FX ← -ARG DONE END
        RETURN
  END
```

Now the main part of the algorithm is given.

```
S ← initial game state
X ← initial game state
V_top ← ∞
BETA_top-1 ← ∞
BETA_top ← ∞
FOR ever DO
  WHILE(X is not a final game state) DO
    BETA ← psh- min(V_top, BETA_top-1)
    V    ← psh-∞
    X    ← o_1(X)
  END
  IF V_top > val(X) THEN
      DO
        V_top ← val(X)
        IF BETA_top ≤ -V_top THEN backup( )
      END
  WHILE(X is the last game state that can be moved to from parent(X)) DO
    [backup( )]
  V_top ← -ARG
    IF BETA_top ≤ -V_top THEN backup( )
  END
  X ← next game state after X that can be moved to from parent(X)
END
```

Notes at the right of the algorithm:

- Beside the lines $V_{top} \leftarrow \infty$ through $BETA_{top} \leftarrow \infty$: The added initializing makes computing the first few psh's possible (see note 4 in section 5.7.3.1).
- Beside the first WHILE loop: Follows from Definition 5.7.2 first child section—updating of BETA added (see note 4).
- Beside the IF ... THEN DO ... END block: This is Algorithm 4.2.6 modified to allow check of V_{top} against BETA for possible cutoff (see note 1).
- Beside "[backup( )]" and "$V_{top} \leftarrow$ -ARG": This is almost the same as Algorithm 4.2.6 backup (see note 2).
- Beside the second "IF $BETA_{top} \leq -V_{top}$ THEN backup( )": This is added test for possible cutoff and backup (see note 3).

Example 5.7.1 shows the algorithm's operation. Here the V list and BETA are traced. V list entries are shown above the corresponding BETA list entries,

that is,

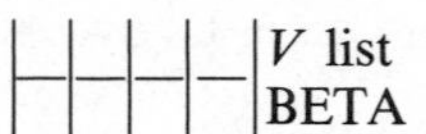

The state for which the V list entry gives the current value to the nonmover is given above entries in the lower part of the figure. The dummy ∞ entries, one in the V and two in the BETA lists, made in initialization, are not shown.

Example 5.7.1

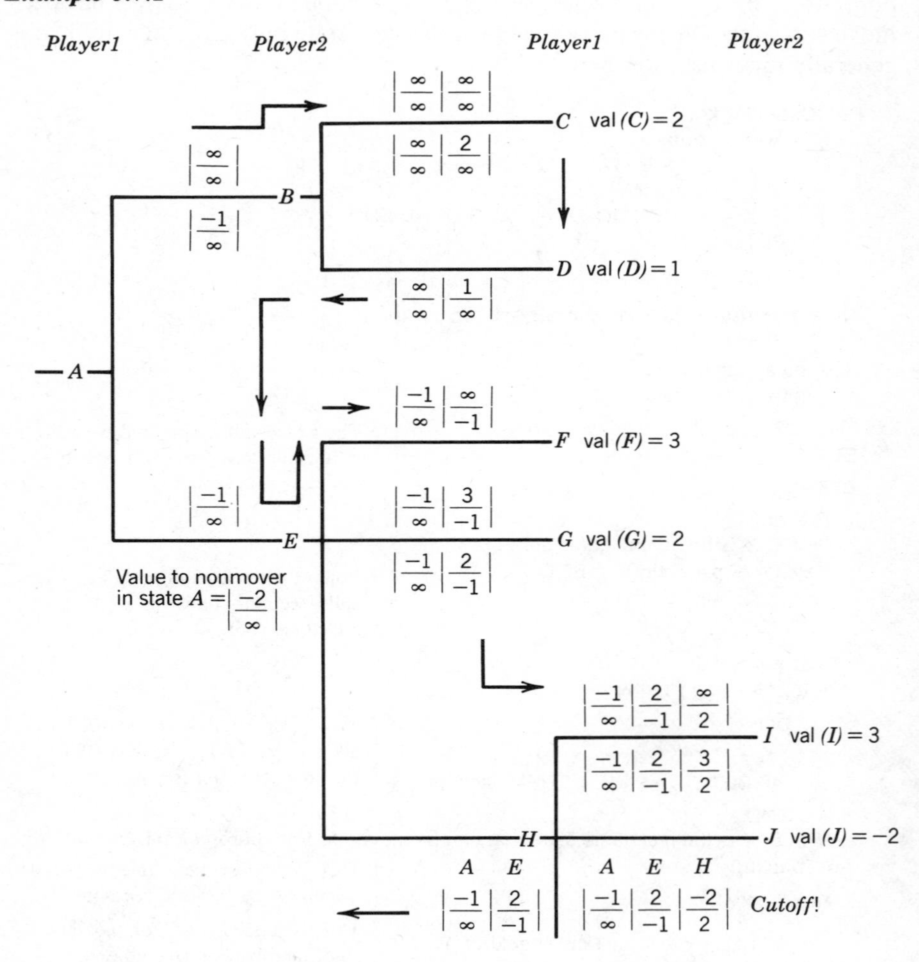

In much of the literature on game analysis, a $(-V)$ list is kept in place of the V list. At every point in the algorithm the $(-V)$ list has entries that correspond one-to-one to the entries in the V list. An entry in one list is the negative of the corresponding entry in the other. The $(-V)$ list is usually called ALPHA, and the entire algorithm written in terms of the ALPHA list is called the ALPHA–BETA algorithm.

If there is a global lower bound b on the value of a game then the minimum part of *nmin* can be evaluated as soon as one of its arguments is b. This is an additional source of pruning beyond that incorporated in the V-BETA algorithm. In the next section we consider an extreme case in which, among other restrictions, such a lower bound occurs. The general case is included as Problem 22 at the end of the chapter.

5.7.4 Evaluation of Game whose only Possible Outcome Is Win or Loss

It is instructive to consider games whose only possible outcome is win (1) or loss (-1). The algorithm in this case is similar but considerably simpler than that for the general zero-sum game.

Instead of trying to adapt the V-BETA Algorithm 5.7.1 to this case, we start from Definition 5.7.1, which only needs modification in the "initially" line to account for the restriction on outcomes. Algorithm 4.2.6 will form the basis for our implementation. We will modify it to take advantage of the properties of the nmin function in combination with restriction of the game values to 1 or -1. We trace the progress of the V list, since the algorithm works on such a game, in search of possible simplifications to Algorithm 4.2.6. This, in fact, is the same route we followed in the development of the V-BETA algorithm. In this case the resultant algorithm is a significant simplification of the V-BETA algorithm. It requires no push-down lists.

We will trace the progress of the interpreted V-list on Example 5.7.2.

Example 5.7.2

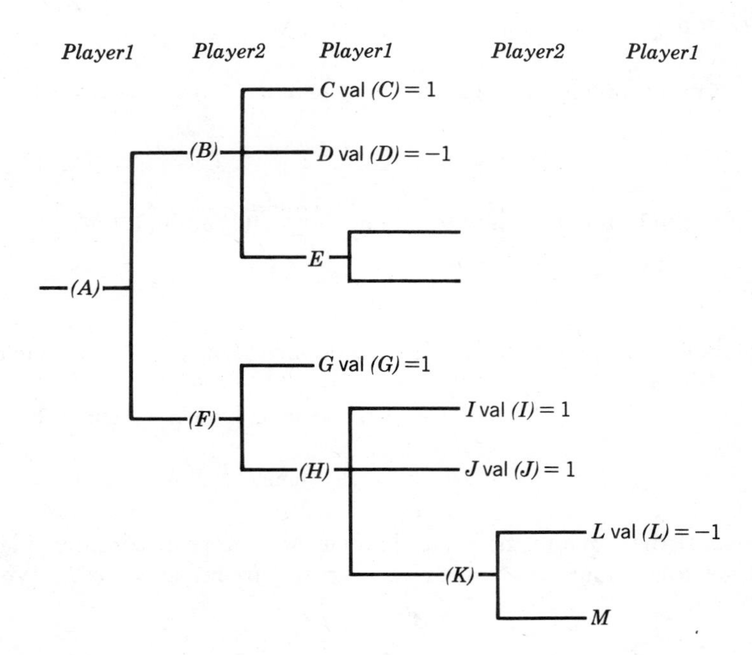

5.7.4.1 Trace of Interpreted V List in Algorithm 4.2.6 Applied to Win–Loss Game

$\text{------} a \text{---}\!\!\rightarrow$ indicates evaluation due to local associativity (taking a minimum).

$\text{------} e \text{---}\!\!\rightarrow$ indicates straightforward evaluation of nmin(x).

$\text{------} t \text{---}\!\!\rightarrow$ indicates evaluation of $\text{nmin}(-1, \ldots) = 1$.

(Once -1 becomes an argument, there can be no smaller argument.) This operation is called truncation.

Start in state A and pass through states A, B, C,

$$\text{nmin}(\infty, \text{nmin}(\infty, 1 \text{ ------ } a \text{ ----}\!\!\rightarrow \text{nmin}(\infty, \text{nmin}(1$$

Go to state D,

$$\text{nmin}(\infty, \text{nmin}(1, -1 \text{ ------ } a \text{ ----}\!\!\rightarrow \text{nmin}(\infty, \text{nmin}(-1 \text{ ------ } t \text{ ----}\!\!\rightarrow \text{nmin}(1$$

The value of the game to the player in state E or beyond need never be considered, since it cannot be lower than the -1 already found at state D. Next pass through states F, G,

$$\text{nmin}(1, \text{nmin}(\infty, 1 \text{ ------ } a \text{ ----}\!\!\rightarrow \text{nmin}(1, \text{nmin}(1$$

Then H and I,

$$\text{nmin}(1, \text{nmin}(1, \text{nmin}(\infty, 1 \text{ ------ } a \text{ ----}\!\!\rightarrow \text{nmin}(1, \text{nmin}(1, \text{nmin}(1$$

and J,

$$\text{nmin}(1, \text{nmin}(1, \text{nmin}(1, 1 \text{ ------ } a \text{ ----}\!\!\rightarrow \text{nmin}(1, \text{nmin}(1, \text{nmin}(1$$

K and L,

$$\text{nmin}(1, \text{nmin}(1, \text{nmin}(1, \text{nmin}(\infty, -1 \text{ ------ } a \text{ ----}\!\!\rightarrow \text{nmin}(1, \text{nmin}(1, \text{nmin}(1, \text{nmin}(-1$$

$$\text{------ } t \,\&\, e \text{ ----}\!\!\rightarrow \text{nmin}(1, \text{nmin}(1, \text{nmin}(1, 1$$

$$\text{------ } a \text{ ----}\!\!\rightarrow \text{nmin}(1, \text{nmin}(1, \text{nmin}(1$$

The value of the game to the player in state M was never considered since it cannot be lower than -1, which was already found at state L. We have returned to K. K is a last sibling. This is tantamount to getting ")" to close the top nmin argument.

$$\text{nmin}(1, \text{nmin}(1, \text{nmin}(1) \dashrightarrow e \dashrightarrow \text{nmin}(1, \text{nmin}(1, -1$$

$$\dashrightarrow a \dashrightarrow \text{nmin}(1, \text{nmin}(-1$$

$$\dashrightarrow t \dashrightarrow \text{nmin}(1, 1$$

$$\dashrightarrow a \dashrightarrow \text{nmin}(1$$

We have backed up to H, then to F, which is a last sibling so we back up to A and the value of the game to A is

$$\text{nmin}(1) \dashrightarrow e \dashrightarrow -1$$

The interpreted V list has a number of critical characteristics useful in an efficient implementation. Each change (except the first) finds the V list interpretation in the form

$$\text{nmin}(\infty, \text{nmin}(\infty, \ldots, \text{nmin}(\infty, \text{nmin}(1, \ldots, \text{nmin}(1, \text{nmin}(1, \text{nmin}(1$$

In what follows, however, the effect of the ∞ in an argument is identical to a 1, because 1 is the highest possible value assignable to a state. So we can assume that after each change the interpreted V list has the form

$$\text{nmin}(1, \text{nmin}(1, \ldots, \text{nmin}(1, \text{nmin}(1, \text{nmin}(1$$

At this point one of the following is possible:

1. 1 is the next outcome recorded in the interpreted V list.
   If 1 is the outcome that results from the next move in the current game state, the interpreted V list becomes

   $$\text{nmin}(1, \text{nmin}(1, \ldots, \text{nmin}(1, \text{nmin}(1, \text{nmin}(1, 1$$

   which becomes

   $$\text{nmin}(1, \text{nmin}(1, \ldots, \text{nmin}(1, \text{nmin}(1, \text{nmin}(1$$

2. If -1 is the outcome that results from the next move in the current game state, the interpreted v list becomes

   $$\text{nmin}(1, \text{nmin}(1, \ldots, \text{nmin}(1, \text{nmin}(1, -1$$

   which becomes

   $$\text{nmin}(1, \text{nmin}(1, \ldots, \text{nmin}(1, \text{nmin}(-1$$

which in one backup becomes

$$\text{nmin}(1, \text{nmin}(1, \ldots, \text{nmin}(1$$

3. The last sibling has been received and the topmost nmin is to be evaluated, in which case the interpreted V list becomes

$$\text{nmin}(1, \text{nmin}(1, \ldots, \text{nmin}(1, \text{nmin}(1, \text{nmin}(1)$$

which in one backup becomes

$$\text{nmin}(1, \text{nmin}(1, \ldots, \text{nmin}(1, \text{nmin}(1, -1$$

which becomes

$$\text{nmin}(1, \text{nmin}(1, \ldots, \text{nmin}(1, \text{nmin}(-1$$

which in a second backup becomes

$$\text{nmin}(1, \text{nmin}(1, \ldots, \text{nmin}(1$$

Notice that all these operations start with the top of the interpreted V list equal to 1 and end with 1 at the top again. Depending on the input received, 1 as in step 1, -1 as in step 2, or last sibling as in step 3, zero, or one, or two backups, respectively, are executed. In the one-backup case V_{top} becomes 1 after backup. During the two-backup case V_{top} becomes -1 after the first, then 1 after the second backup. Notice that each backup also causes a return to a previous game state. If the backup causes return to a game state S and V_{top} has just been made -1, the game is a loss, whereas if V_{top} is 1, the game is a win to the player in state S. (If S is the initial game state, the player in question is the first player.) All this can be done without maintaining any part of the V list except V_{top}. The initial argument structure or game state is recreated (by backup), and that value will determine win or loss.

These consequences for games with only win or loss as an outcome are incorporated in inverse depth-first algorithm, Algorithm 4.2.6, to give Algorithm 5.7.2.

Algorithm 5.7.2: Win–Loss (1, -1) Algorithm
$X \leftarrow$ initial game state
$S \leftarrow$ initial game state

```
FOR ever DO
  WHILE(X is not a final state of game) DO
    X ← o1(X)
  END
```

```
    If val(X) = -1 THEN DO
      X ← parent(X)
      [V ← 1]*
      IF X = S THEN DONE[1]*
      END
    WHILE(X the last state that can be reached by a move from parent(X)) DO
      X ← parent(X)
      [V ← -1]*
      IF X = S THEN DONE[ - 1]*
      X ← parent(X)
      [V ← 1]*
      IF X = S THEN DONE[1]*
    END
    X ← next game state after X that can be moved to from parent(X)
  END
```

Notes. The lines enclosed between [and]* can all be removed provided that at label DONE[1]* the cost is understood to be 1, while at DONE[− 1]* it is understood to be −1.

COMPLEXITY

The game algorithms studied here take advantage of properties common to all two-person zero-sum games. Particular games often have additional properties, such as equal or otherwise comparable states, making even greater simplification possible. The various forms of Nim provide examples of opportunities for extreme simplification, because the same state can be reached in many different ways.

The complexity of game analysis using an implementation of Definition 5.7.1 is related to the number of moves at each game state (the *branching factor*). The savings in using ALPHA–BETA is very sensitive to this number. To explore the savings achievable with ALPHA–BETA, one usually assumes a fixed branching factor B—and a fixed game length n. Under that assumption the total number of possible game states is $O(B^n)$. Assuming a straightforward implementation of Definition 5.7.1, that is, no use of ALPHA–BETA, and constant time to compute each child game state from its parent, then $O(B^n)$ is the complexity of such a game, and is an upper bound on the cost of such a game. On the other hand assume that the game with branching factor B and length n has only two possible values, 1 and -1, and we use Algorithm 5.7.2 to take full advantage of these characteristics to do the maximum number of cutoffs. Our analysis gives a lower bound on the cost of such a game, and thus on the maximum saving to be expected by using ALPHA–BETA techniques.

The two trees shown partially developed in Figure 5-14 each represent complete trees with branching factor B. The tree in Figure 5-14a has depth 2, and would have a total of $B^2 + B + 1 = O(B^2)$ nodes if no pruning had been done.

We have indicated in the tree precisely those nodes which must be visited by an ALPHA–BETA algorithm, such as Algorithm 5.7.2, for a WIN–LOSS

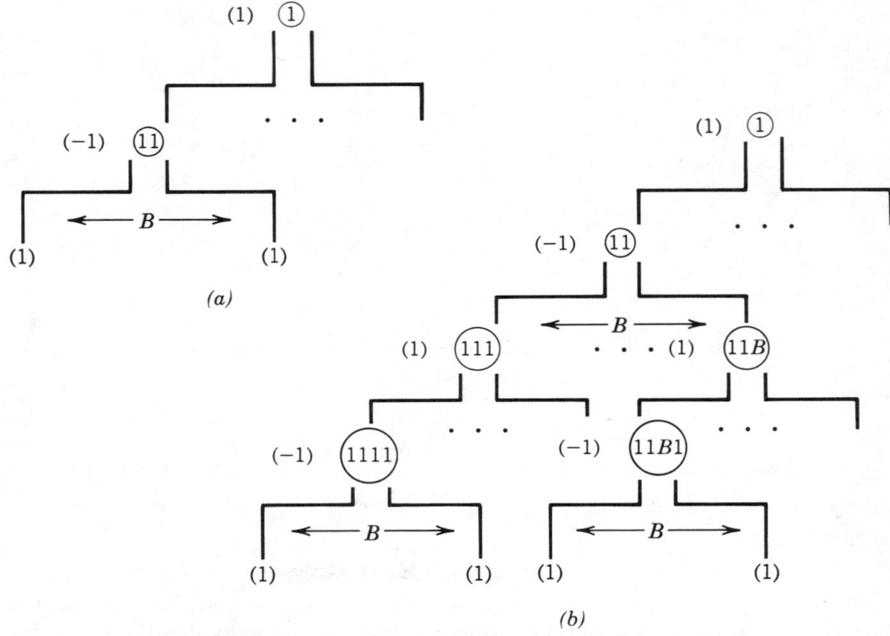

FIGURE 5-14 Best-case win–loss trees for use of ALPHA–BETA.

game, or any game with a priori upper and lower bounds on its value. ($\leftarrow B \rightarrow$) indicates that all B nodes/subtrees must be visited/processed in a similar manner. For depth 2 there are a total of $1 + 1 + B = O(B)$ such nodes.

The tree in Figure 5-14*b*, has depth 4, and if complete would have a total of $1 + B + B^2 + B^3 + B^4 = O(B^4)$ nodes, but only $1 + 1 + B + B + B^2 = O(B^2)$ of these would be visited by Algorithm 5.7.2.

We assume it is clear how these pruned trees can be generalized for any depth n with $n = 2k$, $k \geq 1$. The number of nodes for a pruned tree with depth n is given by $c(n)$.

Definition 5.7.3

$$\begin{bmatrix} c(n) = B + 2 & :n = 2 \\ c(n) = B * c(n-2) + 2 & :n = 2k,\ k > 1 \end{bmatrix}$$

This is satisfied by

$$c(n) = \frac{B^{(n/2+1)} + B^{n/2} - 2}{B - 1} = O(B^{n/2}) \text{ for } n \text{ even.}$$

So a game of n moves with branching factor B having an upper bound on the game value, which without ALPHA–BETA has complexity $O(B^n)$, will, at best, have a complexity of $O(B^{n/2})$ if ALPHA–BETA is used.

5.8 PROBLEMS

1. *Completely Prunable Definition Termination.* Any definition of $f(X)$ which is completely prunable and does not terminate can be made to terminate by adding to its argument a stack of the old argument structures.
   a. Explain how this could be done.
   b. *Simple Pruning Algorithm Generalized.* Prove that a simple pruning algorithm is adequate to implement a completely prunable definition extended to guarantee termination as in part a.
2. *Application Pruning Algorithm.*
   a. Give a completely prunable recursive definition for a function that gives a set of all vertices that can reach a given vertex "goal" in a given digraph G.
   b. Describe an efficient algorithm to find the set of all vertices in a given digraph G that can be reached from a given vertex "begin" and from which a second given vertex "goal" can be reached. Give the worst-case complexity of your algorithm.
3. *Reachability—Exercise, Trace.* This problem uses the adjacency list representation of a graph. Recall that each cell contains two entries—the first giving the identity of a vertex, the second a pointer. There is one row of cells for each vertex in the graph giving that vertex and all its outward

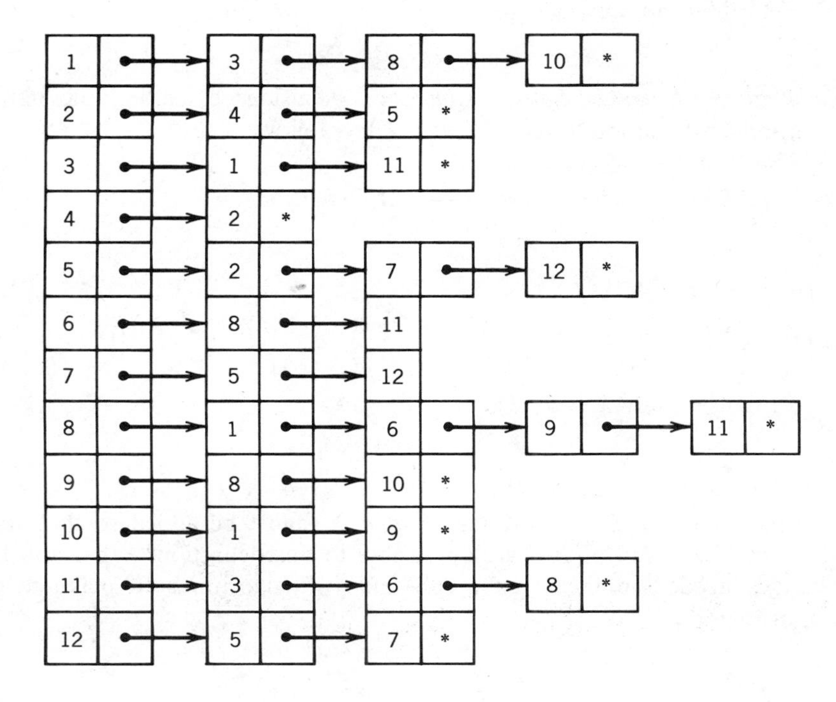

neighbors. The leftmost cell in the row contains the vertex, say x, whose neighbors are given; the pointer in that cell starts the chain leading through the remaining cells in that row which contain the identities of all the neighbors of x.

a. Let vertex 1 be the starting vertex. Trace Algorithm 5.3.1 in the version that uses TBL and does not use OUT on the graph above. Trace X, P, and TBL.

b. Give the order in which vertices are processed when working on the graph above for both Algorithms 5.2.3 and 5.2.1.

4. *Minimum Tour—A Branch and Bound Algorithm.* Write (not for running) a branch and bound program for producing the minimum tour on a given weighted graph, that is, the shortest cycle that passes through every vertex of the graph once. A minimum tour definition is given below in which the vertices are assumed to be 1 through n and the tour is thought of as starting and ending on vertex 1. (Do not worry about any comparability other than that involving the bound computed on the cost of terminal data structures, unless you wish to.)

Minimum Tour (Traveling Salesman)

$$\left[\begin{array}{ll} f(C, p, v, b) = \langle C + c_1(v), p \| \langle v \rangle \| \langle 1 \rangle \rangle & :b = \langle\ \rangle \\ f(C, p, v, b) = \text{MIN-SEL}_{n_i(v) \in b} f(C + c_i(v), p \| \langle v \rangle, n_i(v), & \\ \qquad\qquad b - n_i(v)) & :b \neq \langle\ \rangle \\ \text{Initially } p = \langle\ \rangle, C = 0, v = 1, b = \langle 2, 3, \ldots, n \rangle & \end{array}\right.$$

5. *Generalizing Minimum-Select Definitions.* Consider extending the minimum-select definition to contain $p(c, X)$ as follows.

Definition 5.8.1: Extended Minimum-Select Definitions

$$\left[\begin{array}{ll} f(c, X) = \langle c, t(c, X) \rangle & :T(c, X) \\ f(c, X) = \text{min-sel}(\text{MIN-SEL}_{i=1}^{m(c, X)} [f(u_i(c, X), v_i(c, X))], & \\ \qquad\qquad p(c, X)) & :\sim T(c, X) \\ \text{Initially } c = 0,\ X = d,\ d \in D. & \end{array}\right.$$

In which u has the property $u_i(c, X) \geq c$.

What monotonicity conditions on $p(c, X)$ must be added so that versions of B & B and of B-F are applicable to this definition? What are the changes needed in the given B & B and B-F algorithms to make them applicable?

6. *Best-First Algorithm Using Heap.* Rewrite the general B-F algorithm using a HEAP instead of the list Z.

7. *Shortest-Path Algorithm—Trace Exercise.* Make up a small weighted directed graph (about 5 or 6 vertices and about 10 or 12 branches) and show a trace of the best-first based shorted path algorithm as it works on graph. Trace C, X, Z, and TBL.

8. *Nested Definition for Minimum Tour.* Develop a nested recursive definition for the minimum tour problem starting with a nonnested definition for the problem, adding a bound W, and nesting so as to use W.

9. *(Shortest) Minimum Path Algorithm with Equal Edge Cost.* Develop a (shortest) minimum path algorithm for a graph in which each edge is of the same (positive) weight. It should have a lower complexity bound than when there is no constraint on these weights.

10. *(Shortest) Minimum Path Algorithm for Acyclic Graph with Small Set of Edge Costs.* Develop a minimum path algorithm for acyclic graphs if each edge has one of k (positive) weights.

11. *(Shortest) Minimum Path Algorithm with Small Set of Edge Costs.* Develop a minimum path algorithm for graphs in which each edge has one of k (positive) weights. It should have a lower complexity bound than when there is no constraint on these weights.

12. *Articulation Point—Simple Exercises.* Trace the articulation algorithm Algorithm 4.2.7 on the following graph using b as the root.

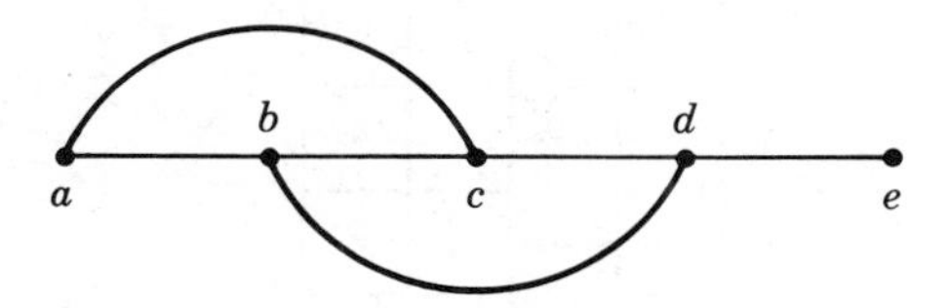

13. *Articulation Point Algorithm Modification.* Describe the alteration necessary in the articulation point algorithm, Algorithm 4.2.7, to print out the articulation points.

14. *Articulation Point in a Directed Graph.* The definition of an articulation point can be extended to directed graphs. y is an articulation point from x to w in G if and only if every path from x to w goes through y. The same definition to detect articulation points used in the text will work for directed graphs. However, the nature of the comparability is different. Describe it in enough detail so that it can be used to improve the basic depth-first implementation.

15. *Bi-connected Components—Exercise.* For the following graph, show all the bi connected components, and give the order in which they would be detected by the text algorithm if 1 is the root.

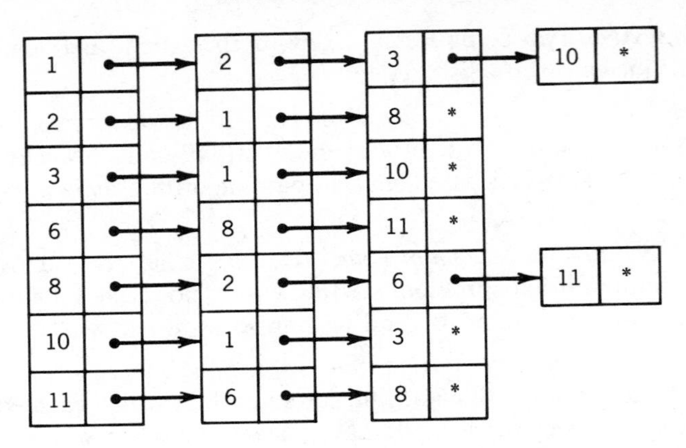

16. *Strongly Connected Component—Exercise.* For the following digraph trace the strongly connected component algorithm, Algorithm 5.6.1, with the "begin" vertex being 1.

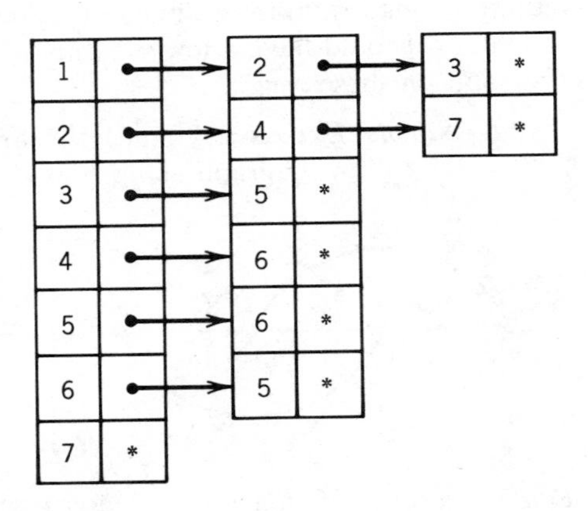

17. *Strongly Connected Component—Proposed Algorithm Complexity.* The strongly connected component which includes vertex x is the set of all vertices reachable from x, intersected with the set of all vertices that can reach x. Since both of these sets can be determined by the reachability algorithm, there is in this approach the basis for a strongly connected algorithm.What is the worst-case complexity of such an algorithm?

18. *Transitive Closure of Undirected Graph.* The transitive closure of a graph G is another graph G'. G' has the same vertices as G, but has an edge from vertex a to b if and only if G has a path from a to b. Describe an algorithm for finding the transitive closure of an undirected graph G based on finding connected components of G.

19. *Game, ALPHA–BETA—Exercise.* Consider the following game played on the graph G.

G = a graph with six vertices numbered 1 through 6 and the following undirected edges: (1, 2) (1, 4) (1, 5) (1, 6) (2, 3) (2, 4) (3, 4) (3, 5) (4, 5). The game rules are

a. The ith move consists of marking an edge from vertex i.

b. A final state is one in which the marked edges form a graph which has a path with three or more vertices, or one in which a move is impossible.

c. The value of the game in the final state to the mover in that state is equal to the number of edges that have been chosen (by either player) in the final state. Note the player who makes the move that puts the game into the final state loses.

Draw the O network and indicate which game states will not be pursued as a result of the ALPHA–BETA algorithm, Algorithm 5.7.1.

Trace the V list in the implementation for this ALPHA–BETA algorithm.

20. *Game with Win–Loss–Draw Values—Algorithm.* Consider a game whose value is 1, 0, or −1. Can a substantially simpler (in space or time usage) algorithm than the general ALPHA–BETA one be designed to take advantage of the restriction on game values as was done when only win–loss values are allowed?

21. *Game with Win–Loss–Draw.* Given a game with a game tree which is a complete trinary tree with n levels whose only terminal values are 1, 0 and −1, we wish to investigate whether it is possible for such a game to have its terminal values assigned so as to have *no* cutoffs, that is, so that every game state must be looked at. If it is possible, give such a worst-case assignment for trees with two and three levels. If not possible, prove it is not.

22. *Game with Upper and Lower Bounds.* Given a game G, suppose the value at every leaf is known to be bounded by a lower bound L and an upper bound U. That is, for every node in the tree, $L \leq v \leq U$. How can Algorithm 5.7.1 be modified to take advantage of these bounds?

CHAPTER SIX

IMPLEMENTATION BY EQUATION SET EVALUATION

In Chapter 5 the significance of comparability for time-efficient implementations was explored. When there is comparability, the basic implementing algorithms can be augmented with a table to record each argument structure X, and, if necessary, $f(X)$, and the record used when X recurs. The form and extent of the table depends on the nature of the comparability. The algorithms in Chapter 5 use a variety of different comparability characteristics. This chapter is devoted to comparability based on complete equality of argument structures (duplicates), and the ESE approach to its exploitation. ESE consists of two steps:

1. The formulation of a set of equations, and then
2. the solution of that set.

The first step, is often unnecessary either because the equations are essentially given by the problem input, or because their form is always the same and can be permanently incorporated in the algorithm. Without the need to formulate equation sets, the running time of the algorithm is entirely devoted to solving the set of equations. Examples are provided by most of the graph problems we have considered, in which the graph representation itself gives the set of equations, and by the Fibonacci number problem explored in Section 2.10.2, where the definition leads to equations whose form is independent of the input. In the simplest case, the second step, equation set solution, is achieved by simple back substitution but often a wider set of techniques is necessary. As an introduction to our study of the implementation of definitions with duplicates, we study the problem of formulating definitions so as to make duplication apparent.

Two definitions for the *same* function may differ widely in comparability. A definition may have no duplicates while a completely equivalent one has many. This chapter starts with the development of *transformations* between equivalent definitions with and without duplicates. The number of different argument structures in the substitutional domain $|\Delta_f(X)|$ gives the number of equations in an ESE solution and is closely related to the complexity of the algorithm. This number usually decreases, often from an exponential to a polynomial function of the input size, in a transformation from a definition to an equivalent. Because of this desired reduction in the size of the substitutional domain, the transformations, though bidirectional, are referred to as reductions.

Transformations of recursive definitions as the basis of a system for design of efficient programs have been studied elsewhere [Burstall 77]. Relatively speaking our transformations involve fairly complex preconditions with relatively large changes, and are intended to stand mainly as a reminder of alternative definitions that can serve as a starting point of an algorithm design.

6.1 TRANSFORMATIONS—SE ↔ ESE STYLE DEFINITIONS

Our first transformation (reduction) has already appeared in Section 2.5, where the following two definitions for the set of all binary n-tuples were proven equivalent.

Definition 6.1.1

$$\left[\begin{array}{ll} f(x, n) = \{x\} & :n = 0 \\ f(x, n) = f(x\|\langle 0\rangle, n-1) \cup f(x\|\langle 1\rangle, n-1) & :n \in N \\ \text{Initially } x = \langle\ \rangle,\ n = d \in N_0 & \end{array}\right.$$

Definition 6.1.2

$$\left[\begin{array}{ll} g(n) = \{\langle\ \rangle\} & :n = 0 \\ g(n) = \langle 0\rangle|*|g(n-1) \cup \langle 1\rangle|*|g(n-1) & :n \in N \\ \text{Initially } n = d \in N_0 & \end{array}\right.$$

The arguments in Definition 6.1.1 are a 0–1 vector x and an integer n, equal to the length of x. In an evaluation, x ranges over all 0–1 vectors from size 0 to d. There are $O(2^d)$ of these. So the substitutional domain of Definition 6.1.1 has $O(2^d)$ members. On the other hand, the substitutional domain of Definition 6.1.2 clearly has only d. This means that using ESE we get $O(2^d)$ equations to solve for one, and only d equations for the other. In this case there is no substantial time saving in implementing Definition 6.1.2 over implementing Definition 6.1.1 with ESE, because the saving in the number of equations is offset by the complexity of repeated implementation of the function $|*|$.

Another pair of definitions illustrating the same basic transformation consists of the shortest path cost, Definitions 5.4.4 and 2.12.4. Both are repeated here.

Repeating Definition 5.4.4.

Definition 6.1.3: Cost of Shortest Path

$$\left[\begin{array}{ll} f(C, v) = C & : v = \text{“goal”} \\ f(C, v) = \text{MIN}_{i=1}^{N(v)} f(C + c_i(v), n_i(v)) & : v \neq \text{“goal”} \\ \text{Initially } C = 0,\ v = \text{a vertex} = \text{“begin” in } G,\ G \text{ is invariant.} & \end{array}\right.$$

Repeating Definition 2.12.4.

Definition 6.1.4: Cost of Shortest Path

$$\left[\begin{array}{ll} g(v) = 0 & : v = \text{“goal”} \\ g(v) = \text{MIN}_{i=1}^{N(v)} [c_i(v) + g(n_i(v))] & : v \neq \text{“goal”} \\ \text{Initially } v = \text{a vertex} = \text{“begin” in } G,\ G \text{ is invariant.} & \end{array}\right.$$

For these definitions, the difference in the size of substitutional domains is large and does affect the ESE complexity. In the first, Definition 6.1.3, the number of argument pairs (C, v) (with v being a vertex in G, and C the cost to reach v from “begin”) gives the size of the substitutional domain. There are generally many ways in which a vertex v can be reached by paths of different cost c in the given graph. Because there can be as many paths to v from “begin” that do not include goal as there are permutations of the remaining $n - 3$ vertices, there can be as many as $O((n - 3)!)$ such paths per vertex, and consequently $O((n - 2)!)$ equations (n being the number of vertices in G). In the second definition the domain contains only those vertices reachable from “begin,” and will yield n equations. Unlike the binary number example, the difference in domain size here results in proportional differences in ESE implementation time. (Neither definition terminates, so straightforward SE is not available without a change in the argument structure.)

6.1.1 Reduction 1—Transforming a Class of Nonnested Definitions

The pairs of equivalent definitions given above are samples from a large class of such pairs. We now delineate that class, first generalizing informally from the examples, then more precisely in a transformation theorem, Theorem 6.1.1.

For the most part the definitions involved have arguments which are either pure information or information–control. Assume that the argument structure of a definition DEF.f has two parts. One of these, X, consists of pure information components. Typically X serves to accumulate a quantity augmented at each parent to child step determined by y, the second part of the argument structure. Each child produces a different augmentation of X. Thus different argument structures on a frontier generally contain different y’s. y is an information–control component. It determines termination and the number of children of a parent, as well as how X is augmented. DEF.f can be transformed to an equivalent, DEF.g, in which only y appears in the argument structure. Each augmentation of X depends on y, as it did before transformation, but now the augmentation appears as the argument of a primitive function *outside* the argument of the nonterminal g of DEF.g. This is not very surprising, given the limited use of X in DEF.f. Starting with any definition, it is easy to produce an equivalent one whose arguments consist of pure information and information–control components, a condition we have said is usually necessary for transformation. However, this is not quite sufficient. Other conditions are given in the following complete statement of the transformation. (The first argument in Definition 1 of Theorem 6.1.1 fails to be pure information only because of its first line. Most applications of the theorem are to definitions without such a line.)

Theorem 6.1.1: Assume that in the following definitions the primitive functions $+$ and $*$ have the properties:

p1. $+$ is associative $\left(\sum_{i=1}^{n} x_i \text{ is short for } x_1 + \cdots + x_n\right)$.

p2. $*$ is associative.

p3. $*$ distributes through $+$.

p4. $0_* * \text{anything} = \text{anything} * 0_* = 0_*$.

For both of the following definitions, $T(y)$ and $\tilde{T}(y)$ are mutually disjoint. Definition 1,

$$\left[\begin{array}{ll} f(X, y) = 0_* & :X = 0_* \\ f(X, y) = X * t(y) & :T(y) \\ f(X, y) = \sum_{i=1}^{m(y)} f(X * d_i(y), o_i(y)) + X * p(y) & :\tilde{T}(y) \\ \text{Initially } y \in D,\ X \text{ is anything.} & \end{array}\right.$$

Definition 2,

$$\left[\begin{array}{ll} g(y) = t(y) & :T(y) \\ g(y) = \sum_{i=1}^{m(y)} [d_i(y) * g(o_i(y))] + p(y) & :\tilde{T}(y) \\ \text{Initially } y \in D. & \end{array}\right.$$

Then $X * g(y)$ satisfies $f(X, y)$.

PROOF:

1. For $X = 0_*$, from Definition 1,

$$f(0_*, y) = 0_*$$

Substituting $X * g(y)$ for $f(X, y)$,

$$0_* * g(y) =? 0_*$$

From p4,

$$0_* =! 0_*$$

2. For $T(y)$, from Definition 1,

$$f(X, y) = X * t(y)$$

Substituting $X * g(y)$ for $f(X, y)$,

$$X * g(y) =? X * t(y)$$

From Definition 2,

$$X * t(y) =! X * t(y)$$

3. For $\tilde{T}(y)$, from Definition 1,

$$f(X, y) = \sum_{i=1}^{m(y)} f(X * d_i(y), o_i(y)) + X * p(y)$$

Substituting $X * g(y)$ for $f(X, y)$,

$$X * g(y) =? \sum_{i=1}^{m(y)} [(X * d_i(y)) * g(o_i(y))] + X * p(y)$$

By p2,

$$X * g(y) =? \sum_{i=1}^{m(y)} [X * (d_i(y) * g(o_i(y)))] + X * p(y)$$

By p3 and p1,

$$X * g(y) =? X * \left(\sum_{i=1}^{m(y)} [(d_i(y) * g(o_i(y))] + p(y) \right)$$

From Definition 2,

$$X * g(y) =! X * g(y) \quad \blacksquare$$

Corollary 6.1.1.1: Theorem 6.1.1 still holds if the two arguments of the $*$ function are interchanged at every appearance of $*$ throughout the theorem. The final conclusion of the theorem would then be that $g(y) * X = f(X)$.

PROOF: To prove the theorem with the interchanged arguments, it is only necessary to interchange the arguments throughout the proof. ■

A number of other points are worth noting.

1. If 1_* is defined so that $1_* *$anything = anything, then Theorem 6.1.1 implies that: $g(y) = f(1_*, y)$ for any y in Δ_g.
2. The theorem still holds if the first line of Definition 1 and the conditions involving 0_* are removed throughout.
3. Definition 2 is satisfied uniquely if and only if Definition 1 is.

Applying Theorem 6.1.1 is sometimes awkward. For example, the equivalence of binary n-tuples Definitions 6.1.1 and 6.1.2, cannot be proven by direct application of the theorem. This is because, for Definition 6.1.1, $\|$ is the natural instantiation for $*$ in the theorem, while $|*|$ in Definition 6.1.2 naturally serves that role. But the theorem allows only one instantiation of $*$. However, Definition 6.1.1 can be replaced by the equivalent Definition 6.1.5.

Definition 6.1.5

$$\left[\begin{array}{ll} f(x, n) = x & : n = 0 \\ f(x, n) = f(x|*|\{\langle 0\rangle\}, n-1) \cup f(x|*|\{\langle 1\rangle\}, n-1) & : n \in N \\ \text{Initially } x = \{\langle\ \rangle\},\ n \in N_0. & \end{array}\right.$$

Now it makes sense for $|*|$ to be instantiated for $*$. To make the theorem applicable it may be necessary to use a primitive operation more powerful than minimally required for the definition. It is possible to restate the theorem so as to avoid this problem, but the necessary increase in complexity of that statement does not seem warranted.

6.1.2 Examples of Application of Reduction 1

The next example also illustrates the occasional need to overcomplicate some primitives. It also contains a line corresponding to the first line in Definition 1 of Theorem 6.1.1.

6.1.2.1 Shortest Path

The following recursive definition gives the shortest path, together with its cost, in a directed graph G with positively weighted edges, from the vertex "begin" to the vertex "goal." It is similar to Definition 6.1.3, except that now the path is explicitly maintained. A growth approach similar to the one used there is employed. As soon as a vertex which duplicates an existing one in path p is added to p, p is assigned infinite cost. This is legitimate since for any such p whose continuation can reach "goal," there is a simple path of smaller cost. Infinite cost causes termination and removal of the corresponding path. All paths with duplicates are thereby removed as candidates for the shortest.

In the definition, argument p represents a vertex path or trail, v a vertex, and C the cost of p. If p is a path, that is, contains no vertex more than once, then the cost of p is the usual sum of edge costs, but if p is a trail with a repeated vertex, its cost is defined to be ∞. So if $p \| \langle v \rangle$ is a path with cost C, then the cost of $p \| \langle v \rangle \| n_i(v)$ is $C +_p c_i(v)$, where

$$
\begin{aligned}
C +_p c_i(v) &= C + c \qquad && \text{if } v \notin p \\
&= \infty && \text{if } v \in p.
\end{aligned}
$$

Definition 6.1.6: Shortest Path and Its Cost

$$
\left[
\begin{array}{ll}
f(C, p, v) = \langle \infty, p \rangle & :C = \infty \\
f(C, p, v) = \langle C, p \| \langle v \rangle \rangle & :v = \text{"goal" and } C \neq \infty \\
f(C, p, v) = \text{MIN-SEL}_{i=1}^{N(v)} f\big(C +_p c_i(v), & \\
\qquad\qquad\qquad\qquad p \| \langle v \rangle, n_i(v)\big) & :v \neq \text{"goal" and } C \neq \infty \\
\text{Initially } C = 0,\ p = \langle\ \rangle,\ v = \text{a vertex} = \text{"begin" in } G,\ G \text{ is invariant.} &
\end{array}
\right.
$$

Unlike Definition 6.1.3, this definition terminates by eliminating paths with repeated vertices.

The theorem applies to a slight modification of this definition. For this we need the function $+_p \|$. Let $A = \langle A_1, A_2 \rangle$ and $B = \langle B_1, B_2 \rangle$, where A_2 and B_2 are themselves vectors,

$$A + \| \ B = \langle A_1 + B_1, A_2 \| B_2 \rangle$$

$$
\begin{aligned}
A +_p \| \ B &= A + \| \ B \text{ if no two components of } A_2 \text{ are the same} \\
&= \langle \infty, \langle\ \rangle \rangle \text{ if there are two components of } A_2 \text{ that are the same}
\end{aligned}
$$

It follows that:

$$
\begin{aligned}
\langle C, p \| \langle v \rangle \rangle &= \langle C + 0, p \| \langle v \rangle \rangle = \langle C, p \rangle + \| \ \langle 0, \langle v \rangle \rangle \\
\langle C +_p c_i(v), p \| \langle v \rangle \rangle &= \langle C, p \rangle +_p \| \ \langle c_i(v), \langle v \rangle \rangle
\end{aligned}
$$

Definition 6.1.7 is revised in terms of $+_p \|$.

Definition 6.1.6 Shortest Path and Its Cost

$$
\left[
\begin{array}{ll}
f(\langle C, p \rangle, v) = \langle \infty, \langle\ \rangle \rangle & :C = \infty \\
f(\langle C, p \rangle, v) = \langle C, p \rangle + \| \langle 0, \langle v \rangle \rangle & :v = \text{"goal" and } C \neq \infty \\
f\big(\langle C, p \rangle, v = \text{MIN-SEL}_{i=1}^{N(v)} f\big(\langle C, p \rangle +_p \| \langle c_i(v), & \\
\qquad\qquad\qquad\qquad \langle v \rangle \rangle, n_i(v)\big) & :v \neq \text{"goal" and } C \neq \infty. \\
\text{Initially } C = 0,\ p = \langle\ \rangle,\ v = \text{a vertex} = \text{"begin" in } G,\ G \text{ is invariant.} &
\end{array}
\right.
$$

This definition satisfies the conditions of Definition 1 of Theorem 6.1.1, with $*$ of the theorem corresponding to $+_p \|$ of Definition 6.1.7, X corresponding to $\langle C, p \rangle$, y to v, $t(y)$ to $\langle 0, \langle v \rangle \rangle$, $m(y)$ to $N(v)$, and Σ to MIN-SEL. These satisfy p1, p2, and p3. 0_* corresponds to the class $\langle \infty, p \rangle$, with p any vertex sequence, so for any pair x, $\langle \infty, p \rangle +_p \| x = \langle \infty, q \rangle$. Thus p4 is satisfied.

This allows us to get the equivalent Definition 6.1.8.

Definition 6.1.8: Shortest Path and Its Cost

$$\left[\begin{array}{ll} g(v) = \langle 0, \langle v \rangle \rangle & : v = \text{“goal”} \\ g(v) = \text{MIN-SEL}_{i=1}^{m(v)}(\langle c_i(v), \langle v \rangle \rangle +_p \| g(n_i(v))) & : v \neq \text{“goal”} \\ \text{Initially } v = \text{a vertex} = \text{“begin” in } G,\ G \text{ is invariant.} & \end{array}\right.$$

As another example of the application of reduction 1, consider the growth definition of permutations.

6.1.2.2 Permutations and Minimum Tour (Traveling Salesman)

Definition 6.1.9: Permutations

$$\left[\begin{array}{ll} f(p, b) = p & : b = \langle\ \rangle \\ f(p, b) = \bigcup_{j=1}^{b} f(p|*|\{\langle b_j \rangle\}, b - b_j) & : b \neq \langle\ \rangle \\ \text{Initially } p = \{\langle\ \rangle\},\ b = \langle 1, 2, \ldots, n \rangle. & \end{array}\right.$$

Note that cross concatenation, $|*|$, is used even though the simpler $\|$ would suffice. This is done to establish the conditions which make the reduction theorem, Theorem 6.1.1, applicable. The reduction yields Definition 6.1.10:

Definition 6.1.10: Permutations

$$\left[\begin{array}{ll} g(b) = \{\langle\ \rangle\} & : b = \langle\ \rangle \\ g(b) = \bigcup_{j=1}^{b} [\{\langle b_j \rangle\}|*|g(b - b_j)] & : b \neq \langle\ \rangle \\ \text{Initially } b = \langle 1, 2, \ldots, n \rangle. & \end{array}\right.$$

A somewhat more complex, related example involves finding the minimum tour, that is, a cycle which contains every vertex of a weighted graph G once. Assume the graph is completely connected, and edge weights are real numbers or ∞. The simple recursive growth formulation is tantamount to enumerating all permutations of vertices other than vertex 1, which is assumed to be the starting and ending vertex of the tour. Assume that the first branch from each vertex is to vertex 1, so $c_1(v)$ is the cost of that branch.

Definition 6.1.11: Minimum Tour (Traveling Salesman)

$$\left[\begin{array}{ll} f(C,p,v,b) = \langle C + c_1(v), p\|\langle v\rangle\|\langle 1\rangle\rangle & :b = \langle\ \rangle \\ f(C,p,v,b) = \text{MIN-SEL}_{n_i(v)\in b}[f(C + c_i(v), p\|\langle v\rangle, & \\ \qquad n_i(v), b - n_i(v))] & :b \neq \langle\ \rangle \\ \text{Initially } p = \langle\ \rangle, C = 0, v = 1, b = \langle 2,3,\ldots,n\rangle. & \end{array}\right.$$

Note that the number of blocks outputs in the O network is at most $n - 1$ in the first stage (the size of set b), $n - 2$ in the next stage, and so on. Thus the total number of block outputs can be as great as $(n - 1)!$ Furthermore since this definition has an inverse, all these outputs are different.

Reduction 1 yields Definition 6.1.12.

Definition 6.1.12: Minimum Tour (Traveling Salesman)

$$\left[\begin{array}{ll} g(v,b) = \langle\ c_1(v), \langle v\rangle\|\langle 1\rangle\ \rangle & :b = \langle\ \rangle \\ g(v,b) = \text{MIN-SEL}_{n_i(v)\in b}[\langle c_i(v), \langle v\rangle\rangle +\| \ g(n_i(v), & \\ \qquad b - n_i(v))] & :b \neq \langle\ \rangle \\ \text{Initially } v = 1,\ b = \langle 2,3,\ldots,n\rangle. & \end{array}\right.$$

Again an O network for this definition can have $(n - 1)!$ block outputs; however, there can only be $n * 2^n$ different outputs because $\langle v, b\rangle$, with v in $\{2,3,\ldots,n\}$ and b a subset of $\{2,3,\ldots,n\}$, can only assume this many values.

6.1.2.3 Knapsack Problem

Another example is provided by the knapsack problem. We are given a vector of numbers, $v = \langle v_1,\ldots,v_n\rangle$, representing sizes of items to be placed in a knapsack whose capacity is the given number s. We are to find all subsets of components of v whose sum is s. A definition is given in which every combination of components of v is grown. The growth continues to a new stage as long as the current combination does not have a sum greater than s. As each component is added to a combination, its value is subtracted from a quantity r, which initially equals s. When $r \le 0$, growth terminates, with a successful combination if $r = 0$, or with an unsuccessful combination if $r < 0$. There are two children for each growing argument-structure at stage i. In the first the combination, x, is unchanged, in the second v_i is concatenated to x. These represent the two possible decisions on inclusion of v_i. In the following definition the combinations x are not actually of components of v, but of the indices of these components.

Definition 6.1.13: Knapsack Problem

$$\left[\begin{array}{ll} f(x,i,r) = \{\ \} & :r > 0 \text{ and } i = 0 \\ f(x,i,r) = \{\ \} & :r < 0 \text{ and } i \geq 0 \\ f(x,i,r) = \{x\} & :r = 0 \text{ and } i \geq 0 \\ f(x,i,r) = f(x,i-1,r) \cup f(x|*|\{\langle i\rangle\}, i-1, r-v_i) & \\ & :r > 0 \text{ and } i > 0 \end{array}\right.$$

Initially $x = \{\langle\ \rangle\}$, $i = n$, $r, v_i \in N$, $v = \langle v_1, \ldots, v_n\rangle$ and is invariant.

The number of block outputs in the first stage of the O network is 2 or fewer. The number of children of every nonterminal parent is 2. So there may be as many as 4 block outputs in the second stage and so on. Since i increases by 1 at each stage and cannot be more than n, the total number of block outputs is not greater than 2^n. Furthermore, even though this definition does not have an inverse, the values of v and r can be chosen to make all 2^n outputs different.

There is no line in this knapsack definition corresponding to the first line of Definition 1 of Theorem 6.1.1. In applying the theorem we let both of the terminal lines in the knapsack definition correspond to the second line of Definition 1 in the theorem. Also 0_* of the theorem corresponds to $\{\ \}$ in the knapsack definition, that is, by convention for all x, $\{\ \}|*|x = \{\ \}$.

Reduction 1 yields Definition 6.1.14.

Definition 6.1.14: Knapsack Problem

$$\left[\begin{array}{ll} g(i,r) = \{\ \} & :r > 0 \text{ and } i = 0 \\ g(i,r) = \{\ \} & :r < 0 \text{ and } i \geq 0 \\ g(i,r) = \{\langle\ \rangle\} & :r = 0 \text{ and } i \geq 0 \\ g(i,r) = \{\langle i\rangle\}|*|g(i-1, r-v_i) \cup g(i-1,r) & :r > 0 \text{ and } i > 0 \end{array}\right.$$

Initially $i = n$, $r, v_i \in N$, $v = \langle v_1, \ldots, v_n\rangle$ and is invariant.

In the substitutional domain of this definition argument structures have the form (i, r) and i can take any value from 1 to n, while r can vary from its initial value to $-\max(v_1, \ldots, v_n)$. Let the initial value of $r = s$. Then the number of different argument structures possible in the substitutional domain is at most $O(n*(s - \max(v_1, \ldots, v_n))$. Any component of v larger than s could be removed by a prepass since it cannot contribute to a combination summing to s. Therefore based on Definition 6.1.14, the complexity of this problem is $O(n*2*s)$. Though this is not exponential in n, it is exponential in the length of the representation of the number s, which is better than n as a measure of the size of the input to this problem, that is, $s \leq 2^l$, where l is the length of the representation of s as a binary number. So the complexity can also be expressed as $O(n * 2^l)$.

For this definition the number of members of the substitutional domain is a good estimate of the complexity only if the number of different ways of fitting the items into the knapsack is small, or only a few of those ways are required. In fact the number of different fits can be exponential in n. The problem is that in solving the equations whose total number of right-side appearances of variables is $O(n * 2 * s)$ we have to substitute *sets*, and such a substitution is O(size of set) rather than $O(1)$.

Definitions of the form of Definition 2 in Theorem 6.1.1 could of course be obtained directly from the problem statement, rather than as the result of a transformation. Usually this would result from the "recursive" approach, as opposed to the "iterative" or growth approach with which definitions in the form of Definition 1 in Theorem 6.1.1 are obtained. (These approaches were described in Chapter 3.) When obtained directly, definitions in the form of Definition 2 are said to have been obtained by *dynamic programming* [Bellman 57], which is a form of "recursive" approach that results in definitions formulated so that comparable argument structures are equal and ESE is applicable.

6.1.3 Reduction 2—Transforming a Class of Nonnested Definitions with a Push-Down List as Part of the Argument

As in reduction 1, the argument structure of the definition to be transformed contains the pure information accumulating part X and the information–control part y, each serving the same function as before. Now, however, the information–control part y does not stand alone, but rather is the top member of a push-down list Y. At each growth step one of two events affecting the next y value occurs. Either y is popped from Y or X is augmented according to the value of y and the index of the child, and y spawns a *set* of new list entries, generally different for *each child*. For each child, new list entries are pushed into Y. These new entries will be processed later in the same fashion as was y. Y serves simply for storage.

Such a definition can be transformed to remove both the accumulated quantity X and all of list Y except its top member y, provided the primitive operations of the definition satisfy the properties given in the theorem below.

Theorem 6.12: Assume that in the following definitions the primitive functions $+$ and $*$ have the properties.

p1. $+$ is associative.

p2. $*$ is associative.

p3. $*$ distributes through $+$.

p4. 0_* has the property that $x * 0_* = 0_* * x = 0_*$.
p5. 1_* has the property that $x * 1_* = 1_* * x = x$.

For both of the following definitions, $Y = \langle\ \rangle \equiv y =$ zilch, $C_1(y)$, $C_2(y)$, and $C_3(y)$ are mutually disjoint.

Definition 1: y is used to refer to Y_1 throughout definition.

$$\left[\begin{array}{ll} f(X,Y) = X & :Y = \langle\ \rangle \\ f(X,Y) = 0_* & :C_1(y) \\ f(X,Y) = f(X * t(y), Y - y) & :C_2(y) \\ f(X,Y) = \sum_{i=1}^{m(y)} \left[f\left(X * d_i(y), \langle o_{i1}(y), \ldots, o_{in(y,i)}(y)\rangle \| Y - y\right)\right] & \\ \qquad + X * p(y) * f(1_*, Y - y) & :C_3(y) \\ \text{Initially } X = \text{anything, typically } \langle\ \rangle,\ Y = \langle y \rangle,\ y \in D. & \end{array}\right.$$

Definition 2

$$\left[\begin{array}{ll} g(y) = 1_* & :y = \text{zilch} \\ g(y) = 0_* & :C_1(y) \\ g(y) = t(y) & :C_2(y) \\ g(y) = \sum_{i=1}^{m(y)} \left[d_i(y) * g(o_{i1}(y)) * \cdots * g\left(o_{in(y,i)}(y)\right)\right] + p(y) & :C_3(y) \\ \text{Initially } y \in D. & \end{array}\right.$$

Assume $Y = \langle Y_1, \ldots, Y_n\rangle = \langle y, Y_2, \ldots, Y_n\rangle$; then

$$X * g(y) * g(Y_2) * \cdots * g(Y_n) \text{ satisfies } f(X, Y).$$

PROOF:

1. For $Y = \langle\ \rangle$ and equivalently $y =$ zilch, from Definition 1,

$$f(X,Y) = X$$

Substituting $X * g(y) * \cdots * g(Y_n)$ for $f(X, Y)$,

$$X * g(y) * \cdots * g(Y_n) = ?\, X$$

and since $Y = \langle\ \rangle$,

$$X * 1_* =?\, X$$

$$X =!\, X$$

2. For $C_1(y)$, in Definition 1: substituting $X * g(y) * \cdots * g(Y_n)$ for $f(X, Y)$,

$$X * g(y) * \cdots * g(Y_n) =?\, 0_*$$

From Definition 2,

$$X * 0_* * g(Y_2) * \cdots * g(Y_n) =?\, 0_*$$

From p4,

$$0_* =!\, 0_*$$

3. For $C_2(y)$, from Definition 1,

$$f(X, Y) = f(X * t(y), Y - y)$$

Substituting $X * g(y) * \cdots * g(Y_n)$ for $f(X, Y)$,

$$X * g(y) * g(Y_2) * \cdots * g(Y_n) =?\, X * t(y) * g(Y_2) * \cdots * g(Y_n)$$

From Definition 2,

$$X * t(y) * g(Y_2) * \cdots * g(Y_n) =!\, X * t(y) * g(Y_2) * \cdots * g(Y_n)$$

4. For $C_3(y)$, from Definition 1,

$$f(X, Y) = \sum_{i=1}^{m(y)} \left[f\left(X * d_i(y), \langle o_{i1}(y), o_{i2}(y), \ldots, o_{in(y,i)}(y) \rangle \| Y\right)\right]$$
$$+ X * p(y) * f(1_*, Y - y)$$

Substituting $X * g(y) * \cdots * g(Y_n)$ for $f(X, Y)$,

$$X * g(y) * \cdots * g(Y_n)$$
$$=?\, \sum_{i=1}^{m(y)} \left[(X * d_i(y)) * g(o_{i1}(y)) * g(o_{i2}(y)) \right.$$
$$\left. * \cdots * g\left(o_{in(y,i)}(y)\right) * g(Y_2) * \cdots * g(Y_n) \right]$$
$$+ X * p(Y) * 1_* * g(Y_2) * \cdots * g(Y_n)$$

By p2 and p5,

$$X * g(y) * \cdots * g(Y_n)$$

$$=? \sum_{i=1}^{m(y)} \left[X * \left(d_i(y) * g(o_{i1}(y)) * g(o_{i2}(y)) * \cdots * g\left(o_{in(y,i)}(y)\right)\right) * g(Y_2) * \cdots * g(Y_n)\right] + X * p(Y) * g(Y_2) * \cdots * g(Y_n)$$

By p3 and p1,

$$X * g(y) * \cdots * g(Y_n)$$

$$=? \sum_{i=1}^{m(y)} \left[X * \left(d_i(y) * g(o_{i1}(y)) * g(o_{i2}(y)) * \cdots * g\left(o_{in(y,i)}(y)\right)\right)\right] * g(Y_2) * \cdots * g(Y_n) + X * p(Y) * g(Y_2) * \cdots * g(Y_n)$$

By p3 and p1,

$$=? X * \sum_{i=1}^{m(y)} \left[d_i(y) * g(o_{i1}(y)) * g(i_{i2}(y)) * \cdots * g\left(o_{in(y,i)}(y)\right)\right] * g(Y_2) * \cdots * g(Y_n) + X * p(Y) * g(Y_2) * \cdots * g(Y_n)$$

By p3 and p1,

$$=? X * \left(\sum_{i=1}^{m(y)} \left[d_i(y) * g(o_{i1}(y)) * g(i_{i2}(y)) * \cdots * g\left(o_{in(y,i)}(y)\right)\right] + p(y)\right) * g(Y_2) * \cdots * g(Y_n)$$

Since by Definition 2,

$$\left(\sum_{i=1}^{m(y)} \left[d_i(y) * g(o_{i1}(y)) * g(i_{i2}(y)) * \cdots * g\left(o_{in(y,i)}(y)\right)\right] + p(y)\right) = g(y)$$

we have

$$X * g(y) * \cdots * g(Y_n) = !\, X * g(y) * g(Y_2) * \cdots * g(Y_n) \quad \blacksquare$$

6.1.4 Examples of Application of Reduction 2

There are a number of approaches to enumerating all the ways of parenthesizing a string of symbols. The following approach provides an illustration of the use of reduction 2, and forms the basis for the subsequent example of a definition for choosing the optimal way of associating, that is, parenthesizing, multiplications of j matrices.

6.1.4.1 Parenthesizing and Association of Matrix Multiplications

The following definition for enumerating all ways of parenthesizing a string of length n is built by the growth method. The parentheses are represented by a list of pairs: $\langle i_k, j_k \rangle$ with k in N^n. Each pair represents a "(" before position i_k and a ")" after position j_k. In the definition, the argument X is the list of such pairs. There is also a list of pairs, L, initially holding the single pair $\langle 1, n \rangle$. Broadly, growth proceeds in the following way. A pair $\langle i, j \rangle$ is popped from list L and concatenated to the sequence X to indicate a pair of parentheses has been added. To determine all possible parentheses to be nested between i and j, the range i to j is partitioned into $\langle i, i+k \rangle$ and $\langle i+k+1, j \rangle$ and pushed onto L as the kth child, with k going from 0 to $j-i-1$. Again the first entry in L, $\langle i, i+k \rangle$, will be used as $\langle i, j \rangle$ was in its parent. Namely, if $i \neq j$, it is concatenated to the X and it spawns a new pair to be pushed onto L. If $i = j$, then the spawning pair becomes infertile; the pair is concatenated on X but no addition is made to L.

Example 6.1.1 gives a partially developed O network in which $y = Y_1$ is initially $\langle 1, 4 \rangle$. The symbols $\langle , \rangle$, as well as commas, are dropped in the example and X and Y are separated by $|$.

Example 6.1.1

```
                                        ┌14,11,24|22,34—14,11,24,22|34—
        X|  Y                           │
       ┌14|11,24——14,11|24——————————————┤
       │                                └14,11,24|23,44—14,11,24,23|22,33,44—*
< >|14─┼14|12,34——14,12|11,22,34—   14,12,11|22,34—14,12,11,22|34————
       │               ┌14,13|11,23,44—   14,13,11|23,44—14,13,11,23|22,33,44—
       └14|13,44———————┤
                       └14,13|12,33,44—   14,13,12|11,22,33,44—**
```

The example is developed only sufficiently to make the scheme clear. For the two * paths the terminal values of X that will be produced by continued development are shown below, together with the parenthesization of the four tokens (x's) each represents.

14, 11, 24, 23, 22, 33, 44	represents	$((x)(((x)(x))(x)))$	*
14, 13, 12, 11, 22, 33, 44	"	$((((\underset{1}{x})(\underset{2}{x}))(\underset{3}{x}))(\underset{4}{x}))$	* *

The pairs (i, i), such as 11 and 22, can be dropped from X, to give a simpler representation of parenthesization. The second line in Definition 6.1.15 effects this simplification; otherwise the definition is modelled on the example above.

Definition 6.1.15: Parenthesizing
In this definition the first member of L is referred to as $\langle i, j\rangle$

$$\left[\begin{array}{ll} f(X, L) = X & : L = \langle\ \rangle \\ f(X, L) = f(X, L - L_1) & : j = i \text{ and } L \neq \langle\ \rangle \\ f(X, L) = \displaystyle\sum_{k=0}^{k=j-i-1} f(X|*|\{\langle i, j\rangle\}, \langle\langle i, i+k\rangle, & \\ \qquad\qquad \langle i+k+1, j\rangle\rangle \| L - L_1) & : j > i \end{array}\right.$$

Initially $X = \{\langle\ \rangle\}$, $i = 1$, $j = n$, $L = \langle\langle i, j\rangle\rangle$ (n is the length of string to be parenthesized).

The result of the application of reduction 2 is given in Definition 6.1.16.

Definition 6.1.16: Parenthesizing

$$\left[\begin{array}{ll} g(\langle i, j\rangle) = \{\langle\ \rangle\} & : j = i \\ g(\langle i, j\rangle) = \displaystyle\bigcup_{k=0}^{k=j-i-1} [\{\langle i, j\rangle\}|*|g(\langle i, i+k\rangle)|*| & \\ \qquad\qquad g(\langle i+k+1, j\rangle] & : j > i \end{array}\right.$$

Initially $i = 1$, $j =$ length of string to be parenthesized.

A related example is the problem of choosing the best way of ordering (associating) multiplications of a series of different size matrices. Although the result is unaffected, the number of element multiplications required is strongly affected by the choice of the order of association. For example, suppose the multiplication is $M_1[4, 2] * M_2[2, 7] * M_3[7, 3]$. The cost in element multiplications is $4*2*7 + 4*7*3 = 140$ if the first and second matrices are multiplied first and the result is multiplied with the third. It is $4*2*3 + 2*7*3 = 66$ if the last two are multiplied first. In the following definition it is assumed that the multiplication is: $M_1 * \cdots * M_n$ and that M_i has r_i rows and c_i columns, with $c_{i-1} = r_i$. It follows that $r_i * c_{i+k} * c_j$ is the cost of multiplying the matrix that results from multiplying matrices i through $i + k$ with the result of multiplying the matrix that results from multiplying matrices $i + k + 1$ through j.

The definition of $f(\langle C, X\rangle, L)$ for producing the minimum cost association order for multiplying the matrices is analogous to the previous definition for parenthesizing. In the argument structure for this definition, there is, along with the record of parenthesization, the cumulative cost of multiplying the matrices in the association order implied by that parenthesization. All the possible orderings are computed, and the minimum cost one is selected with the minimum-select function.

Definition 6.1.17: Association of Matrix Multiplications
In this definition the first member of L is referred to as $\langle i, j\rangle$.

$$\left[\begin{array}{ll} f(\langle C, X\rangle, L) = \langle C, X\rangle & : L = \langle\ \rangle \\ f(\langle C, X\rangle, L) = f(\langle C, X\rangle, L - L_1) & : j = i \text{ and } L \neq \langle\ \rangle \\ f(\langle C, X\rangle, L) = \text{MIN-SEL}_{k=0}^{k=j-i-1}[f(\langle C, X\rangle +\| \langle r_i * c_{i+k} * c_j, \langle i, j\rangle\rangle, & \\ \qquad \langle\langle i, i+k\rangle, \langle i+k+1, j\rangle\rangle \| L - L_1)] : j > i & \\ \text{Initially } \langle C, X\rangle = \langle 0, \langle\ \rangle\rangle,\ i = 1,\ j = \text{number of matrices},\ L = & \\ \langle\langle i, j\rangle\rangle. & \end{array}\right.$$

The result of transformation (reduction) 2 is given in Definition 6.1.18.

Definition 6.1.18: Association of Matrix Multiplications

$$\left[\begin{array}{ll} g(\langle i, j\rangle) = \langle 0, \langle\ \rangle\rangle & : j = i \\ g(\langle i, j\rangle) = \text{MIN-SEL}_{k=0}^{k=j-i-1}[\langle r_i * c_{i+k} * c_j, \langle i, j\rangle\rangle +\| & \\ \qquad g(\langle i, i+k\rangle) +\| g(\langle i+k+1, j\rangle)] & : j > i \\ \text{Initially } i = 1,\ j = \text{number of matrices}. & \end{array}\right.$$

6.1.5 Reduction 3 — Transforming a Class of Nested Definitions

In this section we develop a transformation between nonnested and nested definitions. The following abbreviated notation for nesting first given in

Section 1.2 is used:

$$N_{j=1}^{n(i)} g(p, o_{ij}(y)) \equiv g(g(\dots g(g(p, o_{i1}(y)), o_{i2}(y)) \dots), o_{in(i)}(y))$$

The next lemma is a technical necessity for proving the next transformation; it may be skipped, at least until after the theorem is read.

Lemma 6.1.1: Given the following definition of $g(p, y)$ involving the primitive functions $+$ and t with the properties:

p1. $+$ is associative.

p2. If $T(p, y)$ and $p = p_a + p_b$, then

$$T(p_a, y) \text{ and } T(p_b, y), \text{ and } t(p_a + p_b, y) = t(p_a, y) + t(p_b, y)$$

$$\left[\begin{array}{ll} g(p, y) = t(p, y) & :T(p, y) \\ g(p, y) = \sum_{i=1}^{m(p, y)} N_{j=1}^{n(i)} g(p, o_{ij}(y)) & :\sim T(p, y) \\ \text{Initially } p \in A,\ y \in B. & \end{array}\right.$$

If this definition terminates, then $g(p_a + p_b, y) = g(p_a, y) + g(p_b, y)$ everywhere in Δ_g.

PROOF

1. For $T(p, y)$, by p2 and if $p = p_a + p_b$, then $T(p_a, y)$ and $T(p_b, y)$. From the terminal line of Definition 2,

$$g(p, y) = t(p, t)$$

$$g(p_a + p_b, y) = t(p_a + p_b, y)$$

Substituting $g(p_a, y) + g(p_b, y)$ for $g(p_a + p_b, y)$,

$$g(p_a, y) + g(p_b, y) =?\, t(p_a + p_b, y)$$

$$t(p_a, y) + t(p_b, y) =?\, t(p_a + p_b, y)$$

$$t(p_a, y) + t(p_b, y) =!\, t(p_a, y) + t(p_b, y)$$

2. For $\sim T(p, y)$,

$$g(p_a + p_b, y) = \sum_{i=1}^{m(p, y)} N_{j=1}^{n(i)} g(p_a + p_b, o_{ij}(y))$$

Substituting $g(p_a, y) + g(p_b, y)$ for $g(p_a + p_b, y)$,

$$g(p_a, y) + g(p_b, y)$$

$$=? \sum_{i=1}^{m(p,y)} N_{j=1}^{n(i)}\left[g(p_a, o_{ij}(y)) + g(p_b, o_{ij}(y))\right]$$

$$=? \sum_{i=1}^{m(p,y)} N_{j=2}^{n(i)}\left[g(g(p_a, o_{i1}(y)) + g(p_b, o_{i1}(y)), o_{ij}(y))\right]$$

$$=? \sum_{i=1}^{m(p,y)} N_{j=2}^{n(i)}\left[g(g(p_a, o_{i1}(y)), o_{ij}(y))\right.$$

$$\left. + g(g(p_b, o_{i1}(y)), o_{ij}(y))\right]$$

$$=? \sum_{i=1}^{m(p,y)} N_{j=3}^{n(i)} g(g(g(p_a, o_{1j}(y)), o_{2j}(y))$$

$$+ g(g(p_b, o_{1j}(y)), o_{2j}(y)), o_{ij}(y))$$

$$\vdots$$

$$=? \sum_{i=1}^{m(p,y)} \left[N_{j=1}^{n(i)} g(p_a, o_{ij}(y)) + N_{j=1}^{n(i)} g(p_b, o_{ij}(y))\right]$$

$$=? \sum_{i=1}^{m(p,y)} N_{j=1}^{n(i)} g(p_a, o_{ij}(y)) + \sum_{i=1}^{m(p,y)} N_{j=1}^{n(i)} g(p_b, o_{ij}(y))$$

and

$$g(p_a, y) + g(p_b, y) =! g(p_a, y) + g(p_b, y)$$

So $g(p_a, y) + g(p_b, y)$ satisfies the definition for $g(p_a + p_b, y)$, and because that definition terminates and thus is uniquely satisfied, $g(p_a, y) + g(p_b, y) = g(p_a + p_b, y)$. ■

With this result we can validate a transformation from a definition with a stack argument to a nested definition having simpler arguments. The definition to be transformed is of the type that is typically developed by the growth approach and is similar to Definition 1 in Theorem 6.1.2, with the first argument, p, corresponding to X there; both have a second argument Y and virtually the same recursive lines. They differ most significantly in their C_2

line. In Theorem 6.1.2 the arguments on the right side of this line depend only on Y, whereas in the following theorem they depend on both p and y. So in this respect the next theorem applies more generally but represents a weaker result.

Theorem 6.1.3: Assume that in the following definitions 1 and 2 the primitive functions $+$ and t have the properties:

p1. $+$ is associative ($\Sigma_{i=1}^{n} x_i = x_1 + \cdots + x_n$).

p2. If $T(p, y)$ and $p = p_a + p_b$, then

$$T(p_a, y) \text{ and } T(p_b, y), \text{ and } t(p_a + p_b, y) = t(p_a, y) + t(p_b, y)$$

where $T(p, y) = [C_1(p, y) \text{ or } p = 0_* \text{ or } C_2(p, y)]$.

p3. 0_* has the property that $x * 0_* = 0_* * x = 0_*$.

For both of the following definitions,

$Y = \langle\ \rangle \equiv y = \text{zilch}$.

$\{C_1(p, y) \text{ or } p = 0_*\}, \{C_2(p, y)\}, \{C_3(p, y)\}$ are pairwise disjoint.

Definition 1: y is used in place of Y_1 in this definition.

$$\left[\begin{array}{ll} f(p, Y) = p & : Y = \langle\ \rangle \\ f(p, Y) = 0_* & : C_1(p, y) \text{ or } p = 0_* \\ f(p, Y) = f(t_2(p, y), Y - y) & : C_2(p, y) \\ f(p, Y) = \displaystyle\sum_{i=1}^{m(p,y)} f(p, \langle o_{i1}(y), \ldots, o_{in(p,y,i)}(y)\rangle \| Y - y) & \\ & : C_3(p, y) \\ \text{Initially } p \in D,\ y \in E,\ Y = \langle y \rangle. & \end{array}\right.$$

Definition 2:

$$\left[\begin{array}{ll} g(p, y) = p & : y = \text{zilch} \\ g(p, Y) = 0_* & : C_1(p, y) \text{ or } p = 0_* \\ g(p, y) = t_2(p, y) & : C_2(p, y) \\ g(p, y) = \displaystyle\sum_{i=1}^{m(p,y)} N_{j=1}^{n(p,y,i)} g(p, o_{ij}(y)) & : C_3(p, y) \\ \text{Initially } p \in D,\ y \in E & \end{array}\right.$$

Then $N_{j=2}^{n} g(g(p, y), Y_j) = f(p, Y)$ for $n = |Y|$.

PROOF: First note that Definition 2 can be rewritten as

$$\left[\begin{array}{ll} g(p, y) = t(p, y) & :T(p, y) \\ g(p, y) = \sum_{i=1}^{m(p, y)} N_{j=1}^{n(p, y, i)} g(p, o_{ij}(y)) & :\sim T(p, y) \\ \text{Initially } p \in D,\ y \in E. & \end{array}\right.$$

where

$$\begin{aligned} t(p, y) &= 0_* && \text{if } C_1(p, y) \text{ or } p = 0_* \\ &= t_2(p, y) && \text{if } C_2(p, y) \end{aligned}$$

and thus $T(p, y) = [C_1(p, y)$ or $p = 0_*$ or $C_2(p, y)]$. Therefore Lemma 1 can be applied to Definition 2.

1. For $Y = \langle\ \rangle \equiv y =$ zilch, from Definition 1,

$$f(p, Y) = p$$

Substituting $N_{j=2}^{n} g(g(p, y), Y_j)$ for $f(p, Y)$,

$$N_{j=2}^{n} g\big(g(p, y), Y_j\big) =?\, p$$

$$g(p, \text{zilch}) =?\, p$$

from Definition 2,

$$p =!\, p$$

2. For $C_1(p, y)$ or $p = 0_*$, From Definition 1,

$$f(p, Y) = 0_*$$

Substituting $N_{j=2}^{n} g(g(p, y), Y_j)$ for $f(p, Y)$,

$$N_{j=2}^{n} g\big(g(p, y), Y_j\big) =?\, 0_*$$

From Definition 2,

$$N_{j=3}^{n} g\big(g(0_*, Y_2), Y_j\big) =?\, 0_*$$

and continuing in this fashion, finally

$$g(0_*, Y_n) =?\, 0_*$$

$$0_* =!\, 0_*$$

3. For $C_2(p, y)$, from Definition 1,

$$f(p, Y) = f(t_2(p, y), Y - y)$$

Substituting $N_{j=2}^{n} g(g(p, y), Y_j)$ for $f(p, Y)$, on both left and right sides.

$$N_{j=2}^{n} g\big(g(p, y), Y_j\big) = ?\, N_{j=3}^{n} g\big(g(t_2(p, y), Y_2), Y_j\big)$$

From Definition 2, substituting for $g(p, y)$ on the left,

$$N_{j=3}^{n} g\big(g(t_2(p, y), Y_2), Y_j\big) = ?\, N_{j=3}^{n} g\big(g(t_2(p, y), Y_2), Y_j\big)$$

$$N_{j=3}^{n} g\big(g(t_2(p, y), Y_2), Y_j\big) = !\, N_{j=3}^{n} g\big(g(t_2(p, y), Y_2), Y_j\big)$$

4. For $C_3(p, y)$, in the $C_3(p, y)$ line of Definition 1, substituting $N_{j=2}^{n} g(g(p, y), Y_j)$ for $f(p, Y)$,

$$N_{j=2}^{n} g\big(g(p, y), Y_j\big) = ? \sum_{i=1}^{m(p, y)} N_{j=2}^{n} g\big(N_{k=1}^{n(p, y, i)} g(p, o_{ik}(y)), Y_j\big)$$

But by Lemma 6.1.1,

$$N_{j=2}^{n} g\big(g(p, y), Y_j\big) = ?\, N_{j=2}^{n} g\left(\sum_{i=1}^{m(p, y)} N_{k=1}^{n(p, y, i)} g(p, o_{ik}(Y)), Y_j\right)$$

Substituting from the $\sim T(X)$ line of Definition 2,

$$N_{j=2}^{n} g\big(g(p, y), Y_j\big) = !\, N_{j=2}^{n} g\big(g(p, y), Y_j\big) \quad \blacksquare$$

6.1.6 Example of Application of Theorem 6.1.3 — Recognition and Parsing

Our example of the application of Theorem 6.1.3 is to parsing algorithms for context-free grammars. We assume knowledge of the necessary concepts (see [Aho, Ullman 72*b*] and/or [Sethi, Aho, Ullman 86], where all the concepts as well as the parsing algorithms treated here are covered). In Section 1.2 these concept are briefly reviewed, and a set of related primitive functions is defined. We will make extensive use of these primitives in what follows.

6.1.6.1 Simple Top-Down Parsing Definition

The following definition gives a function whose value is the parse of a string, $s = \langle s_1, \ldots, s_n \rangle$, in a context-free grammar G.

The definition of $f(p, Y, v)$ is based on systematically enumerating all possible leftmost derivations consistent with s in G by the growth approach. p is an index in the input string s. When one of these derivations produces substring $\langle s_1, \ldots, s_k \rangle$, this is recorded by p, which then will equal $k + 1$, indicating that the next input symbol to be matched is s_{k+1}. v is the sequence of rules of G by which $\langle s_1, \ldots, s_k \rangle$ had been derived. Y is that part of the developing sentential form generated by v which has not yet been used in matching input symbols. So the first symbol of Y, designated y, will be used to generate $s_{k+1}, s_{k+2}, \ldots,$ of s. If y cannot produce some subsequence of the input string starting with s_{k+1}, that line of enumeration is abandoned. The arguments at the children of (p, Y, v) depend on whether y is a terminal symbol, TS(y), or not. If TS(y) and $y = s_p$, then p becomes $p + 1$ in the child. If $y \neq s_p$, then (p, Y, v) is a terminal argument structure with the value $\{\ \}$; there is no hope for a parse following the line of derivation given by v. If $\sim$ TS(y), there will be one child for each rule with y on its left, with that rule's right-side replacing y in Y. These and other conditions, which control the generating rule or O function, are detailed in Definition 6.1.19. It is assumed in the definition that s_{n+1} does not exist, that is, that s_{n+1} is matched by no terminal symbol. With the assumption of no left recursion, there is a guarantee that a bounded number ($\leq$ number of NTSs in G) of derivation steps will lengthen the derived string by at least one terminal symbol and thus increase the index p by at least 1. Thus the procedure will have to end. This function $f(p, Y, v)$ can be interpreted as follows.

$$f(p, Y, v) = \{v \| z \mid z \text{ is a rule sequence by which } s_{p+1}, \ldots, s_n \text{ can be obtained in a leftmost derivation from } Y\}$$

In the definition of $f(p, Y, v)$ if (p, Y, v) is an argument structure that actually occurs during evaluation, then $\langle s_1 \ldots s_{p-1} \rangle \| Y$ will be the string of symbols which can be derived from the *distinguished symbol* of G by application of the rule sequence v in a leftmost derivation. From this we can extend the interpretation of $f(p, Y, v)$:

$$f(p, Y, v) = \{v \| z \mid v \| z \text{ is a rule sequence by which } s_1, \ldots, s_n \text{ can be obtained in a leftmost derivation from the } \textit{distinguished symbol} \text{ of } G\}$$

In order to make Theorem 6.1.3 applicable, we will let p represent a set of positions, rather than a single string position. In this definition p will be a set of at most one member. In the transformed definition, however, p can have any number of members. Because p is a set, we need to define the function ad1(p, y) in which y is a terminal symbol,

$$\text{ad1}(p, y) = \{ j + 1 | (j \in p) \text{ and } (y = s_j) |$$

Definition 6.1.19: Simple Top-Down Parse
y is used in place of Y_1 in this definition.

$$f(p,Y,v) = \{\ \} \quad :\mathrm{TS}(y) \text{ and } \sim \exists j \in p \ni y = s_j \text{ and } Y \neq \langle\ \rangle$$
$$f(p,Y,v) = \{\ \} \quad :p \neq \{n+1\} \text{ and } Y = \langle\ \rangle$$
$$f(p,Y,v) = v \quad :p = \{n+1\} \text{ and } Y = \langle\ \rangle$$
$$f(p,Y,v) = f(\mathrm{ad}1(p,y), Y - y, v)$$
$$:\mathrm{TS}(y) \text{ and } \exists j \in p \ni y = s_j \text{ and } Y \neq \langle\ \rangle$$
$$f(p,Y,v) = \bigcup_{i=1}^{N(y)} f(p, \langle R_{i1}(y), R_{i2}(y), \ldots, R_{iN(y,i)}(y)\rangle \| (Y - y),$$
$$v \| P_i(y)) \ :\mathrm{NTS}(y) \text{ and } Y \neq \langle\ \rangle$$

Initially $p = \{1\}$, y is starting symbol of grammar G, $Y = \langle y \rangle$, $v = \langle\ \rangle$, the string $s = \langle s_1, \ldots, s_n \rangle$ and is invariant.

By dropping the rule string v, and other minor modifications, Definition 6.1.19 is converted to a recognition definition. This definition determines how far into the string s each path of a derivation can progress. The value of the recognition function defined is a set of positions in s which can be reached by a derivation of the given grammar. For example, if $\{p_1, \ldots, p_n\}$ is the set of positions, then each of the prefixes from s_1 to s_{p_1} to ... to s_{p_n} of s can be derived in the given grammar. Whether s is recognizable then is easily determined from this set. s is recognizable if and only if the set contains $n + 1$.

Definition 6.1.20: Simple-Top-Down Recognizer
y is used in place of Y_1 in this definition.

$$f(p,Y) = p \quad :Y = \langle\ \rangle$$
$$f(p,Y) = \{\ \} \quad :\mathrm{TS}(y) \text{ and } \sim \exists j \in p \ni y = s_j \text{ and } Y \neq \langle\ \rangle \text{ or } p = \{\ \}$$
$$f(p,Y) = f(\mathrm{ad}1(p,y), Y - y) \ :\mathrm{TS}(y) \text{ and } \exists j \in p \ni y = s_j \text{ and } Y \neq \langle\ \rangle$$
$$f(p,Y) = \bigcup_{i=1}^{N(y)} f(p, \langle R_{i1}(y), R_{i2}(y), \ldots, R_{iN(y,i)}(y)\rangle \| (Y - y))$$
$$:\mathrm{NTS}(y) \text{ and } Y \neq \langle\ \rangle$$

Initially $p = \{1\}$, y is starting symbol of grammar G, $Y = \langle y \rangle$, the string $s = \langle s_1, \ldots, s_n \rangle$ and is invariant.

6.1.6.2 Transformation to Nested Definition for Recognition—Early's Definition

This definition does not have a simple uniform inverse, but modest additions will give it one. Implementation by the depth-first algorithm, Algorithm 4.2.5, would then be appropriate since there does not appear to be any usable form of comparability available (the definition does not produce duplicates). It is easy to see that such an implementation is exponential in the worst case. Applying a reduction transformation may produce a less costly definition. Reduction 2 almost applies; however, because the condition on the second nonterminal line, :TS(y) and $y \neq s_p$ and $Y \neq \langle\ \rangle$, depends on p, that reduction will not apply. On the other hand this definition does fit the form of Definition 1 of reduction 3, Theorem 6.1.3, with the following interpretations:

> y is a symbol, terminal or nonterminal; initially y is the starting symbol of G.
>
> p is a set of positions in the input string s, initially p is $\{1\}$.
>
> Y is a generating string. Initially y is the starting symbol of G.
>
> $o_{ij}(y) = R_{ij}(y)$ is the jth symbol in the right side of the ith rule with y on the left.
>
> $m(p, y) = N(y)$ is the number of rules with y on the left.
>
> $n(p, y, i) = N(y, i)$ is the number of symbols in the right side of the ith rule with y on the left.
>
> Σ = union ($\cup$).
>
> $C_1(p, y) =$:TS(y) and $\sim \exists j \in p \ni y = s_j$ and $Y \neq \langle\ \rangle$.
>
> $C_2(p, y) =$:TS(y) and $\exists j \in p \ni y = s_j$ and $Y \neq \langle\ \rangle$.
>
> $C_3(p, y) =$:NTS(y) and $Y \neq \langle\ \rangle$.
>
> $0_* = \{\ \}$.
>
> $t_2(p, y) = \mathrm{ad}\,1(p, y)$.

This specialization of Definition 1 clearly meets the hypothesis of Theorem 6.1.3. So the theorem applies and Definition 2 of the theorem specialized as above is available for recognizing a given string s proposed as derivable from a context-free grammar G.

Definition 6.1.21: Nested Recursive Recognizer Function (Early's Function)

$$\left[\begin{array}{ll} g(p, y) = \mathrm{ad}\,1(p, y) & :\mathrm{TS}(y) \\ g(p, y) = \bigcup_{i=1}^{N(y)} N_{j=1}^{N(y,i)} g(p, R_{ij}(y)) & :\mathrm{NTS}(y) \end{array}\right.$$

Initially $p = \{1\}$, y is starting symbol of grammar G, the string $s = \langle s_1, \ldots, s_n \rangle$ and is invariant.

DIRECT DEVELOPMENT OF EARLY'S DEFINITION

This definition was obtained by transformation of a straightforward top-down definition based on enumerating all possible derivations. It could also be obtained and justified directly by recursive reasoning. Interpret $g(p, y)$ to be the *set* of all positions e in s such that $y \rightarrow * s_j \ldots s_{e-1}$ is a derivation in G, where j is in p. This is equivalent to saying that $g(p, y)$ is the set of all positions that can be "reached" in the input string starting at a position taken from p with a derivation starting with the grammar symbol y.

For the terminal conditions, that is, when $\mathrm{TS}(y)$, this is clearly a legitimate interpretation of Definition 6.1.21.

If $\mathrm{NTS}(y)$, then any derivation starting with y must start with the right side of a y rule. Say that such a right side was XY. Then by the interpretation of $g(p, y)$, using $y \rightarrow XY$ as the first derivation rule, $g(p, X) =$ the set of all positions that can be "reached" in s starting at a position taken from p with a derivation starting with X. Call this set of positions E. Then $g(g(p, X), Y) = g(E, Y) =$ the set of all positions that can be "reached" in s starting at a position taken from E with a derivation starting with Y. Call this set of positions F. Clearly $F = g(g(p, X), Y)$ is the set of all positions in s that can be "reached" starting at a member of p with a derivation starting with XY, which is equal to "starting with y and using rule $y \rightarrow XY$".

The right side of the recursive line of Definition 6.1.21 is based on similar reasoning for every right side of a y rule.

6.1.6.3 Grammar Recognition Example Using Early's Definition

Example 6.1.2 shows Definition 6.1.21 applied to the grammar $G(S \rightarrow aSb \mid aA \mid a;\ A \rightarrow aA \mid a)$ with the string aab to be recognized. The equations are given in groups. Each group has the same left side. The first equation in each group is the initial form of that equation. New equations are formed as long as there is a $g(j, N)$ on the right which does not appear on the left. When this no longer holds, back substitution starts. The subsequent equations in each group are the result of substitutions. (The set symbols { }, which should appear around the integers, have been dropped.)

Example 6.1.2

$$
\begin{aligned}
g(1, S) &= g(g(g(1, a), S), b) \cup g(1, a) \cup g(g(1, a), A) \\
&= g(g(2, S), b) \qquad \cup\ 2 \qquad \cup\ g(2, A) \\
&= g(3, b) \qquad \cup\ 2 \qquad \cup\ 3 \\
&= 4
\end{aligned}
$$

$$g(1, a) = 2$$

$$g(2, S) = g(g(g(2, a), S), b) \cup g(2, a) \cup g(g(2, a), A)$$
$$= g(g(3, S), b) \qquad \cup 3 \qquad \cup g(3, A)$$
$$g(2, A) = g(g(2, a), A) \cup g(2, a)$$
$$= g(3, A) \cup 3$$
$$g(2, a) = 3$$
$$g(3, S) = \{\ \}$$
$$g(3, A) = \{\ \}$$
$$g(3, b) = 4$$

6.1.6.4 Complexity of Early's Recognition Algorithm

In developing the worst-case complexity for an ESE-based algorithm of Definition 6.1.21, we assume the given grammar is in *Greibach normal form*. In Greibach normal form every rule right side *begins with a terminal symbol. All the other symbols are nonterminal.* Given any CFG grammar G, there is another CFG grammar G' in Greibach normal form with $L_{G'} = L_G$. This assumption allows us to focus on the key points of the complexity argument. Even without this assumption, however, the bounds hold.

In evaluating Definition 6.1.21 by ESE, we first obtain one equation for each string position (p)–grammar symbol (y) pair. For y an NTS, an equation has the form

$$g(p, y) = \mathrm{TM}_1 \cup \mathrm{TM}_2 \cup \cdots \cup \mathrm{TM}_{N(y)}$$

p is a set containing a single position number (Since p is a set containing a single member, we will use p to represent that member.) and TM_i is of the form

$$g(\ldots g(g(p, A_{i1}), A_{i2}), \ldots, A_{in(p, y, i)})$$

corresponding to the grammar rule

$$y \rightarrow A_{i1}A_{i2} \ldots A_{in(p, y, i)}$$

For y a TS, $g(p, y) = p + 1$ (that is, [the member of p] + 1) or $\{\ \}$ terminal symbol.

We are interested in the cost of evaluating the right side of the equation when y is an NTS. In general the value of such equations is a set of positive integers. Assume inductively that the values for $g(q, A)$, with $q > p$ and A any symbol in the grammar, have been determined. This is easy enough to do when $q = n = |s|$. To find the value of $g(p, y)$, each term TM_i on the right of the equation must be evaluated. A term is evaluated by substitution. First $g(p, A_{i1})$ is replaced by its value, which with the Greibach normal form assumption is easily determined. By that assumption A_{i1} is a terminal symbol and need only be compared with s_p to find the value of $g(p, A_{i1})$. Let $p_1 = g(p, A_{i1})$. Next we substitute for $g(p_1, A_{i2})$. Either $p_1 > p$ or $p_1 = \{\ \}$, so, by inductive assumption, we obtain the set of integers which $g(p_1, A_{i2})$ is equal to. Let this set be P_2. It has no more than n members, and thus no more than n substitutions are needed. Next it will be necessary to evaluate $g(p_2, A_{i3})$ for each $p_2 \in P_2$. Each such evaluation may require as many as n substitutions, and there are at most n evaluations, so in total no more than n^2 substitutions are needed. In the end, nevertheless, the total set substituted for $g(p_2, A_{i3})$ for all p_2 in P_2 can have no more than n members. Therefore, repeatedly using an argument like the previous, we conclude that evaluation of the expression $g(\dots g(g(p, A_{i1}), A_{i2}), \dots, A_{in(p,y,i)})$ for each nest level from $j = 3$ to $j = n(p, y, i)$ will require no more than n^2 substitutions. So, if no term TM_i is longer than t, $t * n^2$ substitutions is sufficient to evaluate any term. Therefore evaluation of $g(p, a)$ for any pair (p, a) requires $\leq m * t * n^2$ substitutions, assuming there are $\leq m$ terms on the right of an equation (where m equals the greatest number of rules associated with any NTS). Now if v is the total number of different symbols in the grammar, terminal and nonterminal, the total number of equations will be $\leq v * n$. So the total number of substitutions necessary will be $\leq v * m * t * n^3 = O(n^3)$.

This bound on the number of substitutions necessary for evaluating the recognizing function can be improved if consideration is confined to *unambiguous grammars*. The factor of n^2 required to evaluate each term then becomes n, reducing the evaluation time to $O(n^2)$. This is because, referring to the notation of the previous paragraph, the set of values that $g(p_2, A_{i3})$ can assume for any member p_2 in P_2 must be disjoint from the set of values it can assume if p_2 is any other member of P_2. This must be true since otherwise there are two members of P_2, say r and q, and a position u, such that $g(r, A_{i3}) = u = g(q, A_{i3})$. This would in turn imply that the following two derivations are possible in the given grammar. (We use A_j for A_{ij}, τ for $s_1 \dots s_{p-1}$, and β for a string of terminals and nonterminals. Also, for each nonterminal y in the grammar, we assume there is a P and a derivation $S \overset{*}{\rightarrow} \tau y \beta$ with $|\tau| = p - 1$, since otherwise y can be omitted from the grammar.)

$$S \rightarrow* \tau y \beta \rightarrow \tau A_1 A_2 \dots A_n \beta \rightarrow* \tau \overbrace{s_p \dots s_{r-1}}^{A_1 A_2} A_3 \dots A_n \beta$$

$$\rightarrow* \tau s_p \dots s_r \dots s_{u-1} A_4 \dots A_n \beta$$

Also,

$$S \rightarrow* \tau y \beta \rightarrow \tau A_1 A_2 \dots A_n \beta \rightarrow* \tau s_p \dots s_{q-1} A_3 \dots A_n \beta$$

$$\rightarrow* \tau s_p \dots s_q \dots s_{u-1} A_4 \dots A_n \beta$$

The existence of these two different derivations with the same outcome implies that the grammar is ambiguous—a contradiction. A similar argument shows further that at most n substitutions will be required for each nesting level beyond that at which $j = 3$,

$$g(\ldots g(g(p, A_{i1}), A_{i2}), \ldots, A_{in(p,y,i)})$$

The remaining factors in counting the number of substitutions are the same as in the ambiguous case. Therefore the total number of substitutions required in the nonambiguous case is $O(n^2)$.

6.2 IMPLEMENTATION OF EQUATION SET EVALUATION APPROACH (ESE) — GENERAL

In the previous section we saw how a definition Def.f, with comparability not based on equality can be transformed to an equivalent definition Def.g in which comparability is based on equality. Whereas both Def.f and Def.g can be implemented by SE algorithms developed in Chapter 4, augmented to take advantage of the existence of comparable argument structures, Def.g can also be neatly implemented by the ESE approach. We now study ESE as specialized to nonnested definitions, both terminating and nonterminating. The general ESE process is given in Chapter 2 as Process 2.7.2 and Process 2.7.3. It applies to nested and nonnested definitions and may or may not produce an answer. Here we will identify a class of definitions for which ESE, augmented by a so-called loop-breaking rule, will always produce an answer. The term "loop-breaking rule" is completely defined in subsequent sections. We use it now in our overview of the specialized ESE process to indicate where such a rule is employed. It may be helpful to refer to the example in Section 6.2.2 even at this early stage to see how such a rule is used.

When applied to nonnested definitions, as shown in Lemma 2.7.1, ESE as exemplified by Process 2.7.2, consists of first formulating the equations from the given definition, and second, solving the set by back substitution. A definition is called *equation complete* if the first process ends by producing a finite set of equations. The second process, back substitution, is effective only if the definition terminates. By extending the back substitution process to a complete "Gaussian elimination" process using a loop-breaking rule, we can evaluate many nonterminating definitions. (Because the formulation of the set of equations is often effectively given by the problem statement, we will be concerned mainly with solution of the set of equations rather than with formulation.)

In the remainder of this chapter, our attention will be focused on algorithms for evaluating a class of nonnested, equation complete definitions, including those that terminate and also those that do not terminate but have a loop-breaking rule, by simulating an ESE process. Many definitions of practical interest have significant properties beyond these. They form subclasses with simplified implementation, some of which will be studied. So in addition to simplifications implied by these properties, others may become apparent through consideration of ESE for a number of small examples of the given problem. Such simplifications should then be incorporated in the substitution process.

6.2.1 Loop-Breaking and Gaussian Elimination

Assume a set of equations E of the form

$$\{E_i\colon x_i = W_i(x_{i_1}, \ldots, x_{in_i}) \mid i = 1 \text{ to } n\}$$

where E_i is the label for the ith equation. For each legitimate selection of values in their respective domains assigned to the right side variables $\{x_1, \ldots, x_{n_i}\}$, $W_i(x_{i_1}, \ldots, x_{n_i})$ is constructible and has a single value. (This assumption can be weakened.) The set of values that $W_i(x_{i_1}, \ldots, x_{n_i})$ takes, as the right side variables assume all their legitimate values, has a *partial ordering* PA, inducing the relation $\leq$.

A *substitution transformation* of E, $s(E, i, j)$, for $1 \leq i, j \leq n$, yields the result of substituting the right side of E_i for an occurrence of x_i on the right of equation E_j, $i \neq j$, and simplifying the resultant right side of E_j according to *identities* relating different expressions of w_i. $s(E, i, j)$ differs from E at most in having different E_j equations. $S(E, i, j) = E$ if x_i does not occur nontrivially in W_j.

PROPERTY 1: Each solution of $s(E, i, j)$ is a solution of E and vice versa. The proof of Property 1 of substitutional transformations, as well as that for Properties 2 and 3 given below, are developed in Section 6.9.4.

An equation E_i is said to have a *loop-breaking rule* if there is an equation

$$e_i\colon \quad x_i = w_i(x_{i_1}, \ldots, x_{i_{n_i}}) \qquad (= \text{loop-break}(E_i))$$

for x_i in which:

1. x_i does not occur (nontrivially) on the right of e_i.
2. No variable that wasn't originally on the right is introduced.
3. Every solution to e_i is a solution to E_i.
4. For every solution S to E_i there is a solution of s of e_i so that $s \leq S$.

A *set of equations* E is said to have a loop-breaking rule if for each equation in E and for any equation resulting from E by a sequence of substitutional and loop-breaking transformations, there is a loop-breaking rule.

A loop-breaking transformation of E, $b(E, i)$, $1 \le i, j \le n$, is the result of replacing the equation E_i with loop-break(E_i).

PROPERTY 2: Each solution s of $b(E, i)$ is a solution to E_i, and for every solution S to E_i there is a solution of s of $b(E, i)$ so that $s \le S$.

The loop-breaking rule has a graph-theoretic interpretation. Corresponding to an equation set there is a *dependency digraph*, as defined in Section 2.4 for recursive definitions. This digraph has one vertex for each of the n variables in the equations. It has an edge from vertex (corresponding to) x_a to vertex x_b if x_a appears on the left (of the equal sign) of an equation in which x_b appears on the right.

A set of equations with an acyclic dependency graph can be solved by back substitution. In such a set of equations there must be at least one whose right side is a constant, say, $x_a = c$. c is substituted for every occurrence of x_a on the right of equations in the set, thus eliminating x_a from the right of every equation. After simplification, at least one of the other equations will have only a constant on the right. This process can be continued until the equation for the variable whose value is sought has only a constant on the right. The existence of the finite acyclic dependency graph guarantees that the process will terminate with the solution. If on the other hand the graph has cycles, as it will for nonterminating recursive definitions, something more than back substitution is required. If the back substitutions are done in an equation set with dependency graph cycle(s), an equation will eventually be produced with the same variable appearing on the left and the right of the equality. Back substitution with such an equation will not guarantee elimination of any variables. However, if a rule is available for replacing such an equation with one in which the left variable does not appear on the right, with a guarantee that any solution to this new equation set will satisfy the original, then it becomes possible to move the elimination process forward, though no longer with substitution of constants only. The loop-breaking rule is a rule for cycle breaking.

The Gaussian elimination type of equation set solution we speak about here consists of a sequence of applications of the two transformations (substitution and loop breaking) described above. It follows from Properties 1 and 2 that

PROPERTY 3: If a sequence of applications of these transformations to a set of equations E produces the set E', then the solution to E' is also a solution to E, and if SOL $= \{x_i = S_i | i = 1 \text{ to } n\}$ is a solution to E, then there is a solution sol $= \{x_i = s_i | i = 1 \text{ to } n\}$ of E' with $\{s_i \le S_i | i = 1 \text{ to } n\}$ (we say sol $\le$ SOL).

The following algorithm always yields a solution like E' of Property 3 to a set E with a loop-breaking rule—so such a set always has a unique lowest solution by the partial ordering PA.

Algorithm 6.2.1: Gaussian-Elimination-Type (G-E-T) Algorithm 1

```
    Elimination
FOR I = 1 to N DO
  E ← b(E, I)
  FOR J = I + 1 to N DO
    E ← s(E, I, J)
  END
END

    Back substitution
FOR I = N to 2 DO
  FOR J = I − 1 to 1 DO
    E ← s(E, I, J)
  END
END
```

The complexity of this algorithm is $O(n^3)$.

Lemma 6.2.1: If E has a loop-breaking rule, then the G-E-T1 algorithm, Algorithm 6.2.1, terminates and assigns a unique set of values for x_i, i ranging from 1 to n, and this set of values is the lowest solution to E.

PROOF: For each i from 1 to n G-E-T1 will be able to remove any occurrence of x_i on the right of E_i with the guaranteed applicable loop-breaking rule. Then G-E-T1 would eliminate x_i from all equations E_j, $j > i$. There is nothing to prevent the required substitutions. So the eliminate part of G-E-T1 ends with x_i eliminated from all equations E_j, $j \geq i$. That means that all variables have been eliminated from E_n's right side, leaving only a constant. That constant is the value assigned to x_n. The back substitution part of the algorithm will now be carried out and leave in the end a constant assigned to each variable. There is no reason why the substitution of this part cannot be carried out, so we end G-E-T1 with a single value assigned to each and every variable.

OTHER GAUSSIAN TYPE ALGORITHMS

Algorithm G-E-T1 is quite general. In practice equation sets often arise for which other approaches are more efficient. One such class is determined by the structure of the dependency digraph. The number of edges in a dependency digraph range from as few as $O(1)$ to as many as $O(n)$ per vertex, and these edges may form simple repetitive patterns or display no pattern at all. These characteristics affect the solution complexity. The degree of a vertex affects the number of substitutions that will be necessary in solving for the corresponding variable. If the vertices can be partitioned into blocks with high connectivity within, but light connectivity between blocks, the equations can be efficiently solved by eliminating variables within each block as far as possible before eliminating the remaining variables. As a more detailed example of the effect of structure on the algorithm consider a set of equations in which a subset of variables, say x_1 through x_p, have the property that: In the equations for x_1 through x_p only the variables x_1 to x_{p+1} appear.

Then using only substitutional and loop-breaking transformations we can transform the equations for x_1 through x_p so that each has only one variable, namely, x_{p+1}, on the right. Then by substituting back on the right side of the equation for x_{p+1} for x_1 through x_p, we can eliminate x_1 through x_p from the right of x_{p+1}. The equations for x_1 through x_p may now be set aside, and the remaining $n - p$ equations now contain no occurrence x_1 through x_p, that is, assume we have

$$x_1 = W_1(x_1, \ldots, x_{p+1})$$

$$x_2 = W_2(x_1, \ldots, x_{p+1})$$

$$\vdots$$

$$x_p = W_i(x_1, \ldots, x_{p+1})$$

$$x_{p+1} = W_n(x_1, \ldots, x_n)$$

$$\vdots$$

The following algorithm leaves the x_1 through x_p equations dependent on x_{p+1} only and x_{p+1} only on x_{p+1} through x_n.

Algorithm 6.2.2: Partial Gaussian-Elimination-Type (G-E-T) Algorithm 2

```
        Elimination
FOR I = 1 TO P DO
  E ← b(E, I)
  FOR J = I + 1 TO P DO
    E ← s(E, I, J)
  END
END

    Back substitution
FOR I = P TO 1 DO
  FOR J = I − 1 TO 2 DO
    E ← s(E, I, J)
  END
END
FOR I = 1 to P DO
  E ← s(E, I, P + 1)
END
```

After application of G-E-T2, Algorithm 6.2.2, the equations for x_i through x_p are of the form

$$x_i = w_i(x_{p+1})$$

and the one for x_{p+1} is

$$x_{p+1} = w_n(x_{p+1}, \ldots, x_n)$$

The complexity of this algorithm is $O((p+1)^3)$.

Now suppose that just as the 1st through $(p+1)$th equations can be solved so that the 1st through pth can be set aside, the $(p+2)$th through $(2p+2)$th can be solved to set aside the $(p+2)$th through $(2p+1)$th, and so on for the $(2p+3)$th through $(3p+3)$th equations. This can be done with complexity $O([n/(p+1)](p+1)^3)$, finally leaving a set of $n/(p+1)$ equations which have not been "set aside," consisting of the equations for x_{kp+k}, $1 \le k \le n/(p+1)$. These might again have the structure of the original set of n equation which allowed the solution of groups of $p+1$, but they can always be solved using G-E-T1. Suppose they are solved by G-E-T1, the time cost will be $O(n^3/(p+1)^3)$. The time to substitute the solutions from these equations into those that were set aside is $O(n)$.

$$O\left([n/(p+1)](p+1)^3\right) + O\left(n^3/(p+1)^3\right)$$

If $p+1$ is a constant independent of n, the complexity will still be $O(n^3)$. If, however, $p+1 = n^{1/2}$,

$$O(n^{1/2}n^{3/2}) + O\left(n^3/(n^{1/2})^3\right) = O(n^2) + O(n^{3/2}) = O(n^2)$$

Groupings based on limited dependencies as that above often give efficient algorithms.

In the preceding discussion, groupings of equations with limited dependence of variables were delineated by consecutive numbering. Similar groupings occur in many optimization problems in compiler design, where they are called *intervals* [Aho, Sethi, Ullman 86]. The interval groupings, though not consecutively numbered, can be found efficiently. There are a number of such group finding algorithms, dependent on the nature of the limited dependency that defines the grouping.

In summary, given a set of equations E, for which there is a loop-breaking rule which persists through substitution and loop-breaking transformations, it is legitimate to apply any sequence of substitutional and loop-breaking transformations to get a solution to E.

So it is important to find loop-breaking rules. We now consider a number of examples of equation sets with loop-breaking rules.

STANDARD LINEAR EQUATIONS

$$\{ E_i\text{: } x_i = a_{i1}x_1 + \cdots + a_{in}x_n + c_i) \mid i = 1 \text{ to } n \}$$

a_{ij} and c_i are real numbers.

We are looking for a solution with the x_j's in the reals. Obviously substituting any reals for the x_i's above, the result on evaluation of the right side is unique real.

A standard linear equation, where each x_i's has a value in the reals, has a loop-breaking rule, provided that there is no cycle in its dependency graph along which the product of coefficients equals 1, that is, there is no sequence of variables $\langle y_1, y_2, \ldots, y_p \rangle$ with $a_k y_{k+1}$ on the right of the equation for y_k, in which $y_1 = y_p$ and $y_1 * y_2 * \cdots * y_p = 1$. That loop-breaking rule follows.

If $x = ax + \beta$ (β is the value taken by the sum of all terms other than the x term on the right of this equation), then if $a_i = 0$,

$$\text{loop-break}(E_i) \text{ is } x = \beta$$

and if $a \neq 0$ and $a \neq 1$,

$$\text{loop-break}(E_i) \text{ is } x = \beta/(1 - a_i)$$

$a = 1$ only if there is a cycle in the dependency graph along which the coefficient product is 1, and we do not claim there is a loop-breaking rule in this case. The absence of a loop-breaking rule does not mean that there is no solution to the set E, unique or otherwise. It only means that in our proposed use the loop-breaking rule in G-E-T is not adequate to handle this situation—although there is an augmentation that would.

To be applicable to an equation set, a loop-breaking rule must continue to be applicable after each use. The key to continued applicability in linear equations is that after loop-breaking, the resultant right side remains a sum of terms, each of which is a constant times a variable. This is so because the operations * and + are associativity and * distributes over +. Since the loop-breaking rule depends on this form, it will continue to hold.

Since loop-break(E_i) given above is the result of applying simple algebraic equivalences whenever it is applied, any solution satisfying an equation E_i also satisfies loop-break(E_i) and vice versa. So the solution we obtain using G-E-T1 is unique. Consider a second example

UNION–INTERSECTION "LINEAR" EQUATIONS

$$\{ E_i\text{: } x_i = a_{i1} \cap x_1 \cup \cdots \cup a_{in} \cap x_n \cup c_i) \mid i = 1 \text{ to } n \}$$

for a_{ij}, c_i finite subsets of a given universal finite set D.

We are looking for a solution with the x_j's being subsets of D. Obviously if we substitute any finite sets for the x_i's on the right of the equation above, the result on evaluation will be a unique subset of D. The loop-breaking rule follows.

If E_i: $x = a \cap x \cup \beta$, then for any finite set, including the empty set,

$$\text{loop-break}(E_i) \text{ is } x = \beta$$

The loop-breaking rule is applicable in all cases. In Section 6.7, which is devoted to developing a large number of such rules, we show that any solution satisfying the equation e_i = loop-break(E_i) also satisfies E_i, and for every solution S to E_i there are solutions of s of e_i so that $s \leq S$.

The key to the continued applicability of loop-breaking in this case is that the right side remains a union of terms each of which is a constant set intersected with a variable. This is so because of the laws of associativity and distributivity obeyed by the operations $\cap$ and $\cup$. Since the loop-breaking rule justification depends crucially on this form, it will continue to hold.

As another similar example consider the following.

INTERSECTION–UNION "LINEAR" EQUATIONS

$$\{E_i\colon x_i = (a_{i1} \cap x_1 \cup c_{i1}) \cap \cdots \cap (a_{in} \cap x_n \cup c_{in}) \cap C_i | i = 1 \text{ to } n\}$$

a_{ij}, c_{ij}, C_i are finite subsets of a given universal finite set D.

We are looking for a solution in which the x_j's are subsets of D. Obviously if we substitute any finite sets for the x_i's on the right of the equation above, the result on evaluation is a unique subset of D. The loop-breaking rule is given as follows.

If $x = (a \cap x \cup c) \cap \beta$,

$$\text{loop-break}(E_i) \text{ is } x = c \cap \beta$$

This is shown in a way analogous to that for showing the union–intersection loop-breaking rule.

Again applicability continues because of the laws of associativity and distributivity obeyed by the operations $*$ and $+$. The loop-breaking rule continues to hold after each transformation.

Now we consider a (sub)class of definitions to which ESE is applicable.

p1. *Nonnested recursive definitions* which are

p2. *Equation complete*, and for which

p3. *There is a loop-breaking rule for the equation set generated by the definition.*

Equation sets which arise from such definitions can always be solved by an analog of the Gauss (or Gauss–Jordan) algorithm. This algorithm is illustrated by application to a shortest path problem. Later the general algorithm will be outlined. Loop-breaking rules are established for a considerable number of different kinds of equation sets in Section 6.7.

6.2.2 Elimination Methods

Example 6.2.1: Shortest Path—Gaussian Elimination
Here Definition 6.1.8 for the shortest path problem is repeated.

Definition 6.2.1: The Minimum Path

$$\left[\begin{array}{ll} g(v) = \langle 0, \langle v \rangle \rangle & : v = \text{“goal”} \\ g(v) = \text{MIN-SEL}_{i=1}^{N(v)}(\langle c_i(v), \langle v \rangle \rangle +\| \ g(n_i(v))) & : v \neq \text{“goal”} \\ \text{Initially } v = \text{“begin.”} & \end{array}\right.$$

This definition can be evaluated using ESE for any digraph G. First a set of equations is developed. For each vertex v in G there will be an equation with $g(v)$ on the left. Lowercase letters are used for vertices. In the following example, V is used to abbreviate $g(v)$. ($\downarrow$ is short for minimum-select, $+$ is short for $+\|$.)

Consider the digraph shown in Figure 6-1. The shortest path definition gives the equation set

$$A = (\langle 1, a \rangle + B) \downarrow (\langle 4, a \rangle + C)$$

$$B = (\langle 1, b \rangle + D)$$

$$C = (\langle 1, c \rangle + E)$$

$$D = (\langle 1, d \rangle + C) \downarrow (\langle 4, d \rangle + E)$$

$$E = (\langle 1, e \rangle + B) \downarrow (\langle 5, e \rangle + F)$$

$$F = (\langle 0, f \rangle)$$

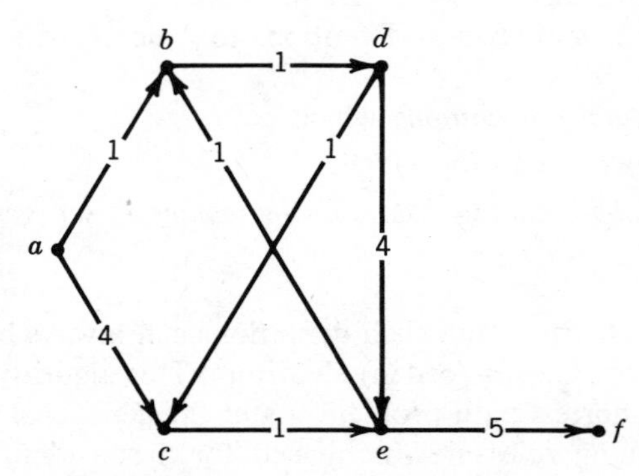

FIGURE 6-1 Graph example for shortest path.

We begin to solve by substitution of constants where possible. We can substitute $\langle 0, f\rangle$ for F, eliminating F to obtain

$$A = (\langle 1, a\rangle + B)\downarrow(\langle 4, a\rangle + C)$$

$$B = (\langle 1, b\rangle + D)$$

$$C = (\langle 1, c\rangle + E)$$

$$D = (\langle 1, d\rangle + C)\downarrow(\langle 4, d\rangle + E)$$

$$E = (\langle 1, e\rangle + B)\downarrow(\langle 5, ef\rangle)$$

Now there are no more constants on the right and we progress by the substitution of entire expressions for variables. Substitute for B in A and E, eliminating B.

$$A = (\langle 1, a\rangle + \langle 1, b\rangle + D)\downarrow(\langle 4, a\rangle + C)$$

$$= (\langle 2, ab\rangle + D)\downarrow(\langle 4, a\rangle + C)$$

$$C = (\langle 1, c\rangle + E)$$

$$D = (\langle 1, d\rangle + C)\downarrow(\langle 4, d\rangle + E)$$

$$E = (\langle 1, e\rangle + \langle 1, b\rangle + D)\downarrow(\langle 5, ef\rangle)$$

$$= (\langle 2, eb\rangle + D)\downarrow(\langle 5, ef\rangle)$$

Eliminate C,

$$A = (\langle 2, ab\rangle + D)\downarrow(\langle 4, a\rangle + \langle 1, c\rangle + E)$$

$$= (\langle 2, ab\rangle + D)\downarrow(\langle 5, ac\rangle + E)$$

$$D = (\langle 1, d\rangle + \langle 1, c\rangle + E)\downarrow(\langle 4, d\rangle + E)$$

$$= (\langle 2, dc\rangle + E)\downarrow(\langle 4, d\rangle + E)$$

$$= (\langle 2, dc\rangle + E)$$

$$E = (\langle 2, eb\rangle + D)\downarrow(\langle 5, ef\rangle)$$

Eliminate D,

$$A = (\langle 4, abdc\rangle + E)\downarrow(\langle 5, ac\rangle + E)$$

$$= (\langle 4, abdc\rangle + E)$$

$$E = (\langle 4, ebdc\rangle + E)\downarrow(\langle 5, ef\rangle)$$

To eliminate E, substitution is not sufficient; we need a loop-breaking rule. As will be shown in our general discussion of loop-breaking rules, there is such a rule for this case.

An equation of the form

$$X = (v + X) \downarrow Q$$

appearing in a set of equations may be replaced with

$$X = Q \qquad (\text{since } Q = (v + Q) \downarrow Q)$$

and the solution to the altered set will also satisfy the original set.

So we conclude that $E = \langle 5, ef \rangle$, and thus substituting,

$$A = \langle 9, abdcef \rangle$$

Notice that the first components of each coefficient in the initial equation set are simply the entries that would appear in an adjacency matrix for a weighted graph. The second component is the vertex represented by the left side of the equation in which the coefficient occurs. So the set of equations is given directly by the definition of the graph in the problem specification.

This illustrates Gaussian elimination. A general description of elimination methods is given next. We can eliminate x_j from the right of equation E_j by applying the loop-breaking rule. Call this process ELIM(j, j).

We can eliminate x_j from the right of equation E_i by the following process:

1. ELIM(j, j) (loop-breaking).
2. Substituting $r(E_j)$ for all appearances of x_j in $r(E_i)$ and simplifying the resultant $r(E_j)$.

Call this process ELIM(i, j). If we apply ELIM(i, j) in any order to pairs $\langle i, j \rangle$, taking care that once having eliminated a variable on the right of an equation we never reintroduce it in eliminating some other variable, we will eventually have all solutions. This can always be done. In fact it can be done systematically in a variety of ways, some of which will be explored later.

Summarizing

Process 6.2.1: Elimination Methods

```
WHILE the equation whose solution is desired is not solved DO
  [a, b is a pair of integers, each ≤ n]
  IF r(E_b) contains no variable already eliminated from r(E_a) THEN ELIM(a, b)
END
```

By constraining the order in which ELIM(a, b) is applied we get different elimination methods, including, among others, the methods described by Process 6.2.2.

Process 6.2.2: Gauss and Gauss–Jordan Elimination
[E^* initially is the set of equations to be solved. In the operation of the algorithm, equations will be continually removed from E^* as the variable on the left of the equation is eliminated. B^* is a set, initially empty, containing equations for all variables which have been eliminated from E^*. Assume the object is to solve for every variable in E^*]

```
WHILE (E* is not empty) DO
    FOR E_a ∈ E* DO
        FOR ALL E_j ∈ E* [and B*]^1 with j ≠ a
            ELIM(a,j)
        END                                              [Elimination]
    END
     B* ← B* ∪ E_a
     E* ← E* − E_a
END
WHILE (B* is not empty) DO
    FOR (E_b ∈ B* and r(B*) = constant) DO
        FOR ALL E_j ∈ B* with j ≠ b DO
            ELIM(b,j)
        END                                              [Back Substitution]
         print E_b
         B ← B − B*
    END
END
```

Notes

1. For Gaussian elimination exclude, and for Gauss–Jordan include, the bracketed part at superscript 1. The back substitution part will be unnecessary in the Gauss–Jordan case.
2. If the object is to solve for only one variable, say x_d, then the set B^* is unnecessary provided that whenever the *Elimination part* is entered, E_d is checked first for a constant right side. The entire procedure would terminate as soon as this happens — that constant is the answer. On the other hand, whenever the *Elimination part* is entered, the equation E_d is not considered unless the application of the loop-breaking rule is guaranteed to leave a constant right side, in which case the entire procedure terminates with that constant as answer.

Next we consider more efficient versions of this elimination algorithm, and the class of recursive definitions to which they apply.

Elimination approaches generally require the substitution of entire expressions involving variables as well as constants, though on occasion only constants need be substituted with consequent saving in work. Such an occasion is present when the dependency graph has no cycles. Such occasions also arise for a significant class of equation sets which include many sets with dependency cycles, namely those cases to which the *minimum-constant principle* applies.

6.3 MINIMUM-CONSTANT PRINCIPLE

The minimum-constant principle applies to equation sets, satisfying p1 and p2 (nonnested and equation complete), and in which each variable x_i is given as a function of a subset of the other variables and a constant in the following way:

f1. $\{x_i = w_i(x_{i_1}, \ldots, x_{i_{n_i}}) \downarrow c_i \mid i \in N^n\}$ ($\downarrow$ is min or minimum-select).

Any equation with numerical variable values can be put into this form by making $c_i = \infty$. Furthermore each equation in the set must satisfy the condition that the function w_i on the right has a value greater than that of at least one variable in the set of equations; more precisely,

p4. For each $i \in N^n$ there is a $k \in N^n \ni w_i(x_{i_1}, \ldots, x_{i_{n_i}}) > x_k$.

This condition does not require determination of k for a given i, only the existence of some k for each i. For example, in one formulation the equation for the cost of the shortest path would be

$$x_i = (a_{i1} + x_1) \downarrow (a_{i2} + x_2) \downarrow \cdots \downarrow (a_{in} + x_n) \downarrow c_i, \qquad \text{with } a_{ij} > 0$$

Each term $a_{ij} + x_j$ is greater than x_j; therefore even though we do not know k, we do know that

$$(a_{i1} + x_1) \downarrow (a_{i2} + x_2) \downarrow \cdots \downarrow (a_{in} + x_n)$$

is greater than x_k for some $k \in N$.

If p4 holds, then we can draw the following important conclusion.

Theorem 6.3.1: Let E^* be a set of equations of form f_1 satisfying $p4$. Let a, $1 \leq a \leq n$ be such that $c_a = \mathrm{MIN}_{i=1}^{i=n}[c_i]$. If the equation for x_a in E^* is replaced with $x_a = c_a$ to obtain $E^{*\prime}$, then any solution of E^* is a solution of $E^{*\prime}$, and vice versa.

PROOF: Taking minima on both sides of the set of equations E^*, and using the commutativity and associativity of the minimum function,

$$\mathrm{MIN}_{i=1}^{i=n}[x_i] = \mathrm{MIN}_{i=1}^{i=n}\left[w_i(x_{i_1}, \ldots, x_{i_{n_i}})\right] \downarrow \mathrm{MIN}_{i=1}^{i=n}[c_i] \tag{1}$$

By p4, for each i,

$$w_i(x_{i_1}, \ldots, x_{i_{n_i}}) > x_k \geq \text{MIN}_{i=1}^{i=n}[x_i] \qquad \text{for some } 1 \leq k \leq n$$

$$\text{MIN}_{i=1}^{i=n}\left[w_i(x_{i_1}, \ldots, x_{i_{n_i}})\right] > \text{MIN}_{i=1}^{i=n}[x_i]$$

Therefore by Eq. (1),

$$\text{MIN}_{i=1}^{i=n}[x_i] = \text{MIN}_{i=1}^{i=n}[c_i] = c_a$$

Since $x_a \geq \text{MIN}_{i=1}^{i=n}[x_i]$, $x_a \geq c_a$. But by definition, $x_a \leq c_a$. So

$$x_a = c_a$$

This shows that any solution to $E^{*\prime}$ satisfies E^*. Furthermore $E^{*\prime}$ can only be satisfied with $x_a = c_a$ because it is, as in E^*, the minimum constant. ■

If p4 continues to hold after substitutions, we have p5.

p5. p4 holds for a set of equations E, and after substitution by the minimum constant, simplification, and the elimination of variables whose value is the minimum constant, *p4 continues to hold for the remaining set*. This continues to be true for repeated elimination of minimum constants. We say the *minimum-constant principle is applicable to E*.

If the minimum-constant principle is applicable to a set of equations, E^*, then the following adaptation of Gaussian elimination, involving the *substitution of constants only*, will solve E^*. Where one would have had to substitute an expression involving variables, because all equation right sides involve variables, one may now find the equation with the minimum constant and eliminate everything but that constant from its right side. So one need never substitute an expression having any variables.

Process 6.3.1: Minimum-Constant Elimination
[E^* initially is the set of equations. In the operation of the algorithm, equations will be continually removed from E^*. K^* is a set of equations, initially empty, holding the solution to the set of equations. Assume the object is to solve for every variable in E^*.]

```
Main routine
  WHILE (E* is not empty) DO
    elim-const-rght-side(E*) or
    change-eq-with-min-const( )
  END
```

```
Subroutines
  PROC: elim-const-rght-side(E*)
    IF (r(E_a) for any E_a in E* is a constant) THEN DO
      FOR ALL E_j ∈ E* with j ≠ a DO
          ELIM(a,j)
      END
      K* ← K ∪ E_a
      E* ← E* − E_a
    END
    WHILE (r(E_j) for any E_j in E* contains an occurrence of x_a) DO
      Subst c_a for all occurrences of x_a in E_j and
      Simplify result (gather terms)
    END
  END:elim-const-rght-side
  PROC: change-eq-with-min-const( )
    c_a ← minimum constant (c_i) in all equations in E*
    Make E_a in E* = "x_a = c_a"
  END:change-eq-with-min-const
```

Notes. By altering the algorithm to halt when the equation for a variable, say x, is solved, that is, entered into K, an algorithm that gives the solution for x, rather than for all variables is obtained.

Whether seeking the solution for one variable or of all, the nondeterminism in Process 6.3.1 may be resolved in a number of ways. The *elim-const-rght-side* procedure could be repeated until its preconditions were no longer satisfied, followed by execution of the *change-eq-with-min-const* procedure. This pattern would be repeated until the solution is complete. Alternately the procedure *changing-eq-with-min-const* could be executed and *elim-const-rght-side* used to substitute the minimum constant (even though more than one constant is available). This pattern could be repeated until the solution was obtained. This has the advantage of avoiding substitution into expressions whose constant term is minimum. On the other hand, it may require more minimum operations to determine the minimum constant than the alternative.

The minimum-constant principle serves as a special form of loop-breaking rule. If the equation with the minimum constant has a "loop," it will be removed when the right side is reduced to a constant.

The equation set generated by the minimum path definition in Example 6.2.1 satisfies the minimum-constant principle. A typical equation is

$$X = (\langle c_1, x \rangle + y) \downarrow (\langle c_2, x \rangle + z) \downarrow \langle c_3, xw \rangle$$

As long as c_1 and c_2 are positive, the terms $(\langle c_1, x \rangle + y)$ and $(\langle c_2, x \rangle + z)$ must have cost components c_i greater than those of y and z respectively. Therefore the cost component of $(\langle c_1, x \rangle + y) \downarrow (\langle c_2, x \rangle + z)$ is greater than at least one of the costs of y and z. Also, condition p4 is met. (Notice that no constraint on c_3 is required.) In general, the cost component of a term T, such as $(\langle c_1, x \rangle + y)$, has the form "constant + variable," so with the assurance that *these constants are greater than* 0, we can conclude that the cost component of T is greater than that of a least one variable. Thus the *minimum of all terms* is greater than some variable. So p4 applies, and since the equation forms are unchanged after constant substitution and simplification, p5 also holds.

The operation of Process 6.3.1 is illustrated on the same minimum path problem we solved with Gaussian elimination, Example 6.2.1.

Example 6.3.1: Shortest Path—Minimum Constant
We pick up the solution after the first substitution of a constant, namely, $\langle 0, f\rangle$ for F,

$$A = (\langle 1, a\rangle + B)\downarrow(\langle 4, a\rangle + C)$$

$$B = (\langle 1, b\rangle + D)$$

$$C = (\langle 1, c\rangle + E)$$

$$D = (\langle 1, d\rangle + C)\downarrow(\langle 4, d\rangle + E)$$

$$E = (\langle 1, e\rangle + B)\downarrow(\langle 5, ef\rangle)$$

Now there is no remaining right side that is constant. Previously we progressed by the substitution of an entire expression for variable B. This time we invoke the minimum-constant principle. The only equation with an explicit constant term is that for E, all the others have an understood constant of ∞. Thus we can conclude that $E = \langle 5, ef\rangle$ and substitute $\langle 5, ef\rangle$ for E as the next elimination step.

$$A = (\langle 1, a\rangle + B)\downarrow(\langle 4, a\rangle + C)$$

$$B = (\langle 1, b\rangle + D)$$

$$\begin{aligned} C &= (\langle 1, c\rangle + \langle 5, ef\rangle) \\ &= (\langle 6, cef\rangle) \end{aligned}$$

$$\begin{aligned} D &= (\langle 1, d\rangle + C)\downarrow(\langle 4, d\rangle + \langle 5, ef\rangle) \\ &= (\langle 1, d\rangle + C)\downarrow(\langle 9, def\rangle) \end{aligned}$$

Now the equation for C has a constant on the right, so C is eliminated next. (C is also, trivially, the minimum constant variable.)

$$\begin{aligned} A &= (\langle 1, a\rangle + B)\downarrow(\langle 4, a\rangle + \langle 6, cef\rangle) \\ &= (\langle 1, a\rangle + B)\downarrow(\langle 10, acef\rangle) \end{aligned}$$

$$B = (\langle 1, b\rangle + D)$$

$$\begin{aligned} D &= (\langle 1, d\rangle + \langle 6, cef\rangle)\downarrow(\langle 9, def\rangle) \\ &= (\langle 7, dcef\rangle)\downarrow(\langle 9, def\rangle) \\ &= (\langle 7, dcef\rangle) \end{aligned}$$

D is a constant. It is eliminated; then B will be a constant $= \langle 8, bdcef \rangle$, and finally,

$$A = (\langle 9, abdcef \rangle)$$

As illustrated by this example, use of the minimum-constant principle results in substituting constants only.

Even when a graph has negative edges, it can sometimes be transformed in $O(n)$ time to one with all positive edges and thus one to which the minimum-constant principle applies. The solution to the transformed graph can then be applied to the original graph in constant time. Assume there is in graph G, and a set of edges S, exactly one of which must be in any minimum path (for example, all edges out of vertex "begin," or all vertices into "goal.") Assume further that all negative edges are in S, and that the most negative has value $-r$. Then if to each edge cost in S we add $r+1$, all edges become nonnegative and the minimum-constant principle is applicable. The minimum path of the altered graph is the same as that of the original graph, while its cost is $r+1$ greater. If not all negative edges lie in S, the process can be iterated.

The minimum-constant principle can also be extended to be applicable in the minimum path problem, even if coefficients of zero are allowed on any edge.

More generally, the minimum-constant principle holds under slightly less stringent conditions than those required for Theorem 6.3.1. A form of the principle applies even to definitions in which the value of the w function on the right side of an equation is less than or equal to any of its arguments, as long as the equations have the form required for Theorem 6.3.1. Assume that instead of p4 we have

p4′. For each $i \in N^n$ there is a $k \in N^n \ni w_i(x_{i_1}, \ldots, x_{i_{n_i}}) \geq x_k$.

Under this condition we cannot guarantee a unique value for the equation with minimum constant, but we can conclude the following.

Theorem 6.3.2: Let E^* be a set of equations of form f_1 satisfying $p4$. Let a, $1 \leq a \leq n$ be such that $c_a = \mathrm{MIN}_{i=1}^{i=n}[c_i]$. If the equation for x_a in E^* is replaced with $x_a = c_a$ to obtain $E^{*\prime}$, then the maximum solution of E^* is a solution of $E^{*\prime}$, and any solution of $E^{*\prime}$ is also a solution of E^*.

PROOF: Taking minima on both sides of the set of equations E^*, gathering terms using commutativity and associativity of minimum,

$$\mathrm{MIN}_{i=1}^{i=n}[x_i] = \mathrm{MIN}_{i=1}^{i=n}\left[w_i(x_{i_1}, \ldots, x_{i_{n_i}})\right] \downarrow \mathrm{MIN}_{i=1}^{i=n}[c_i] \tag{1}$$

By p4′ for each i,

$$w_i(x_{i_1}, \ldots, x_{i_{n_i}}) \geq \text{MIN}_{i=1}^{i=n}[x_i]$$

because each is $\geq$ than some x_i. If $>$ held, then the argument of Theorem 6.3.1 would hold, so assume equality holds for some of the i's. Therefore there exists $1 \leq m \leq n \ni x_m = \text{MIN}_{i=1}^{i=n}[x_i]$, which together with the equality assumption implies

$$\text{MIN}_{i=1}^{i=n} w_i(x_{i_1}, \ldots, x_{i_{n_i}}) = x_m$$

So letting $c_a \equiv \text{MIN}_{i=1}^{i=n}[c_i]$
Eq. (1) becomes $x_m = x_m \downarrow c_a$ (2)
As is easy to show by substitution and as we show in Lemma 6.7.2, Eq. (2) is satisfied by any x_m for which $x_m \leq c_a$. Therefore the largest value of x_m that will satisfy this equation is given by $x_m = c_a$. This shows that any solution to E^* satisfies $E^{*\prime}$. In $E^{*\prime}$ c_a is, as in E^*, the minimum constant, so $x_a \leq c_a$ and by p4, $x_a \leq w_a(x_{a_1}, \ldots, x_{a_{n_a}})$. Therefore,

$$x_a = w_a(x_{a_1}, \ldots, x_{a_{n_a}}) \downarrow c_a = c_a \quad \blacksquare$$

If p4′ continues to hold after substitution, we have p5′.

p5′. p4′ holds for a set of equations E, and after substitution by the minimum constant, simplification, and the elimination of the variable whose value is the minimum constant, p4′ continues to hold for the remaining set. This continues to be true for repeated elimination of minimum constants. We say the minimum-constant principle (with equality) is applicable to E.

Example 6.3.2: Minimum–Maximum Path—Minimum Constant
Consider the problem of determining the maximum weighted edge on each path from vertex "begin" to vertex "goal" and then choosing the path with the smallest of these edges. With $\uparrow$ replacing $+$, the recursive definition for this problem becomes identical to that for the shortest path, Definition 6.2.1. Consider a situation in which an edge weight represents the degree of risk associated with traversing that edge. Then this problem has the interpretation: Find a path from "begin" to "goal" which minimizes the greatest risk. This problem arises in planning, for example, for medical procedures or rescues.

Definition 6.3.1: Minimum–Maximum Path

$$\left[\begin{array}{ll} g(v) = \langle 0, \langle v \rangle \rangle & : v = \text{"goal"} \\ g(v) = \text{MIN-SEL}_{i=1}^{m(v)}(\langle c_i(v), \langle v \rangle \rangle \uparrow\| \, g(n_i(v))) & : v \neq \text{"goal"} \\ \text{Initially } v = \text{"begin."} & \end{array}\right.$$

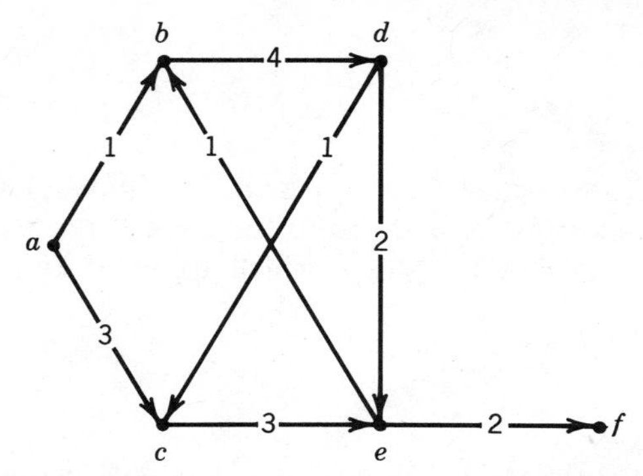

FIGURE 6-2 Graph example for minimum–maximum path.

An example of the application of the minimum-constant principle to the problem in Figure 6-2 is now given. We pick up the solution after the first substitution of a constant, namely, $\langle 0, f\rangle$ for F, ($\uparrow$ is short for $\uparrow\|$)

$$A = (\langle 1, a\rangle \uparrow B) \downarrow (\langle 3, a\rangle \uparrow C)$$

$$B = (\langle 4, b\rangle \uparrow D)$$

$$C = (\langle 3, c\rangle \uparrow E)$$

$$D = (\langle 1, d\rangle \uparrow C) \downarrow (\langle 2, d\rangle \uparrow E)$$

$$E = (\langle 1, e\rangle \uparrow B) \downarrow (\langle 2, ef\rangle)$$

There are no more equations having only a constant on the right. However, we may invoke the minimum-constant principle p5′. The terms have the form "constant (coefficient) $\uparrow$ variable," so they are each clearly greater or equal to a variable, and so is the minimum of all these terms. This is true *no matter what the signs of the coefficients are*.

The only equation with an explicit constant term is that for E; all the others have an understood constant of ∞. Thus we can conclude that $E = \langle 2, ef\rangle$ and substitute $\langle 2, ef\rangle$ for E throughout as the next elimination step,

$$A = (\langle 1, a\rangle \uparrow B) \downarrow (\langle 3, a\rangle \uparrow C)$$

$$B = (\langle 4, b\rangle \uparrow D)$$

$$C = (\langle 3, c\rangle \uparrow \langle 2, ef\rangle)$$

$$= (\langle 3, cef\rangle)$$

$$D = (\langle 1, d\rangle \uparrow C) \downarrow (\langle 2, d\rangle \uparrow \langle 2, ef\rangle)$$

$$= (\langle 1, d\rangle \uparrow C) \downarrow (\langle 2, def\rangle).$$

Now the equation for C has only a constant on the right, namely, $\langle 3, cef \rangle$, which is substituted for C throughout. (Note that $\langle 3, cef \rangle$ is not the minimum constant. D has the minimum constant on its right, $\langle 2, def \rangle$, and could be set equal to that constant. This requires finding the minimum constant, which is unnecessary since we already have a constant to substitute.)

$$A = (\langle 1, a \rangle \uparrow B) \downarrow (\langle 3, a \rangle \uparrow \langle 3, cef \rangle)$$

$$= (\langle 1, a \rangle \uparrow B) \downarrow (\langle 3, acef \rangle)$$

$$B = (\langle 4, b \rangle \uparrow D)$$

$$D = (\langle 1, d \rangle \uparrow \langle 3, cef \rangle) \downarrow (\langle 2, def \rangle)$$

$$= (\langle 3, dcef \rangle) \downarrow (\langle 2, def \rangle)$$

$$= (\langle 2, def \rangle)$$

D is a constant. It may be eliminated, then B will be a constant $= \langle 4, bdef \rangle$, and finally,

$$A = (\langle 3, acef \rangle)$$

6.4 "LINEAR" EQUATIONS

An important class of equation sets is that which includes and is modeled on the familiar sets of linear equations, formulated in terms of the primitive functions addition $(+)$ and multiplication $(*)$. These were used in school to solve problems like: determine Jack's age given that it is two times Mary's age minus twelve years, and that Jack's plus three times Mary's age is 38 years; and later to find the voltages in an electrical network. Such *a linear* set arises from a nonnested, equation complete definition with a loop-breaking rule, in which each equation is of the form

p6. $x_i = \Sigma_{k=1}^{n}[a_{ik} * x_k] + c_i$

The operations $+$ and $*$ need not be addition and multiplication, but they are required to have some of the properties of those functions, namely, those necessary to allow Gaussian elimination to work.

p7. $+$ is commutative (for gathering terms).

p8. $+$ and $*$ are associative (for recovering the same form of equation after substitution).

p9. $*$ distributes through $+$ (for multiplying a coefficient through a substituted set of terms).

p10. There is a 0, with $x * 0 = 0 * x = 0$ and $x + 0 = 0 + x = x$ (technically useful to represent each right side as involving every variable, with 0 coefficients to eliminate those not actually present).

We propose to solve these equations by Gaussian elimination techniques. This involves the systematic elimination of variable occurrences on the right of the equations. We repeatedly substitute for an occurrence of a variable x_j on the right of an equation with the right side of the equation for x_i, multiply ($*$) through with the coefficient of that occurrence, and then gather terms. In order for this to be effective in eliminating x_i from the x_j equation, however, the equation for x_i must have no loop, that is, no occurrence of x_i on its right. If it does have such an occurrence, it will be removed with the loop-breaking rule. In general this removal will result in changing the coefficients, but not the form, of the equation for x_i. Given properties p7 through p10, this process is guaranteed to be *reproducible* in the sense that after substitution of one equation into another, the result will be an equation of the same form. The existence of reproducibility implies that once the loop-breaking rule is justified for the original equation form, it is justified throughout the process.

It is convenient to use vector notation in developing the algorithm. In vector notation $\cdot$ is the vector dot product, $*$ is used to indicate the multiplication of a scalar by a vector, and $+$ is a vector addition. Given that

$$\mathbf{A}_i^u = \langle \overbrace{0, \ldots, 0}^{u-1}, a_{iu}, a_{iu+1}, \ldots, a_{in} \rangle \qquad (\mathbf{A}_i^1 = \mathbf{A}_i)$$

$$\mathbf{X} = \langle x_1, x_2, \ldots, x_n \rangle$$

p6 can be expressed in vector notation as

$$x_i = \mathbf{A}_i^1 \cdot \mathbf{X} + c_i \equiv x_i = \mathbf{A}_i \cdot \mathbf{X} + c_i$$

The algorithm repeatedly substitutes for x_i on the right side of the equation for x_j with $j > i$. Its effect can now be developed easily using the vector notation. When the algorithm is ready to eliminate x_i in the equation for x_j, the first through $(i-1)$th variables will already have been removed from this equation. Before substitution, the x_i term is eliminated from the x_i equation by the use of the loop-breaking rule, after which we have:

$$x_i = \mathbf{A}_i^{i+1} \cdot \mathbf{X} + c_i$$

At this time the equation for x_j, $j > i$, can be expressed as

$$x_j = a_{ji} * x_i + \mathbf{A}_j^{i+1} \cdot \mathbf{X} + c_j$$

Now substituting for x_i in this equation,

$$x_j = a_{ji} * (\mathbf{A}_i^{i+1} \cdot \mathbf{X} + c_i) + \mathbf{A}_j^{i+1} \cdot \mathbf{X} + c_j$$

Using p7 through p10 to simplify,

$$x_j = (\mathbf{A}_j^{i+1} + a_{ji} * \mathbf{A}_i^{i+1}) \cdot \mathbf{X} + (c_j + a_{ji} * c_i)$$

When expanded this becomes

$$x_j = \langle (a_{ji+1} + a_{ji} * a_{ii+1}), \ldots, (a_{jk} + a_{ji} * a_{ik}), \ldots, (a_{jn} + a_{ji} * a_{in}) \rangle \cdot \mathbf{X}$$
$$+ (c_j + a_{ji} * c_i)$$

This substitution adds $a_{ji} * a_{ik}$ to the kth coefficient a_{jk} on the right of the x_j equation—with k ranging from $i + 1$ to n.

The components of the updated A_j^{i+1}, that result from substituting for x_i in a *single* equation for x_j still designated $a_{ji+1}, a_{ji+2}, \ldots, a_{jn}$, are then computed as follows. ($A[JK] = a_{jk}$ and $C[J] = c_j$.)

Algorithm 6.4.1: Innermost Loop of Gaussian Elimination

```
FOR K = I + 1 TO N DO
  A[JK] ← A[JK] + A[JI] * A[IK]      { Eliminates x_i on the
                                       right of the jth equation
END
C[J] ← C[J] + A[JI] * C[I]
```

The algorithm for doing this operation for all x_j with $j > i$, and thus eliminating all occurrences of x_i in all equations beyond the ith, is

Algorithm 6.4.2: Middle Loop of Gaussian Elimination

```
FOR J = I + 1 TO N DO
  FOR K = I + 1 TO N DO
    A[JK] ← A[JK] + A[JI] * A[IK]    { Eliminates x_i on the right of the jth
                                       equation, with j ranging from i + 1 to n.
  END
  C[J] ← C[J] + A[JI] * C[I]
END
```

Eliminating x_i for all i from 1 to $n - 1$ is then accomplished by the ELIMINATION phase of the following algorithm. On completion, it will have solved for one variable, namely, x_n, and generally triangularized the equations, making each x_i dependent only on x_j, $j > i$. ELIMINATION includes removal of variables x_i from the right of its own equation using the procedure *loop-break*(i). The back substitution phase of the algorithm computes the value of all other variables. The value of x_j is in C_j at termination.

Algorithm 6.4.3: Gaussian Elimination

```
            Elimination
FOR I = 1 TO N - 1 DO
  loop-break(I)**
  FOR J = I + 1 TO N DO
    FOR K = I + 1 TO N DO

      A[JK] ← A[JK] + A[JI] * A[IK]   { For i ranging from 1 to n, eliminates x_i
                                       { on the right of the jth equation, with j
                                       { ranging from i + 1 to n.
    END
    C[J] ← C[J] + A[JI] * C[I]
  END
END

            Back Substitution
loop-break(N)
FOR I = N TO 2
  FOR J = I - 1 TO 1 DO
    C[J] ← C[J] + A[JI] * C[I]
  END
END
Answer: The value of X[I] is in C[I] at termination.
```

In *elimination*, there is a triply nested set of FOR loops, each with range $O(n)$, inside of which is an $O(1)$ assignment. The outer two loops enclose another $O(1)$ assignment, and the outer loop itself encloses the loop-breaking procedure which we assume is also $O(1)$. The triply nested operation is dominant and results in a time of $O(n^3)$.

In *back substitution*, first $x_n = c_n$ is eliminated in all equations with index lower than n. As a result the equation for x_{n-1} has only a constant on its right. Then x_{n-1} is eliminated in all equations of index lower than $n - 1$. In general, when elimination of x_i in equations of index lower than i is about to start, the equation for x_i will have only a constant on its right. The total number of right sides into which constant substitutions are made, and thus the worst-case complexity, is $O(n^2)$.

The elimination part of Algorithm 6.4.3 computes the value of a single variable, with has complexity $O(n^3)$. Nothing is added to the order of complexity by applying back substitution to find all the other variables, so there is only a modest time saving in getting one rather than all variables using this technique. The following Gauss–Jordan algorithm evaluates all variables in a more integrated, though slightly more costly, way than Algorithm 6.4.3.

The Gauss–Jordan algorithm, Algorithm 6.4.4, eliminates successive variables from *all* equations rather than only those of higher index. It terminates with x_i's value in the C_i terms, as in the Gaussian algorithm.

Algorithm 6.4.4: Gauss–Jordan Elimination

```
FOR I = 1 TO N DO                          {Eliminating each variable.
  loop-break(I)**
  FOR J = 1 TO N but not J = I DO          {Equations into which variables are sub-
                                           {stituted.
    FOR K = I + 1 TO N DO                  {Updating each coefficient in an equation
      A[JK] ← A[JK] + A[JI] * A[IK]
    END
    C[J] ← C[J] + A[JI] * C[I]             {Updating the constant in an equation.
  END
END
Answer: The value of X[I] is in C[I] at termination.
```

Further simplification is possible if

p11. For all i loop-break(i) is: Remove $a_{ii} * x_i$, on the right of the equation for x_i.

If p11 holds, the loop-break(I)** in Algorithms 6.4.3 and 6.4.4 can be removed. Because this usually results in zeroing a coefficient at a point in the algorithm at which that coefficient has no further effect on changing values of A or C. Since it is the C's that will ultimately contain the answers at the end of the algorithm, there is no need to actually zero these coefficients.

Property p11 holds quite often—it holds in the minimum path and min–max path problems *if the edge weights are all* ≥ 0. But in these cases the minimum-constant principle also holds, and a more efficient algorithm than Gaussian elimination is applicable. If, however, negative edges are inappropriately located, the minimum-constant principle will not apply, but Gaussian elimination still will, provided that the sum of the edges on every cycle is non-negative. (This also guarantees that there will be an answer other than $-\infty$.) For such problems elimination algorithms, Gaussian or Gauss–Jordan, with their $O(n^3)$ complexity may be reasonable choices.

6.4.0.1 Linear Equations and Minimum-Constant Principle

If, in addition to the linear equation conditions, those necessary for the application of the minimum-constant principle, p4 and p5, also hold, then a specialization of Algorithm 6.4.4 produces a particularly efficient algorithm. In this algorithm the variable whose equation has the minimum constant will always be the next one to be eliminated. Because a constant is substituted for a single variable in an equation of the linear form, only the constant term need be updated at each substitution step.

The details of the minimum-constant algorithm depend on the method used for getting the minimum of a changing set of constants.

The algorithm for the first case, Algorithm 6.4.5, keeps the indices of all unsolved equations in a list(Z). It also has the index of the current minimum constant (MINCON) and the variable, say x, which has that value. It substitutes the current minimum constant for x on the right of each equation whose index is in Z (even if the coefficient is 0), updating the constant of that equation, and using it to update the computation and the new minimum. If it is the smallest new constant so far, it, and its index, are saved (MIN/MININD) and replaced in Z by the previous occupants of MIN/MININD. The complexity of one such pass through Z is $O(n)$. This complexity could not be decreased even if we avoided the 0 coefficient substitutions altogether—since all $O(n)$ indices in Z must be scanned to compute the new minimum. It is necessary, in the worst case, to compute n new minimum constants over the course of the algorithm, so the complexity is $O(n^2)$.

Consider on the other hand the use of a HEAP to get the minima. Assume the HEAP is initialized with all n constants—$O(n)$. Then the minimum constant is obtained from the top of the HEAP and the HEAP readjusted—$O(\log n)$. This happens at most n times altogether, giving a total complexity of $O(n \log n)$. Every time we get a new constant, say for index x_i, it is substituted for all right-side occurrences of x_i *where the coefficients are not* 0. (In order to assure that we will not do any substitutions where there are 0 coefficients, it will be necessary to represent the graph appropriately, that is, by an inverted adjacency list). The new constant computed as a result of the substitution ($\leq$ to previous ones)—$O(1)$— is entered into the HEAP—$O(\log n)$. Say there are $r(i)$ such substitutions or right side occurrences for the ith variable. Then the cost to substitute and re-heap for each minimum constant is $r(i) * O(\log n)$. Over the run of the algorithm, this will give $\Sigma_{i=1}^{i=n}[r(i) * \log n]$, or letting $\Sigma_{i=1}^{i=n}[r(i)] = e$, $e * \log n$. The total complexity then is $O((e + n) * \log n)$, $e \leq O(n^2)$. In the worst case this is worse than the previous approach. However, if the right side coefficients are limited in various ways (see problems, Section 6.10), this approach may prove the better of the two.

If instead of the standard binary HEAP, an F-HEAP is used, the amortized cost to adjust the F-HEAP when a one of its entries is decreased is $O(1)$ instead of $O(\log n)$, so the overall complexity drops to $O(e + n \log n)$.

Algorithm 6.4.5: "Linear" Equation—Minimum-Constant Elimination
Z is a list containing the indices of all as yet unsolved equations.

Initially Z has the index of every equation. The minimum constant from the previous pass is in MINCON, and the index of the equation with that constant is in MINCONIND. MINCON is used to update equations. The smallest constant on the current pass through the equations is in MIN, and its index is in MININD.

[The first part of the algorithm simply finds the first minimum constant. Often this information is implicit in the problem (minimum path), so this part is unnecessary.]

```
MININD ← pop- Z
MIN ← C(MININD)
FOR DUMMY  = 2 TO N DO
  I ← pop- Z
  IF C[I] < MIN THEN DO
    Z ← queue- MININD
    MININD ← I
    MIN ← C[I]
  END
  ELSE Z ← queue- I
END
MINCONIND ← MININD
MINCON ← MIN
```

[The next part finds each subsequent minimum constant and substitutes it for the corresponding variable throughout the right sides. The substitution is done in the inner FOR loop.]

```
FOR K = 2 TO N − 1
  MININD ← pop- Z
  MIN ← C(MININD)
  FOR J = K TO N DO
    I ← pop- Z
    C[I] ← min(A[I, MINCONIND] * MINCON, C[I])   ⎧ Combines
    IF C[I] < MIN THEN DO                        ⎪ substitution
      Z ← queue- MININD                          ⎪ of current minimum
      MININD ← I                                 ⎨ constant with
      MIN ← C[I]                                 ⎪ computation of
    END                                          ⎪ new minimum
  ELSE Z ← queue- I                              ⎩ constant.
  END
  MINCONIND ← MININD
  MINCON ← MIN
END
Answer: The value of X[I] is in C[I] at termination.
```

We have already given an example of a minimum path problem to which "linear" equation algorithms are applicable. Next we consider a special case of the "linear" equation class.

6.4.1 Coherent Linear Equations

Consider an equation set which satisfies the properties for linear equations, and has the following property.

p12. The constant term in each equation is itself a vector with the same number of components as there are variables in the equation.

Assume c_j is $\langle c_{j1}, \ldots, c_{jn} \rangle$. Then Gauss–Jordan algorithm, Algorithm 6.4.4, is equivalent to the following assuming p11 holds.

Algorithm 6.4.6: Modification 1 in Algorithm 6.4.4—Simple Loop-Breaking, Coherence Using Gauss–Jordan Elimination

```
FOR I = 1 to N DO
  FOR J = 1 TO N but not J = I DO
   ⎧ FOR K = I + 1 TO N DO
  A⎨   A[JK] ← A[JK] + A[JI] * A[IK]
   ⎩ END
   ⎧ FOR K = 1 TO N DO
  C⎨   C[JK] ← C[JK] + A[JI] * C[IK]
   ⎩ END
  END
END
```
Answer: The value of $X[I]$ is in $C[I]$ at termination.

We wish to jam the inner loops, A, which updates coefficients, and C, which updates constants. We first extend the range of loop A from FOR $K = I + 1$ TO N DO, to FOR $K = 1$ TO N DO.

The revised loop A' causes, for a given I and J, values to be assigned to $A[J1], A[J2], \ldots, A[JI]$ different from those given by A. In particular, $A[JI]$ changes, which in turn affects the computation of $A[JI + 1], A[JI + 2], \ldots, A[JN]$, and eventually that of $C[J1], C[J2], \ldots, C[JN]$ as computed by loop C. Ultimately, it is the $C[JK]$ we are interested in. We want them to be unchanged by the change from A to A'. Thus the change to $A[JI]$ cannot be allowed to propagate. We save $A[JI]$ in a variable AJI on entry to loop A', and use the saved value in place of $A[JI]$ within A' and C. Thus extending from A to A' will not change the values used for computation in A and C. Thus the modified loops can be jammed, with the result shown in Algorithm 6.4.7.

Algorithm 6.4.7: Modification 2 in Algorithm 6.4.4—Simple Loop-Breaking, Coherence Using Gauss–Jordan Elimination

```
FOR I = 1 TO N DO
  FOR J = 1 TO N but not J = I DO
    AJI ← A[JI]
    FOR K = 1 TO N DO
      A[JK] ← A[JK] + AJI * A[IK]        ... [1]
      C[JK] ← C[JK] + AJI * C[IK]
    END
  END
END
```
Answer: The value of $X[I]$ is in $C[I]$ at termination.

The set of equations is also *coherent* if the vector constants c_j satisfy the condition,

p13. For all jk, $a_{jk} = c_{jk}$.

With this condition met, $C[IJ] = A[IJ]$ initially, and also after each passage through the inner loop. Therefore we can replace $A[JI]$ with $C[JI]$ at the entry to the inner loop and drop line 1 in Algorithm 6.4.7. The Gauss–Jordan algorithm becomes

Algorithm 6.4.8: Gauss–Jordan Elimination—Simple Loop-Breaking, Coherence

```
FOR I = 1 TO N DO
  FOR J = 1 TO N but not J = I DO
    AJI ← C[JI]
    FOR K = 1 TO N DO
      C[JK] ← C[JK] + AJI * C[IK]
    END
  END
END
Answer: The value of X[I] is in C[I] at termination.
```

This algorithm clearly has time complexity $O(n^3)$. It is Warshall's algorithm and is, as shown next, applicable to the computation of the *transitive closure*, that is, *all* vertices reachable from each and every vertex in a graph, as well as *all* shortest paths or *all* min–max paths between pairs of vertices in a graph.

Assume that the vertices of digraph G are designated with integers 1 through n. Consider the reachability function $F(i)$ whose value is a 0–1 vector with 1 in position j if and only if there is a path from vertex i to vertex j in G. Let $c_{ik} = 1$ if and only if vertex k is a neighbor of vertex i. Let $C_i = \langle c_{i1}, \ldots, c_{in} \rangle$, that is, C_i is the ith row in an adjacency matrix representation of G. In the definition we give for $F(i)$, + is vector "or," while & is the "and" of a scalar with each component of a vector. We give a schematic description of the sense of the definition followed by the definition itself.

All the vertices reachable from the ith vertex [i.e., $F(i)$ represented as a 0–1 vector]

= All those reachable from the 1st [i.e., $F(1)$ represented as a vector] if there is an edge from the ith to 1st vertex (i.e., if $c_{i1} = 1$)

\+ All those reachable from the 2nd [i.e., $F(2)$ represented as a vector] if there is an edge from the ith to 2nd vertex (i.e., if $c_{i2} = 1$)

\+ $\cdots$ + All those reachable from the nth vertex [i.e., $F(n)$ represented as a vector] if there is an edge from the ith to nth vertex i.e., if $c_{in} = 1$)

\+ All those reachable directly by an edge from the ith vertex, C_i.

The definition is an abbreviation of this description. It does not terminate but is equation complete, has a loop-breaking rule, and satisfies p12 and p13.

Definition 6.4.1: All Reachable Vertices or Transitive Closure—Nonterminating Definition

$$\left[\begin{array}{l} F(i) = \sum_{k=1}^{k=n} [c_{ik} \,\&\, F(k)] + C_i \\ \text{Initially } i \text{ is vertex in } G,\ G \text{ is an invariant argument.} \end{array}\right.$$

As another example of an application of Algorithm 6.4.8, consider a definition for the cost of the shortest path function, $F(i)$. $F(i)$ is a nonnegative vector $c_i = \langle c_{i1}, c_{i2}, \ldots, c_{ik} \rangle$ in which c_{ij} is the minimum of the costs of all paths from vertex i to vertex j. The definition for $F(i)$ given below does not terminate, but will produce a finite set of equations and has a loop-breaking rule. In the definition, min or $\downarrow$ is vector minimization, that is,

$$\min(\langle a_1, \ldots, a_n \rangle, \langle b_1, \ldots, b_n \rangle) = \langle \min(a_1, b_1), \ldots, \min(a_n, b_n) \rangle.$$

and + is addition of a scalar to each component of a vector.

Definition 6.4.2: Cost of Minimum Path between Every Pair of Vertices

$$\left[\begin{array}{l} F(i) = \mathrm{MIN}_{k=1}^{k=n} [c_{ik} + F(k)] \downarrow C_i \\ \text{Initially } i = \text{vertex in } G,\ G \text{ is an invariant argument.} \end{array}\right.$$

There is a simple loop-breaking rule with property p11 for both Definitions 6.4.1 and 6.4.2, so Algorithm 6.4.8 without inclusion of loop-break(I) is applicable to both. With the interpretation for c_{ij} given for these definitions, the algorithm requires no special plug in. It will work as written for either, although it can be refined slightly for some applications. This is done for reachability, Definition 6.4.1, taking advantage of the fact that all c_{ij}'s are either 0 or 1 and properties of & and +.

Algorithm 6.4.9: Warshall's Algorithm

```
FOR I = 1 TO N DO
  FOR J = 1 TO N but not J = I DO
    IF C[J1] = 1 THEN
      FOR K = 1 TO N DO
        C[JK] ← C[JK] + C[IK]
      END
    END
END
```

If +* "ors" two vectors of length N, then the inner loop can be replaced with $C[J] \leftarrow C[J] +^* C[I]$

Answer: The value of the $X[I]$ is in $C[I]$ at termination.

Unfortunately the minimum-constant principle does not apply; basically because the vector minimum of a number of vectors need not be any of those vectors.

6.5 ARGUMENTS DEPENDING ON THE STRUCTURE OF THE SOLUTION

In the minimum path problem the solution has two parts: the cost of the minimum path, and the path itself. In the ESE solution each vertex in the minimum path corresponds to the variable in the selected minimum term on the right of an equation (see Example 6.2.1). This is true even if the path is removed as an explicit argument in the equation set. We can in fact simplify the definition by removing the path argument and computing the minimum cost only, in an ESE implementation. Then we can reconstruct the path by noting the terms selected in each equation solution, eliminating the need to carry around a partially developed path as an argument. This kind of partition of a solution is not uncommon. Recursive definitions formulated for ESE can often be simplified because part of the value of the function is dependent on the structure of the solution. Essentially the same solution would be obtained with a simpler definition. After obtaining the solution, using the simpler definition, the structure of the solution can be used to produce the part of the value of the original definition which depended only on that structure. We next develop a class of definitions with such solution structure dependent values.

Consider how a pure information component in a definition in which w is a selection function applies when using ESE. Definition 6.5.1 gives a class of such definitions. (This continues a discussion started in Section 2.3.3.1; applications to B-F implementations were then considered in Section 5.4.2.2. There we showed the application to an expanded form of the minimum-select monotonic definition.)

Definition 6.5.1: Transformable Pure Information Select Definition

$$\left[\begin{array}{ll} f(C, Y, X) = \langle C + tc(X), ty(X, Y)\rangle & :T(X) \\ f(C, Y, X) = \mathrm{SEL}_{i=1}^{i=m(X)}[f(C + c_i(X), y_i(X, Y), o_i(X))] & :\sim T(X) \\ \text{Initially } C = d0 \in D0,\ X = d1 \in D1,\ Y = d3. & \end{array}\right.$$

Definition 6.5.2 has essentially the same solution structure as Definition 6.5.1. The pure-info component Y is replaced by a queue V, which starts with $d3$, the initial value of Y. The functions v_i and tv of Definition 6.5.2 can be designed so that at termination V has the information needed to compute $ty(X, Y)$ of Definition 6.5.1 assuming they start with the same initial argument value. The value of C at termination will obviously be the same in both definitions. The functions v_i and tv necessary to obtain the relation between the two definitions will vary considerably depending on ty and y_i. This variation will be illustrated in example.

Definition 6.5.2 can now be transformed to an equivalent Definition 6.5.3 which is natural for ESE (Definition 6.5.1 was designed so this would happen). With ESE the solution structure, V will be simple to extract, and from V the corresponding value of Y can be easily computed.

Definition 6.5.2: Transformable Pure Information Select Definition

$$\left[\begin{array}{ll} f(C,V,X) = \langle C + tc(X), V \| tv(X,V)\rangle & :T(X) \\ f(C,V,X) = \mathrm{SEL}_{i=1}^{i=m(X)}\left[f(C + c_i(X), V \| \langle v_i(X)\rangle, o_i(X))\right] & :\sim T(X) \\ \text{Initially } C = d0 \in D0,\ X = d1 \in D1,\ V = \langle\ \rangle. & \end{array}\right.$$

V is a pure information component, and assume the SEL selects on the first components of the generated pairs, $+$ and $\|$ are associative and both distribute through SEL, and $0_+ + a = a + 0_+ = a$. Given these properties we can transform this definition using reduction 1, Theorem 6.1.1, identifying $+$ and $\|$ here with $*$ in the theorem and minimum-select with $+$ in the theorem.

Definition 6.5.3: Transformable Pure Information Select Definition—ESE Form

$$\left[\begin{array}{ll} G(X) = \langle tc(X), tv(X)\rangle & :T(X) \\ G(X) = \mathrm{SEL}_{i=1}^{i=m(X)}\left[\langle c_i(X), v_i(X)\rangle + |{*}|\ G(o_i(X))\right] & :\sim T(X) \\ \text{Initially } X = d1 \in D1. & \end{array}\right.$$

The solution to this definition, which is guided completely by the value of X, will give a set of pairs, each of the form $\langle C, V\rangle$. The pairs will have been selected because of C, while V is a queue of $v_i(X)$ values on the path from the root of the O network to the selected pair. The selected queues constitute the *solution structure*. If the solution is obtained by ESE, the child indices that would appear in the queues associated with the final solution will be those of the terms $[\langle c_i(X), v_i(X)\rangle + |{*}|\ G(o_i(X))]$ which are SELected in each equation of the set. Thus the solution of the equation set is equivalent to solving the simplified definition below, noting the indices of the selected terms, and reconstructing the queues or solution structure.

Definition 6.5.4: Transformable Pure Information Select Definition—ESE Form

$$\left[\begin{array}{ll} g(X) = tc(X) & :T(X) \\ g(X) = \mathrm{SEL}_{i=1}^{i=m(X)}\left[c_i(X) + g(o_i(X))\right] & :\sim T(X) \\ \text{Initially } X = d1 \in D1. & \end{array}\right.$$

In the ESE solution for Definition 6.5.4 one or more right side terms (g calls) will be selected in each equation for the final solution. Let

$r(X) = \{i \mid g(o_i(X))$ is a selected term in the solution of the $g(X)$ equation$\}$

The solution structure or set of queues, $V(X)$, which consists of second components of the solution pairs for Definition 6.5.3, can be recreated from the solution to Definition 6.5.4 by evaluating the definition below. (Examples are given later in this section.)

Definition 6.5.5: Solution Structures—Definition 1

$$\left[\begin{array}{ll} V(X) = \langle\langle tv(X)\rangle\rangle & :T(X) \\ V(X) = \langle\ \langle\langle v_i(X)\rangle\rangle|*|V(o_i(X)) \mid i \in r(X)\ \rangle & :\sim T(X) \\ \qquad \text{or if } v_i(X) = v(X) \text{ is independent of } i, & \\ V(X) = \langle\langle v(X)\rangle\rangle|*|\langle V(o_i(X)) \mid i \in r(X)\rangle & :\sim T(X) \\ \text{Initially } X = d1 \in D1 = \text{the initial component from Definition 6.5.4.} & \end{array}\right.$$

Given $i \in r(X)$, computing $o_i(X)$ may require time which increases with i; it cannot be bounded above with a constant. If, for example, $o_i(X)$ is the ith neighbor of vertex X in a graph represented by an adjacency list, it would take time $O(i)$ to find it. In such a case, rather than keep the child#, it is advisable to keep the argument structure or the applicable part of the argument structure associated with the right side terms used in the solution of each equation. This alternative is expressed as follows.

Definition 6.5.6: Solution Structures—Definition 2
Let $R(X) = \{o_i(X) | i \in r(X)\}$.

$$\left[\begin{array}{ll} V(X) = \langle\langle tv(X)\rangle\rangle & :T(X) \\ V(X) = \langle\ \langle\langle v_i(X)\rangle\rangle|*|V(Y) \mid Y \in R(X)\ \rangle & :\sim T(X) \\ \qquad \text{or if } v_i(X) \text{ is independent of } j, & \\ V(X) = \langle\langle v(X)\rangle\rangle|*|\langle V(Y)| Y \in R(X)\rangle & :\sim T(X) \\ \text{Initially } X = d1. & \end{array}\right.$$

The solution structure computed by $V(X)$, in one of the two forms above, is usually sufficient for our needs. However, we can produce the set of second components of selected pairs as given in Definition 6.5.1 [$ty(X, Y)$'s] from the $r(X)$'s obtained from a solution to Definition 6.5.4. The next definition of $Q(X, Y)$ does that.

Definition 6.5.7

$$\left[\begin{array}{ll} Q(X, Y) = ty(X, Y) & :T(X) \\ Q(X, Y) = \{Q(o_i(X), y_i(X, Y)),\ i \in r(X)\} & :\sim T(X) \\ \text{Initially } X \in D1 = d1,\ Y \in D2 = d2. & \end{array}\right.$$

Computing the solution structure is particularly easy if $r(X)$ [or $R(X)$] contains a small number, preferably only one, of solution structures, or if a small expression, which compactly expresses all possible solution structures, without listing each separately, is adequate. In general the cost to compute $V(X)$ is $O(k)$ per structure, where k is less than or equal to the number of equations [$= |\Delta_f(d)|$].

In summary, a recursive definition, such as Definition 6.5.3, may be simplified by removing all parts that involve the solution structure(s), to give a definition of the form of Definition 6.5.4. Generally a simplified algorithm is adequate to implement the simplified definition. Then the solution structure $V(X)$ incorporated in Definition 6.5.3 can be recreated by use of Definition 6.5.5 or 6.5.6. (The complexity of the algorithm may even be of *lower* order than that of computing $V(X)$.) Then $V(X)$ can be computed to recapture the parts removed from the original definition. If obtaining $V(x)$ takes no more than the simplified algorithm, as will usually occur if there are only a few solution structures, the complexity of the algorithm implementing the simplified definition is governing.

For example, instead of using Definition 6.2.1 for finding the shortest path, we could use the definition for the minimum cost, which does not involve the path at all, keeping track of the terms that gave the minimum in each equation, and reproduce the winning path.

Definition 6.5.8: Minimum Path—Simplified Definition

$$\left[\begin{array}{ll} g(v) = 0 & : v = \text{"goal"} \\ g(v) = \text{MIN}_{i=1}^{N(v)}(c_i(v) + g(n_i(v)) & : v \neq \text{"goal"} \\ \text{Initially } v = \text{"begin."} & \end{array}\right.$$

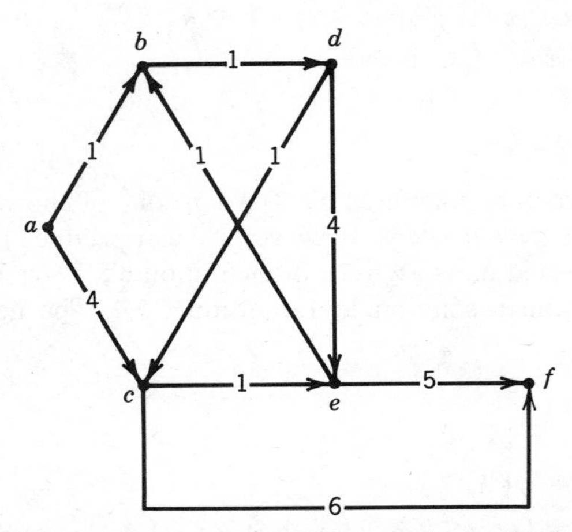

FIGURE 6-3 Graph example for shortest path, simplified definition.

TABLE 6-1. Set of equations for shortest path example, simplified definition.

Child#					
	1	2	arg $= v$	$r(v)$	$R(v)$
$A =$	$(1 + B)$	$\downarrow (4 + C)$	a	1	b
$B =$	$(1 + D)$		b	1	d
$C =$	$(1 + E)$	$\downarrow (6 + F)$	c	1,2	e, f
$D =$	$(1 + C)$	$\downarrow (4 + E)$	d	1	c
$E =$	$(1 + B)$	$\downarrow (5 + F)$	e	2	f
$F =$	0				

As previously, we evaluate this definition for a digraph, slightly different from that in Example 6.2.1 used previously to illustrate the ESE approach. The change is just sufficient to give two minimum paths. First the set of equations: for each vertex v in the given digraph there will be an equation with $g(v)$ on the left. As before, lowercase letters represent vertices, and if x represents a vertex, its uppercase form X represents $g(x)$.

For the digraph of Figure 6-3 the shortest path definition is of the form of Definition 6.5.4, and gives the equation set in Table 6-1.

Gaussian elimination, or more efficiently, the minimum-constant principle algorithm, can be used to solve this equation set. In either case, assuming we record the minimum term in each equation, as illustrated above, we can reconstruct the path(s), using a simplified version of Definition 6.5.6. Let $R(v) = \{n_i(v) \mid i \in r(v)\}$.

$$\left[\begin{array}{ll} V(v) = \langle\langle v \rangle\rangle & : v = \text{“goal”} \\ V(v) = \langle\langle v \rangle\rangle |*| \langle V(Y) | Y \in R(v) \rangle & : v \neq \text{“goal”} \\ \text{Initially } v = \text{“begin.”} & \end{array}\right.$$

$$\begin{aligned} V(a) &= \langle\langle a \rangle\rangle |*| \langle V(b) \rangle \\ &= \langle\langle a \rangle\rangle |*| \langle \langle\langle b \rangle\rangle |*| \langle V(d) \rangle \rangle \\ &= \langle\langle a \rangle\rangle |*| \langle \langle\langle b \rangle\rangle |*| \langle \langle\langle d \rangle\rangle |*| \langle V(c) \rangle\rangle\rangle \\ &= \langle\langle a \rangle\rangle |*| \langle \langle\langle b \rangle\rangle |*| \langle \langle\langle d \rangle\rangle |*| \langle \langle\langle c \rangle\rangle |*| \langle V(e), V(f) \rangle \rangle\rangle\rangle \\ &= \langle\langle a \rangle\rangle |*| \langle \langle\langle b \rangle\rangle |*| \langle \langle\langle d \rangle\rangle |*| \langle\langle\langle c \rangle\rangle |*| \langle\langle\langle e \rangle\rangle |*| \langle V(f) \rangle, V(f) \rangle \rangle\rangle\rangle \\ &= \langle\langle a \rangle\rangle |*| \langle \langle\langle b \rangle\rangle |*| \langle \langle\langle d \rangle\rangle |*| \langle \langle\langle c \rangle\rangle |*| \langle\langle\langle e \rangle\rangle |*| \langle\langle\langle f \rangle\rangle\rangle, \langle\langle f \rangle\rangle\rangle \rangle\rangle\rangle \\ &= \langle\langle abdcef \rangle, \langle abdcf \rangle\rangle \end{aligned}$$

There are two minimum paths. If one is sufficient, then the following definition for $V(v)$ would be adequate. Let $R(v) = \{n_i(v) \mid i \in r(v)\}$

$$\left[\begin{array}{ll} V(v) = \langle v \rangle & : v = \text{“goal”} \\ V(v) = \langle v \rangle \| V(Y) \text{ for any single } Y \in R(v) \rangle & : v \neq \text{“goal”} \\ \text{Initially } v = \text{“begin.”} & \end{array}\right.$$

and in the example instead of $\langle V(e), V(f)\rangle$ either $\langle V(e)\rangle$ or $\langle V(f)\rangle$ would be calculated.

As a second example consider the following simplified version of the traveling salesman definition, Definition 6.1.11.

Definition 6.5.9: Traveling Salesman Simplified

$$\left[\begin{array}{ll} g(v,b) = c_1(v) & :b = \langle\ \rangle \\ g(v,b) = \mathrm{MIN}_{n_i(v)\in\{b\}}[c_i(v) + g(n_i(v), b - n_i(v))] & :b \neq \langle\ \rangle \\ \text{Initially } v = 1,\ b = \langle 2,3,\ldots,n\rangle. & \end{array}\right.$$

We can obtain the same result as Definition 6.1.11 if we solve this definition with ESE and keep a record of $r(v, b)$, the index of the selected terms on the right of the equation for $g(v, b)$. Finally, the path is obtained by evaluating the following definition.

Let $R(v,b) = \{\langle n_i(v), b - n_i(v,b)\rangle | i \in r(v,b)\}$.

$$\left[\begin{array}{ll} V(v,b) = \langle\langle v\rangle\rangle & :b = \langle\ \rangle \\ V(v,b) = \langle\langle v\rangle\rangle|*|\langle V(v',b') \mid \langle v',b'\rangle \in R(v,b)\rangle & :b \neq \langle\ \rangle \\ \text{Initially } v = 1. & \end{array}\right.$$

The technique developed for simplified evaluation of the class of definitions given by Definitions 6.5.1 and 6.5.3 is not restricted to definitions in that form.

We can also handle cases in which the terms on the right are more complex, in which each term depends on more than one G call. Consider, for example, a definition containing the recursive line below and otherwise like Definition 6.5.4.

$$G(X) = \mathrm{SEL}_{i=1 \text{ to } m(X)}\big[\langle c_i(X), v_i(X)\rangle +\| G(o_i(X)) +\| \cdots +\| G(o_{ik_i}(X))\big] \quad :\sim T(X)$$

Such a definition can be replaced with a simpler definition with the recursive line below and otherwise like Definition 6.5.6:

$$G(X) = \mathrm{SEL}'_{i=1 \text{ to } m(X)}\big[c_i(X) + G(o_{i1}(X)) + \cdots + G(o_{ip_i}(X))\big] \quad :\sim T(X)$$

The simplified definition can then be solved with ESE, and the solution for the original definition obtained using the solution structure of Definition 6.5.10 below. Assume the solution to this simplified definition involves a single right side term, whose first index is $i = r(X)$.

Definition 6.5.10: Solution Structures—Definition 3

$$\left[\begin{array}{ll} V(X) = \langle tv(X) \rangle & :T(X) \\ V(X) = \langle v(X) \rangle \| V\big(o_{r(X)j}(X)\big) \| \cdots \| V\big(o_{r(X)p_{r(X)}}(X)\big) & :\sim T(X) \\ \text{Initially } X = e. & \end{array}\right.$$

An example of such a use is given in the next section for a simplified version of Definition 6.1.18, association of matrix multiplications.

6.6 ACYCLIC DEPENDENCY GRAPHS — GENERAL APPROACHES

ESE algorithms for recursive definitions involving acyclic dependency graphs require setting up equations, then solving them by back substitution. In the most general case, setting up equations requires evaluating the recursive definition at the initial argument structure to get the first equation, and for each argument structure appearing on the right of that first equation, setting up new equations. In general, for each new argument structure on the right of a newly set up equation, add a new equation, until there is an equation for each argument structure that appears on the right of any equation. This much exertion is, however, seldom necessary. In fact setting up equations is not usually a major part of the algorithm because one of the two situations below holds.

1. *Form given by initial argument.* The form is dependent on the initial argument structure, so that the initial argument structure itself must contain information that gives the equation set. In fact the input in many instances gives the equation set in detail directly. Most of the graph problems, such as minimum path or min–max path, acyclic or not, as well as maximum path on an acyclic graph, provide examples. In these cases the weighted graph representation, adjacency matrix, or list gives the essentials of the equations. The weights of edges are the coefficients of the equations, and they appear in the relative locations in the representations that they occupy in the equations.
2. *Almost fixed form.* The form of the equation is fixed independent of, or it depends in only simple parameterized ways on, the argument structure.

In this section we study the implementation of definitions involving acyclic dependency graphs in which setting up equations is unnecessary and the major emphasis is on back substitution.

Like the minimum-constant principle, an acyclic dependency graph allows solution of the resultant equation set by substitutions of constants only. Because complex expressions need never be substituted, the cost of simplification is almost eliminated. No cycles in the dependency graph means no loops in the equation set, so loop breaking is also unnecessary. That only constants need be substituted is a key consequence of acyclicity. If each such substitution costs $O(1)$, as is often the case, the complexity is proportional to $O(e)$, where e is the total number of f calls = variables on the right of *all* equations. If on the other hand the equations have loops, it may be necessary to substitute an entire right side of an equation for each of the e variable occurrences on the right of the equations. This typically has cost $O(m)$, where m is the maximum possible number of terms on the right of an equation. To eliminate all variables, we may have to do $O(n^2)$ substitutions of equation right sides, and the total cost is typically $O(mn^2)$. Note that $e \leq mn$.

Given an acyclic set of equations, there must be an equation whose right side is a constant. Assume that equation is $x = c$. We then have a solution for the variable x. Substitute for x on the right of all other equations with c and simplify. Eliminate the equation $x = c$. Repeat the process on the remaining equations, solving for and eliminating another variable. Continue solving for and eliminating variables until the one with the constant on the right is the variable for which a solution is sought. The process terminates at that point. This is a straightforward specialization of Process 6.2.2.

Central to any algorithm for carrying out this process is a means for identifying equations whose right side has become a constant. An equation cannot be finally evaluated to a constant until all the variables on which it depends have been evaluated. When it has been evaluated, its value must be stored (or recomputed) until it has been substituted in all the right sides that depend on it. There are various ways the equation solution can be completely ordered consistent with these constraints. Which is chosen depends in part on how the equations arise from the problem statement. If the details of their form and content are known when the initial argument structure is given (situation 1 above), then a topological-sort or a depth-first inverse type of algorithm is applicable.

6.6.1 Form Given by Initial Argument Structure — Topological-Sort Ordering, Depth-First Inverse

The variables of a set of equations without loops can be partially ordered according to their dependency relations. In this *topological ordering*, a variable is of higher order than any of the variables on which it depends. Variables whose value does not depend on any other variable are of lowest order. The order of substitutions prescribed above for solving acyclic equations sets follows this topological ordering. We substitute the right side (a constant) of the equation for the variable with lowest topological ordering first, the right side of the variable with the next lowest topological ordering next, and so on.

Algorithm 6.6.1: Topologically Ordered Algorithm
For equation j, for which DEP(j) gives the number of distinct variables on which x_j depends. Assume $C[J]$ contains the value of the constant term on the right of the equation for x_j.

```
PUT into TOPLST all J for which DEP(J) = 0.
FOR I = 1 TO N DO
  IND ← pop- TOPLST
  FOR each equation of the form "x_J = E" with x_IND in E DO
    FOR each occurrence of X[IND] in E
      SUBSTITUTE C[IND] and GATHER terms (including updating C[J])
      DEP[J] ← DEP[J] − 1.
      IF DEP(J) = 0 then add J at end of TOPLST
    END
  END
END
```

Because the form of the equations is usually embedded in the initial argument structure, for example, a graph representation, the details of the algorithm for finding DEP(J), for determining which equations of the form "$x_j = E$" contain variable x_{IND} in E, and so on, will depend on and in part determine the data structure for representing the initial argument structure. The same comments apply to the following alternative approach to ordering back substitution.

For the same ESE-oriented recursive definition with acyclic dependencies to which the topologically ordered algorithm applies a standard SE depth-first inverse algorithm with the same order of complexity also applies. For this we require that the algorithm be augmented to take advantage of comparability based on equality, which probably suggested an ESE formulation initially. The fact that the definition has the acyclic dependency property implies that there is a uniform inverse.

It is also worth noting that the definitions developed for ESE solution seldom have w functions which are associative, so the more generally applicable depth-first implementations, Algorithm 4.2.4 or, if w is locally associative, Algorithm 4.2.6, are likely to be appropriate.

EXAMPLES

The form for the set of equations for graph problems is often congruent to the adjacency structure of the graph and thus is given by the graph representation. Thus the path problem definitions minimum and minimum–maximum, given previously, if applied to acyclic graphs, can be implemented efficiently using the techniques of this section. The following two definitions of functions on a graph are easily implemented on an acyclic graph using either the topological-sort or the depth-first inverse algorithm for acyclic equation sets. In the case of Definition 6.6.2, remembering the variable on the right of each equation which is in the maximum term can guide the generation of the maximum path as described in Section 6.5.

Definition 6.6.1: Topological Partial Ordering

$$\left[\begin{array}{ll} g(v) = 1 & : v = \text{"goal"} \\ g(v) = 1 + \text{MAX}_{i=1}^{i=N(X)}[g(n_i(v))] & : v \neq \text{"goal"} \end{array}\right.$$

Initially v is a vertex = "begin" in acyclic digraph G. Every vertex reachable from "begin" is on a path to "goal."

Definition 6.6.2: Maximum Path in Network

$$\left[\begin{array}{ll} g(v) = 0 & : v = \text{"goal"} \\ g(v) = \text{MAX}_{i=1}^{i=N(X)}(c_i(v) + g(n_i(v))) & : v \neq \text{"goal"} \end{array}\right.$$

Initially v is a vertex = "begin" in acyclic digraph G. Every vertex reachable from "begin" is on a path to "goal."

6.6.1.1 Almost Fixed Form ESE—Dynamic Programming Ordering

[Bellman 62] [Dreyfus 77] [Nemhauser 77] [Held 67] [Puterman 77]

There is an alternative way of ordering variables to guide a back substitution algorithm, which is particularly useful when the equation form is "almost fixed," as described in situation 2 above. If the definition is acyclic, then the n equations can be ordered in such a way that the equation for x_i with i in N^n depends only on variables x_j, with $j > i$ and j in N^n. Then the right side of the equation for x_n must be a constant—it is solved. Instead of substituting for all right side occurrences of x_n at this point, note that x_{n-1} can depend only on x_n. Substitute on the right of the equation for x_{n-1} if necessary and obtain the solution for x_{n-1}. In general when it comes time to solve for x_j, the solution for $x_n, x_{n-1}, \ldots, x_{j+1}$ will be available, and x_j will depend only on these variables. So x_j can be solved with the substitution of constants only.

Algorithm 6.6.2: Dynamic Programming Algorithm

[Assume the equations are of the form $\{x_i = w_i(x_{i+1}, x_{i+2}, \ldots, x_n) \mid i \in N\}$. Let the variable whose value we wish be x_1.]

```
FOR I = N TO 1 DO
  X[I] ← w_i(X[I + 1], X[I + 2],..., X[N])
END
```

This algorithm requires the same number of substitutions of constants for variables as the topological-sort and depth-first inverse algorithms for acyclic dependency graphs; it simply makes those substitutions in a different order. Only storage requirements are affected by this difference in ordering. Generally in Algorithm 6.6.1 partially evaluated equations are maintained, even though once a variable has been substituted for, its constant value is not again needed in solving the equations. In this algorithm, Algorithm 6.6.2, on the other hand, whenever a substitution is made, it is made for all variables on the right so only a constant remains. Thus there is at most one partially evaluated equation necessary at any time, namely the one being evaluated, though all evaluated variables are maintained for possible future use. This dynamic programming order is convenient when the equation forms are sufficiently fixed so that we can delay producing an equation until we wish to evaluate it. If there is no effort required to put the equations in the order required for the application of this algorithm, it is usually the preferable one, evaluated equations usually requiring less space than partially evaluated ones. In cases in which the equation forms are given by the initial argument structure, it may even pay to do a prepass through that argument structure to establish the ordering.

The Fibonacci number and the optimal matrix association definitions, are examples for which the equation form is "almost" independent of the details of the input. Thus the dynamic programming order is easily achieved for these. Several examples of this kind are covered in detail in the next section.

6.6.2 Acyclic Dependency Graphs — Dynamic Programming Examples

A large class of definitions involving dynamic programming are formed as a generalization of the k-dimensional definitions studied in Chapter 2. The examples given here are organized into subclasses involving increasingly more constrained forms of recursive definitions of the k-dimensional form. [Bellman 57]

6.6.2.1 Substitutional Domain Maps to k-Dimensional Space

Consider definitions whose general form is

Definition 6.6.3: Generalized k-Dimensional Definitions

$$f(V,Q) = t(V,Q) \qquad : V \le \vec{0}$$

$$f(V,Q) = w\big(f(V - r_1(V,Q),Q),\ldots,f(V - r_{m(V,Q)}(V,Q),Q)\big) \qquad : V > \vec{0}$$

Initially $Q \in D$, $V = \langle n_1, \ldots, n_k \rangle$, $r_i(V,Q) = \langle r_{i1}(V,Q),\ldots,r_{ik}(V,Q)\rangle$, Q is invariant.

A vector $V \le \vec{0}$ if one or more components are ≤ 0, and $> \vec{0}$ if all components are positive.

All such definitions can be implemented using nested FOR loops having k (dimensionality of V) levels of nesting. This kind of implementation can be viewed as a dynamic programming implementation, since the outer indices of the nested loops track which equations are being solved. The other index moves through terms of the equation, and the solution of each equation depends only on those previously evaluated. The exact range of the FORs depends upon the terminal conditions. Within the FOR loops the operation is a variant of

$$F(V) \leftarrow w\big(F(V - r_1(V,Q)), \ldots, F(V - r_{m(V,Q)}(V,Q))\big)$$

where F is a k-dimensional array.

If the time per argument for w is $O(1)$, the complexity is

$$O(n_1 * n_2 * \cdots * n_k * \text{maximum value of } m(V,Q))$$

Now consider

CASE 1:

1. $r_i(V,Q) = r_i =$ a constant independent of V and Q.
2. $m(V,Q) = m =$ a constant independent of V and Q.

This implies that the recursive line can be written

$$f(V) = w(f(V - r_i), \ldots, f(V - r_m))$$

This case has been covered extensively in Chapter 2. Its main distinction is that both the vector distances $|r_i|$ and the number m of places upon which $f(V)$ depends are bounded, and thus the temporary memory requirement is significantly less than the number of equations $= |\Delta_f(V)|$.

In general, if time per argument is $O(1)$, its complexity is implementation

$$O(n_1 * n_2 * \cdots * n_k)$$

There are often faster algorithms for definitions in this class. Definitions in which V is one-dimensional and w is a linear function form the class of linear recurrences. For this class there is a procedure for finding a closed-form solution. Other Case 1 definitions also have a closed form, or other more efficient solutions than the nested FOR loop algorithm. Of course w is not necessarily $O(1)$ per argument, so sometimes the best algorithm is considerably slower than the complexity given above. An example is provided by the recursive line for $f(n) =$ set of all binary numbers of length n, which can be written

$$f(n) = \{\langle 0,1\rangle\}|*|f(n-1)$$

The operation $|*|$ here takes time proportional to

$$|f(n-1)| * 2 = 2^{n-1} * 2 = 2^n$$

Less restrictive than Case 1 is

CASE 2:

1. $r(V, Q) = r_i(V)$ is a function of V and independent of Q.
2. $m(V, Q) = m(V)$ is a function of V and independent of Q.

This implies that the recursive line can be written

$$f(V) = w\big(f(V - r_i(V)), \ldots, f\big(V - r_{m(V)}(V)\big)\big)$$

In this case the function value at position V may depend on its values at *any* vector distance $r_i(V)$ from V, constrained only by the size of the k-dimensional space. A value computed at any place in the k-dimensional substitutional domain may have to be saved until the final value $f(n_1, \ldots, n_k)$ is computed.

Case 2 is similar to Case 1 in that, in general, it can be implemented by a nested FOR loop program with nesting of depth k with the line

$$F(V) \leftarrow w\big(F(V - r_1(V)), \ldots, F\big(V - r_{m(V)}(V)\big)\big)$$

at the center of the nest.

This case is special in that neither r_j nor m depend on Q.

An example is provided by the following simplified version of the association of multiplication of matrices, Definition 6.1.18, in which we have removed the record of the optimal association.

Definition 6.6.4: Association of Matrix Multiplications, Simplified

$$\left[\begin{array}{ll} g(\langle i, j\rangle) = 0 & : j = i \\ g(\langle i, j\rangle) = \text{MIN-SEL}_{k=0}^{k=j-i-1}\big[c_i * r_{i+k} * c_j + g(\langle i, i+k\rangle) & \\ \qquad\qquad + g(\langle i+k+1, j\rangle)\big] & : j > i \\ \text{Initially } i = 1,\ j = \text{number of matrices.} & \end{array}\right.$$

In the example evaluation of this definition in Table 6-2, the equations are simplified. Instead of constants $r_i * r_{i+k} * r_j$ consistent with the actual matrix dimensions, we use arbitrarily chosen values. On the right of each equation for $f(i, j)$, the value of $f(i, j)$ is given in column f, and the child# of the minimum right side term in column i. (We also drop commas between arguments of f.)

In a dynamic programming algorithm, the equations, here all of the same form, are each completely evaluated in an order in which no equation is evaluated before all of its variable values are available. We need solutions for $f(i, j)$ for every pair (i, j), with $1 \le i,\ j \le 4$ and $i < j$. These can be obtained

TABLE 6-2. Set of equations for example matrix association definition.

Child# $= i$				
1	2	3	f	i
$f(14) = 6 + f(11) + f(24)$	$0 + f(12) + f(34)$	$2 + f(13) + f(44)$	8	3
$f(24) = 2 + f(22) + f(34)$	$4 + f(23) + f(44)$		5	1
$f(13) = 5 + f(11) + f(23)$	$0 + f(12) + f(33)$		6	2
$f(34) = 3 + f(33) + f(44)$			3	1
$f(23) = 2 + f(2)) + f(33)$			2	1
$f(12) = 6 + f(11) + f(22)$			6	1
$\{f(ii) = 0 \mid i \in \{1,2,3,4\}\}$			0	

in different orders. In ordering by diagonals, all $f(i, j)$ with $j - i < k$ are computed before those in which $j - i = k$. Alternatively, in ordering by columns, all $f(i, j)$ with $j = k_1$ and $i = k_2$ are computed after all those with $j < k_1$, and those with $j = k_1$ but $i > k_2$ have been computed. Other orderings are possible. The different orderings can affect the amount of temporary memory necessary. Examples of two such orderings are given in Algorithm 6.6.3.

Algorithm 6.6.3: Optimal Matrix Multiplication—Dynamic Programming Algorithm

```
Ordered by columns
FOR I = 1 TO N DO
  X[I, I] ← 0
END
FOR J = 2 TO N DO *
  FOR I = J − 1 TO 1 BY −1 DO *
    X[I, J] ← ∞
    FOR K = 0 TO J − I − 1
      X[I, J] ← min(X[I, J], C[I] * R[I + K] * C[J] + X[I, I + K] + X[I + K + 1, J])
    END
  END
END

Ordered by diagonals (D = J − I)
The same as above, except the * lines are replaced with
FOR D = 1 TO N − 1 DO
  FOR I = 1 TO N − D
    J ← D + I
```

The solution to the set of equations is given in Figure 6-4. $f(i, j)$ is given at position ij in the matrix, as well as the index of a right side term, which gives the minimum for $f(ij)$.

i ↑				
4				0
3			0	1 3
2		0	1 2	1 5
1	0	1 6	2 6	3 8
	1	2	3	4
		j →		

FIGURE 6-4 Evaluation of set of equations for example matrix association.

FINDING THE ASSOCIATION FOR OPTIMAL MATRIX MULTIPLICATION
As described in Section 6.5, we can use the details of this solution to actually get the solution structure that would have resulted if Definition 6.1.18 had been implemented.

Let $r(i, j)$ be the child# for the right-hand term which solves $f(i, j)$, and let $o_{r(i,j)}(i, j)_1$ be the argument of the first and $o_{r(i,j)}(i, j)_2$ that of the second function occurrence in the $r(i, j)$th term. We use Definition 6.5.10 to generate the description of the association which gives the minimum cost.

$$\left[\begin{array}{ll} V(i, j) = \langle\langle i, j\rangle\rangle & :i = j \\ V(i, j) = \langle\langle i, j\rangle\rangle \| \langle V\big(o_{r(i,j)}(i, j)\big)_1 \| V\big(o_{r(i,j)}(i, j)\big)_2\rangle & :i < j \end{array}\right.$$

We want $V(\langle 1, 4\rangle)$. It is obtained as follows;

$$\begin{aligned} V(1,4) &= \langle\langle 14\rangle\rangle \| V_3(1,4)_1 \| V_3(1,4)_2 \\ &= \langle\langle 14\rangle\rangle \| V(13) \| V(44) \\ &= \langle\langle 14\rangle\rangle \|(\ \langle\langle 13\rangle\rangle \| V(12) \| V(33)\ ) \| V(44) \\ &\ \ \vdots \\ &= \langle\langle 14\rangle\rangle \|(\ \langle\langle 13\rangle\rangle \|(\ \langle\langle 12\rangle\rangle \| V(11) \| V(22)\ ) \| V(33)\ ) \| V(44) \\ &= \langle\langle 14\rangle\rangle \|(\ \langle\langle 13\rangle\rangle \|(\ \langle\langle 12\rangle\rangle \| \langle\langle 11\rangle\rangle \| \langle\langle 22\rangle\rangle\ ) \| \langle\langle 33\rangle\rangle\ ) \| \langle 44\rangle \\ &= \langle\langle 14\rangle, \langle 13\rangle, \langle 12\rangle, \langle 11\rangle, \langle 22\rangle, \langle 33\rangle, \langle 44\rangle \end{aligned}$$

Case 3:

1. $r_i(V, Q)$ is a function of both V and Q, and Q is invariant.
2. $m(V, Q) = m$ is independent of V and Q.

Q contains information given in the initial argument structure. Though it helps determine how the function value at one point depends on its value at other points in the k-dimensional space, it does not affect the overall size and character of the space. Thus the nested FOR loop implementation is still valid. The quantity of temporary storage required is determined by the nature of $r_i(V, Q)$, but an upper bound is fixed by the boundaries of the k-dimensional space.

It is helpful to have an efficient way of determining, for any point U in the space, every pair $(i, V) \ni V - r_i(V) = U$ [or $r_i(V) = V - U$].

This case is illustrated by the knapsack problem, a simplified version of which is given in Definition 6.1.14, below. The sequence of indices which gives the solution has been removed. It will be recreated by using the solution structure. In Definition 6.6.5, $\langle i, r\rangle$ corresponds to V, and $v = \langle v_1, v_2, \ldots, v_n\rangle$ corresponds to Q of the generic definition.

Definition 6.6.5: Knapsack Problem, Simplified

$$\left[\begin{array}{ll} g(i, r) = \{\ \} & :r > 0 \text{ and } i = 0 \\ g(i, r) = \{\ \} & :r < 0 \text{ and } i \geq 0 \\ g(i, r) = \{\langle\ \rangle\} & :r = 0 \text{ and } i \geq 0 \\ g(i, r) = g(i-1, r-v_i) \cup g(i-1, r) & :r > 0 \text{ and } i > 0 \\ \multicolumn{2}{l}{\text{Initially } i = n, r, v_i \in N, v = \langle v_1, \ldots, v_n\rangle \text{ and is invariant.}} \end{array}\right.$$

Consider an example: initially $v = \langle 1, 3, 4, 2\rangle$, $r = 5$, $i = 4$. We may need $g(i, r)$ for any pair i, r with $0 \leq i \leq 4$ (number of items) and $0 \leq r \leq 5$ (knapsack size), but we will not have to compute all such arguments. To see how to avoid computing $g(i, r)$ at as many arguments as possible, we consider the example in some detail. Since we wish finally to have no items left to consider for assignment and no space remaining in the knapsack, we want solutions that involve $g(0, 0)$. We then have as a first equation $g(0, 0) = \{\langle\ \rangle\}$. Now move to $g(1, v_1) = g(1, 1)$, which is the only equation with $i = 1$ using $g(0, 0)$ in its right side. It follows that $g(1, x) = \{\ \}$ if $x \neq 1$. Next consider those values of x for which the equation for $g(2, x)$ has $g(0, 0)$ or $g(1, 1)$ on the right,

$$g(2, v_2) = g(1, 0) \cup g(1, v_2) = g(1, 0)$$

$$g(2, v_1 + v_2) = g(1, 1 = v_1) \cup g(1, v_1 + v_2) = g(1, 1)$$

In Figure 6-5 we develop a number of relevant $g(i, r)$'s in this way. In doing so we take advantage of the fact that the equations in which $g(i, r)$ can be a right side term are $g(i + 1, r)$ in which $g(i, r)$ is the 2nd term, and $g(i + 1, r + v_{i+1})$ in which $g(i, r)$ is the 1st term.

The entire solution is shown in Table 6-3, in which position (i, j) contains $g(i, j)$ as well as the index of the right side term, 1 or 2, that provided the

child# 1	child# 2		f	i
⋮				
$g(2,0) = g(1,-3)$	$\cup\ \underline{g(1,0)}$		$\{\langle\ \rangle\}$	2
$g(2,3) = \underline{g(1,0)}$	$\cup\ g(1,3)$		$\{\langle\ \rangle\}$	1
$g(2,1) = g(1,-2)$	$\cup\ \underline{g(1,1)}$		$\{\langle\ \rangle\}$	2
$g(2,4) = \underline{g(1,1)}$	$\cup\ g(1,4)$		$\{\langle\ \rangle\}$	1
$g(1,0) = g(1,-1)$	$\cup\ \underline{g(0,0)}$		$\{\langle\ \rangle\}$	2
$g(1,1) = \underline{g(0,0)}$	$\cup\ g(0,1)$		$\{\langle\ \rangle\}$	1
$g(0,0) = \{\langle\ \rangle\}$				

$v = \langle 1\ 3\ 4\ 2 \rangle$

$i \quad\ 1\ 2\ 3\ 4$

FIGURE 6-5 Set of equations for example knapsack problem definition.

solution. Position $(0,0)$ is filled first. Then in row 1, a 2, indicating that $g(0,0)$ is the 2nd term in the equations for $g(1,0)$, is entered in column 0. Also a 1 is entered in row 1 at a position $v_1 = 1$ unit to the right of an entry in row 0, indicating that $g(0,0)$ is the first term on the right of the equation for $g(1,1)$. In general a 2 is entered in row i just above each entry in row $i-1$, and a 1 is entered in each position in row i which is v_i to the right of an entry in row $i-1$.

This solution implies that some of the entries in the matrix must be empty, In row 1, for example, all entries $g(1,x)$ for $x > v$, will be empty, so there are $r - v_1$ empty entries. For row 2, at least $r - v_1 - v_2$ will be empty, etc. On the

TABLE 6.3. Evaluation of set of equations for example knapsack problem definition.

i	$r=0$	$r=1$	$r=2$	$r=3$	$r=4$	$r=5$		*empty* = ↔
4	←———	———	———	———	———→	12 ⟨{ }⟩	−0 or 2 →	$r = 5$
3	←———	———→		2	12 ⟨{ }⟩	1 ⟨{ }⟩	−0 or 4 →	$r - v_n = 3$
2	2	2		1 ⟨{ }⟩	1 ⟨{ }⟩	←——→	−0 or 3 →	$r - v_1 - v_2 = 1$; $r - v_i - v_{n-1} < 0$
1	2	1 ⟨{ }⟩	←———	———	———	———→	−0 or 1 →	$r - v_1 = 4$
0	⟨{ }⟩	←———	———	———	———	———→		$r = 5$

other hand, since there will be only one entry in row n, namely $g(n, r)$, corresponding to a valid solution, there will be r empty locations in row n. This in turn has consequences for row $n - 1$, and so on. In general, with $j \ni n - v_1 \cdots - v_j > 0$ and $n - v_1 \cdots - v_{j+1} \leq 0$ and $k \ni n - v_n \cdots - v_k > 0$ and $n - v_n \cdots - v_{k+1} \leq 0$.

In row 0: r	In row n: r
In row 1: $r - v_1$	In row $n - 1$: $r - v_n$
In row 2: $r - v_1 - v_2$	In row $n - 2$: $r - v_n - v_{n-1}$
⋮	⋮
In row j: $r - v_1 \cdots - v_j$	In row $n - k$: $r - v_n \cdots - v_{n-k-1}$
$(j + 1)r - (j)v_1 - \cdots - v_j$	$(k + 1)r - kv_n - \cdots - v_{n-k-1}$

$$\text{grand total} = n(j + k + 2) - (jv_1 + kv_n) - \cdots - (v_j + v_{n-j-1})$$

We can reorder the components of v any way we like. It is well to do this so as to minimize T. Algorithm 6.6.4 implements a row ordered passage through the equations for this example.

Algorithm 6.6.4 gives a row order construction of a matrix like that in Table 6-3 for Definition 6.6.5. The matrix which we call CS, gives in $CS(I, J)$ the child# for the right-side terms which solve $g(i, r)$. The algorithm uses a queue, Q, to record the entries in row i when building row i of $CS(I, R)$. RINIT is the size of the knapsack.

Algorithm 6.6.4: Knapsack-Dynamic Programming

```
CS ← { }
Q ← queue- 0
Q ← queue- #
FOR I = 1 TO N DO
  REPEAT
  R ← pop- Q
  Q ← queue- R
  IF R ≠ # THEN DO
            CS[I, R] ← CS[I, R] ∪ {2}
            IF R + V[I] ≤ RINIT THEN DO
                          Q ← queue- R + V[I]
                          CS[I, R + V[I]] ← CS[I, R + V[I]] ∪ {1}
                          END
            END
  UNTIL R = #
END
END
```

FINDING THE INDICES OF THE ENTRIES IN V WHICH SUMS TO R

Definition 6.6.5 now finds all sequences of indices whose corresponding items fill the knapsack. Let $O(\text{cs}(ir)) = \{i\}$ if $\text{cs}(ir)$ includes 1, and $= \{\ \}$ otherwise. Let $N1(ir) = r - v_i$ if cs(ir) contains 1 and $N2(ir)$ if it contains 2.

$$\left[\begin{array}{ll} V(i,r) = \langle\langle\ \rangle\rangle & :r = 0 \\ V(i,r) = \langle\ \langle\langle i\rangle\rangle|*|V(i-1, r-v_i), & \\ \qquad\quad V(i-1,r)\ \rangle & :r > 0, cs(ir) = \langle 1,2\rangle \\ V(i,r) = \langle\langle i\rangle\rangle|*|V(i-1, r-v_i) & :r > 0, cs(ir) = \langle 1\rangle \\ V(i,r) = V(i-1,r) & :r > 0, cs(ir) = \langle 2\rangle \end{array}\right.$$

We want $V(4,5)$. It is obtained as follows;

$$\begin{aligned} V(4,5) &= \langle\ \langle\langle 4\rangle\rangle|*|V(3,3), V(3,5)\rangle \\ &= \langle\ \langle\langle 4\rangle\rangle|*|\langle\langle 3\rangle\rangle|*|V(2,0), V(3,5)\ \rangle \\ &= \langle\ \langle\langle 3,4\rangle\rangle, V(3,5)\rangle \\ &= \langle\ \langle\langle 3,4\rangle\rangle, V(2,5)\rangle \\ &= \langle\ \langle\langle 3,4\rangle\rangle, \langle\langle 2\rangle\rangle|*|V(1,1)\ \rangle \\ &= \langle\ \langle\langle 3,4\rangle\rangle, \langle\langle 2\rangle\rangle|*|\langle\langle 1\rangle\rangle|*|V(0,0)\ \rangle \\ &= \langle\ \langle\langle 3,4\rangle\rangle, \langle\langle 2,1\rangle\rangle\ \rangle \end{aligned}$$

To get all solutions takes time proportional to the number of solutions. In many instances one solution is sufficient, with consequent time saving.

6.7 LOOP-BREAKING RULES

The assumption of a loop-breaking rule has been critical in the algorithm development in this chapter. Up to now we have asserted the existence of such rules but not justified our assertion nor considered how such rules can be developed. That problem will be addressed next.

First we consider some cases in which there is a simple loop-breaking rule. Consider a set equation involving the operations union ($\cup$) and intersection ($\cap$) of the form

Lemma 6.7.1: Let $x = (q \cap x) \cup w$, where w and q are each sets given by any expressions involving variables and constants. Then $x = (q \cap x) \cup w$ is satisfied by any set s which contains w and is contained in $w \cup q$, and by no other.

PROOF: If W satisfies the equation, then

$$W = (q \cap W) \cup w \tag{1}$$

Let $W \supseteq w$ and $W \subseteq w \cup q$, that is, $W = w \cup z$, where $z \subseteq q$, and see whether

TABLE 6-4. Loop-Breaking Rules

Given an equation of the form:

$$x = q \cdot x + w$$

where q and w may contain variables including the variable x.

	Operation interpretations		Solutions x is		
	$\cdot$	$+$	$\geq$	$\leq$	
1	union	inter-section	$q \cap w$	w	
2	inter-section	union	w	$q \cup w$	
3	max	min	$\min(q, w)$	w	
4	min	max	w	$\max(q, w)$	
5	or	and	q and w	w	
6	and	or	w	q or w	
7	sum	min	w	w	$q > 0$
8	$\vert * \vert$	union	$q^* \vert * \vert w$	$q^* \vert * \vert w$	$\langle\, \rangle \notin q,\ q, w \neq \{\ \}$

Eq. (1) holds,

$$\begin{aligned} w \cup z &=? \, q \cap (w \cup z) \cup w \\ &=? \, q \cap z \cup w \\ &=? \, z \cup w \qquad \text{because } q \text{ includes } z \\ &=! \, w \cup z \end{aligned}$$

On the other hand, if y is in W but in neither w nor q, then y is not in the set given by the right of Eq. (1), so Eq. (1) cannot hold. Furthermore if y is in w but not in W, then y is in the set on the right of Eq. (1) but not that on the left, and again equality cannot hold. ■

Thus there is a range of solutions satisfying the equation in $(\cup, \cap)$. Similar proofs apply to other pairs of operators. These include (maximum, minimum) and (and, or). For pairs of operators labelled $\cdot$ and $+$, with appropriate properties, x will be a solution of $x = q \cdot x + w$ if and only if $w \leq x \leq q + w$. In Table 6-4, rows 1 through 6 give pairs of operations $(\cdot, +)$ having the necessary properties.

Now consider another equation with very different operators, namely, addition $(+)$ and minimum $(\downarrow)$. With these operators and the domain restricted to numbers greater than 0, we get a unique solution rather than a range of solutions.

Lemma 6.7.2: Let $x = (q + x) \downarrow w$, where q and w are both > 0. Then $x = (q + x) \downarrow w$ is uniquely satisfied by $x = w$.

PROOF: First, we show that w satisfies the equation.

$$w = ?\, q + w \downarrow w$$

$$w = ?\, w$$

On the other hand, let $w^{-} < w$. If w^{-} satisfies $x = q + x \downarrow w$, then

$$w^{-} = q + w^{-} \downarrow w$$

which is impossible because since $q > 0$, $q + w^{-} \downarrow w > w^{-}$. If w^{+}, with $w^{+} > w$, satisfies $x = q + x \downarrow w$, then

$$w^{+} = q + w^{+} \downarrow w$$
$$= w$$

which also is impossible. ■

Consider still another group of operators, namely, cross concatenation ($|*|$) and union ($\cup$), and closure (*), where closure is defined as follows. If q is a set of vectors or strings, then q^{+} is all strings that can be formed by concatenating ($\|$) any number of members of q,

$$q^{*} = q^{+} \cup \{\langle\ \rangle\}$$

We get a unique solution for an equation involving these operations, as shown in the next lemma, though here, unlike the previous case, it involves an operation that does not appear in the original equation.

Lemma 6.7.3: (*Arden's Rule*): Let $x = (q|*|x) \cup w$, where q and w are nonempty sets of strings (vectors) and q does not include $\langle\ \rangle$. Then $x = (q|*|x) \cup w$ is uniquely satisfied by $x = q^{*}|*|w$.

PROOF: If W satisfies the equation, then

$$W = q|*|W \cup w \qquad (1).$$

Let $W = q^{*}|*|w$,

$$(q^{*}|*|w) = ?\, q|*|(q^{*}|*|w) \cup w$$
$$= ?\, (q^{+}|*|w) \cup \langle\ \rangle|*|w$$
$$= !\, (q^{*}|*|w)$$

On the other hand, let y, which is not a vector in $q^*|*|w$, be in W. In fact, let it be a smallest length vector in W but not in $q^*|*|W$. Then although y is in W, that is, left of Eq. (1), it is not in the set formed on the right. It is not in w, so it could only be in $q|*|W$. In order for y to be a member of $q|*|W$, y must equal $q^1\|z$, with $q^1 \in q$, and z in W. But this could only be true if z were not in $q^*|*|w$ (otherwise $q^1\|z$ must be in $q^*|*|w$). But z is shorter than y (since q^1 cannot be $\langle\ \rangle$), a contradiction.

Similarly let y be a smallest member of $q^*|*|w$ which is absent from W. If y is in w, then the right of Eq. (1) would include y and Eq. (1) would not hold. So for no member z of W and q^1 in q can $q^1\|z = y$. But if y is a member of $q^*|*|w$, then there is a z in $q^*|*|w$ and q^1 in q such that $q^1\|z = y$ (y cannot be $\langle\ \rangle$ because q cannot contain $\langle\ \rangle$). So z is smaller than y and cannot be in $q^*|*|w$: a contradiction. ■

The loop-breaking rules derived above, and others which can be derived similarly, are summarized in Table 6-4. (The pairs of operations given on the left of the table define a lattice, and the inequalities on the right represent the extreme points of the sublattice of solutions.)

Loop-Breaking in Nonlinear Equations

Although the loop-breaking rules in Table 6-4 have been used in this text almost exclusively with linear equations, they also apply to nonlinear equations. Examples of the application of the rules to a linear and to a nonlinear equation are given next.

Example 6.7.1: The rules are applicable to linear equations. Assume the equations are of the form

$$x_i = \mathrm{MIN}_{k=1}^{n}[a_{ik} + x_k] \downarrow C_i$$

which, because of commutativity of min, can be written

$$x_i = (a_{ii} + x_i) \downarrow \mathrm{MIN}_{k=1}^{n \text{ but not } i}[a_{ik} + x_k] \downarrow C_i$$

It can be seen that Lemma 6.7.2 is applicable with $q = a_{ii}$, and

$$w = \mathrm{MIN}_{k=1}^{n \text{ but not } i}[a_{ik} + x_k] \downarrow C_i$$

Example 6.7.2: The rules are also applicable to equations which are not linear. Assume we are given equations of the form

$$x_i = \bigcup_{k=1}^{n-1} [a_{ik} \cap x_k \cap x_{k+1}] \cup C_i$$

which, because of properties of operators $\cap$ and $\cup$, can be written

$$x_i = x_i \cap (a_{i-1,i} \cap x_{i-1} \cup a_{i,i+1} \cap x_{i+1})$$

$$\cup \bigcup_{k=1}^{n-1,\text{ but neither } i \text{ nor } i-1} [a_{ik} \cap x_k \cap x_{k+1}] \cup C_i$$

It can be seen that Lemma 6.7.3 is applicable with

$$q = x_i \cap (a_{i-1,i} \cap x_{i-1} \cup a_{i,i+1} \cap x_{i+1})$$

$$w = \bigcup_{k=1}^{n-1,\text{ but neither } i \text{ nor } i-1} [a_{ik} \cap x_k \cap x_{k+1}] \cup C_i$$

6.8 NONTERMINATING RECURSIVE DEFINITIONS AND THE MINIMUM STATE SOLUTION

Many loop-breaking rules do not give unique solutions. In practice an *extreme solution*, either the minimum or the maximum, is usually of interest. (With appropriate choice of ordering relations, the extreme of interest can, and here will, be forced to be the minimum.) To better understand why this is so, to introduce iterative algorithms and their incremental properties, as well as to establish conditions for the existence and determination of loop-breaking rules, and therefore of the applicability of Gaussian type algorithms, a physical interpretation of extreme solutions is now developed.

Consider a general nonnested recursive definition.

Definition 6.8.1: Standard Form Again

$$\begin{bmatrix} f(X) = t(X) & :T(X) \\ f(X) = w\big(p(X), f(o_1(X)), \ldots, f(o_{m(x)}(X))\big) & :\sim T(X) \\ \text{Initially } X = d \in D. & \end{bmatrix}$$

Let the substitutional domain of $f = \Delta_f = \{d_1, \ldots, d_n\}$, and let $o_j(d_i) = d_{ij}$, $f(d_i) = x_i$, and $f(d_{ij}) = x_{ij}$. Then the set of equations generated by the above definition is

$$\{x_i = w\big(p(d_i), x_{i1}, x_{i2}, \ldots, x_{in_i}\big) \mid i \text{ in } N\}$$

or in the abbreviated notation we will use

$$\{x_i = w_i(x_{i1}, x_{i2}, \ldots, x_{in_i}) \mid i \text{ in } N\}$$

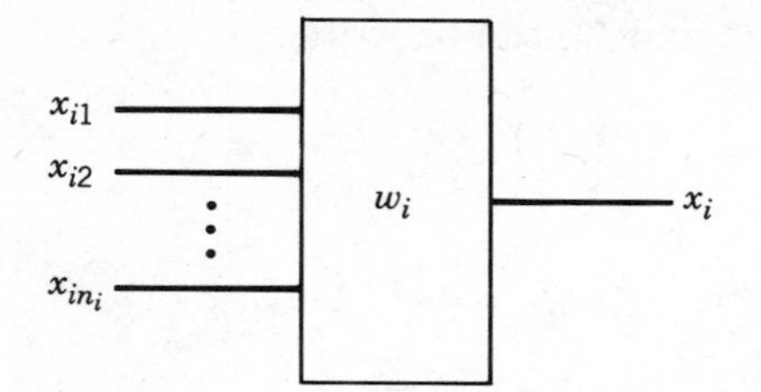

Figure 6-6 Typical block for Definition 6.8.1.

It is easy to see what the dependency graph and the closely related folded network for Definition 6.8.1 look like. The folded network will consist of blocks of the form shown in Figure 6-6. For example, the folded network for the *cost* of the shortest path is given in Figure 6-7. The relevant definitions are Definition 2.12.4, or Definition 6.1.6 with the path part removed, and the graph is that in Figure 6-1.

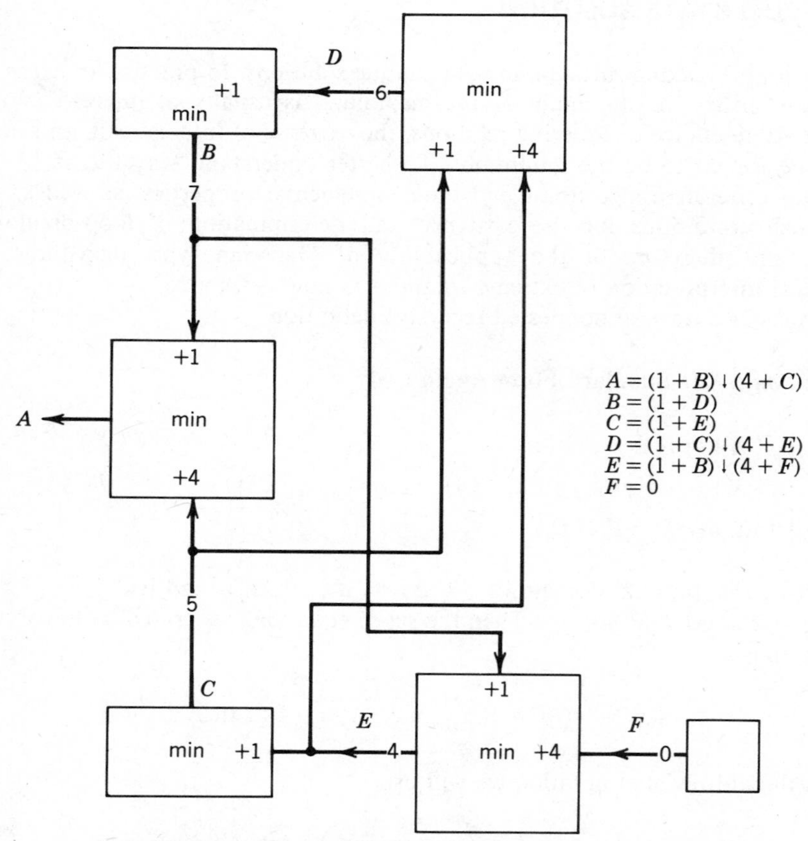

FIGURE 6-7 Folded Network, Definition 2.12.4, Graph of Figure 6-1.

If everything except the letters identifying the vertices were removed from each block, we would have the dependency digraph for the given set of equations—which of course is Definition 2.12.4 applied to a specific digraph, say G. In fact the dependency digraph is G with the direction of each edge reversed (the converse graph of G). In Figure 6-7 the output of block x is the cost of the shortest path from x to f. The vector of all block outputs of a network is called the *state* of the network. Viewing a network as a physical configuration of blocks which perform the indicated functions, the final state is the block outputs when the network is *stable*. A network NET is stable in a given state X if, once established, that state does not change. In that state every block input is itself an output in X. So block inputs form a subset of the components of X, say $i(X)$. If X is a state of NET, and the output computed for NET with input $i(X)$ is X, then NET is stable in state X. Such a state is also a solution to the corresponding set of equations, and thus to the recursive definition from which that set was derived. A stable state is a *fixed point* of the equations, the network, and the recursive definition which it solves.

The algorithmic approach for simulating the passage of such a network from an initial to a stable state to find its solution is *iteration*. In Section 6.9.2 such algorithms are developed, and applied, for example, to the network of Figure 6-7.

The network in Figure 6-7 has only one stable state. In general, however, a network may have many stable states; thus, the equations and the definition which generated them have many solutions. But the definition is formulated as a solution to a particular problem, and thus in expectation of a single solution with certain properties; there must be some additional unstated condition. Our experience is that the condition usually implies that the minimum of the several stable states is the desired solution. In order to appreciate this, and as a guide in selecting the correct stable state in particular cases, we must understand the significance of the minimum stable state. As an aid to reaching such an understanding, we develop a physical interpretation of minimum stable state.

6.9 INTERPRETATION OF THE MINIMUM STABLE STATE OF A NETWORK — BASIC DEFINITIONS AND PROPERTIES; LIMITS

We consider how a fixed point or stable state of a network X can be found by iteration of the function W, by the network—given that W is monotonic—on a partially ordered set of possible network values Q. The development here differs somewhat from lattice developments [Birkhoff 67; Hecht 77; Aho, Sethi, Ullman 86].

We think of a network, like that in Figure 6-9, possibly arising from a recursive definition or set of equations, as evaluating a function W on the vector of block inputs to give a vector of block outputs. In equation set terms, W generates the values X_i (outputs) on the left of the equations when values for X_i (inputs) are plugged in on the right. That is, the equations are of the form

$$\{ X_i = w_i(X_{i1}, \ldots, X_{in_i}) | i \in N^n \}, \quad \text{where } X_{ij} \in \{ X_i | i \in N^n \}.$$

Because inputs and outputs of a network (the variables on the left and right of the equations) are the same, W has the same set Q as both range and domain. When Q has certain order properties, W can exhibit interesting and useful behavior. This motivates the following definition.

Q is a *bounded set* if

1. Q is *partially ordered* by the relation $\leq$.
2. Q has a *least element* designated $\underline{0}$, that is, for all $X \in Q$, $\underline{0} \leq X$.

Usually Q is a set of n-vectors, and W is determined by a system of n equations in n unknowns. The relation as well as the zero of a bounded set arise from those of the components of these vectors, and the components are usually of the same type, for example, all real numbers or all sets. But the results we develop do not require any of these restrictions.

The function W behaves nicely if it has properties related to those of its domain Q. A function W is called a *bounded function* if

1. W: $Q \to Q$ and Q is a bounded set.
2. W is *monotonic*, that is, if A and B are in Q and $A \leq B$ then $W(A) \leq W(B)$.

We represent n applications of W as follows:

$$W^0(X) = X$$

$$W^n(X) = W(W^{n-1}(X))$$

[$W^1(X)$ is also written $W(X)$].

Notice that monotonicity *does not* mean that $W(X)$ is necessarily $\geq X$, for example, let $W(X) = x^{1/2}$. On the other hand it does imply that

Lemma 6.9.1: If W is monotonic, then for all $j \in N_0$; $W^{j+1}(\underline{0}) \geq W^j(\underline{0})$.

PROOF: $W^1(\underline{0}) \geq \underline{0}$, so $W^2(\underline{0}) \geq W^1(\underline{0})$, and so for all $j \in N_0$, $W^{j+1}(\underline{0}) \geq W^j(\underline{0})$. ■

(In fact if, for any X, $W(X)$ is $\geq$ or $\leq X$, $W^{j+1}(X)$ will be similarly related to $W^j(X)$ and to X.

Unless otherwise specified, *the function W referred to subsequently in this section is bounded*. W is *stable* at X, or X is a *stable state* (of W), or X is a *fixed point* of W, if and only if $X = W(X)$.

Lemma 6.9.2: If W is a bounded function and W is stable at X, then for all $j \in N_0$, $X \geq W^j(\underline{0})$.

PROOF: For any X, $\underline{0} \leq X$. So by monotonicity, for all $j \in N_0$, $W^j(\underline{0}) \leq W^j(X) = X$. ■

Definition 6.9.1: Definition of Network—Finite, Infinite
A *network* NET consists of *a pair* (Q, W), where Q is a bounded set and W a bounded function. (We will sometimes speak of the "network W," meaning the network whose bounded function is W.) Given a network NET $= (Q, W)$, if $X \in Q$,

X is stable (in NET) if W is stable at X.
X is infinite (in NET) if for all $j \in N_0$, $W^j(\underline{0}) < X$.
X is finite (in NET) if there exists $j \in N_0$, $W^j(\underline{0}) \geq X$.

Intuitively, $X \in Q$ is finite if for some finite j $X \in Q | 0 \leq X \leq W^j(\underline{0})$; X is infinite if X is unreachable by W, that is, no such j exists and $X > W^j(\underline{0})$ for all j. Notice that a member of Q may have no relation to W^j and be neither finite or infinite.

Definition 6.9.2: Definition of Limit, Finite and Infinite
We say that L is a limit of W with respect to X, $\lim_w(X) \to L$, if W is stable at L and either

1. *Finite limit*: There exists $j \in N_0 \ni L = W^j(X)$.
2. *Infinite limit*: [Effectively we want L to equal $W^\infty(\underline{0})$; we take $L =$ least upper bound $\{W^j(\underline{0})\}$.]
   a. L is infinite.
   b. For $X \in Q$, if X is infinite, then $X \geq L$. (This condition assures the uniqueness of the infinite limit.)

L is a limit of W if L is a limit of W with respect to $\underline{0}$, that is, if $\lim_W(\underline{0}) \to L$.

The following examples illustrate the definitions above.

Example 6.9.1: For $Q =$ sets, $\leq$ given by $\subseteq$, and $\underline{0} = \{\ \}$, let a and b be two fixed sets in sets, and

$$W(X) = X \cap a \cup b$$

Then $W^0(\underline{0}) = \{\ \}$; $W^2(\underline{0}) = W^1(\underline{0}) = b$, so $L = b$ is the finite limit of this network. Any subset of b is finite, and any superset of b is infinite; any other set is neither finite nor infinite.

Example 6.9.2: The set of equations for the shortest path network in Figure 6-7 provide an example in which the limit is finite but not given by the first application of W.

Example 6.9.3: Let G be a directed graph with vertices V and edges E. Let Q be the power set of $V \times V$, ordered by inclusion, with $\underline{0} = \{\ \}$. Let

$$A \square B = \{(x, z) | \exists y\ (x, y) \in A \text{ and } (y, z) \in B\}$$

and

$$W(X) = X \square E \cup E$$

It is easy to verify that L is the edge set of the transitive closure of G, and that it requires k = diameter of G (the maximum length of a shortest non-trivial path) iterations to reach the stable state. Thus, if G is a finite graph, L is a finite limit.

Example 6.9.4: Let Q = sets of vectors (or strings in a language), ordered by set inclusion, with $\underline{0} = \{\langle\ \rangle\}$. Let $A \subset Q$, and

$$W(X) = A|*|X \cup A$$

We know that $L = A^*$. This is an infinite limit, since A^* is not reached by a finite number of applications of W.

Example 6.9.5: Let $f(x)$ be a real-valued function with some fixed point in an interval in which $0 < f'(x) \leq r < 1$. Let $Q = \mathbf{R}$ with the usual order, and let $\underline{0} = x_0$ = some initial guess to the left of the fixed point. Then the usual fixed point iteration $W(x) = f(x)$ will in general have the fixed point as its infinite limit.

Stability alone is a strong condition, as seen from the following.

Lemma 6.9.3: If X is stable, then either

1. X is a finite limit or
2. X is infinite.

PROOF: $\forall j\ X \geq W^j(\underline{0})$, X is stable and Lemma 6.9.2 holds. So either

$$\exists j\ X = W^j(\underline{0}) \text{ and } X \text{ is a finite limit}$$

or

$$\forall j\ X > W^j(\underline{0}) \text{ and } X \text{ is infinite.}$$

An alternative characterization of limit follows directly from this lemma, together with the definition of infinite limit. The following lemma gives that characterization.

Lemma 6.9.4: L is a limit of W if and only if the following two conditions are satisfied:

1. L is stable.
2. If $Y \neq L$ and for all j, $Y \geq W^j(\underline{0})$, then $Y \geq L$.

6.9.0.1 Limits and the Minimum Stable State

We give some simple cases in which limits can be determined and the relation of limit to the minimum stable state. This forms the basis for subsequent development, ranging from the insensitivity of iterative algorithms to operation sequencing, to incremental iterative algorithms.

Lemma 6.9.5: If L is a finite limit of W and $X \leq L$ (in particular, if X is finite), then L is a finite limit of W with respect to X.

PROOF: First we show that if L is a finite limit of W and X is finite

$$\exists j \in N_0,\ W^j(\underline{0}) \geq X \qquad X \text{ is finite}$$

$$\forall j \in N_0,\ W^j(\underline{0}) \leq L \qquad L \text{ is stable and Lemma 6.9.2 holds}$$

Therefore if X is finite, $L \geq X \geq \underline{0}$.

$$W^k(L) \geq W^k(X) \geq W^k(\underline{0}) \text{ for all } k \in N_0 \qquad \text{monotonicity}$$

But there is a z so that when $k = z$, $L = W^z(\underline{0})$, and

$$L = W^z(L) \geq W^z(X) \geq W^z(\underline{0}) = L. \quad \blacksquare$$

In some cases a network NET $= (W, Q)$ has a finite stable state which is the only stable point in a finite domain Q. In such a case, a finite number of applications of W to any X in Q produces that limit. The minimum path network will generally have this property, as will most networks which arise from a set of equations whose dependency graph has no cycles.

Lemma 6.9.6: Let NET $= (Q, W)$ such that

1. Q is finite.
2. L is the only stable member of Q.

Then

1. NET has a finite limit.
2. If $X \in Q$ and there is a $Y \geq X$ in Q which is "order related" to $W(Y)$, that is, either $Y \geq W(Y)$ or $Y \leq W(Y)$, it follows that L is a finite limit with respect to X.

PROOF: Since L is stable, L is either a finite limit or is infinite by Lemma 6.9.3. But since L is the only stable state, it can only be infinite if there is no finite limit. If there is no finite limit, then for all $k \in N_0$, $W^{k+1}(\underline{0}) > W^k(\underline{0})$, so Q has an infinity of members. By hypothesis this cannot be so. Thus L is a finite limit. Assume $L = W^z(\underline{0})$, and $Y \geq X \geq \underline{0}$. For all $k \in N_0$,

$$W^k(Y) \geq W^k(X) \geq W^k(\underline{0}).$$

But if $k = z$, $L = W^z(\underline{0})$,

$$W^z(Y) \geq W^z(X) \geq W^z(\underline{0}) = L$$

But $Y \geq W(Y) \geq W^2(Y) \geq \cdots W^q(Y) \geq \cdots$ or $Y \leq W(Y) \leq W^2(Y) \leq \ldots W^q(Y) \leq \ldots$. But $>$ in the first case and $<$ in the second case cannot hold for all $q \in N_0$, because Q is finite. So there must be a $u \ni W^u(Y) = L$. Thus for $j \ni j + z \geq u$,

$$W^{j+z}(Y) = L \geq W^{j+z}(X) \geq L$$

So there exists $j + z \in N_0 \ni W^{j+z}(X) = L$. ■

Notice that if Q has a largest element $\underline{1}$ which is $\geq$ every member of Q, then $\underline{1}$ will serve in place of Y in the lemma above for any X in Q. Also X itself would serve in place of Y if it were guaranteed that either $X \geq W(X)$ or $X \leq W(X)$.

Even if Q is not finite, as long as sequences of increasing and decreasing members of Q are of finite length, the limit will be reached by a finite number of applications of W. This is stated more precisely as follows.

A partial order satisfies the *ascending chain condition* if any sequence of its members $a_1 < a_2 < \ldots < a_p$ is finite.

A partial order satisfies the *descending chain condition* if any sequence of its members $a_1 > a_2 > \ldots > a_p$ is finite.

Lemma 6.9.7: Suppose NET $= (Q, W)$ meets the same conditions as in Lemma 6.9.6, except that condition 1 is replaced with

1. Q satisfies both the ascending and the descending chain condition.

Then if $X \in Q$ and there is a Y in Q so that either $Y \geq W(Y)$ or $Y \leq W(Y)$ for $Y \geq X$, it follows that L is a finite limit with respect to X.

PROOF: The proof is identical to that of Lemma 6.9.6, except that the finite chain conditions rather than finiteness is used to argue that neither $>$ nor $<$ can hold for all $q \in N_0$.

The next two lemmas establish the close relation of limit and minimum stable set.

Lemma 6.9.8: If a network has a finite stable state Z, then Z is the unique finite limit and the smallest stable state.

PROOF: Because Z is stable and finite, Z is a finite limit (Lemma 6.9.3).

If Y is stable, then it is either a finite limit or infinite (Lemma 6.9.3).

If it is finite, then there exists a $j \in N_0$, $W^j(Y) = Z$ (Lemma 6.9.5). If Y is also a limit, it is stable, and $Y = W^j(Y) = Z$. If Y is an infinite limit, then $Y > Z$, since Z is finite.

So all stable states in the network are $\geq Z$. ■

Lemma 6.9.9: If a network has an infinite limit and no finite limit, then its limit is unique and is its minimum stable state.

PROOF: Proof follows from the definition of infinite stable state. ■

From the last two lemmas it follows immediately that

Theorem 6.9.1: The limit of a network is its minimum stable state.

Notice, however, that a network with a number of stable states, none of which are finite, may not have a limit. This occurs when the state which would be the infinite limit is not stable. An example of this is given as problem 12 in the exercises, Section 6.10.

6.9.1 Incremental Changes in a Network

This section is optional; it may be skipped without loss in continuity.

Generally, W, the network function is the composition of primitive functions operating on variables and parameters, in which only the output and not the form of the algorithm is affected by changes in the parameters. For example, in the minimum path example, Figure 6-7, addition and minimum are the primitive functions, while the coefficients representing edge costs are the parameters. Suppose W runs to its limit, after which one or more of its parameters are changed, so that W becomes W', and then W' is allowed to run on. Will W' reach a limit, supposing it has one?

To develop the answer to this question, we need some preliminary results.

Assume two networks W_A and W_B, have the same bounded domain or set of outputs Q, with the minimum element $\underline{0}$, and are both monotonic. Assume that for all $X \in Q$, $W_B \geq W_A$.

Lemma 6.9.10: If for all $X \in Q$, $W_B(X) \geq W_A(X)$, then for all $j \in N_0$, $W_B^j(\underline{0}) \geq W_A^j(\underline{0})$.

PROOF: By induction:

$\underline{0} = \underline{0}$	obvious
$W_B^1(\underline{0}) \geq W_A^1(\underline{0})$	condition of theorem
$W_B^{j-1}(\underline{0}) \geq W_A^{j-1}(\underline{0})$	inductive hypothesis
$W_A\left(W_B^{j-1}(\underline{0})\right) \geq W_A^j(\underline{0})$	monotonicity
$W_B^{j-1}(\underline{0}) = W_B^{j-1}(\underline{0})$	obvious
$W_B^j(\underline{0}) \geq W_A\left(W_B^{j-1}(\underline{0})\right)$	condition of theorem
$W_B^j(\underline{0}) \geq W_A^j(\underline{0}))$	transitivity ∎

We now show that if the network W_A (which may or may not have a limit)

1. Runs for any time, reaching a value, say X_A, immediately after which
2. the parameters of W_A are changed to give network W_B with $W_B \geq W_A$, which does have a finite limit L_B, following which
3. W_B is run starting at X_A,

then W_B will reach its limit L_B.

Theorem 6.9.2: Assume for all $X \in Q$, $W_B(X) \geq W_A(X)$ and

1. $X_A = L_A$ or
2. $X_A = W_A^j(\underline{0})$ for some j, or more generally X_A is finite for W_A. Then $\lim_{W_B}(X_A) \to L_B$ if and only if W_B has a limit L_B.

PROOF: (*if*) Assume that W_B has a limit L_B. If there exists $z \in N_0 \ni X_A \leq W_A^z(\underline{0})$, then by Lemma 6.9.10,

$$W_B^z(\underline{0}) \geq W_A^z(\underline{0}) \geq X_A$$

So X_A is finite for W_B, and the result follows by Lemma 6.9.5.

On the other hand, if $X_A = L_A$ is an infinite limit, then for all $z \in N_0$,

$$L_B \geq W_B^z(\underline{0}) \geq W_A^z(\underline{0})$$

so $L_A \leq L_B$. Thus either $L_A = L_B$, or L_A is finite for B.

(*only if*) If $\mathrm{Lim}(W_B^j(X_A)) \to L_B$, then we can show that the conditions of Lemma 6.9.4 are met by L_B. First note that $W_B(X_A) \geq W_A(X_A) \geq X_A$. Then

1. L_B is stable by assumption.
2. $L_B \geq X_A \geq \underline{0}$, so for all j, $L_B = W_B^j(L_B) \geq W_B^j(X_A) \geq W_B^j(\underline{0})$ by monotonicity, and finally
3. If for all $j \in N_0$, $Y \geq W_B^j(\underline{0})$ then for all $j \in N_0$, $Y \geq W_A^j(\underline{0})$, which in turn implies for all $j \in N_0$, $Y \geq L_A$ by Lemma 6.9.4 applied to W_A, and thus that $Y \geq W_A^j(X_A)$, so $Y \geq L_B$. ■

Not only will W_B reach its limit, it is also easy to show that it will always reach the limit at least as rapidly as if after the conversion it is again started with an input of $\underline{0}$. On the other hand, if instead of $W_B \geq W_A$, $W_B \leq W_A$, no such guarantee can be given. This is shown by the following example.

Example 6.9.6: Let $W_A(\langle x_1, x_2 \rangle)$ be the network function given by the following equations characterized by the operations $\cap$ and $\cup$, and the parameters $A = \{\{1,2\},\{2,4\},\{1,3\},\{3,4\}\}$. The minimum element in Q is $\{1,2,3,4\}$. Q is all subsets of this set. The partial ordering is $x \supseteq y$ corresponding, unintuitively, to $x \leq y$.

W_A:

$$A = \{\{1,2\},\{2,4\},\{1,3\},\{3,4\}\}$$

$$x_1 = \{1,2\} \cap x_1 \cup \{2,4\} \cap x_2$$

$$x_2 = \{1,3\} \cap x_1 \cup \{3,4\} \cap x_2$$

Starting at $X^0 = \langle \{1,2,3,4\}, \{1,2,3,4\} \rangle$,

$$X^1 = \langle \{1,2,4\},\{1,3,4\} \rangle$$

$$X^2 = \langle \{1,2,4\},\{1,3,4\} \rangle = L_A$$

W_B:

$$B = \{\{1,2,3\},\{2,4\},\{1,3\},\{3,4\}\}$$

[For $X =$ any subset of $\{1,2,3,4\}$, $W_B(X) \supseteq W_A(X)$ and so $W_B(X) \leq W_A(X)$.]

$$x_1 = \{1,2,3\} \cap x_1 \cup \{2,4\} \cap x_2$$

$$x_2 = \{1,3\} \cap x_1 \cup \{3,4\} \cap x_2$$

Starting at $X^0 = \langle\{1,2,3,4\},\{1,2,3,4\}\rangle$,

$$X^1 = \langle\{1,2,3,4\},\{1,3,4\}\rangle$$

$$X^2 = \langle\{1,2,3,4\},\{1,3,4\}\rangle = L_B$$

However, if we assumed we changed from W_A to W_B after the former stabilized at $\langle\{1,2,3\},\{1,3,4\}\rangle$, we get the following. Starting at $X^0 = \langle\{1,2,3\},\{1,3,4\}\rangle$ in W_B,

$$X^1 = \langle\{1,2,3\},\{1,3,4\}\rangle = \text{ a stable state in } W_B = X_A$$

$$X_A \neq L_B$$

If certain strong, but still realistic, conditions hold, we can always continue from a solution of one network, L_A of W_A, to that of a second one, L_B of W_B, by applying W_B to L_A a finite number of times, no matter whether $W_A \geq W_B$ or not. Such conditions are given in the following theorem.

Theorem 6.9.3: Let $\text{NET}_A = (Q, W_A)$, $\text{NET}_B = (Q, W_B)$.

1. Q is a finite set or, more generally, Q meets the ascending and descending chain conditions.
2. Both NET_A and NET_B have (finite) limits L_A and L_B, respectively.
3. NET_A and NET_B each have only one stable state.

Then if there is a Y in Q so that either $Y \geq W_B(Y)$ or $Y \leq W_B(Y)$ for $Y \geq X$, it follows that there exists $j \in N_0 \ni W_B^j(L_A) = L_B$.

PROOF: Follows directly from Lemma 6.9.6. ■

The conditions for Theorem 6.9.3 are satisfied by the minimum path network. In fact, stable points in acyclic networks are almost invariably unique and finite, so unique limits almost always exist. Even in the presence of cycles, it is sometimes feasible and usually desirable to partially solve the corresponding equation set symbolically, in terms of its parameters, to obtain an equivalent acyclic network, which then has a unique solution.

6.9.1.1 Variation in Sequencing of Iteration Operations— Timing

Thus far our study of network behavior has implicitly assumed that all network blocks responded to an input change with equal speed. It is physically more realistic to assume that blocks do not respond with the same but with different and even varying speed. We will remove the restriction on response time and consider conditions guaranteeing that a network will nevertheless reach its limit. The removal of response time restrictions results in fewer restrictions on the order of evaluation in algorithms simulating the corresponding networks.

To do this we need to modify our definitions. Consider a set of vectors, each of the form $\langle x_1, \ldots, x_n \rangle$, drawn from the domain D, in which $x_i \in D_i$, where D_i is partially ordered and has a minimum element $\underline{0}_i$. The vectors in D are partially ordered by the relation $\langle x_1, \ldots, x_n \rangle \geq \langle z_1, \ldots, z_n \rangle$ if $x_i \geq z_i$ in the partial ordering on D_i. Although the results given in this section only require that each x_i come from its own domain D_i, the x_i's usually come from the same domain.

We need to distinguish block inputs from outputs so we extend the network function W's definition to include both as arguments.

Definition 6.9.3: Extended Network Function Definition

Let

x_i = output of the ith block.

$X \in D = \langle x_1, \ldots, x_n \rangle$ = a vector of all n block outputs in the network.

$y_{i,j}$ = jth input to block i.

$Y_i \in D = \langle y_{i,1}, \ldots, y_{i,n_i} \rangle$ = a vector of all n_i inputs to block i.

$Y = \langle Y_1, \ldots, Y_n \rangle$ = the vector of all vectors of inputs to the network blocks.

Then

$w_i(X) = x$ with X as above and $x \in D_i$.

$W(X, Y) = \langle \langle w_1(X), \ldots, w_n(X) \rangle, Y \rangle$ with X and Y as above.

$W_i(X, Y) = \langle \langle x_1, \ldots, x_{j-1}, w_j(X), x_{j+1}, \ldots, x_n \rangle, Y \rangle$ with X and Y as above.

We consider only networks for which $W(X, Y)$ is monotonic. Therefore $w_i(X)$ and $W_i(X)$ must also be monotonic. So if $X' \geq X$, then $w_j(X') \geq w_j(X)$ and $W_j(X') \geq W_j(X)$.

These definitions, as well as the next two, are illustrated in Figure 6-8.

out(i, k) = the block (number) whose output feeds input k of block $i(y_{i,k})$.

$C_{i,k}(X, Y) = \langle X, Y$ with $y_{i,k}$ set to x_j, where $j = \text{out}(i, k)$, all other components unaltered$\rangle$.

Consider a network NET $= (D, W, C)$. As before, this is a network because D is partially ordered and has a lowest member, $\langle \underline{0}_1, \ldots, \underline{0}_n \rangle = \underline{0}^n$. Finite and infinite limits are defined as before. In previous sections limit L was defined in terms of the output vector X only, since it was implicitly assumed that the transmission time from an output $x_{\text{out}(i,k)}$ and the corresponding input $y_{i,k}$

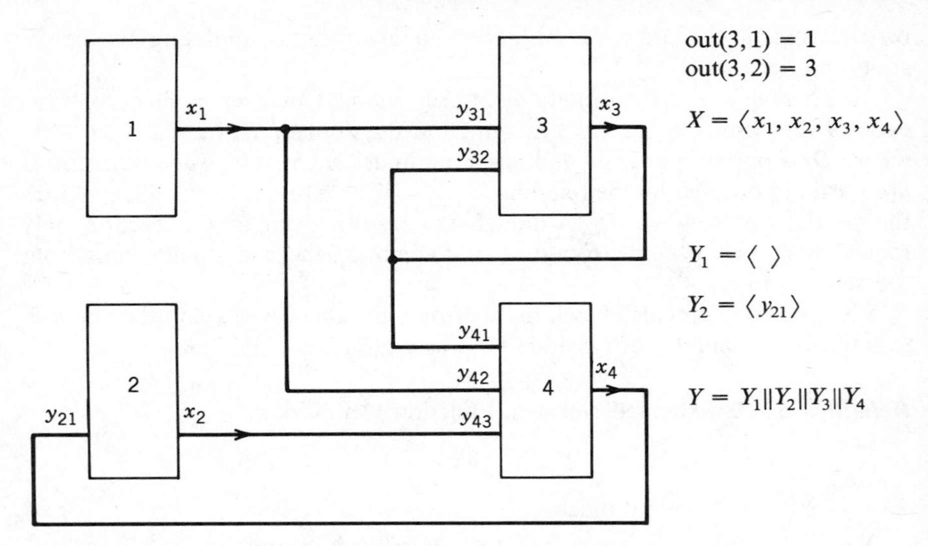

$W_4(X, Y) = \langle X$ with $x_4 =$ output of block 4 when input is $\langle y_{41}, y_{42}, y_{43}, Y_j \rangle$

$C_{4,2}(X, Y) = \langle X$ unchanged, Y with $y_{42} = x_1 \rangle$

FIGURE 6-8 Input–output definitions.

was instantaneous, making both equal in every state. This is no longer generally so, but it is nevertheless true once the output X reaches its limit L. In general we will eventually assume that the value of a block output will affect each input to which it is connected. So in the limit $X = L$ and $x_{\text{out}(i,k)} = y_{i,k}$. Because our discussion now must maintain the input–output distinction, we will think of the limit here as a pair $\langle L, L' \rangle$, with L being the output vector X in the limit and L' the vector of input vectors Y when all block inputs have taken the same value as the limit outputs to which they are connected.

We say that a block *input* Y_{ij} *is irritable* if it has a different value than the block output X_k which feeds it. Otherwise it is *relaxed*.

We say *block i is irritable* if its output does not equal $w_i(X)$, X being the vector of inputs of the network.

Our development will continue under the assumption that the following hypothesis holds.

Hypothesis 6.9.1: Whenever a set of one or more blocks or inputs is irritable, at least one such irritability will relax within a bounded time b.

Notice that this leaves the possibility that relaxation will be confined to the same block or group of blocks, and some irritable block may never relax. In fact this can happen, as will be illustrated, so this property by itself is not sufficient to ensure that a network with different timings reaches its limit. First, however, we consider some of the consequences of Hypothesis 6.9.1.

From the untimed case we know that if the sequence of events is all blocks relax, all inputs relax, all blocks relax, all inputs relax, and so on, then the limit will be reached if the network is finite, and if it is infinite, for any j, a value greater than $W^j(\underline{0}^n)$ will be reached. But will it do so if there is no fixed order of block response?

We first show that

Lemma 6.9.11: Given Hypothesis 6.9.1, if a network NET starts with $\underline{0}$ at all inputs and outputs, then no matter in what order relaxations occur, NET's output vector and/or its input vector either increases or remains unchanged.

PROOF: Assume that the output and input vectors after $p - 2$ relaxations are X' and Y', and after $p - 1$, are X and Y. The inductive hypothesis is if the pth relaxation is an input or a block relaxation, $X \geq X'$ and $Y \geq Y'$, and that $x_{\text{out}(i,k)} \geq y_{i,k}$. With all inputs and outputs initially $\underline{0}$ (on the 0th relaxation), this will clearly be true after the first relaxation. The inductive part of the argument covers the second and successive relaxations.

We consider three cases:

1. The pth relaxation is an input relaxation.
2. The pth relaxation is a block relaxation and the $(p - 1)$th relaxation is an input relaxation.
3. The pth relaxation is a block relaxation and the $(p - 1)$th relaxation is a block relaxation.

Assume the pth relaxation is in an input, namely, $y_{i,k}$ that is, $C_{i,k}(X, Y)$ is applied. Since by the inductive assumption $x_{\text{out}(i,k)} \geq y_{i,k}$ for any (i, k), this relaxation causes the new X and Y vectors to be greater than or equal to X and Y before the relaxation and also maintains the property $x_{\text{out}(i,k)} \geq y_{i,k}$. Suppose the pth relaxation is in a block, namely, block j, so that $W_j(X, Y)$ is applied, and that the $(p - 1)$th was an input relaxation, say $C_{i,k}(X, Y)$. Because of the application of $C_{i,k}(X, Y)$ and inductive assumption, $Y_j \geq Y_j'$ and $X = X'$ since an input relaxation has no effect on the outputs.

$W_j(X, Y) \geq W_j(X, Y')$	monotonicity
$W_j(X, Y') \geq X$	inductive assumption
$W_j(X, Y) \geq X$	transitivity and $X = X'$.

Outputs increase while inputs are unchanged, so outputs remain larger than inputs $(x_{\text{out}(i,k)} \geq y_{i,k})$.

Finally, suppose the pth relaxation is in a block, namely, block k, that is, $W_k(X)$ is determined, and that the $(p-1)$th was also a block relaxation, say block i. Since by the inductive assumption $\langle X, Y \rangle = W_i(X', Y)$,

$$W_j(X', Y) \geq \langle X', Y \rangle \qquad \text{for any } j$$

so

$$W_i(X', Y) \geq \langle X', Y \rangle \text{ and} \tag{1}$$

$$W_k(X', Y) \geq \langle X', Y \rangle \tag{2}$$

$$\begin{aligned}
&W_k(W_i(X', Y)) \geq W_k(X', Y) && \text{by monotonicity on Eq. (1)} \\
&W_i(W_k(X', Y)) \geq W_i(X', Y) && \text{by monotonicity on Eq. (2)} \\
&W_i(W_k(X', Y)) = W_k(W_i(X', Y)) && \text{by commutativity of } W_i \text{ and } W_k \\
&W_k(W_i(X', Y)) \geq W_i(X', Y) &&
\end{aligned} \tag{3}$$

and since $\langle X, Y \rangle = W_i(X', Y)$, substituting in Eq. (3),

$$W_k(X, Y) \geq \langle X, Y \rangle$$

Outputs increase and inputs do not change, so outputs remain larger than inputs ($x_{\text{out}(i,k)} \geq y_{i,k}$). ■

Lemma 6.9.12: If the network NET starts with $\underline{0}$ at all inputs and outputs, then no matter in what order relaxations occur, NET's output and input vectors remain *less than or equal to* their value at the network limit.

PROOF: This is because $L \geq \underline{0}^n$, and thus if R is a relaxation operator, W_i or $C_{i,k}$, on a block or input, then by monotonicity of R, $R(L, L') \geq R(\underline{0}^n, \underline{0}^m)$, so $\langle L, L' \rangle \geq R(\underline{0}^n, \underline{0}^m)$. This argument can be extended to any number of R operations. ■

Consider applying the relaxation operators to irritable inputs and blocks in any order, starting with all inputs at $\underline{0}$. Assuming Hypothesis 6.9.1 will this guarantee reach a limit if one exists? The answer is no, as shown by the following example.

Example 6.9.7: Consider the set equations

$$\begin{bmatrix} x1 = a|*|x1 + x2 + \{\langle\ \rangle\} \\ x2 = a* \end{bmatrix}$$

In the network interpretation $x1$ and $x2$ on the left represent block outputs, while $x1$ and $x2$ on the right of Eq. (1) are inputs to block 1, namely y_{11} and y_{12}, respectively. Block 2 has no inputs. We assume x_1, x_2, y_{11}, and y_{12} are all $\{\langle\ \rangle\}$ initially.

Block 1 is irritable—apply $W_1(X, Y)$

$$W_1(X, Y) = W_1(\langle\{\langle\ \rangle\}, \{\langle\ \rangle\}\rangle, \langle\langle\{\langle\ \rangle\}, \{\langle\ \rangle\}\rangle, \langle\ \rangle\rangle)$$
$$= \langle\langle\{\langle\ \rangle, a\}, \{\langle\ \rangle\}\rangle, \langle\langle\{\langle\ \rangle\}, \{\langle\ \rangle\}\rangle, \langle\ \rangle\rangle\rangle$$

That is

$$\begin{bmatrix} x1 = a|*|\{\langle\ \rangle\} + \{\langle\ \rangle\} + \{\langle\ \rangle\} = \{\langle\ \rangle, a\} \\ x2 = a^* \end{bmatrix}$$

gives an output $X = \langle\{\langle\ \rangle, a\}, \{\langle\ \rangle\}\rangle$.

Input y_{11} is irritable—apply $C_{11}(X, Y)$

$$C_{11}(X, Y) = C_{11}(\langle\{\langle\ \rangle, a\}, \{\langle\ \rangle\}\rangle, \langle\langle\{\langle\ \rangle\}, \{\langle\ \rangle\}\rangle, \langle\ \rangle\rangle)$$
$$= (\langle\{\langle\ \rangle, a\}, \{\langle\ \rangle\}\rangle, \langle\langle\{\langle\ \rangle, a\}, \{\langle\ \rangle\}\rangle, \langle\ \rangle\rangle)$$

The only change is that now $y_{11} = \{\langle\ \rangle, a\}$.

Block 1 is irritable—apply $W_1(X, Y)$

$$W_1(X, Y) = W_1(\langle\{\langle\ \rangle, a\}, \{\langle\ \rangle\}\rangle, \langle\langle\{\langle\ \rangle, a\}, \{\langle\ \rangle\}, \langle\ \rangle\rangle)$$
$$= \langle\{\langle\ \rangle, a, a^2\}, \{\langle\ \rangle\}\rangle, \langle\langle\{\langle\ \rangle, a, a^2\}, \{\langle\ \rangle\}\rangle, \langle\ \rangle\rangle)$$

$$\begin{bmatrix} x1 = a|*|\{a, \langle\ \rangle\} + \{\langle\ \rangle\} + \{\langle\ \rangle\} = \{\langle\ \rangle, a, a^2\} \\ x2 = a^* \end{bmatrix}$$

gives an output $X = (\{\langle\ \rangle, a, a^2\}, \{\langle\ \rangle\})$.

If we continue applying C_{11} and W_1 alternately to irritable input 1 at block 1, and to block 1, the limit will never be reached. However, the three successive operations W_2, C_{12}, W_1 will get $X = \langle a^*, a^*\rangle$, which is the limit. ■

Thus we need a hypothesis stronger than Hypothesis 6.9.1 to guarantee reaching a limit. The following property suffices.

Hypothesis 6.9.2: Whenever any block or input is irritable, that block or input will eventually (within a finite time) be relaxed.

Lemma 6.9.13: If Hypothesis 6.9.2 holds, NET will reach a finite limit in finite time, or if the limit is infinite, it will reach a value $\geq W^j(\underline{0}^n)$ for any $j \in N_0$.

PROOF: We have established that in any network both inputs and outputs below the limit increase when relaxed. It follows that once a block is irritable, it will remain so until it is relaxed. Therefore when $\underline{0}$ is actually applied at the network inputs, all blocks whose output is not $W(\underline{0}^n)$ are irritable, and if relaxed will have the output of that block in $W(\underline{0}^n)$. Each of these blocks will eventually relax, and their inputs can only increase, so eventually the output will be $\geq W(\underline{0}^n)$. Inductively assume the outputs are $\geq W^{j-1}(\underline{0}^n)$. Then these or higher values must eventually appear as inputs relaxing irritable inputs, and with these new inputs, irritable blocks will all eventually be relaxed with these or even larger inputs. Therefore eventually outputs are $\geq W^j(\underline{0}^n)$. ■

Hypothesis 6.9.3: The network domain Q satisfies the ascending chain condition. Whenever any block or input is irritable, at least one such irritability will relax within a bounded time b.

Hypothesis 6.9.3 incorporates Hypothesis 6.9.1 and guarantees that Hypothesis 6.9.2 will hold. This often provides an easy way of guaranteeing that limits will be reached.

Once the ascending chain condition holds, the network NET must have a finite limit, and we have

Lemma 6.9.14: Given Hypothesis 6.9.3, NET will reach its finite limit L in finite time.

PROOF: As long as the output of the network is less than L, it must increase whenever a block is relaxed. A block must eventually be relaxed because once an input is relaxed, it cannot become irritable again until some block is relaxed, so eventually (after nb), if no block is relaxed, all inputs are relaxed and only blocks remain irritable. So after each period of at most nb the output must increase. Thus there must be a sequence of increasing output states generated as long as the output is less than the limit. Because of the ascending chain condition, this sequence must eventually end and the limit be reached. ■

In a similar way we can show that

Lemma 6.9.15: If in Hypothesis 6.9.3 ascending is replaced with the condition that NET has a finite limit L, it will reach L in finite time.

In summary, if a network has a finite limit L, then L is the minimum stable state, and the network will actually reach and be stable at L if started at $\underline{0}^n$. If the limit is infinite, the state of the network will continually increase and eventually exceed $W^j(\underline{0}^n)$ for any finite j, while remaining less than L, which in turn is less than any other state greater than $W^j(\underline{0}^n)$ for all finite j. As we have shown, these conclusions hold virtually *independent of the block response times*. The common techniques for evaluation of sets of equations usually give this minimum solution. This is true for the Gaussian elimination approach, as will be verified in Section 6.9.4.

6.9.1.2 Simple Conditions for the Existence of a Limit

We need techniques for determining whether a network has a limit. One simple condition in which a limit can be guaranteed is given by

Lemma 6.9.16: If the network domain Q is finite, then W, and thus the network, has a limit.

PROOF: $W^{j+1}(\underline{0}^n) \geq W^j(\underline{0}^n)$ by Lemma 6.9.1, but $>$ cannot hold for all j because that would make Q infinite. Thus there exists $j \in N$, $W^{j+1}(\underline{0}^n) = W^j(\underline{0}^n) =$ limit $=$ minimum stable state of W. ■

In Definition 5.2.1, for the vertices reachable from a given vertex in graph G, $Q =$ the domain $=$ the range consists of all subsets of vertices of G. Thus Q is finite. Q is bounded and W is monotone, so the W function has a limit. The same is true for the cost of the minimum path definition provided there is some finite cost path from "begin" to "goal."

The following lemma shows that, even if a network domain is not finite, as long as sequences of increasing values are finite, a finite limit is guaranteed.

Lemma 6.9.17: If the network domain Q satisfies the ascending chain condition, then W and thus the network has a limit.

PROOF: $W^{j+1}(\underline{0}^n) \geq W^j(\underline{0}^n)$ by 6.9.1, but $>$ cannot hold for all j because then there would be an infinite ascending chain. Thus there exists $j \in N_0$, $W^{j+1}(\underline{0}^n) = W^j(\underline{0}^n) =$ limit $=$ minimum stable state of W. ■

Lemmas 6.9.16 and 6.9.17 can help establish the existence of a limit. It is still necessary to find that limit.

We now consider techniques for computing the minimum stable state, given that there is one. This will eventually result in some other tests for the existence of a limit.

6.9.2 Iterative Algorithms—For Definitions with a Finite Limit

In the following discussion assume the function W is given by a set of equations.

Definition 6.9.4: The Network Function W as a Set of Equations

$$\{x_i = w_i(x_{i1}, x_{i2}, \ldots, x_{in_i}) \mid i \in N\}, \text{ or equivalently,}$$

$$\{x_i = w_i(x_1, \ldots, x_n) \mid i \in N\}.$$

Thus if

$$X = \langle x_1, x_2, \ldots, x_n \rangle$$
$$W(X) = \langle w_1(x_{11}, x_{12}, \ldots, x_{1n_1}), \ldots, w_n(x_{n1}, x_{n2}, \ldots, x_{nn_n}) \rangle.$$

Iteration starts by substituting $\underline{0}$'s for all variable occurrences on the right of all equations and obtaining the solution. Let $X^0 = \underline{0}^n$; then $X^1 = W(X^0)$ is the result of this first substitution. Next substitute the variable values in X^1 for all variable occurrences on the right of all equations and solve to get $X^2 = W(X^1)$. In general $X^{k+1} = W(X^k)$. Provided the limit is finite, this process, which simulates the progress in time of the outputs in the corresponding network with equal response times, continues until substitution of X^k on the right and evaluation results in X^k on the left, that is, until $X^k = W(X^k)$. This is the minimum stable state.

Algorithm 6.9.1: Iterative Algorithm I—Solving Equations in Definition 6.9.4

```
X ← 0^n
FLG ← 1
TX ← 0^n
WHILE(FLG = 1) DO
  X ← TX
  FLG ← 0
  FOR I = 1 TO N DO
    TX[I] ← w_I(X[1],..., X[n])
    IF X[I] ≠ TX[I] THEN FLG ← 1
  END
END
```

In Algorithm 6.9.1 X^k is substituted on the right in all equations to compute the new components which form X^{k+1}. This is not generally necessary. The new values of components can be computed one or more at a time in any order. The designation X^{k+1} can be given to the vector which contains any subset of these changed components and is otherwise identical to X^k. Then the new input vector X^{k+1} can be applied on the right again to whichever equations we wish,, with any or all changes included in X^{k+2}. This nondeterministic procedure is justified by Lemmas 6.9.13, 6.9.14, and 6.9.15, which ensure a unique result even if one block responds more rapidly than another, and the delay of an individual block varies within bounds.

In particular let $X^0 = \underline{0}^n$. Let I start at 1 and k at 0. If, on substituting with X^k on the right of the equations, the Ith variable $X[I]$ either

1. Does not change, in which case I is incremented (modulo n) and again $X[I]$ is examined for change, or
2. Does change, in which case k is incremented and X^k becomes the vector with the new $X[I]$. Then the new X^k is substituted on the right, I is incremented modulo n and again $X[I]$ is examined for change,

then I is incremented until either there is a change in $X[I]$ and the substitution of a new X^k, or termination is signaled by n successive failures to find such a change. This procedure corresponds to a network in which the ith block responds more slowly than the $(i + 1)$th, but in which the nth block responds in less than twice the time of the first.

We give the algorithm for this variation of the iterative approach.

Algorithm 6.9.2: Iterative Algorithm II—Solving Equations in Definition 6.9.4

```
X ← 0^n
I ← 1
COUNT ← 0
WHILE (COUNT < N) DO
  T ← X[I]
  X[I] ← w_I(X[1],..., X[n])
  IF T = X[I] THEN COUNT ← COUNT +1
                ELSE COUNT ← 0
  I ← (I + 1) MOD (N)
END
```

If this algorithm is run on the graph of Figure 6-1 for minimum cost (but not the path), the values of variable v's start at $\underline{0}^n$ and progress with each passage through the algorithm loop (passage number or time $= t$) are as follows.

Example 6.9.8: Result of Algorithm 6.9.2 Run on Graph in Figure 6-1 Starting with $\underline{0} = 0$

Variable	t	v	t	v	t	v	t	v	t	v	t	v	t	v
$A = x_1$	1	1	7	2	13	4	19	6	25	8	31	9	37	9
$B = x_2$	2	1	8	3	14	5	20	7	26	8	32	8		
$C = x_3$	3	1	9	3	15	5	21	6	27	6	33	6		
$D = x_4$	4	2	10	4	16	6	22	7	28	7	34	7		
$E = x_5$	5	2	11	4	17	5	23	5	29	5	35	5		
$F = x_6$	6	0	12	0	18	0	24	0	30	0	36	0		

For this problem the lowest element, $\underline{0}$, of the bounded set is 0 or ∞, depending on how we define the ordering in Q. Working the same problem with the lowest element ∞ yields

Example 6.9.9: Result of Algorithm 6.9.2 Run on Graph in Figure 6-1 Starting with $\underline{0} = \infty$

Variable	t	v	t	v	t	v	t	v	t	v	t	v
$A = x_1$	1	∞	7	∞	13	∞	19	10	25	9	31	9
$B = x_2$	2	∞	8	∞	14	∞	20	8	26	8		
$C = x_3$	3	∞	9	∞	15	6	21	6	27	6		
$D = x_4$	4	∞	10	∞	16	7	22	7	28	7		
$E = x_5$	5	∞	11	5	17	5	23	5	29	5		
$F = x_6$	6	0	12	0	18	0	24	0	30	0		

Rather than continually increasing modulo n, other fixed orders of testing variables can be used. However an implementation with lower average time than all of these can be obtained if, after a change in x_i, changes are next considered for x_j's which depend on x_i. In terms of equations, these are the x_j's with x_i on the right. To do this we need:

DEP(i): a set of indices of equations which have x_i on the right.

CHK: a subset of the indices to be reevaluated; j is in CHK if it appears in DEP(i) for some i that has been changed.

$\ni (S)$, when S is a set, returns an arbitrary element in S.

Algorithm 6.9.3: Iterative Algorithm III—Solving Equations in Definition 6.9.4

```
X ← 0ⁿ
CHK ← {1,2,3,...,n}
WHILE (CHK ≠ { }) DO
  I ← ∋ (CHK)
  CHK ← CHK − {I}
  T ← X[I]
  X[I] ← w_I(X[1],...,X[n]
  IF T ≠ X[I] THEN CHK ← CHK ∪ DEP (I)
END
```

Iterative algorithms are widely applicable. Whether the equation set or network is linear or nonlinear, as long as the function is monotone and the limit is finite, these algorithms apply. Furthermore these algorithms require just enough storage for the current variable values, and the set of equations W. Each new value required for x_i, is completely recomputed using current values of variables on the right of its equation. Thus time cost is large. Under some conditions, however, it is possible to speed up iteration by storing relevant results of previous iterations. More specifically we can save the evaluation of parts of equations which depend on variables no longer change. The value of a variable will no longer change once all the variables on the right of its equation (excluding itself) can no longer change. Note a special but common case. If a variable, say x, becomes fixed, then any term whose only variable is x as in "linear" equations, also is fixed. Furthermore, if a change in variable x does occur, then only variables whose value depends on x can change. This can be used to compute new components systematically in an order in which they are likely to change.

These time reductions are applicable, for example, when the dependency graph is acyclic or when the minimum-constant principle holds. In such cases iteration can be applied so that a new variable will become fixed on each iteration of Algorithm 6.9.1. Of course finding the fixed variable takes time.

6.9.3 A Path Interpretation of Minimum Solution for a Class of Recursive Definitions

The interpretation given here is a modest generalization of that given in [Aho, Hopcroft, Ullman 75], although the development is different.

We have seen that the minimum stable state solution is the one that is first reached when a network representing the function W is initialized with the minimum value $\underline{0}$ on all its inputs and allowed to iterate until it stabilizes. This physical interpretation of the minimum solution distinguishes it from other stable states of W, but does not determine if the minimum stable state is appropriate. This depends on having a satisfying interpretation for the folded network for the given equation set. We develop such an interpretation for a subclass of the definitions considered in Section 6.9, more detailed and thus more useful in determining whether the minimum stable state is the desired solution. (It often is.)

Consider the class of recursive definitions which give rise to equation sets of the form

Definition 6.9.5: Interpretable Equation Set

$$\{x_i = [a_{i1} * x_1] + \cdots + [a_{in} * x_n] + c_i \mid i \in N^n\}$$

The operations + and * need not be addition and multiplication, but must have some of the properties of those functions,

1. * distributes through +.
2. + is commutative and associative.
3. a_{ij}, c_i and $x_i \in$ domain D, and D contains $0 \ni$ for $x \in D$,

$$x * 0 = 0 = 0 * x \qquad \text{and} \qquad x + 0 = x = 0 + x$$

4. The constants c_i have the property that $c_i + c_i = c_i$.

Consider the n-vertex graph G, with vertices $i \in N^n$. Vertex i is associated with variable x_i, and there is an edge from x_i to x_k, weighted with $a_{i,k}$ for precisely those x_k with $a_{i,k} \neq 0$. G is isomorphic to the dependency graph and bears a close resemblance to the network corresponding to the set of equations. If this set of equations was derived from a problem on a graph, that graph is probably also isomorphic to G (or its converse graph). Minimum path, min–max path problems, domination, and a large number of problems which can be formulated on a graph give rise to such a set of equations. The term *genpath* refers to a path in a graph in which vertices can be repeated (see Section 1.2). The solution to the set of equations in Definition 6.9.5 can be related to genpaths in G. (For the following, define the length of a genpath to be the length of the sequence of vertices representing it.)

If $p = \langle i, j, k, \ldots, z, w \rangle$ is a vertex genpath from i, then we define the $*$-evaluation of p by

$$
\begin{aligned}
*\text{-evaluation of } p &= 0 && :|p| = 0 \\
&= c_i = c_w && :|p| = 1 \\
&= a_{ij} * a_{jk} * \cdots * a_{zw} * c_w && :|p| \geq 2
\end{aligned}
$$

If P is a set of vertex genpaths, then define the $+*$-evaluation of P by

$$+*\text{-evaluation of } P = \sum_{p \in P} *\text{-evaluation of } p.$$

Finally, if P_t is the set of all genpaths of length t starting at i, let

$$\text{ev}(i, t) = +*\text{-evaluation of } P_t$$

It follows that for all $1 \leq i \leq n$

$$
\begin{aligned}
\text{ev}(i, t) &= 0 && :t = 0 \\
\text{ev}(i, t) &= c_i && :t = 1 \\
\text{ev}(i, t) &= a_{i1} * \text{ev}(1, t-1) + \cdots + a_{in} * \text{ev}(n, t-1) && :t \geq 2
\end{aligned}
$$

The first two lines follow directly from the $*$-evaluation definitions. The last, recursive line follows from the definitions of $*$- and $+*$-evaluation, and the partition of the set of all paths out of i of length $t \geq 2$ into subsets whose second vertex is k, for $k = 1, \ldots, n$, using the properties of $*$ and $+$, including distributivity.

Next we define the function EV, which will be shown to be a solution to Definition 6.9.5 given as a computation for a set of paths.

Definition 6.9.6: Path Evaluation
If $P^t = \bigcup_{k=0}^{t} P_t$ is the set of all genpaths of length $\leq t$ starting at i, then for $1 \leq i \leq n$

$$
\begin{aligned}
\text{EV}(i, t) &= 0 && :t = 0 \\
\text{EV}(i, t) &= c_i && :t = 1 \\
\text{EV}(i, t) &= +*\text{-evaluation of } P_t + \text{EV}(i, t-1) && :t \geq 2
\end{aligned}
$$

Thus for $t \geq 1$

$$\begin{aligned}
\mathrm{EV}(i,t) &= \sum_{k=0}^{t} \mathrm{ev}(i,k) \\
&= \sum_{k=2}^{t} [a_{i1} * \mathrm{ev}(1, k-1) + \cdots + a_{in} * \mathrm{ev}(n, k-1)] + c_i + 0 \\
&= \sum_{k=1}^{t-1} [a_{i1} * \mathrm{ev}(1, k) + \cdots + a_{in} * \mathrm{ev}(n, k)] + c_i \\
&= a_{i1} * \sum_{k=1}^{t-1} \mathrm{ev}(1,k) + \cdots + a_{in} * \sum_{k=1}^{t-1} \mathrm{ev}(n,k) + c_i \\
&= a_{i1} * \mathrm{EV}(1, t-1) + \cdots + a_{in} * \mathrm{EV}(n, t-1) + c_i
\end{aligned}$$

This gives an alternate way of representing EV.

Definition 6.9.7: Path Evaluation Recursive Definition
For $1 \leq i \leq n$

$$\begin{aligned}
\mathrm{EV}(i,t) &= 0 && :t = 0 \\
\mathrm{EV}(i,t) &= c_i && :t = 1 \\
\mathrm{EV}(i,t) &= a_{i1} * \mathrm{EV}(1, t-1) + \cdots + a_{in} * \mathrm{EV}(n, t-1) + c_i && :t \geq 2
\end{aligned}$$

This recursive formulation will be useful in showing that EV evaluates Definition 6.9.5. In summary:

Theorem 6.9.4: Definitions 6.9.6 and 6.9.7 are equivalent.

To show that EV does evaluate the equations of Definition 6.9.5, we will show that $\mathrm{EV}(i, j)$ is equivalent to $[W^j(\underline{0})]_i$, the jth iterate for x_i.

Referring to Definition 6.9.5, the iterative function W is defined as follows.

Definition 6.9.8: Iterative Evaluation
For $1 \leq i \leq n$

$$\begin{aligned}
[W^t(\underline{0}^n)]_i &= 0 && :t = 0 \\
[W^t(\underline{0}^n)]_i &= a_{i1} * [W^0(\underline{0}^n)]_1 + \cdots + a_{in} * [W^0(\underline{0}^n)]_n + c_i = c_i && :t = 1 \\
[W^t(\underline{0}^n)]_i &= a_{i1} * [W^{t-1}(\underline{0}^n)]_1 + \cdots + a_{in} * [W^{t-1}(\underline{0}^n)]_n + c_i && :t \geq 2
\end{aligned}$$

where $W^t(\underline{0}^n) = \langle [W^t(\underline{0}^n)]_1, \ldots [W^t(\underline{0}^n)]_n \rangle$.

Letting $\underline{0}^n$ in Definition 6.9.8 be the n-tuple $(0, 0, \ldots, 0)$, we have

Theorem 6.9.5: $[W^t(\underline{0}^n)]_i$ in Definition 6.9.8 and $\mathrm{EV}(i, t)$ have the same value for all i and t.

Thus $[W^t(\underline{0}^n)]_i$ has a finite limit L_i attained when $t = T$ if and only if $\mathrm{EV}(i, t) = L_i$ for all $t \geq T$. But then the $+ *$-evaluation along all paths from i of length $\leq T$ in the dependence graph gives L_i. The minimum solution is thus the solution found by $+ *$-evaluations along paths of length at most T, where T is the smallest integer such that the same solution results if the path length is extended. This may not be the only solution but the path interpretation may suggest whether the minimum fixed point is appropriate.

The path interpretation is particularly simple when c_1 through c_{n-1} are each $\underline{0}$ and c_n is not. In that case, the only paths whose $*$-evaluation is not $\underline{0}$ are those ending at vertex n.

$\mathrm{EV}(i, t)$ then becomes the sum $(+)$ of $*$-evaluations of the paths from i to n. This further simplifies if $c_n = \underline{1}$.

Consider an example. In solving for the cost of the least costly genpath, we want the minimum $(+)$ of the sum $(*)$ of the edge costs along each genpaths from vertex i to vertex n. In this case we have shown that this solution is the minimum stable state (when edge weights are constrained to be non-negative) for the set of equations for the least cost path problem.

Another simplification that often occurs and makes path interpretation particularly useful is when $x * x = x$ (as when $*$ is minimum). In this case, $*$-evaluations of genpaths with cycles can be ignored, and only simple paths need be considered. In other situations, only genpaths with no vertex repeated more than k times need be considered. In such cases, practical algorithms are sometimes based on $*$-evaluation of genpaths.

6.9.4 Justification for Elimination Algorithms for Bounded Functions

The iterative algorithms of Section 6.9.2 are applicable to equation sets with finite limits (compare Section 6.9.1). Gaussian elimination is also applicable to such equation sets—as well as to those with infinite limits. The basic results for that technique were developed in Section 6.2.1. We will now show that, as with iteration from an input of $\underline{0}^n$, the elimination algorithm yields the minimum state solution. Such elimination algorithms consist of two transformations: *substitution* and *loop-breaking*. The appropriate application of these transformations to a set of equations E will leave a smaller set with the same solution. Ultimately, repeated applications of these transformations result in the right side of each equation being a constant. In network terms, given a network NET′, there are simple transformations, equivalent to substitutions and loop-breaking in the corresponding equation set, which result in a network NET, simpler than NET′, but with the same limit. Ultimately, after repeated applications a network results, in which every variable is the output of a block whose inputs are all constant. In order to show the relation of elimination and iteration, we develop these transformations in network terms.

6.9.4.1. Substitutional Transformation—Network Interpretation

Substitution is illustrated in Figure 6-9. X is the output of block $b1$, and the ith input to block $b2$. A substitution transformation eliminates X as an input to $b2$, by replacing that $b1'$ to $b2$ with a copy of $b1$. $b1'$ is incorporated in an enlarged block $b2'$, which replaces $b2$. In equation terms, this is equivalent to

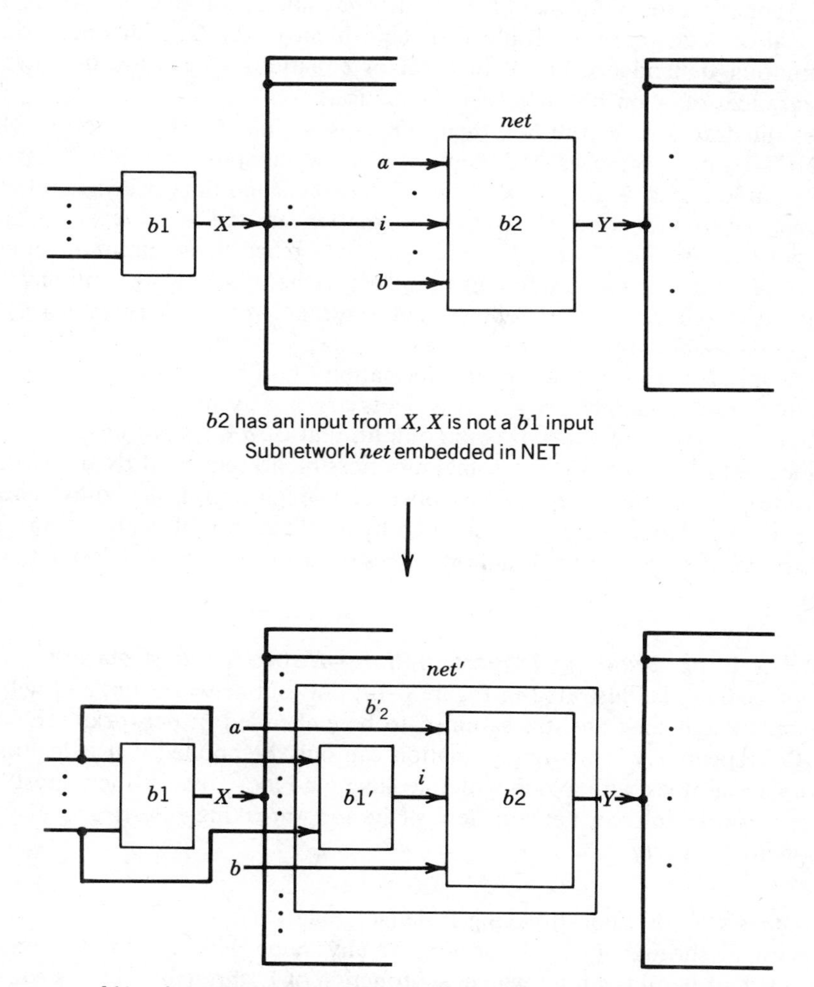

$b2$ has an input from X, X is not a $b1$ input
Subnetwork *net* embedded in NET

$b1'$ performs the same function as $b1$, $b2'$ does not have an input from X
Subnetwork *net'* embedded in NET'

FIGURE 6-9 Elimination transformation 1—substitution. The part of NET' excluding net' is the same as the part of NET excluding net.

substituting the expression giving the output of $b1$ in terms of its inputs for input $b1$ in the expression for the function of $b2$. That expression for $b2$ is then independent of $b1$ provided $b1$'s output does not return directly to its input.

The substitution transformation (Figure 6-9) results decreasing by one the number of blocks to which X goes, provided that no input to $b1$ is itself X. If this proviso is met, the transformation can be repeated once for each block which receives X in the original network until no block output depends on X. The substitution transformation is equivalent to substitution of the right side of the equation for X for that of X in the right side of the equation for Y. The removal of X as input to all blocks is equivalent to the familiar operation of eliminating dependence on X in a set of equations by substitution for all appearances of X on the right of the equations.

As illustrated in Figure 6-9, then, the substitution transformation of NET to NET′ simply renames and duplicates some computation. NET′ has the same stable states, if any, as NET. If $b1'$ has the same time delays as $b1$, and in fact all blocks with the same designation in the two networks have the same delay, then both NET and NET′ will reach their minimum state in the same time. Even with different time delays, the two networks still have the same stable states. Thus it follows that *substitutional transformation will not change the stable states of the network.*

In order for the substitution transformation to produce progress toward the elimination of a variable, say X, it is necessary that X not be an input to the same block, say b, of which it is the output, that is, b does not have a *loop*. If a block does have a loop, it is sometimes possible to remove it by transformation into an essentially equivalent network. Though useful, the equivalence is not generally as strong as is maintained by a substitution transformation. The possibility of such a transformation hinges on the existence of a *loop-breaking rule*.

6.9.4.2 Loop-Breaking Transformation—Network Interpretation

Loop-breaking is illustrated in Figure 6-10; only subnetworks net and net′ are shown, though they are still assumed to be embedded in networks NET and NET′, respectively. This transformation can only be applied if it is legitimate to replace a block with a loop with one having no loop, that is, there must be a loop-breaking rule. In network terms, the loop-breaking rule is given by the following property.

Hypothesis 6.9.4: Loop-Breaking Rule (see Figure 6-10)

With i_1 through i_n held constant at any value within their domain, the output X of net has a limit which is a function of i_1 through i_n and is equal to $w'(i_1, i_2, \ldots, i_n)$.

NET and NET′ as described in Figure 6-10 will be referred to repeatedly in the subsequent development.

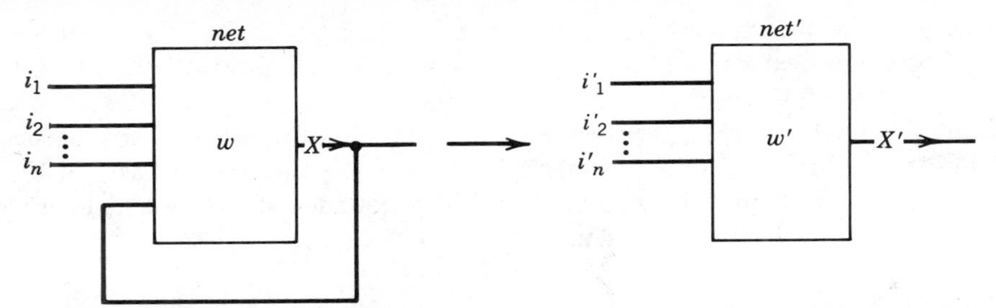

Subnetwork *net* embedded in NET Subnetwork *net'* embedded in NET'

FIGURE 6-10 Elimination transformation 2—Loop Breaking. The part of NET′ excluding net′ is the same as the part of NET excluding net.

In finding the limit, NET′ will serve in place of NET, NET′ containing net′ in place of net. The number of blocks to which the net′ output goes is one less than for net. So NET′ is closer to the ultimate solution in which no block has an output to any block. This then is a useful transformation. In equation set terms this is the same loop breaking as considered previously (Section 6.2.1). In the following lemmas, assume that NET is bounded.

Lemma 6.9.18: If network NET′ is stable in state Z' then network NET is also stable in state Z'.

PROOF: In stable state Z' in NET′ let X have the value $\$X$, and for all $j \in N^n$, let i_j have the value $\$i_j$. Certainly $\$X = w'(\$i_1, \ldots, \$i_n)$. Assume that in NET, the value at X is set to $\$X$, at each i_j to $\$i_j$, and each other output line of NET to the same value as the corresponding line in NET′ in state Z'. Then by Proposition 6.9.4, $w'(\$i_1, \ldots, \$i_n) = \$X = w(\$i_1, \ldots, \$i_n, \$X)$, and given the assigned values in NET, $w'(\$i_1, \ldots, \$i_n) = \$X =$ the value at $X = w(\$i_1, \ldots, \$i_n, \$X)$. Thus net within NET is stable under the assigned outputs, with the same external inputs and outputs as net′ within NET′, and the remainders of the two networks are identical. Therefore NET is stable when assigned the outputs on the corresponding lines in NET′ in state Z'. ■

Thus any solution of NET′ is also a solution of NET. Now we show that the minimum solution is included among those solutions; in fact, it equals the limit of NET′. First we establish that when NET reaches its limit, net, with its simple loop, also is at its limit.

Lemma 6.9.19: If Z is the limit of network NET, then the output at X is the limit of the subnetwork net with i_1 through i_n held constant at the values they have in Z.

PROOF: In limit state Z of NET let X have the value $\$X$ and for all $j \in N^n$, i_j have the value $\$i_j$. Let L be the limit of net with $i_j = \$i_j$, and held constant, that is, L is the smallest value such that $L = w(\$i_1, \ldots, \$i_n, L)$. There are three possibilities: $\$X < L$, $\$X > L$, or $\$X = L$. We eliminate the first two. If $\$X < L$, then $\$X < w(\$i_1, \ldots, \$i_n, \$X)$ and $\$X$ cannot be the limit. So $\$X < L$ cannot hold. If $\$X > L$, then, in the limit, as the output of NET approaches $\$X$, the output at X must become greater than L. But as long as $i_j \leq \$i_j$, $w(i_1, \ldots, i_n, X)$ cannot be greater than L (by boundedness of w and Lemma 6.9.2). So $\$X > L$ also is not true. ■

Finally we have

Lemma 6.9.20: If Z is the limit of network NET, then it is also the limit of NET′.

PROOF: As in the proof of the previous lemma, assume that when NET is its limit state Z, X has the value $\$X$ and for all $j \in N^n$, i_j has the value $\$i_j$. By Lemma 6.9.19, $\$X$ is the limit of net with $i_j = \$i_j$, and held constant, that is, $\$X$ is the smallest value such that $\$X = w(\$i_1, \ldots, \$i_n, \$X)$. Therefore by the definition of w', $\$X = w'(\$i_1, \ldots, \$i_n)$. So if in NET′, the value at X is set equal to $\$X$, the value at each i_j is set equal to $\$i_j$, and each other output line of NET′ is set to the same value as the corresponding line in NET in state Z, then NET′ will be stable. So NET′ is stable in the limit state Z of NET. Finally, Lemma 6.9.18 implies that NET′ is stable only in states in which NET is also stable, and is stable in the limit state of NET if there is one. So the limit state of NET′ is the same as that of NET. ■

In summary, assume block b has output X. If X does not loop back to its input, it can be eliminated as an input to blocks other than b by *substitution transformations*. Now assume there is a loop-breaking transformation for any block function w in the original network or one which arising as a result of a sequence of loop-breaks and substitutions. The loop-breaking transformation allows elimination of output X as an input to its own block b. So all input occurrences of X can be eliminated. This can be done for every block output or variable X so that finally each block output is dependent only on constant block inputs. This solves the network. Algorithm 6.9.1 is an implementation of the systematic application of these transformations. The development here justifies the algorithm and its specializations to any bounded function with a limit. In its application, loop-breaking rules applicable to individual blocks, as opposed to entire networks, are required. A number of these were justified, although not derived, in Section 6.7.

6.9.4.3 Derivation of Loop-Breaking Rules

The development above suggests a way in which such rules may be derived. The limit of the subnetwork in the transformation of Figure 6-10 consisting of block net with all its inputs other than X held constant is given by $X = w^j(i_1, \ldots, i_n, \underline{0})$ as j increases to ∞, or, letting $\langle i_1, \ldots, i_n \rangle = C$, $X = w^j(C, \underline{0})$ as j increases to ∞.

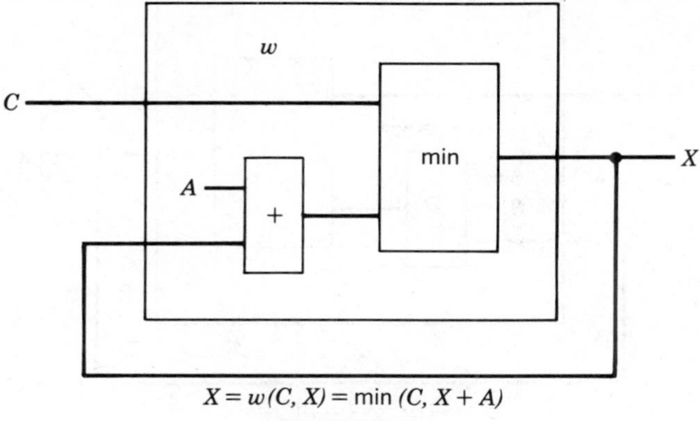

$X = w(C, X) = \min(C, X + A)$

FIGURE 6-11.

Now consider particular w functions which occur in practice and find the value of $w^j(C, \underline{0})$ when $j = 1, 2, \ldots$. Generalize these results to get a value at any j. This is not an algorithmic process, but a correct generalization is often provable by induction. A few examples will illustrate this approach.

First consider the network of Figure 6-11. min is minimum, + is ordinary addition and A and C are real numbers.

Example 6.9.10

$$w^1(C, \underline{0}) = \min(C, \underline{0} + A)$$

Then

$$w^2(C, \underline{0}) = \min(C, \min(C, \underline{0} + A) + A)$$

$$= \min(C, \min(C + A, \underline{0} + 2A))$$

$$\text{because } \min(X, Y) + Z = \min(X + Z, Y + Z)$$

$$= \min(C, C + A, \underline{0} + 2A)$$

$$\text{because } \min(X, \min(Y, Z)) = \min(\min(X, Y), Z)$$

$$= \min(C, \underline{0} + 2A)$$

Generalizing we obtain

$$w^j(C, \underline{0}) = \min(C, \underline{0} + jA)$$

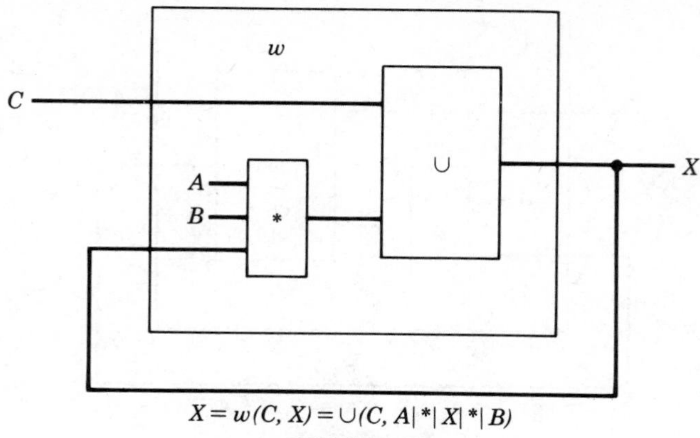

$X = w(C, X) = \cup(C, A|*|X|*|B)$

FIGURE 6-12.

whose validity we establish by the following induction

$$\begin{aligned} w^{j+1}(C,\underline{0}) &= \min(C, \min(C, \underline{0} + jA) + A) \\ &= \min(C, \min(C + A, \underline{0} + (j+1)A)) \\ &= \min(C, \underline{0} + (j+1)A) \end{aligned}$$

It follows that if $A > 0$ then there exists at j such that $(j+1)A > C - \underline{0}$, so that $\lim_{j \to \infty} w(C, \underline{0}) = C$. If $\underline{0} = 0$, it will reach that limit C in j steps, where j is the smallest integer such that $C/A \leq j$. If $\underline{0} = \infty$, then $w(C, \underline{0}) = \min(C, \infty + A) = C$. This reaches the limit in one step.

Example 6.9.11: As another example, consider Figure 6-12. A, B, and C are sets of vectors or strings, $\cup$ is union, $|*|$ cross-concatenation, and $0 = \{\langle\ \rangle\}$.

Let $w^1(C, \{\langle\ \rangle\}) = C \cup A|*|\{\langle\ \rangle\}|*|B$. It is convenient to define a function *surround* as follows

$$\operatorname{surr}^0(X, A, B) = X$$

$$\operatorname{surr}^j(X, A, B) = A^j|*|X|*|B^j \text{ for } j > 0,$$

$$A^j = A|*| \ldots |*|A \; (j \text{ factors})$$

and $\operatorname{surr}(X, A, B) \equiv \operatorname{surr}^1(X, A, B)$. Then

$$w^1(C, \{\langle\ \rangle\}) = C \cup \operatorname{surr}(\{\langle\ \rangle\}, A, B)$$

$$w^2(C, \{\langle\ \rangle\}) = C \cup \operatorname{surr}(C \cup \operatorname{surr}(\{\langle\ \rangle\}, A, B), A, B)$$

Since surr distributes through $\cup$, that is, $\text{surr}^j(X \cup Y, A, B) = \text{surr}^j(X, A, B) \cup \text{surr}^j(Y, A, B)$.

$$w^2(C, \{\langle\ \rangle\}) = C \cup \text{surr}(C, A, B) \cup \text{surr}(\text{surr}(\{\langle\ \rangle\}, A, B), A, B)$$

$$= C \cup \text{surr}(C, A, B) \cup \text{surr}^2(\{\langle\ \rangle\}, A, B)$$

$$w^3(C, \{\langle\ \rangle\}) = C \cup \text{surr}(C, A, B) \cup \text{surr}^2(C, A, B) \cup \text{surr}^3(\{\langle\ \rangle\}, A, B)$$

$$\vdots$$

$$w^j(C, \{\langle\ \rangle\}]) = \text{surr}^0(C, A, B) \cup \text{surr}^1(C, A, B)$$

$$\cup \text{surr}^2(C, A, B) \cup \cdots \cup \text{surr}^j(\langle\ \rangle, A, B)$$

This can again be established inductively. We therefore have

$$\lim_{j\to\infty} w^j(C, \{\langle\ \rangle\}) = \bigcup_{j\geq 0} \text{surr}^j(C, A, B)$$

$$= \bigcup_{j\geq 0} A^j|{*}|C|{*}|B^j$$

$$\equiv \text{surr}^*(C, A, B) \qquad \text{(by definition)}$$

$\text{surr}^*(C, A, B)$ is infinite since it contains $\text{surr}^j(C, A, B)$ for all $j \in N_0$. w is stable at $\text{surr}^*(C, A, B)$, that is, it is a fixed point of w because $w(C \cup \text{surr}^*(C, A, B)) = C \cup \text{surr}(\text{surr}^*(C, A, B), A, B) = \text{surr}^*(C, A, B)$.

If X is any set of strings made by concatenating those in A, B, and C and $X \supset \text{surr}^j(C, A, B)$ for all $j \in N_0$, then X is clearly infinite.

The two examples of generalization given above are extremes in a spectrum of possibilities. In the first, the limit of $w^j(\underline{0})$ occurs when $j = 1$, in the second when $j = \infty$. There are intermediate cases in which the limit is finite but in general reached for $j > 1$. (Compare Examples 6.9.1–6.9.5 and Problem 10 in Section 6.10).

6.10 PROBLEMS

1. *Minimum Covering—Transformation.* Apply the transformation given in Theorem 6.1.1 to the following definition, which is a growth-derived solution to the minimum covering problem.

   For $j = 1$ to n, M_j is a 0–1 vector of length m. If the kth component of M_j is 1, then k is "covered" by M_j. The object is to find all smallest

sets of j's such that the corresponding set of M_j's covers all m components.
"or_v" is a vector "or" operation.
$\text{ONE}(v) = 1$ if every component of vector v is 1, it is 0 otherwise.

$$\left[\begin{array}{ll} f(C,p,v,j) = \langle C, p\rangle & : j \le n+1 \text{ and } \text{ONE}(v) \\ f(C,p,v,j) = \langle \infty, p\rangle & : j = n+1 \text{ and } \sim \text{ONE}(v) \\ f(C,p,v,j) = \text{MIN-SEL}(f(C+1, & \\ \quad p \| \langle j \rangle, v \text{ or}_v M_j, j+1), & \\ \quad f(C,p,v,j+1)) & : j < n+1 \\ \text{Initially } C = 0,\ p = \langle\ \rangle,\ v = \langle 0,0,\dots,0\rangle;\ |v| = n,\ j = 1. & \end{array}\right.$$

2. *Partition Cost—Transformation.* Here is a recursive definition designed to find the cost of the least costly partition of the integer D, where the cost of a partition $D = v_1 + v_2 + \cdots + v_k$ is $P(v_1) + P(v_2) + \cdots + P(v_k)$. $P(i)$ for all i = integers from 1 to D is an integer greater than 0, and is part of the input to the definition, that is, an invariant argument.

$$\left[\begin{array}{ll} f(c,v) = c & : v = \langle\ \rangle \\ f(c,v) = f(c + P(v_1), v - v_1) & : v_1 = 1 \\ f(c,v) = \text{MIN}(f(c + P(v_1), v - v_1), & \\ \quad f(c, \langle 1\rangle \| \langle v_1 - 1\rangle \| v - v_1), \dots, & \\ \quad f(c, \langle \lfloor v_1/2 \rfloor\rangle \| \langle v_1 - \lfloor v_1/2 \rfloor\rangle \| v)) & : v_1 > 1 \\ \text{Initially } c = 0,\ n \in N = D,\ v = \langle n \rangle. & \end{array}\right.$$

a. This definition enumerates all partitions of D so its 0-tree has a number of terminal argument-structures equal to the number of partitions of D, which is exponential in D. So $|\Delta_f(0, D)|$ is exponential in D. Give an alternative definition in which $|\Delta_f(0, D)|$ is subexponential in D.

b. Describe a dynamic programming to implement the definition obtained in the previous question. Show its complexity is polynomial in D.

c. If $P(i) = \infty$ for $i > k$, describe an improved algorithm to implement the definition obtained in the previous question. Give its complexity.

3. *Parse Exercise.* Given the grammar G, $A \xrightarrow{1} a$; $B \xrightarrow{2} b$; $C \xrightarrow{3} AZ$; $D \xrightarrow{4} BZ$

$$Z \rightarrow C\overset{5}{A} \mid D\overset{6}{B} \mid A\overset{7}{A} \mid B\overset{8}{B} \mid$$

Give a leftmost derivation if there is one (sequence of rule numbers) for the following input strings: *abaaba* and *bbaabb*.

4. *Early Parse Nested-Definition Exercise.* Evaluate functions in Definitions 6.1.20 and 6.1.21 for the grammar in Problem 3 and the string *baab*.
5. *Cocke–Kasami–Younger Parse—Transformation.* Consider the following definition for recognition. Assume that there are two kinds of NTSs, those whose right sides consist of pairs of NTSs, and those whose right sides consist solely of one terminal symbol. If an NTS y is of the first type, then NTSR(y) is true. If y is of the second type, then TSR(y) is true. If TSR(y), then y has only one right side referred to as $r(y)$. A grammar in this form is in *Chomsky normal form*. Given any CFG, there is another CFG that produces the same language and is in Chomsky normal form. [Aho, Ullman 72a]

   Given the grammar G, and the input string $s = \langle s_1, \ldots, s_n \rangle$ to be recognized we proceed as follows.

   In the definition the argument consists of a single list Y. Each entry in Y is a three-component vector. The first two components are numbers (the second $\geq$ the first) representing a range of positions in the input string. The third is always an NTS, of one of the two types, from G. The top member of Y is symbolized $\langle \ell, u, y \rangle$.

   The final value of the defined function $f(Y)$ is 1 if for every triple $\langle \ell_\ell, u_k, y_k \rangle$ in y the substring s_{ℓ_k} through s_{u_k} can be derived from the TS or NTS y_k. Otherwise $f(Y)$ will be 0. Thus the $f(\langle 1, n, \text{starting symbol} \rangle)$ $= 1$ if the input string is recognized as a string of G.

Definition 6.10.1: Transform of Cocke–Kasami–Younger Definition

$$\left[\begin{array}{ll}
f(Y) = 1 & :Y = \langle\ \rangle \\
f(Y) = 0 & :l = u \text{ and } r(y) \neq s_u \text{ and } Y \neq \langle\ \rangle \\
 & \text{or } l < u \text{ and TSR}(y) \text{ and } Y \neq \langle\ \rangle \\
f(Y) = f(Y - \langle l, u, y \rangle) & \\
 & :l = u \text{ and } r(y) = s_u \text{ and } Y \neq \langle\ \rangle \\
f(Y) = \text{OR}_{i=1}^{N(y)} \text{OR}_{j=l}^{u-1} f(\langle\langle l, j, R_{i1}(y)\rangle, & \\
\qquad \langle j+1, u, R_{i2}(y)\rangle\rangle \| (Y - \langle l, u, y \rangle)) & \\
 & :l < u \text{ and NTSR}(y) \text{ and } Y \neq \langle\ \rangle \\
\text{Initially } Y = \langle 1, n, \text{starting symbol} \rangle, \text{ grammar } G \text{ is invariant.} &
\end{array}\right.$$

   a. Apply reduction 2, Section 6.1.3, to Definition 6.10.1.
   b. The algorithm that results from a straightforward equation set implementation of the transformed version of Definition 6.10.1 was the first general CFG recognizer known to have polynomial complexity. It was discovered independently by Cocke, Kasami, and Younger. Show that the complexity of this algorithm is $O(n^3)$.

c. Give a definition analogous to Definition 6.10.1, or to the transformed version whose application is not restricted to Chomsky normal form grammars, but is directly applicable to any CFG.

6. *Domination—Obtain ESE Definition.* Consider a digraph G with a single begin vertex and a number of end vertices, called a flow graph. We are interested in what happens along paths from begin to end vertices. All vertices in G are assumed to be reachable from begin. Vertex x is said to be *dominated* by a vertex y if y appears along every path from begin to x. By definition x dominates itself. Let $f(x)$ = the set of all vertices which dominate x. Note that besides itself, x is dominated by a vertex y if and only if y is in the set of dominators of every inward neighbor of x. Based on this fact,

a. Give a recursive definition for $f(x)$.

b. Give an algorithm which implements that definition.

(Let $\text{in}_i(x)$ be the ith inward neighbor of x and $\text{IN}(x)$ the number of x's inward neighbors.)

7. *Maximum–Minimum Edge on All Paths from a to b—Weighted Vertices.* Consider a digraph G, each of whose vertices x has an associated weight $w(x)$. We wish to find the minimum weighted vertex on each path from a to b, and then choose the largest of these minima. For any vertex x in G, let $f(x)$ give this value on all paths from x to b. A proposed recursive line for a definition of $f(x)$

$$f(x) = \text{MAX}_{i=1}^{N(x)}[w(x) \downarrow f(n_i(x))]$$

a. Give a recursive definition for $f(x)$.

b. Give an algorithm which implements that definition.

8. *The kth Shortest Path Algorithm [Hu 82].* Develop a minimum path algorithm to find the cost of each of the k shortest paths in a given graph between two given vertices. Also elaborate the algorithm to give the k shortest paths in addition to their costs.

9. *Loop-Breaking Rule-Proof.* Prove the following lemma.

Lemma 6.10.1: Let $x = x \cup w$ where w is any set. Then $x = x \cup w$ is satisfied by any x which contains w, and by only such sets.

10. *Aliasing—Loop-Breaking Rule.* In the problem of aliasing of the parameters of procedures we get equations of the form

$$x_j = a_1 \Box x_1 \cup \cdots \cup a_n \Box x_n$$

where the x_i's and a_k's are finite sets of ordered pairs and,

$$X \Box Y = \{\langle a, b\rangle \mid \exists c \ \langle a, c\rangle \in X, \langle c, b\rangle \in Y\}$$

a. Find the loop-breaking rule.
b. What is the interpretation of the rule if the sets of pairs are interpreted as sets of edges of a graph?

11. *Limits.* Suppose we have the following two network functions.

a. $W_1(X) = X \| \langle 1 \rangle$
b. $W_2(X) = \langle 1 \rangle \| X$

Suppose $X \in Q$ = set of all strings of 1,s of zero or greater, including infinite length. The strings are ordered by length, the minimum = $\underline{0}$ = $\langle \ \rangle$.

a. The functions are both monotone. Why?
b. Which of the functions, if either, has a limit? Explain.

12. *Nonexistence of Limit.* Let Q = the set of nonempty closed subintervals of the real interval $[0, 2]$ ordered by $[a, b] \geq [c, d]$ if and only if $a \geq c$ and $b \geq d$. Then $0_Q = [0, 0]$. Let W be defined as follows:

$$W[a, b] = \begin{cases} [a, (b + 2)/3] & b < 1 \\ [a, (b + 2)/2] & b \geq 1. \end{cases}$$

Show:

a. $[0, 1] \geq W^{(j)}[0, 0]$ for all j.
b. $[0, 1]$ is the least upper bound of $\{W^{(j)}[0, 0]\}$.
c. $[0, 1]$ is not a fixed point of W.
d. $[0, 2]$ is the least fixed point of W. Thus we can conclude W does not have a limit.

13. *Finite Limits and Acyclic Dependency Graphs.* In Section 6.9 it is asserted that a finite limit is virtually assured for a problem with an acyclic dependency graph. Justify this statement.

14. *Iterative Algorithm—Definition of Q.* In Examples 6.9.8 and 6.9.9, Algorithm 6.9.2 is applied to a minimum path cost problem. Fewer iterative steps are required when the iteration begins with variables initialized to ∞ than if initialization is to 0. Is this always true for the minimum cost problem?

CHAPTER SEVEN

IMPLEMENTATIONS DEPENDENT ON MANY AND / OR VERY STRONG PROPERTIES

In previous chapters we developed a framework for "explaining" the origin of good algorithms by identifying properties of recursive definitions that lead to good algorithms. These properties are simple, easy to detect, and applicable to a reasonable number of diverse examples. There remain many functions, however, for which a disappointingly large complex of properties is required to explain their good implementation. Furthermore, each class of functions with such a large collection of properties is usually small, so that many such classes will be needed to explain any reasonable range of good algorithms. Of course the framework we have chosen is not the only choice. Algorithms need not be developed from simple recursive definitions, and the fact that some algorithms do not fit so neatly into this framework suggests that at least some should not. In any case we will describe a variety of algorithms in this section which fit our chosen framework only awkwardly.

7.1 A CLASS OF SIMPLE GAMES—DEFINITIONS WHOSE RANGE AND DOMAIN ARE FINITE

Here is an example which fits our framework well enough, but for which the properties of the recursive definition giving the good algorithm, in this case with complexity $O(1)$, are so restrictive as to apply to only a very narrow class of definitions.

In the game of NIM two players alternate, removing one or two counters from a pile which initially contains n counters. The loser (-1) is the player who first finds, when about to move, that no counters remain. $f(n)$ as given in Definition 7.1.1 gives the value of the game to the player who moves when there are n counters present. It is a direct adoption of Definition 5.7.1 to NIM.

Definition 7.1.1: NIM

$$\begin{cases} f(n) = -1 & :n = 0 \\ f(n) = 1 & :n = 1 \\ f(n) = -\min(f(n-1), f(n-2)) & :n > 1 \end{cases}$$

Clearly there is comparability. So ESE seems appropriate. With such an approach we can get an $O(n)$ algorithm analogous to that for Fibonacci numbers, which has a similar definition. However, since the arguments of $-\min$ can only be 1 or -1, we can do better. (Actually, we could also do better for Fibonacci, because its w is sum.)

If we record $f(n)$ for a series of value of n, we can see the cyclic behavior, which is general.

n:	0	1	2	3	4	5	6	...
$f(n)$:	-1	1	1	-1	1	1	-1	...

In such a recording of a sequence of $f(n)$ values, $f(n)$ depends on the previous two values, $(f(n-1)$ and $f(n-2))$. There are only four possible pairs, and given a pair, the next pair is uniquely determined. So eventually there must be a cycle, that is, considering the sequence of overlapping pairs, one value must repeat after at most four such pairs. This is shown in two different forms in Figure 7-1*a*. From this we can see how starting in state (-1), that is, with $(f(0), f(1))$, as is provided by Definition 7.1.1, $f(n)$'s value for $n > 1$ is given by tracing the second component of state diagram Figure 7-1*a*. The result is shown in Figure 7-1*b*.

From this we conclude that the game is a loss to the first player if it starts with n a multiple of three, otherwise it is a win. So evaluating $f(n)$ is $O(1)$.

The result of this example can be generalized somewhat beyond game definitions. Consider the following definition of $f(n)$, where f is a function whose domain is the positive integers, and whose range is a known finite set F. (If w is $-\min$ and $t(n)$ either 1 or -1, this gives a definition for a NIM game m in which players may remove $a_i, 1 \le i \le m$, counters from the pile.)

Definition 7.1.2: Generalization of One-Dimensional NIM

$$a_1 < a_2 < \cdots < a_m < L, a_i \in N.$$

$$\begin{cases} f(n) = t(n) & :1 \le n < L \\ f(n) = w(f(n-a_1), \ldots, f(n-a_m)) & :n \ge L \\ \text{Initially } n \in N. & \end{cases}$$

F is a finite set, $|F| = r$; $t(n) \in F$ [so $f(n) \in F$ when $1 \le n < L$]. $w(x_1, \ldots, x_m) \in F$ if all the x_i's are in F [so $f(n) \in F$ when $n \ge L$].

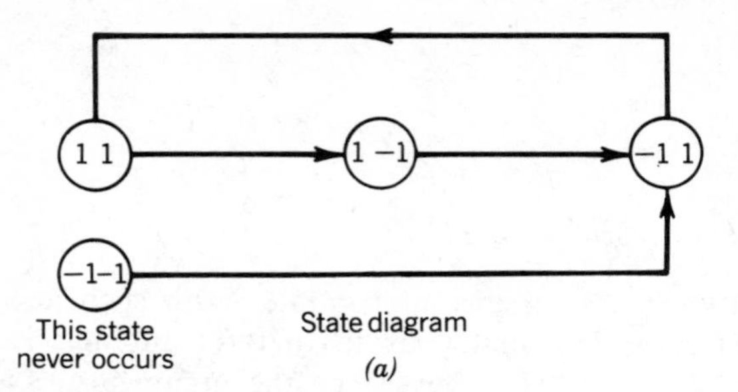

(a)

Present state		Next state	
$i-1$	i	i	$i+1$
1	1	1	−1
1	−1	−1	1
−1	1	1	1
−1	−1	−1	1

State table

(c)

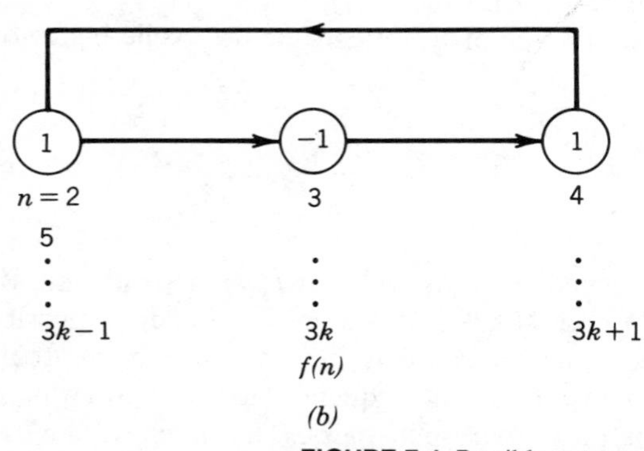

(b)

FIGURE 7-1 Possible outcomes of NIM.

Given an instance of Definition 7.1.2 (that is, particular choices of t, F, L and w), we can compute $f(n)$ for any $n > 0$. For simplicity, assume $L = a_m + 1$ and that w is extended to a function W on a_m-tuples,

$$W(x_1, x_2, \dots, x_{a_m}) = w(x_{a_1}, x_{a_2}, \dots, x_{a_m}).$$

For $n \le a_m$, $f(n)$ is given by the terminal values $t(n)$, and for $n > a_m$, $f(n)$ depends on $f(n - a_1), \dots, f(n - a_m)$. Thus if we knew $f(x)$ for each of the a_m values of x: $n-1, n-2, \dots, n-a_m$, we could easily determine $f(n)$. This motivates the following definitions.

Call an a_m-tuple of values of F a *state*. There are r^{a_m} states, since $|F| = r$. Let state $X = \langle x_{-a_m}, x_{-a_m+1}, \dots, x_{-1} \rangle$.

Then $\delta(X)$, the next state of X, is given by

$$\delta(X) = \langle x_{-a_m+1}, \dots, x_{-1}, f \rangle, \text{ where } f = W(x_{-a_m}, x_{-a_m+1}, \dots, x_{-1}).$$

But this again gives a state diagram, this time with m^{a_m} states and an edge from state X to state Y if and only if $\delta(X) = Y$. Thus no state has more than one successor.

Let the initial state be $\langle f(1), f(2), \ldots, f(a_m) \rangle$. Starting in that state and tracing the path given by the next-state function, the process eventually (after no more than r^{a_m} steps) enters a cycle, giving the solution trace for the problem. Numbering the initial state 1 and each successive state one greater than its predecessor, we assign a state to every $n \in N$ (where each state in the cycle will receive an infinity of integer labels). Let $f(a_m + n) =$ the last component of the state to which n is assigned. The solution can now be formulated in terms of the value of n modulo the length of the cycle, and is thus computable in $O(1)$ time.

This development is pleasing, but it only applies to a small set of definitions. A natural way to generalize further is to consider analogous definitions with two or more integer arguments. Definition 7.1.3 does this for two arguments, where the tuple (a_i, b_i) represents a legitimate combination of moves.

Definition 7.1.3: Generalized 2-Dimensional NIM

$$a_1 \leq, \ldots, \leq a_m < L_1, a_i \in N, \max b_i < L_2, \text{ and if } a_i = a_{i+1} \text{ then } b_i < b_{i+1}$$

(that is, the tuples (a_i, b_i) are in lexicographic order).

$$f(n, p) = t(n, p) \qquad :1 \leq n < L_1 \text{ or } 1 \leq p < L_2$$
$$f(n, p) = w(f(n - a_1, p - b_1), \ldots, f(n - a_m, p - b_m)) \qquad :n \geq L_1 \text{ and } p \geq L_2$$

Initially $n \in N$, $p \in N$.
F is a finite set, $|F| = r$; $t(n, p) \in F$ [so $f(n, p) \in F$ when $1 \leq n < L_1$ or $1 \leq p < L_2$]. $w(x_1, \ldots, x_m) \in F$ if all the x_i's are in F [so $f(n, p) \in F$ when $n \geq L_1$ and $p \geq L_2$].

There are a number of alternative formulations for k-dimensional NIM, each with its own set of properties and difficulties. For our formulation, note that there are an infinite number of terminal cases (which probably follow some recursive pattern themselves), that a player is required to remove counters from each pile every time, and that the game halts when either pile enters the terminal region. Development of a good algorithm for any of these definitions, even with the ambiguities resolved, is not nearly as simple as for the one argument case.

7.2 THE ORIGIN OF THE TOWER OF HANOI RECURSIVE DEFINITION

The recursive definition for the Tower of Hanoi first given in Chapter 3, Definition 3.1.1, is not really the most straightforward solution to that problem. It already incorporates insight which goes beyond that necessary for a first solution. A more straightforward solution arises from thinking in terms of enumeration/selection, enumeration of all disk move sequences, and then, selection of the shortest of those that end with all disks on peg 3. This leads directly to a recursive definition like that of Definition 7.2.1, developed either by backtracking or by sample growth trees.

Definition 7.2.1: Tower of Hanoi
$sz(i)$ = size of the disk on top of peg i. $\delta(S, i \to j)$ = resulting state if move $i \to j$ is made with disks in state S. F = final state = state in which all disks are on peg 3.

$$\left[\begin{array}{ll} f(n, v, S) = \langle n, v \rangle & :S = F \\ f(n, v, S) = \text{MIN-SEL}_{i, j \in \{1,2,3\}, sz(i) < sz(j)} f(n+1, v \| \langle i \to j \rangle, & \\ \qquad \delta(S, i \to j)) & :S \neq F \\ \text{Initially } S = \text{state in which all disks are on peg 1, } v = \langle\ \rangle, n = 0. & \end{array}\right.$$

Definition 7.2.1 does not terminate. This is easily remedied by augmenting the definition to assure that no state appears more than once in any move sequence. In any case it is unimportant for this discussion. The largest disk d_n must be moved to peg 3 eventually, at which time peg 3 must be empty and d_n must be alone on either peg 1 or 2. Since d_n starts on peg 1 and must be alone there at some time, it might as well be moved to peg 3 when this first occurs. After all, to move d_n at all, either peg 2 or peg 3 must be empty. By symmetry, emptying either requires the same number of moves from the starting state. This means that the state in which d_n is alone on peg 1, peg 3 is empty, and all other disks are on peg 2 must be in the shortest move sequence. This fact is the basis for a recursive definition (Definition 3.1.1), which generates only the optimal move sequence rather than generating all feasible ones and selecting one of those.

Ideally this property, which leads to the good definition, should be inferable from Definition 7.2.1 in its abstract form *without using or even knowing its Tower of Hanoi interpretation*. How this might be done is considered in the Artificial Intelligence literature [Nilsson 80].

7.3 MINIMUM SPANNING TREE — GREEDY DEFINITION

The minimum spanning tree problem provides another example in which the original recursive definition has strong properties which validate a much simpler definition, called a greedy definition. This in turn leads to simple implementations. However, as with the Tower of Hanoi, these properties are hard to formulate abstractly. The minimum spanning problem is a member of a well-known class of problems that lead to good algorithms. These are the *matroids* in [Papadimitriou, Steiglitz 82, Chap. 12; Lawlor 76]. The class is not, however, usually defined in terms of properties on recursive definitions.

We give a possible development from a "straightforward" to a "greedy" definition for the minimum spanning tree, to illustrate the kind of properties which allow such simplifications.

A spanning tree of a *connected* graph G is a subgraph of G; a tree containing every vertex of G.

The weight or cost of a spanning tree of a weighted connected graph G is the sum of the weights or costs of its edges. A spanning tree of minimum cost is a *minimum spanning tree*. Concrete versions of the problem of finding the minimum spanning tree involve minimizing the total length of an electrical (fluid, gas) connection between points at given distances from each other.

As a start in developing an efficient algorithm to compute the minimum spanning tree, we develop a definition that enumerates all spanning trees of G each with its cost, and minimum-select selects all $\langle$cost, tree$\rangle$ pairs with minimum cost.

In the definition, G initially is the given graph and T is a tree, initially a single vertex. Edges, e, are removed from G and placed into T in a fashion which ensures that T remains a tree and is eventually a spanning tree. We can guarantee this if only edges with exactly one end in T are ever placed in T.

Definition 7.3.1: Minimum Spanning Tree—Enumeration Selection Definition 1

$P(T, G)$ is the vector of all (p) edges still in G with exactly one vertex in T. $c(e)$ is the weight of edge e.

$$\left[\begin{array}{ll} f(c, T, j, G) = \langle c, T \rangle & : j = n \\ f(c, T, j, G) = \text{MIN-SEL}_{e \in P(T,G)}[f(c + c(e), T \cup \{e\}, j + 1, G - e)] & \\ & : j < n \\ \text{Initially } G = \text{an } n\text{-vertex graph, } v = \text{vertex in } G,\ c = 0,\ T = \{v\},\ j = 1. & \end{array}\right.$$

Only edges in $P(T, G)$ are candidates for movement from G to T, so only edges with exactly one vertex in T can be moved to G. So T remains a subtree of the original graph. This property related to the earlier loop-detecting Definition 3.6.5 for growing a tree, means that even this initial definition is not entirely straightforward.

Two trees with the same edges are the same independent of the order in which edges are accumulated. A definition need only enumerate one such representative of each tree. We order the edges available for T by size. So $P(T, G) = \langle e_1, e_2, \ldots, e_p \rangle$, with $c(e_1) \le c(e_{i+1})$. First e_1 is added to T and removed from G, then e_2 (but not e_1) is added to T and removed with e_1 from G, ... Finally e_p is added to T while $e_1, e_2, \ldots, e_p$ (designated $e_{1:p}$) are removed from G. This removal of edges can result in cases in which G has no edges before T has $n - 1$. An additional terminal line accommodates this condition in the next definition.

***Definition** 7.3.2:* Minimum Spanning Tree—Enumeration Selection Definition 2
$P(T, G)$ is vector of all (p) edges still in G with exactly one vertex in T.

$$\left[\begin{array}{ll} f(c,T,j,G) = \langle c, T\rangle & : j = n \\ f(c,T,j,G) = \langle \infty, T\rangle & : j < n \text{ and } P(T,G) = \{\ \} \\ f(c,T,j,G) = \text{MIN-SEL}_{i=1}^{p}[f(c + c(e_i), T \cup \{e_i\}, j+1, G - e_{1:i})] & \\ & : j < n \text{ and } P(T,G) \neq \{\ \} \\ \text{Initially } G = \text{an } n\text{-vertex graph, } v = \text{a vertex in } G,\ c = 0, T = \{v\},\ j = 1. & \end{array}\right.$$

Even though Definition 7.3.2 has a commutative w function (MIN-SEL), the order in which the children f calls are executed is constrained to eliminate duplication of spanning trees. The result is that the edge in the first f call, e_1, is the minimum member of $P(T, G)$. Luckily it turns out that all other f calls beyond the first can be dispensed with. The definition that results gives the same result as Definition 7.3.2.

***Definition** 7.3.3:* Minimum Spanning Tree—Greedy Formulation 1
$P(T, G)$ is the vector of all (p) edges in G with one vertex in T.
The first r components of $P(T, G)$ have the same cost.

$$\left[\begin{array}{ll} g(c,T,j,G) = \langle c, T\rangle & : j = n - 1 \\ g(c,T,j,G) = \langle \infty, T\rangle & : j < n \text{ and } P(T,G) = \{\ \} \\ g(c,T,j,G) = \text{MIN-SEL}_{i=1}^{r}[g(c + c(e_i), T \cup \{e_i\}, j+1, G - e_{1:i})] & \\ & : j < n \text{ and } P(T,G) \neq \{\ \} \\ \text{Initially } G = \text{an } n\text{-vertex graph, } v = \text{a vertex in } G,\ c = 0, T = \{v\},\ j = 0. & \end{array}\right.$$

This formulation is termed *greedy* because it is based on the expectation that the minimum cost edges in $P(T, G)$ are the correct ones to put into T. If this expectation is fulfilled, a cheap algorithm will suffice. The combination of "hope" and "cheap" accounts for the word "greedy." On the other hand, in many definitions returning the minimum of an enumerated set, for example, the minimum of all possible paths in a graph from vertex a to b, the same expectation is frustrated. This may be another, moral, justification for "greedy." The term should not, however, discourage one from testing the expectation. Even if the greedy algorithm does not implement the desired function, it is often useful as an approximation algorithm.

The fact that Definition 7.3.3 is adequate will require proof. We need to prove that

Theorem 7.3.1: If the function $f(c, T, j, G)$ as defined in Definition 7.3.2 produces all minimum cost trees built by augmenting the tree T of cost c with edges drawn from G, then the solution using Definition 7.3.2 equals that using Definition 7.3.3.

PROOF: The proof is easily established for the terminal lines. For the recursive line we need to establish the initial tentative equality. For $j < n$ and $P(T, G) \neq \{\ \}$, referring to Definition 7.3.2, the right side of the recursive line can be partitioned so

$$f(c, T, j, G) = \text{min-sel}\big(\text{MIN-SEL}_{i=1}^{r}\big[f(c + c(e_1), \text{T} \cup \{e_i\}, j + 1, G - e_{1:i})\big],$$

$$\text{MIN-SEL}_{i=r+1}^{p}\big[f(c + c(e_i), T \cup \{e_i\}, j + 1, G - e_{1:i})\big]\big),$$

If $\text{MIN-SEL}_{i=r+1}^{p}[f(c + c(e_i), T \cup \{e_i\}, j + 1, G - e_{1:i})]$ has a greater cost than $\text{MIN-SEL}_{i=1}^{r}[f(c + c(e_i), T \cup \{e_i\}, j + 1, G - e_{1:i})]$ then *we say fact (1) holds*. If fact (1) holds, then it follows that:

$$f(c, T, j, G) = \text{MIN-SEL}_{i=1}^{r}\big[f(c + c(e_i), T \cup \{e_i\}, j + 1, G - e_{1:i})\big]$$

Which with g replacing f is the recursive line of Definition 7.3.3, so

$$f(c, T, j, G) = g(c, T, j, G)$$

It remains only to show that fact (1) holds. Fact (1) is equivalent to "The trees in $f(c + c(e_b), T \cup \{e_b\}, j + 1, G - e_{1:b})$ are all more costly than the trees in $f(c + c(e_a), T \cup \{e_a\}, j + 1, G - e_{1:a})]$, where a is an index in the range 1 through r, and b is an index in the range $r + 1$ through p." Let F be a minimum spanning tree in $f(c + c(e_b), T \cup \{e_b\}, j + 1, G - e_{1:b})$ for any $b > r$. T, which is a subgraph of F, is represented on the left of Figure 7.2; on the right, T' represents the subgraph of F on the vertices not in T. Unlike T, T' may be a forest rather than a single tree. There is at least one edge in F, say e_b, between a vertex v_{b1} in T and another v_{b2} in T'. There may be two or more, represented pictorially by e_b and e_b'. For $a \leq r$, all edges in F between T and T' cost more than e_a, whose ends are v_{a1} in T and v_{a2} in T'. e_a is not in F since its cost is less than that of the current choice, and all such edges are removed before each future choice. There must be a path entirely in T' from v_{a2} to a first vertex in T, since v_{a2} is a vertex in the minimum spanning tree F. Without loss of generality, assume this path runs from v_{a2} to $v_{b2'}$ in T', and then uses the edge e_b' to $v_{b1'}$. Since T is connected, there is a path in T from $v_{b1'}$ to v_{a1}. Thus if e'_b were replaced with e_a, a less costly spanning tree would result, since $c(e_a) < c(e_b')$. Therefore for every spanning tree F in $f(c + c(e_b), T \cup \{e_b\}, j + 1, G - e_{1:b})$, there is a less costly one in $f(c + c(e_a), T \cup \{e_a\}, j + 1, G - e_{1:a})$, which produces the minimum cost spanning trees containing $T \cup \{e_a\}$. ■

Definition 7.3.3 will generate all minimum spanning trees. Unfortunately it also sometimes produces spanning trees which are not minimum. (See Problem 1, Section 7.5.) However, a simple modification of the definition yields one

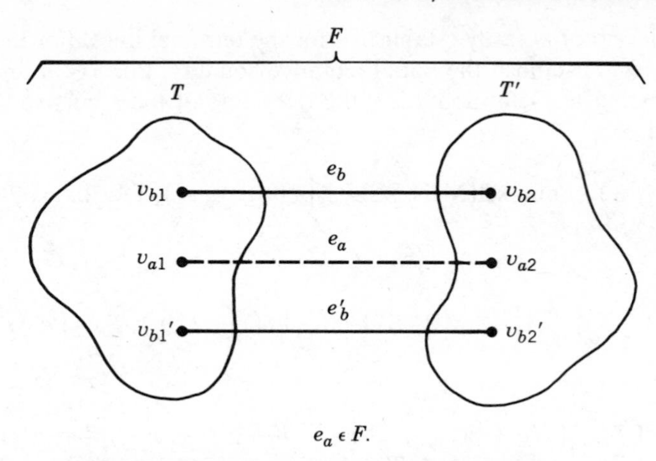

FIGURE 7-2 Partially developed minimum spanning tree.

that gives a single minimum spanning tree. This is achieved by setting the upper limit on the MIN-SEL to 1. The result is Definition 7.3.4.

Definition 7.3.4: Minimum Spanning Tree—Greedy Formulation 2

$$\left[\begin{array}{ll} g(c,t,j,G) = \langle c,T\rangle & :j=n \\ g(c,T,j,G) = g(c+c(e_1), T\cup\{e_1\}, j+1, G-e_1) & \\ & :j<n \text{ and } P(T,G)\neq\{\,\} \\ \text{Initially } G = \text{an } n\text{-vertex graph}, v = \text{a vertex in } G, c=0, T=\{v\}, j=1 & \end{array}\right.$$

Clearly Definition 7.3.4 has one solution, so G cannot become empty before T has $n-1$ edges. That is why the second line of Definition 7.3.3 is not needed. We have shown that definition 7.3.3 gives all minimum spanning trees, but it can also give some which are not minimum. Definition 7.3.4 gives one of the trees returned by 7.3.3, which we need to verify is one of the minimum ones. If at each application of the recursive line $P(T,G)$ has only one edge, which will be the case if all edge weights are different, then Definition 7.3.3 is the same as 7.3.4 and obviously gives the minimum spanning tree. We now sketch an argument which handles nonunique weights. Assume we run Definition 7.3.4 on a graph G in which some edges have equal weight. Each time Definition 7.3.3 would give a choice of more than one equal weight edges, $\langle e_1,\ldots,e_r\rangle$, reduce the weight of e_1 by $\delta>0$, and alter G accordingly. This makes e_1 the unique choice of the resultant graph G'. Finally for G' the minimum spanning *tree* has a weight, w', less than or equal to w, the weight for G, minus δm, where m is the number of edge weights reduced to produce G'. But G' differs from G only in that it has m edges of lower weight than G. Thus w' for any $\delta>0$ is a lower bound on w. But in fact Definition 7.3.4 would have chosen the same edges as were chosen for G' with a resultant weight of w' for $\delta = 0$. So, since each w' is a lower bound on w, we have $w' = w$.

7.3.1.1 Update Function to Compute Candidate Edges for Addition to Minimum Spanning Tree (P(T, G))

In the unary definition, Definition 7.3.4, only one smallest edge of $P(T, G)$, called SMALL1($P(T, G)$, is chosen. Consider the O network for that definition. It consists of a sequence of O blocks, each of which has one output. There is one t block. It has constant complexity. There are no w blocks. The complexity is equal to the sum of complexities for computing O block outputs. Computing an O block output requires computing SMALL1($P(T, G)$). If there is an edge between every pair of vertices (x, y) with $x \in T$, $y \in T'$, and T has t vertices, then there may be as many as $t(n - t)$ edges in $P(T, G)$. To find the minimum of these edges requires as many as $t(n - t) - 1$ comparisons. The number of O blocks is $n - 1$, and the total complexity could be as much as

$$\sum_{t=1}^{n-1} [t(n - t)] = O(n^3)$$

This complexity can be decreased by replacing the computation of SMALL1($P(T, G)$) with an update of information from which SMALL1($P(T, G)$) can be easily inferred.

Let T be the tree, represented by both edges and vertices, as before; G the n vertices and e edges of the given graph. Let P_v be all vertices in G but not in T, from which there is an edge to a vertex in T; P a set containing one edge for each vertex in P_v, namely, the minimum cost edge from v to any vertex in T. Clearly SMALL1($P(T, G)$) = MIN(P).

The update function for these quantities follows:

Let v_{initial} be the single vertex which forms T initially.
Let e_1 = the smallest edge in P.
Let v_l = the vertex in P_v incident on e_1.

$$\begin{bmatrix} P_v \text{ initially} = \text{All vertices in } G \text{ with an edge to } v_{\text{initial}} \\ P_v\text{@child} = P_v\text{@parent} - v_1 \cup \{ y | (v_1, y) \in G, y \notin T \} \\ = ud(P_v) \end{bmatrix}$$

$$\begin{bmatrix} P \text{ initially} = \text{For each vertex } x \text{ in } P_v \text{ initially the minimum edge} \\ \text{from } x \text{ to } v_{\text{initial}} \\ P\text{@child} = P\text{@parent} - e_1 \cup (\text{for each vertex } x \text{ in } P_v\text{@child}, \\ \text{with an edge } (x, v,) \text{ the minimum of } (x, v,) \text{ and the existing} \\ \text{edge from } x \text{ in } P\text{@parent})] \\ = ud(P) \end{bmatrix}$$

This update version of P is incorporated in Definition 7.3.5.

Definition 7.3.5: Minimum Spanning Tree—Greedy Formulation
SMALL1(P) = e_1

$$g(c, T, j, P_v, P) = \langle c, T\rangle \qquad : j = n$$
$$g(T, c, j, P_v, P) = \langle \infty, T\rangle \qquad : j < n \text{ and } P = \{\,\}$$
$$g(c, T, j, P_v, P) = g(c + c(e_1), T \cup \{e_1\}, j+1, ud(P_v), ud(P) \qquad : j < n \text{ and } P \neq \{\,\}$$

Initially G = an n-vertex graph, v_{initial} = a vertex in G, $T = \{v_{\text{initial}}\}$, $c = 0$, $j = 0$, P_v = all vertices whose neighbor is v_{initial}, P = all edges incident on v_{initial}.

To determine the complexity of the update functions for P_v and P note that both require finding all edges (v_1, y) such that $(v_1, y) \in G$, with $y \notin T$, after v_1 is moved into T. This involves, at most, looking at all edges incident on v_1. Since each vertex in G is "v_1" at some time, this can involve all e edges in G. Thus the updates have total complexity $O(e)$, assuming appropriate data-structures. To complete the update of P, $|P| \leq n - t$ minimums must be found, where t is the number of vertices in T. So for $n - 1$ updates the complexity is $\Sigma_{t=1}^{n-1}[n - t] = O(n^2)$. A computation of the same complexity determines the minimum edge, e_1, in P. These are the major costs of the algorithm, so the complexity is $O(e + n^2)$. By use of additional principles and the data-structures (F-Heap) of $O(e + n \log n)$ has in fact been found. [Tarjan 83].

7.4 OPTIMAL STATIC BINARY SEARCH TREE

In this example we obtain a recursive definition in a natural way whose ESE yields an $O(n^3)$ algorithm. However, recognizing additional, rather subtle properties, we can get an improvement to an $O(n^2)$ algorithm. The abstract formulation of the necessary properties in terms of the recursive definition is not trivial (see [Yao 80; Yao 82; Mehlhorn 84]. The justification of this property, even confined to its search tree application (See Section 7.4.1.1), is not simple.

A key is stored at each node of a binary tree in such a way as to minimize average number of comparisons required to access a key. The following assumptions are made.

1. The keys are $a_1 < a_2 < \cdots < a_n$. The subtree containing $a_1 < a_2 < \cdots < a_j$ is T_{ij}.
2. The keys are arranged, as in any binary search tree, so that if a_i precedes a_j in an "inorder" sequence through the tree, the $a_i < a_j$.

3. In searching for a key X it is assumed, letting P = probability, that

$$P(X = a_i) = p_i$$

$$P(a_i < X < a_{i+1}) = q_i$$

BASIC RECURSIVE DEFINITION

The average cost to search a binary tree with keys at each internal node, and whose external nodes are failure nodes, is the sum over all nodes in the tree of path length (number of nodes in the path) from root to node times the probability of the event represented by that node.

$c(i, j)$ is the average cost to search the binary tree T_{ij} of minimum cost computed as above and using the following assumptions.

1. $$P(X = a_k) = p_k$$

$$P(a_k < X < a_{k+1}) = q_k \qquad \text{for all } i \le k \le j - 1$$

2. the a_k, $i \le k \le j$, are arranged by "inorder" as described above.

In general, the probabilities in 1 are not correct; $c(i, j)$ computed as prescribed is not the actual average cost of a search within T_{ij}. On the other hand, for searches in the full tree (containing a_1 through a_n), $c(i, j)$ gives a correct average over all searches in T_{1n} in which the cost of moving within T_{ij} is assigned to each X which ends up being in T_{ij} (including falling on either side of a key in T_{ij}), and *a cost of* 0 is assigned to each case in which X ends up outside of T_{ij}. It follows that for $c(1, n)$ all searches are in T_{1n}, so $c(1, n)$ is the average cost of moving in T_{1n}.

Note that in considering the different binary trees which contain $a_i, a_{i+1}, \ldots, a_j$, any a_k, $i \le k \le j$, could be the root. The recursive definition for $c(i, j)$ is built on this observation.

Figure 7.3 illustrates the key principle of the recursive definition. For keys X found in the left subtree T_{ik-1}, the cost of $c(i, k - 1)$ is accrued for movement in T_{ik-1} only. In addition since X has to move from the root of T_{ij} into T_{ik-1}, X moves 1 + depth (r) to move to node r in T_{ik-1}. T_{ik-1} contains nodes $a_i, \ldots, a_{k-1}$, with corresponding success and failure $p_i, \ldots, p_{k-1}$ and $q_{i-1}, \ldots, q_k$. The cost contributed by T_{ik-1} is

$$\begin{gathered} c(i, k - 1) + \\ p_i + \cdots + p_{k-1} + \\ q_{i-1} + \cdots \quad \cdots \quad \cdots + q_{k-1} \end{gathered}$$

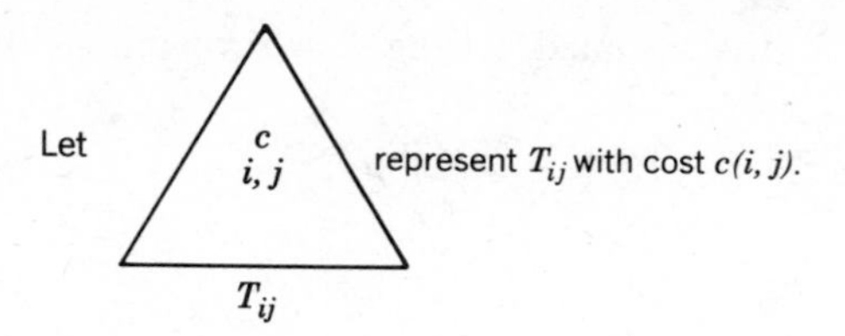

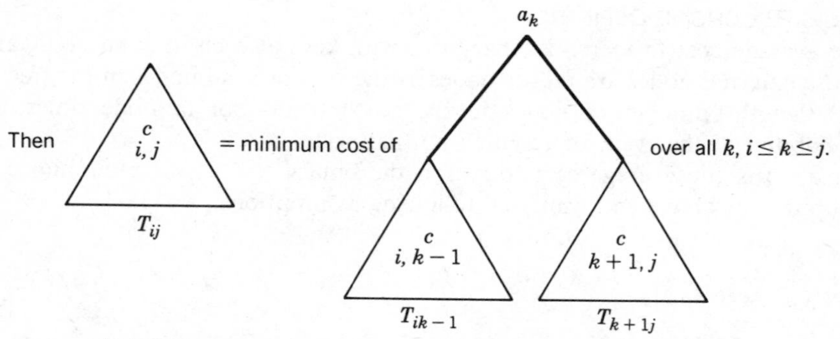

FIGURE 7-3 Optimum static binary search tree-recursive formulation.

In a similar way the right subtree, T_{k+1j} contributes

$$c(k+1, j) +$$

$$p_{k+1} + \cdots + p_j +$$

$$q_k + \cdots \quad \cdots \quad \cdots + q_j$$

and the cost contributed by the root a_k is $1 * p_k$. The total is $c(i, k-1) + c(k+1, j) + p_i + \cdots + p_j + q_{i-1} + \cdots + q_j$ which, letting $w_{i,j} = p_i + \cdots + p_j + q_{i-1} + \cdots + q_j$ becomes $w_{i,j} + c(i, k-1) + c(k+1, j)$. See figure 7.4

Now adding a terminal condition and value we get

Definition 7.4.1: Cost of Optimal Binary Search Tree

$$\left[\begin{array}{ll} c(i, j) = 0 & : i > j \\ c(i, j) = w_{i,j} + \text{MIN}_{k=i}^{k=j}[c(i, k-1) + c(k+1, j)] & : i \leq j \\ \text{Initially } i = 1, j = n. & \end{array}\right.$$

This definition obviously terminates and will have repeated argument structures, so it will be implemented by ESE simulation, setting up the set of equations and back substituting. Termination implies no loops and no need of

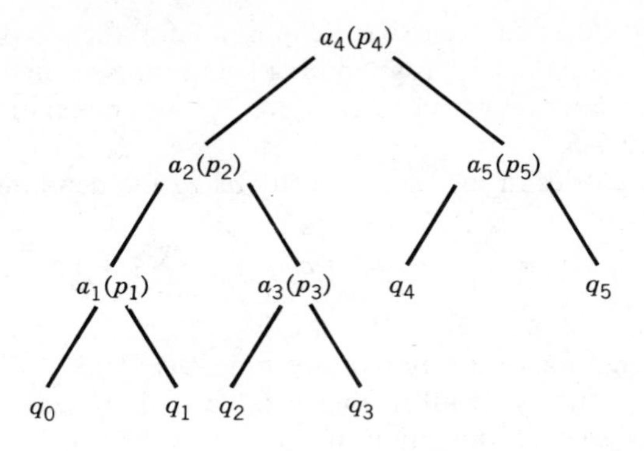

average cost (one comparison per internal node)

$$c(1,5) = 1*p_4 + (2*p_2 + 3*p_1 + 3*p_3 + 3*q_0 + 3*q_1 + 3*q_2 + 3*q_3)$$
$$+ (2*p_5 + 2*q_4 + 2*q_5)$$

and this is equal to:
Left subtree (T_{13})

$$c(1,3) = 1*p_2 + 2*p_1 + 2*p_3 + 2*q_0 + 2*q_1 + 2*q_2 + 2*q_3$$
$$+ (p_2 + p_1 + p_3) + (q_0 + q_1) + (q_2 + q_3)$$

$+$

Right subtree (T_{55})

$$c(5,5) = 1*p_5 + 1*q_4 + 1*q_5$$
$$+ (p_5) + (q_4 + q_5)$$

$+$

Root

p_4

FIGURE 7-4 Optimum static binary search tree example.

w

i \ j	1	2	3	4
4				3
3			3	5
2		7	9	12
1	8	12	14	16

FIGURE 7-5 Optimum static binary search tree example, w values.

a loop-removal rule. The recursive definition and the accompanying ESE implementation is a dynamic programming solution. As a first step in designing the implementation consider computing w_{ij} by systematically filling in a table as in Figure 7-5.

We can make entries in the table directly using the definition,

$$w_{ij} = q_{i-1} + \cdots + q_j + p_i + \cdots + p_j$$

Done in this straightforward way requires time $O(n^3)$.

Here is an $O(n^2)$ way to fill in the w table. It is based on the following easily justified recursive formulation of w_{ij}.

Let

$$w_{ij} = 0 \qquad :i < 0 \text{ or } j < i \text{ and}$$

$$w_{ij} = q_{i-1} + q_i + p_i \qquad :i > 0,\ j = i$$

$$w_{ij} = w_{ij-1} + q_j + p_j \qquad :j > i > 0$$

Algorithm 7.4.1: Algorithm to Compute w_{ij}

```
FOR j = 1 TO n DO
  s ← q_j + p_j
  w_jj ← s + q_j-1
  FOR i = 1 TO j - 1 DO
    w_ij ← w_ij-1 + s
  END
END
```

Next consider an example of the ESE simulation for Definition 7.4.1. The root chosen in finding the minimum of $c(i, j)$ is designated k_{ij}, that is, k_{ij} is the root of the subtree containing $a_i, \ldots, a_j$. Let $(p_1, \ldots, p_4) = (3, 3, 1, 1)$ and $(q_0, q_1, \ldots, q_4) = (2, 3, 1, 1, 1)$.

The equations set up on these values, using Definition 7.4.1, are shown in Figure 7-6. The results can be stored in a matrix. From the equation we can see that we can evaluate $c(i, j)$ starting at the lowest value of i and j and increasing i and j toward their final values. This can be represented as filling in values in a matrix, as in other dynamic programming examples. For this example Figure 7-7 gives the matrix of $c(i, j)$ values. It also shows how the matrix generalizes and the derivation of the complexity of $O(n^3)$ for this algorithm. The algorithm can be improved significantly by using a principle which allows equations to be solved without evaluating all terms on the right of every equation by use of the following principle.

Diagonal	Calculation	
$j - i = 1$	$c(1,2) = \min[c(1,0) + c(2,2) + w_{12},$	$:k = 1$
	$c(1,1) + c(3,2) + w_{12}]$	$:k = 2$
	$= \min[0 + 7 + 12,$	$:k = 1$
	$8 + 0 + 12]$	$:k = 2$
	$= 19$	$:k = 1$
$j - i = 2$	$c(1,3) = \min[c(1,0) + c(2,3) + w_{13},$	$:k = 1$
	$c(1,1) + c(3,3) + w_{13},$	$:k = 2$
	$c(1,2) + c(4,3) + w_{13}]$	$:k = 3$
	$= \min[0 + 12 + 14,$	$:k = 1$
	$8 + 3 + 14,$	$:k = 2$
	$19 + 0 + 14]$	$:k = 3$
	$= 25$	$:k = 2$
$j - i = 3$	$c(1,4) = \min[c(1,0) + c(2,4) + w_{14},$	$:k = 1$
	$c(1,1) + c(3,4) + w_{14},$	$:k = 2$
	$c(1,2) + c(4,4) + w_{14},$	$:k = 3$
	$c(1,3) + c(5,4) + w_{14}]$	$:k = 4$

FIGURE 7-6 Optimum static binary tree example, equations for c values.

Definition 7.4.2: Knuth's Principle

$$k_{i,j-1} \leq k_{ij} \leq k_{i+1,j}.$$

In terms of matrix computations, the Knuth principle has the form shown in Figure 7-8. As a result the cost of each diagonal can be decreased from $O(n^2)$ to $O(n)$, and the overall complexity from $O(n^3)$ to $O(n^2)$. We show this now.

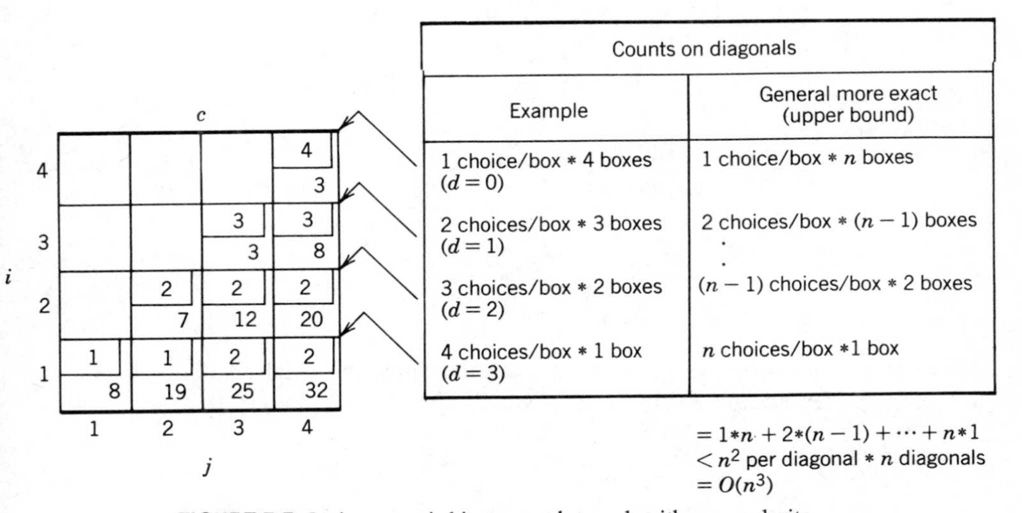

Counts on diagonals	
Example	General more exact (upper bound)
1 choice/box ∗ 4 boxes ($d = 0$)	1 choice/box ∗ n boxes
2 choices/box ∗ 3 boxes ($d = 1$)	2 choices/box ∗ $(n - 1)$ boxes ⋮
3 choices/box ∗ 2 boxes ($d = 2$)	$(n - 1)$ choices/box ∗ 2 boxes
4 choices/box ∗ 1 box ($d = 3$)	n choices/box ∗ 1 box

FIGURE 7-7 Optimum static binary search tree algorithm, complexity.

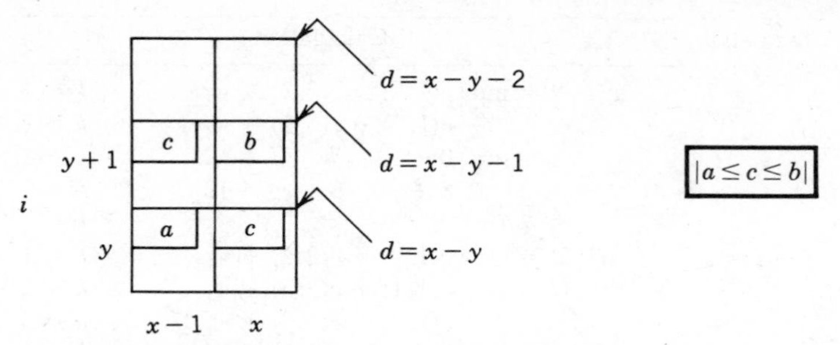

FIGURE 7-8 Optimum static binary search tree, Knuth's principle.

Instead of considering the number of possible root values in *each box* of the matrix (Figure 7-7) as the basis for computing the complexity, we now consider the number of possible values of the root that needs to be considered on an *entire diagonal*. This will depend on the diagonal just below the one for which we are doing the computation. The lowest possible value k can have in the entire diagonal is 1, the highest is n. In each box the value of k must lie in between.

Let the diagonal in which $j - i = d$ contain root $k_{d,j}$ in column j. In Figure 7-7, $k_{0,1} = 1, k_{0,2} = 2, \ldots, k_{0,4} = 4$; $k_{1,2} = 1, k_{1,3} = 2, k_{1,4} = 3$; $k_{2,3} = 2, k_{2,4} = 2$. In general clearly, j goes from $d + 1$ to n and $1 \le k_{d,d+1} \le k_{d,d+2} \le \cdots \le k_{d,n} \le n$.

Then in diagonal $d + 1$, j goes from $d + 2$ to n and $1 \le k_{d+1,d+2} \le k_{d+1,d+3} \le \cdots \le k_{d+1,n} \le n$. By Knuth's principle $k_{d+1,d+2}$ is constrained by

$$1 \le k_{d,d+1} \le k_{d+1,d+2} \le k_{d,d+2} \le n \text{ so}$$

$$k_{d+1,d+2} \text{ has one of } \underline{k_{d,d+2}} - k_{d,d+1} + 1 \text{ values.}$$

Similarly,

$$k_{d+1,d+3} \text{ has one of } k_{d,d+3} - \underline{k_{d,d+2}} + 1 \text{ values,}$$

$$\vdots$$

$$\text{and } k_{d+1,n} \text{ has one of } k_{d,n} - k_{d,n-1} + 1 \text{ values.}$$

Because of cancellations between succesive lines the total number of values obtained by adding the number of values above is

$$k_{d,n} - k_{d,d+1} + n - d + 1,$$

For all $1 \le d \le n - 1$, this is $\le n - 2 + n = 2n - 2 = O(n)$.

7.4.1.1 Refining the Optimum Binary Search Tree Algorithm

Knuth's principle (Definition 7.4.2) asserts that there is a relation between the optimal root of a tree and those of trees containing subsets of consecutive nodes, and allows us to reduce the number of candidates for the optimal choice. Now we give a proof of this principle using the definitions of the previous section; this proof closely follows his own [Knuth 71, Knuth 73].

Lemma 7.4.1: If $q_n = p_n = 0$, an optimum binary tree may be obtained by relacing the rightmost terminal (failure) node of the optimum tree for $a_1, \ldots, a_{n-1}$ by the subtree with a_n at the root and two new terminal nodes, as shown in Figure 7-9, and thus with no change in the roots.

PROOF: (Based on [Knuth 73, p. 673, answer to exercise 24]. Let P be the weighted path length of a tree obtained by deleting the a_n'th key of an optimum tree. Then

$$c(1, n-1) \leq P \leq c(1, n) - q_{n-1}$$

since the deletion moves the failure node up one level. Also, with $q_n = p_n = 0$,

$$c(1, n) \leq c(1, n-1) + q_{n-1}$$

Simple algebraic manipulation of these two inequalities shows that

$$c(1, n-1) = P = c(1, n) - q_{n-1}$$

Thus $c(1, n) = c(1, n-1) + q_{n-1}$, which is the weighted path length after the operation proposed by the lemma. ■

Theorem 7.4.1: When adding key $a_n \geq a_i$ for $i = 1, \ldots, n-1$ to the optimum tree, it is always possible to find a new optimum tree without moving the root left. In other words, there is always a solution to the dynamic programming equations such that $k_{1,n-1} \leq k_{1,n}$ when $n \geq 2$.

PROOF: The proof is by induction on n. The argument is vacuous when $n = 1$. Assume the theorem is true for all trees with fewer than n keys. The optimum

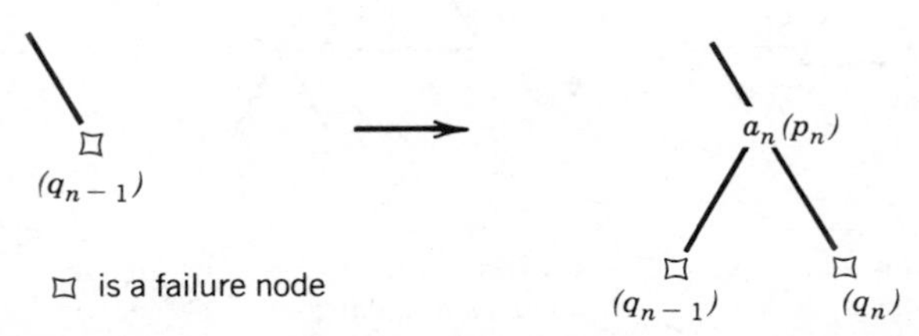

FIGURE 7-9 Replacement that maintains optimum tree. (□–failure node.

tree is a function of $q_n + p_n$, so we assume without loss of generality that $p_n = 0$. When $q_n = 0$, the theorem is true by Lemma 7.4.1. We will show that the theorem remains true as q_n gets arbitrarily large. If the theorem is false for some q_n, but true when $q_n = 0$, then there must be some critical value, say q, above which it first fails. Thus let the optimum tree be T when $q_n = q - \epsilon$, but T' when $q_n = q + \epsilon$ for all sufficiently small $\epsilon > 0$. Assume that the root of T' is less than, that is, to the left of, the root of T. We will show that if this is the case, there is another tree T'' whose root is the same as T's, and which is optimum where T' is optimum. Thus we never need consider moving the root to the left.

The weighted path length $c(1, n)$ of T is a linear expression;

$$l(q_0)q_0 + l(q_1)q_1 + \cdots + l(q_n)q_n + l(p_1)p_1 + \cdots + l(p_n)p_n$$

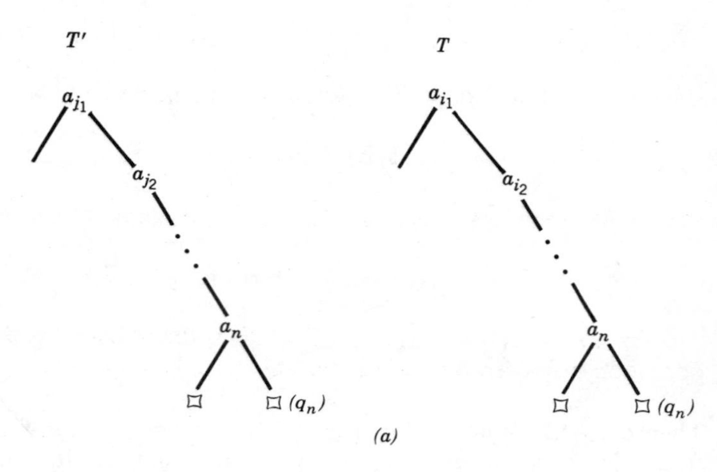

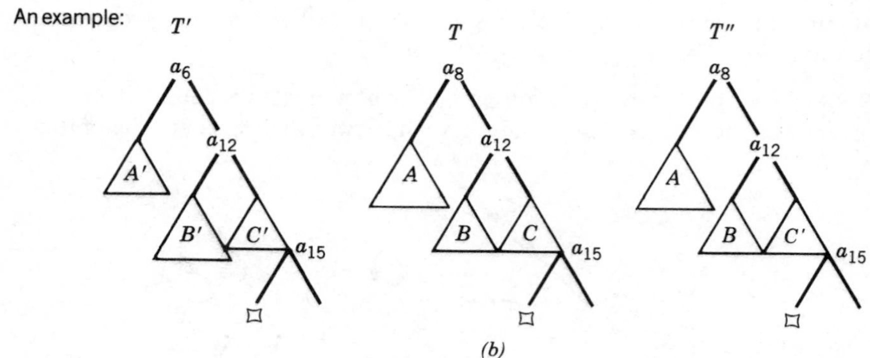

A' contains a_1 to a_5; B' contains a_7 to a_{11}; C' contains a_{13} to a_{14};
A contains a_1 to a_7; B contains a_9 to a_{11}; C contains a_{13} to a_{14}.

FIGURE 7-10 Trees T, T' and T''.

where $l(x)$ is the level of a_n if $x = p_n$, or 1 less than the level of the nth failure node if $x = q_n$ in the optimum tree. Similarly, $c'(1, n)$ for T' is

$$l'(q_0)q_0 + l'(q_1)q_1 + \cdots + l'(q_n)q_n + l'(p_1)p_1 + \cdots + l'(p_n)p_n$$

By assumption, $c'(1, n) < c(1, n)$ whenever $q_n > q$, so $l'(q_n)q_n$ must grow more slowly than $l(q_n)q_n$. After all, nothing else in the respective linear expressions changes as q is varied. Hence $l'(q_n) < l(q_n)$. Consider the partial pictures of T and T' presented in Figure 7-10(a).

By assumption, $j_1 < i_1$, $i_{l(q_n)} = j_{l'(q_n)} = n$. All keys larger than a_{j_1} and less than or equal to a_{i_1} must be somewhere in the subtree rooted at a_{j_2} in T'. Thus the subtree rooted at a_{j_2} contains the same keys as that rooted at a_{i_2}, plus one or more additional keys, *each of which must be smaller than any key in the subtree rooted at* a_{i_2}. Recall that the induction hypothesis applies to the subtrees of T and T'. Also, note that the principle of the theorem is left–right symmetric. Hence when this sequence of smaller keys is added to the subtree rooted at a_{j_2}, its root cannot move to the *right*; hence $j_2 \le i_2$. Suppose $j_2 < i_2$; then, similarly, we have that $j_3 \le i_3$. But we also know that $l'(q_n) < l(q_n)$, which implies that $j_{l'(q_n)} = n > i_{l'(q_n)}$, so j_p must equal i_p for some p. Recall that every subtree of an optimal tree is optimal. Thus we can replace the right subtree of a_i in T by the similar subtree in T', obtaining a binary tree T'' whose weighted path length is equal to that of T' for all q_n (see Figure 7-10(b)). Since T'' has the same root as T, we need never move the root to the left. ■

Corollary 7.4.1.1: There is always a solution to the dynamic programming conditions satisfying

$$k_{i,j-1} \le k_{i,j} \quad \text{and} \quad k_{i,j} \le k_{i+1,j}, \quad \text{for } 0 \le i < j - 1 \le n$$

PROOF: This is simply the result of the theorem applied to all subtrees, and using left–right symmetry. ■

Thus when we build the algorithm, we will not have to search the entire range $i \le k \le j$ when determining k_{ij}. Instead, only $k_{i+1,j} - k_{i,j-1} + 1$ cases need to be examined when k_{ij} is selected, which as we have shown implies that the total amount of work is $O(n^2)$.

This proof depends on an interpretation of the set of equations that arise from the recursive definition, Definition 7.4.1. "Trees" and "weighted path lengths" are interpretations of this definition, not obvious from properties of primitive operations that make up that definition, in the absence of knowledge of its origin. A proof that depended more directly on the properties of the definition would focus on those properties that are essential to the existence of Knuth's principle, and would thereby extend the applicability of that principle. Such a proof is given by Yao. [Yao 80, 82]. Alternatively one could identify the properties in the definition which allow the "tree" and "weighted path length" interpretation.

7.5 PROBLEMS

1. *Two-Dimensional Nim.* [Berlekamp Conway Guy 82]. Give one or more alternative (inequivalent) definitions for two-dimensional Nim.
2. *Two-Dimensional Nim.* Generalize the argument of Section 7.1 to Definition 7.1.2. This may require additional data structures.
3. *Minimum Spanning Tree—Nonminimum Spanning Trees.* When Definition 7.3.3 is applied to a graph, in addition to all minimum spanning trees, it may produce some which are not minimum. Give such a graph.
4. *Minimum Spanning Tree—Cost is a Two-Component Vector.* Write the best recursive definition you can for finding a "minimum" spanning tree for a graph which has two numbers assigned to each edge. The spanning tree found should have the sum of its first numbers' minimum and among such spanning trees be one whose second number sum is smallest.
5. *Minimum Spanning Tree vs. Shortest Path Tree.* Determine where the arguments of Section 7.3 break down in deriving the shortest path tree, or show how to modify the development to obtain a definition for shortest path trees.
6. *Knuth's Principle.* Following the discussion in the last paragraph of Section 7.4, give a proof of Knuth's principle which does not depend on the tree interpretation, or show how to relate this interpretation to properties of recursive definitions in general.

REFERENCES

[Aho, Hopcroft, Ullman 75]
A. V. Aho, J. E. Hopcroft, and J. D. Ullman, *Design and Analysis of Computer Programs*. Addison-Wesley, Reading, MA, 1975.

[Aho, Hopcroft, Ullman 85]
A. V. Aho, J. E. Hopcroft, and J. D. Ullman, *Data Structures and Algorithms*. Addison-Wesley, Reading, MA, 1985.

[Aho, Ullman 72*a*]
A. V. Aho and J. D. Ullman, *The Theory of Parsing, Translation, and Compiling*. I. Prentice-Hall, Englewood Cliffs, NJ, 1972.

[Aho, Ullman 72*b*]
A. V. Aho and J. D. Ullman, *The Theory of Parsing, Translation, and Compiling*. II. Prentice-Hall, Englewood Cliffs, NJ, 1972.

[Auslander 78]
M. A. Auslander and H. R. Strong, Systematic Recursion Removal, *Commun. ACM on Programming Languages and Systems*, **2**(21): 127–134, 1978.

[Baase 78]
S. Baase, *Computer Programs—Introduction to Design and Analysis*. Addison-Wesley, Reading, MA, 1978.

[Bellman 57]
R. Bellman, *Dynamic Programming*. Princeton University, Princeton, NJ, 1957.

[Bellman 62]
R. Bellman and S. E. Dreyfus. *Applied Dynamic Programming*. Princeton University Press, Princeton, NJ, 1962.

[Bird 77]
R. S. Bird, Notes on Recursion Elimination, *Commun. ACM*, **11**(20):434–439, 1977.

[Birkhoff 67]
G. Birkhoff, *Lattice Theory*, American Mathematical Society, Colloquium Publications, Providence, R.I., 1967.

[Burge 82]
W. H. Burge, *Recursive Programming Techniques*. Addison-Wesley, Reading, MA, 1982.

[Berlekamp Conway Guy 82]
Berlekamp, E., Conway, J., Guy, R., *Winning Ways for Your Mathematical Plays*, Academic, New York, 1982.

[Burstall 77]
R. M. Burstall and J. Darlington, A Transformation System for Developing Recursive Programs, *J. ACM* **1**:44–67, 1977.

[Cohen 79]
N. H. Cohen, "Characterization and Elimination of Redundancy in Recursive Programs," *6th Annual Conf. Principles of Programming Languages*, pp. 143–157, 1979.

[Darlington 76]
J. Darlington and R. M. Burstall, "A System Which Automatically Improves Programs," *Acta Informatica*, **6**:41–60, 1976.

[Dreyfus 77]
S. E. Dreyfus and A. M. Law, *The Art and Theory of Dynamic Programming*. Princeton University Press, Princeton, NJ, 1977.

[Edelen, Kydoniefs 72]
D. G. B. Edelen and A. D. Kydoniefs, *An Introduction to Linear Algebra*. Elsevier, New York, 1972.

[Even 79]
S. Even, *Graph Algorithms*. SIAM, Rockville, MD, 1979.

[Fredman 84]
M. L. Fredman and R. E. Tarjan, "Fibonacci Heaps and Their Uses in Improved Network Optimization Algorithms," in *FOCS*, pp. 338–346, IEEE Computer Society, 1984.

[Fredman et al. 86]
Fredman, M. L., Sedgewick, R., Sleator, D. D. Tarjan, R., "The Pauing Heaping: A New Form of Self-Adjusting Heap," *Algorithmica*, 1986, pp. 111–129.

[Goodman, Hedetniemi 79]
S. E. Goodman and S. T. Hedetniemi, *Introduction to Design and Analysis of Algorithms*. McGraw-Hill, New York, 1979.

[Hecht 77]
M. S. H. Hecht, *Flow Analysis of Computer Programs*. Elsevier North-Holland, New York, 1977.

[Held 67]
M. Held and R. M. Karp, *Finite-State Processes and Dynamic Programming*. SIAM, 1967.

[Henderson 83]
P. Henderson, *Functional Programming: Application and Implementation*. Prentice-Hall, Englewood Cliffs, NJ, 1983.

[Horowitz, Sahni 78]
E. Horowitz and S. Sahni, *Fundamentals of Computer Algorithms*. Computer Science Press, Rockville, MD, 1978.

[Hu 82]
T. C. Hu, *Combinatorial Algorithms*, Addison-Wesley, Reading, MA., 1982.

[Knuth 71]
D. E. Knuth, "Optimum Binary Search Trees," *Acta Informatica*, **1**:14–26, 1971.

[Knuth 73]
D. E. Knuth, *The Art of Computer Programming, vol. 3: Sorting and Searching*. Addison-Wesley, Reading, MA, 1972.

[Knuth 75]
D. E. Knuth and R. W. Moore, "An Analysis of Alpha–Beta Pruning," *Acta Informatica*, **1**:293–326, 1975.

[Lawlor 76]
E. Lawlor, *Combinatorial Optimization*. Holt, Reinhart and Winston, New York, 1976.

[Manna 80]
Z. Manna, *The Logic of Computer Programming*. SIAM, Phil. PA., 1980.

[Mehlhorn 84]
K. Mehlhorn, *Graph Algorithms and NP-Completness*. Springer-Verlag, Berlin, 1984.

[Nadel 1986]
B. A. Nadel,, "The Consistent Labelling Problem and its Algorithms; Towards Exact-Complexities and Theory Based Heuristics," Computer Science Dept., Rutgers University, Ph.D. Thesis, April 1986.

[Nemhauser 77]
G. L. Nemhauser, *Dynamic Programming and Its Applications*, Wiley, New York, 1977.

[Nilsson 80]
N. J. Nilsson, *Principles of Artificial Intelligence*. Tioga, Palo Alto, CA, 1980.

[Paige 81]
R. Paige, *Formal Differentiation*. UMI Research Press, Ann Arbor, MI, 1981.

[Papadimitriou, Steiglitz 82]
C. H. Papadimitriou and K. Steiglitz, *Combinatorial Optimization: Algorithms and Complexity*. Prentice-Hall, Englewood-Cliffs, NJ, 1982.

[Peter 51]
R. Peter, *Rekursive Funktionen*. Akadémiai Kiadó, Budapest, Hungary, 1951.

[Polya 45]
G. Polya, *How to Solve It: A New Aspect of Mathematical Method*. Doubleday, New York, 1945.

[Purdom 85]
P. W. Purdom and C. A. Brown, *The Analysis of Algorithms*. Holt, Rinehart and Winston, New York, 1985.

[Puterman 77]
M. L. Puterman, *Dynamic Programming and Its Applications*. Academic, New York, 1977.

[Reingold, Nievergelt, Deo 77]
E. M. Reingold, J. Nievergelt, and N. Deo, *Combinatorial Algorithms: Theory and Practice*. Prentice-Hall, Englewood Cliffs, NJ, 1977.

[Roberts 86]
E. Roberts, *Thinking Recursively*. Wiley, New York, 1986.

[Rogers 67]
H. Rogers Jr., *Theory of Recursive Functions and Effective Computability*. McGraw-Hill, New York, 1967.

[Rosser 53]
J. B. Rosser, *Logic for Mathematicians*. McGraw-Hill, New York, 1953.

[Sedgewick 83]
R. Sedgewick, *Algorithms*. Addison-Wesley, Reading, MA, 1983.

[Aho, Sethi, Ullman 86]
A. V. Aho, R. Sethi, and J. D. Ullman, *Compilers: Principles, Techniques and Tools*. Addison-Wesley, Reading, MA, 1986.

[Strong 71]
H. R. Strong, "Translating Recursive Equations into Flowcharts," *J. Cmputer. Sys. Sci.*, **5**:254–285, 1971.

[Tarjan 72]
R. E. Tarjan, Depth-First Search and Linear Graph Algorithms, *SIAM J. Computing*, **1**(2):146–159, 1972.

[Tarjan 83]
R. E. Tarjan, *Data Structures and Network Algorithms*, SIAM, Philadelphia, PA, 1983.

[Wells 71]
M. B. Wells, *Elements of Combinatorial Computing*. Pergamon, Elmsford, NY, 1971.

[Yao 80]
F. F. Yao, "Efficient Dynamic Programming Using Quadrangle Inequality," *Proc. 12th Ann. ACM Proceedings on Theory of Computing*, pp. 429–435, Apr. 1980.

[Yao 82]
F. F. Yao, "Speed-Up in Dynamic Programming," *SIAM J. ALG. DISC. MATH*, **3**(4):532–539, Dec. 1982.

[Yasuhara 71]
A. Yasuhara, *Recursive Function Theory and Logic*. Academic, New York, 1971.

INDEX

Numbers in *italics* indicate pages on which the complete reference appears.